Introduction to Nuclear Reactions

Introduction to Nuclear Reactions

Second Edition

Carlos Bertulani
Pawel Danielewicz

CRC Press
Taylor & Francis Group
Boca Raton London New York

CRC Press is an imprint of the
Taylor & Francis Group, an **informa** business

Second edition published [2021]
by CRC Press
6000 Broken Sound Parkway NW, Suite 300, Boca Raton, FL 33487-2742

and by CRC Press
2 Park Square, Milton Park, Abingdon, Oxon, OX14 4RN

© 2021 Taylor & Francis Group, LLC

First edition published by Taylor & Francis 2004

CRC Press is an imprint of Taylor & Francis Group, LLC

Library of Congress Cataloging-in-Publication Data

Names: Bertulani, Carlos A., author. | Danielewicz, Pawel, author.
Title: Introduction to nuclear reactions / Carlos Bertulani, NSCL and
Department of Physics and Astronomy, Michigan State University, USA,
Pawel Danielewicz, NSCL and Department of Physics and Astronomy,
Michigan State University, USA. Description: Second edition. |
Boca Raton : CRC Press, 2021. | Series: Graduate student series in physics |
Includes bibliographical references and index. Identifiers: LCCN 2020050052 |
ISBN 9780367353629 (hardback) | ISBN 9780429331060 (ebook)
Subjects: LCSH: Nuclear reactions.
Classification: LCC QC794 .B48 2021 | DDC 539.7/6--dc23 LC record
available at https://lccn.loc.gov/2020050052

ISBN: 9780367353629 (hbk)
ISBN: 9780429331060 (ebk)

Typeset in Computer Modern font
by KnowledgeWorks Global Ltd.

To Eliete
and Ewa.

Contents

Foreword

This book has its origin in the lecture notes of a course on nuclear reactions that the authors have taught at the Physics and Astronomy Department of the Michigan State University. Since that time, the original text was enlarged substantially and it contains an amount of material, somewhat larger than the usually taught in a one semester course.

The general idea of the book is to present the basic information on scattering theory and its application to the atomic nucleus. Although there is reference to experiments or measurements when we find it necessary, there was no attempt to describe equipments and methods of experimental nuclear physics in a systematic and consistent way. Also, practical applications of nuclear reactions are mentioned sporadically, but there is no commitment of giving a general panorama of what exists in this area.

In the ordering of the subjects, we chose to begin with a study of the basic scattering theory, followed by its applications in the study of nuclear reaction processes.

In Chapters 1-3, the low-energy scattering theory is described in details. Strong absorption and applications to the formation and decay of a compound nucleus is described in Chapter 4.

The following two chapters work with the nuclear reactions at low energies, starting with the general study of fusion reactions followed by the description of direct reactions.

Chapter 7 is dedicated to nuclear reactions occurring in stars. To discuss in details, this important branch of astrophysics we use, much of the knowledge developed in Chapters 1–6.

Chapters 8 and 9 embrace the second great block of study in nuclear reactions which are the high-energy nuclear collisions. In Chapter 8, we only treat applications of the eikonal wave function to nuclear reactions in general. The important branch of nuclear transport equations is studied in Chapter 9.

The adequate level for the complete understanding of this book corresponds to a student at the end of a graduation study in physics, what includes, besides the basic physics, a course of modern physics and a first course of quantum mechanics. But, students of exact sciences, and of technology in a general way, can take profit of a good part of the subjects presented in this book.

September 2020

Carlos A. Bertulani (Texas A&M University-Commerce, Commerce, Texas)
Pawel Danielewicz (Michigan State University, East Lansing, Michigan)

1

Classical and quantum scattering

1.1   Experiments with nuclear particles

Figure 1.1 represents schematically a typical scattering experiment; in fact this is a sketch of
E. Rutherford experiment in 1910 [1]. Projectiles (here, α-particles) from a source go through a
collimator and collide with a target (gold foil in Rutherford's experiment). Some projectiles are
scattered by the target and reach the detector (here, a fluorescent screen). Rutherford expected
them to go straight through, with perhaps a minor deflection. Most did go straight through,
but to his surprise some particles bounced directly off the gold sheet! What did this mean?
Rutherford hypothesized that the positive alpha particles had hit a concentrated mass of positive
particles, which he termed the nucleus. We shall describe the Rutherford scattering theoretically
later in this chapter.

Instead of a radioactive nuclear source, as in Rutherford's experiment, one could use particles
accelerated in a beam of particles, which could be protons, electrons, positrons, pions, ions,
ionized molecules, etc. According to the nature of the projectile, to the required collision energy,
resolution, beam intensity, etc., a variety of *accelerators* may be used. Ideally, the beam should
be sharp in momentum space, i.e., should have good energy resolution and be well collimated.
Usually, good experiments require high count rates. This is attained with an intense beam and/or

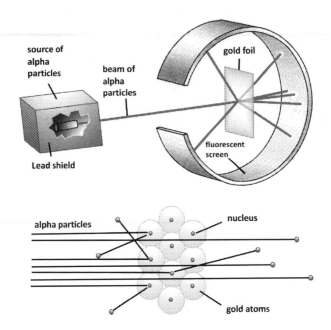

FIGURE 1.1   Schematic representation of Rutherford's scattering experiment.

a thick target. However, if the beam it so intense that the projectiles are close enough to interact, considerable complications appear in the theoretical description. Theoretical difficulties also arise when the target is so thick that multiple scattering becomes relevant.

In Figure 1.2 we show the concept of the most popular of the particle accelerators, the *cyclotron*. The cyclotron is a particle accelerator conceived by Ernest O. Lawrence in 1929. It is composed of two parts called Ds, or Dees, because of their D-shape. The two Dees were placed back-to-back with their straight sides parallel but slightly separated. An oscillating voltage was applied to produce an electric field across this gap. Particles injected into the magnetic field region of a Dee trace out a semicircular path until they reach the gap. The electric field in the gap then accelerates the particles as they pass across it. The particles now have higher energy so they follow a semi-circular path in the next Dee with larger radius and so reach the gap again. The electric field frequency must be just right so that the direction of the field has reversed by their time of arrival at the gap. The field in the gap accelerates them and they enter the first Dee again. Thus the particles gain energy as they spiral around. The trick is that as they speed up, they trace a larger arc and so they always take the same time to reach the gap. In this way a constant frequency electric field oscillation continues to always accelerate them across the gap. The limitation on the energy that can be reached in such a device depends on the size of the magnets that form the Dees and the strength of their magnetic fields.

The *detectors* can simply acknowledge the arrival of a particle or they can give additional information like the particle charge or energy. In the Rutherford experiment a fluorescent screen was used to detect particles. Nowadays, several particle detection techniques are available. One of the simplest particle detectors is the cloud chamber. Cloud chambers were first developed by Charles T.R. Wilson around 1911 for experiments on the formation of rain clouds. The principle of a cloud chamber is shown in Figure 1.3. A volume of saturated vapor contained in a vessel is made supersaturated through a sudden adiabatic expansion. When ionizing radiation passes through such a supersaturated vapor the ionization produced in the vapor serves as condensation

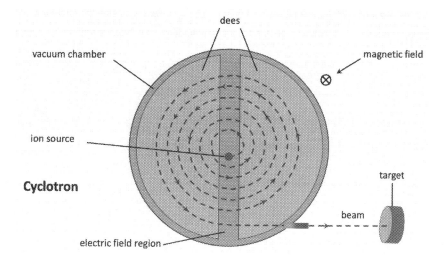

FIGURE 1.2 A cyclotron consists of two large dipole magnets designed to produce a semi-circular region of uniform magnetic field. In the figure, the magnetic field is perpendicular to the page. Between the two Dees, an electric field changes sign when the beam changes from the top to the lower part of the trajectory.

seeds. As a result small droplets of liquid can be observed along the path of the radiation. These condensation tracks have a lifetime of less than a second and can be photographed through the chamber window. The density of the condensation depends on the ionization power of the projectile as well as on the nature of the vapor, which is often an alcohol or water. Cloud chamber photographs are shown in Figure 1.4 for the reaction

$$\mathrm{^{14}_{7}N} + \mathrm{^{4}_{2}He} \longrightarrow \mathrm{^{18}_{9}F^{*}} \longrightarrow \mathrm{^{17}_{8}O} + \mathrm{^{1}_{1}H},$$

which was the first nuclear transmutation produced in a laboratory. This experiment was done by Rutherford in 1919.

Nowadays, nuclear experiments are performed with a high degree of sophistication. Large and expensive accelerators and detectors using different techniques are in use worldwide. A detailed account of the physics of accelerators and detection techniques is found in numerous sources in the literature. Here, we will concentrate on a study of the theoretical tools needed to understand the experiments involving atomic nuclei. Discussion of experimental techniques will be brief and schematic. Comparison between theory and experiments will be shown when opportunities arise.

1.2   Theories and experiments

To investigate the properties of complex microscopic systems (atoms, nuclei, molecules etc.) it would be necessary to solve the Schrödinger equation with a Hamiltonian with many interacting degrees of freedom, which is a hopeless task. To handle this situation, approximate models have been developed. These models must reproduce the main features of the system, while leading to simpler equations of motion. Calling ξ the set of coordinates of the model and $h(\xi)$ the corresponding Hamiltonian, the stationary states, $\varphi_{\alpha}(\xi)$, are obtained by solving the Schrödinger equation

$$h(\xi)\varphi_{\alpha}(\xi) = \varepsilon_{\alpha}\varphi_{\alpha}(\xi). \tag{1.1}$$

Neglecting the projectile's intrinsic degrees of freedom, the projectile-target system is described by the set of coordinates $\{\xi, \mathbf{r}\}$, where $\mathbf{r}$ is the projectile-target separation vector, and the total

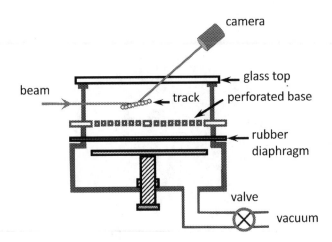

FIGURE 1.3   Principle of a cloud chamber.

Hamiltonian is

$$H(\mathbf{r}, \xi) = T_{\mathbf{r}} + h(\xi) + V(\mathbf{r}, \xi). \tag{1.2}$$

Above, $T_{\mathbf{r}}$ is the kinetic energy operator associated with $\mathbf{r}$ and $V(\mathbf{r}, \xi)$ is the coupling interaction.

Before the collision takes place, the target is in its ground state $\varphi_0(\xi)$. As the projectile approaches the target, the action of $V(\mathbf{r}, \xi)$ leads to excitations, such that there are finite probabilities for finding the target in different final states $\varphi_\alpha(\xi)$. These probabilities are measured and compared with the theoretical predictions based on the approximation used for the target dynamics. These predictions depend strongly on matrix elements of the interaction between pairs of intrinsic states φ_α, φ_β. In this way, the approximate model can be tested.

Therefore, scattering experiments play a fundamental role in the investigation of the interactions between elementary particles and in assessing the validity of approximations for the dynamics of the complex system. Of course, this testing procedure requires the capacity to go from the model wave functions to observable quantities. This is provided by quantum scattering theory, which we will discuss in details in this book.

1.3   Reactions channels

When one of the collision partners (or both) have relevant internal degrees of freedom, the intrinsic states of the projectile-target system may change from $\varphi_0(\xi)$ to a different final state $\varphi_\alpha(\xi)$. Therefore, the characterization of the final state of the system requires the knowledge of the relative wave function and also of the intrinsic quantum numbers expressed by the label α. Such states are called reaction channels ("channel-α"). The most important reaction channel is the *elastic* or *entrance* channel, corresponding to the ground state, $\alpha = 0$. Let us list below some frequently studied channels:

a) *Elastic channel* – In this channel, the collision does not produce intrinsic excitation. Since the collision process conserves the total energy, the kinetic energy of the relative motion is also unchanged. A simple example of elastic scattering is

$$^{16}\mathrm{O} + \mathrm{n} \rightarrow\ ^{16}\mathrm{O} + \mathrm{n},$$

with $^{16}\mathrm{O}$ remaining in the ground state. Above, we used the standard Nuclear Physics notation where $^{16}\mathrm{O}$ stands for an oxygen nucleus with mass number 16.

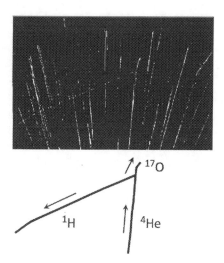

FIGURE 1.4   Cloud chamber photograph of the Rutherford experiment for the reaction $^{14}_{7}\mathrm{N} + ^{4}_{2}\mathrm{He} \longrightarrow$ $^{17}_{8}\mathrm{O} + ^{1}_{1}\mathrm{H}$.

b) *Inelastic channels* – Inelastic channels are those where projectile and target maintain their identities but one (or both) collision partner(s) is (are) excited to a higher intrinsic energy ε_α. For example,

$$^{16}\mathrm{O} + \mathrm{n} \rightarrow\ ^{16}\mathrm{O}^* + \mathrm{n},$$

where the notation A^* is used to indicate excitation. In inelastic collisions, part of the kinetic energy of the relative motion is used up in intrinsic excitation. *Inelastic scattering* can be characterized by a reduction of the kinetic energy of the emerging projectile or detection of the de-excitation product (usually photons).

c) *Reaction channels* – The name reaction channel is used for states of the system where the projectile or/and the target change identity. The population of such channels is called a *reaction*. A few examples are given below:

$$^{16}\mathrm{O} + \mathrm{n} \rightarrow\ ^{16}\mathrm{N} + \mathrm{p}$$
$$e^+ + \mathrm{H} \rightarrow\ e^+ + \mathrm{p} + e^-$$
$$\mathrm{p} + \mathrm{p} \rightarrow\ \mathrm{p} + \mathrm{n} + \pi^+ + \pi^0.$$

It is essential to highlight that the resulting products of a nuclear reaction are not determined univocally: starting from two or more reactants there can exist dozens of possibilities of composition of the final products with an unlimited number of available quantum states. As an example, the collision of a deuteron with a nucleus of $^{238}\mathrm{U}$ can give place, among others, to the following reactions:

$$\mathrm{d} + ^{238}\mathrm{U} \rightarrow\ ^{240}\mathrm{Np} + \gamma$$
$$\rightarrow\ ^{239}\mathrm{Np} + \mathrm{n}$$
$$\rightarrow\ ^{239}\mathrm{U} + \mathrm{p}$$
$$\rightarrow\ ^{237}\mathrm{U} + \mathrm{t}.$$

In the first of them the deuteron is absorbed by the uranium, forming an excited nucleus of $^{240}\mathrm{Np}$, that de-excites by emitting a γ-ray. The following two are examples of what one calls *stripping reactions*, where a nucleon is transferred from the projectile to the target. The last one exemplifies the inverse process: the deuteron captures a neutron from the target and it emerges of the reaction as a ^{3}H (tritium). This is denominated as a *pick-up reaction*. Another possibility would be, in the first reaction, that the nucleus ^{240}Np fissions instead of emitting a γ-ray, contributing with dozens of possible final products for the reaction.

The probability of a nuclear reaction taking place through a certain exit channel depends on the energy of the incident particle and is measured by the cross section for that channel. The theory of nuclear reactions must besides elucidating the mechanisms that determine the occurrence of the different processes, evaluate the cross sections corresponding to all exit channels.

1.4   Conservation laws

Several conservation laws contribute to restrict the possible processes that take place when a target is bombarded with a given projectile.

Baryonic number – There is no experimental evidence of processes in which nucleons are created or destroyed without the creation or destruction of corresponding antinucleons. The application of this principle to low-energy reactions is still more restrictive. Below the threshold for the production of mesons ($\sim$140 MeV), no process related to the nuclear forces is capable to transform a proton into a neutron and vice-versa, and processes governed by the weak force (responsible for the β-emission of nuclei) are very slow in relation to the times involved in nuclear reactions ($\sim$10^{-22} to 10^{-16} s). In this way, we can speak separately of proton and neutron conservation, which should show up with same amounts in both sides of a nuclear reaction.

Charge – This is a general conservation principle in physics, valid in any circumstances. In purely nuclear reactions charge is computed summing the atomic numbers for each side of the reaction.

Energy and linear momentum - Those are two of the most applied principles in the study of the kinematics of reactions. Through them, angles and velocities are related to the initial parameters of the reaction.

Total angular momentum – Is always a constant of motion. In the reaction

$$^{10}\text{B} + \,^4\text{He} \rightarrow \,^1\text{H} + \,^{13}\text{C}, \tag{1.3}$$

for example, ^{10}B has $I = 3$ in the ground state, whereas the α-particle has zero angular momentum. If it is captured in an s-wave ($l_i = 0$), the intermediate compound nucleus is in a state with $I_c = 3$. Both final products have intrinsic angular momenta equal to $1/2$, thus their sum is 0 or 1. Therefore the relative angular momentum of the final products will be $l_f = 2, 3$, or 4.

Parity – Is always conserved in reactions governed by the nuclear interaction. In the previous example, ^{10}B, ^{4}He and the proton have even parities, while ^{13}C has odd parity. Therefore, if $l_i = 0$, we necessarily have $l_f = 3$. Thus, the orbital momentum of the final products of (1.3) is determined by the joint conservation of the total angular momentum and of the parity.

Isospin – This is an approximate conservation law that is applied to light nuclei, where the effect of the Coulomb force is small. A nuclear reaction involving these nuclei not only conserves the z-component of the isospin (a consequence of charge and baryonic number conservation) but also the total isospin **T**. Reactions that populate excited states not conserving the value of **T** are strongly inhibited. A well mentioned example is the reaction d + ^{16}O $\rightarrow \alpha$ + ^{14}N, where the excited state 0$^+$ of ^{14}N, with 2.31 MeV, is about a hundred times more populated than the 1$^+$ ground state. Conservation of energy, angular momentum, and parity do not impose any prohibition for that channel, whose low occurrence can only be justified by isospin conservation; the 1$^+$ ground state of ^{14}N has **T** = 1, and the other states of the four participants in the reaction have all **T** = 0.

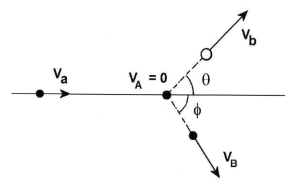

FIGURE 1.5   Nuclear reaction a + A → b + B as seen in the laboratory system.

1.5   Kinematics of nuclear reactions

We shall study the kinematics of a typical reaction, where the projectile a and the target A give rise to two products, b and B, respectively. This can also be expressed in the notation that we used so far

$$a + A \rightarrow b + B, \tag{1.4}$$

or even in a more compact notation,

$$A(a,b)B. \tag{1.5}$$

Figure 1.5 exhibits the parameters related to the reaction in (1.4).

In the most common situation a and b are light nuclei and A and B, heavy ones; the nucleus b has its emitting angle θ and its energy registered in the laboratory system and the recoil nucleus B has short range and cannot leave the target. Thus, we can for convenience eliminate the parameters of B in the system of equations that describe the conservation of energy and momentum:

$$
\begin{aligned}
E_a + Q &= E_b + E_B \\
\sqrt{2m_a E_a} &= \sqrt{2m_b E_b}\cos\theta + \sqrt{2m_B E_B}\cos\phi \\
\sqrt{2m_b E_b}\sin\theta &= \sqrt{2m_B E_B}\sin\phi,
\end{aligned}
\tag{1.6}
$$

where the *Q-value of the reaction* measures the energy gained (or lost) due to the difference between the initial and final masses:

$$Q = (m_a + m_A - m_b - m_B)c^2. \tag{1.7}$$

Eliminating E_B and ϕ from (1.6) we can relate Q to the parameters of collision that can be measured in a reaction:

$$Q = E_b\left(1 + \frac{m_b}{m_B}\right) - E_a\left(1 - \frac{m_a}{m_B}\right) - \frac{2}{m_B}\sqrt{m_a m_b E_a E_b}\cos\theta. \tag{1.8}$$

A useful relation for the analysis of a nuclear reaction is obtained by observing that (1.8) is an equation of the second degree in $\sqrt{E_b}$, whose solution is:

$$\sqrt{E_b} = \frac{1}{m_b + m_B}\left\{\sqrt{m_a m_b E_a}\cos\theta \pm \sqrt{m_a m_b E_a \cos^2\theta + (m_b + m_B)[E_a(m_B - m_a) + Qm_B]}\right\}. \tag{1.9}$$

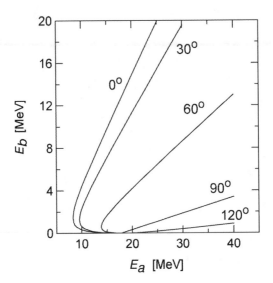

FIGURE 1.6   Energy E_b of boron nuclei in the reaction $^{12}C + ^{14}N \rightarrow {}^{10}B + {}^{16}Al$, as function of the energy E_a of the incident ^{12}C nuclei, for several scattering angles.

If we plot the energy E_b of the emitted particles in a graph, observed at an angle θ, as a function of the energy E_a of the incident particle, we obtain a set of curves, one for each value of θ. Figure 1.6 exhibits the curves obtained for reaction $^{12}C + ^{14}N \rightarrow {}^{10}B + {}^{16}O$, where $Q = -4.4506$ MeV.

Two things are evident when observing Figure 1.6. In the first case, since Q is negative for that reaction, an energy threshold exists for the incident particle, $E_a = E_t$, as a function of the angle θ, below which nuclei b are not observed in that angle. At these energies the square root of (1.9) vanishes, and this condition yields:

$$E_t = \frac{-Qm_B(m_B + m_b)}{m_a m_b \cos^2 \theta + (m_B + m_b)(m_B - m_a)}. \tag{1.10}$$

These thresholds are just due to the nuclear force, they could be smaller when the Coulomb repulsion is taken into account.

The smallest value of E_t in (1.10),

$$E_t = \frac{-Q(m_B + m_b)}{m_B + m_b - m_a}, \tag{1.11}$$

occurs for $\theta = 0$, and is the absolute threshold of the reaction, i.e., the smallest value of the incident energy E_a for which the reaction can occur. If $Q > 0$ (*exothermic* reactions), the threshold is negative and the reaction occurs for any incident energy. If $Q < 0$ (*endothermic* reactions), an incident energy $E_a = E_t$ begins to produce events for $\theta = 0°$. With the increase of E_a starting from E_t, other angles become accessible.

The second important observation to be done in Figure 1.6 is with respect to the unique relation between E_b and E_a. For each energy E_a exists only one value of E_b for each angle θ, except in the region of energy between 8.26 MeV and 17.82 MeV. The first number corresponds to the energy threshold (1.11). The second can be determined setting the numerator of (1.9) to zero; this implies

$$E_a' = \frac{-m_B Q}{m_B - m_a}, \tag{1.12}$$

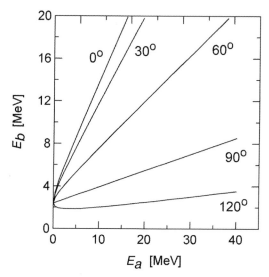

FIGURE 1.7 Energy of the nucleus ^{12}C emitted in the reaction ^{10}B $+^{16}$ O $\rightarrow^{12}$ C $+^{14}$ N.

independent of the value of the angle θ. Thus, the curves for all θ angles cut the horizontal axis at the same point.

In a scattering experiment with collision energy E, a channel with $-Q > E$ can never be excited. The reason is that the maximal kinetic energy available to be transformed into intrinsic excitation is E, which is not enough. As an example, the inelastic process

$$e + H \rightarrow e + H^*,$$

with an electron beam of $E = 10$ eV is forbidden by energy conservation, since the lowest excited state in H has the energy $\varepsilon_1 = 10.2$ eV. The lowest collision energy to allow the excitation of an endothermic channel is called *threshold*. In the above example, $E_{threshold} = 10.2$ eV. When the collision energy is below the threshold for a given channel, one says that this channel is *closed*. Otherwise, it is said to be *open*.

The region of double values of E_b only exists for endothermic reactions; when $Q > 0$, the correspondence between E_b and E_a is also unique for all energies and any value of θ. This is a clear consequence of (1.12) and is illustrated in Figure 1.7, where the inverse reaction to the one of the Figure 1.6 is plotted.

One also sees from (1.10) and (1.12) that the energy region where double values of E_b happen is wider when the participants of the reaction have comparable masses.

It is worthwhile to observe, that only values of θ within $0°$ and $90°$ admit a region of double values for the energy; for $\theta > 90°$ the first part of (1.9) is negative and is not possible to have more than one value of E_b.

1.6   Cross sections, center of mass, and laboratory frames

Let us consider an experiment as that represented in Figure 1.1, measuring the count rate of events leading to the population of channel-α, $N_\alpha(\Omega, \Delta\Omega)$. A few considerations should be made:

1. If the detector aperture is small, $N_\alpha(\Omega, \Delta\Omega)$ should be proportional to $\Delta\Omega$.

2. Assuming that multiple scattering can be neglected, $N_\alpha(\Omega, \Delta\Omega)$ should be proportional to the number of target particles in the region attained by the beam, n.

3. Assuming that interactions between beam particles can be neglected, the count rate $N_\alpha(\Omega, \Delta\Omega)$ should be proportional to the incident flux J.

$N_\alpha(\Omega, \Delta\Omega)$ can then be written

$$N_\alpha(\Omega, \Delta\Omega) = \Delta\Omega \cdot n \cdot J \cdot \frac{d\sigma_\alpha(\Omega)}{d\Omega}.$$

The constant of proportionality,

$$\frac{d\sigma_\alpha(\Omega)}{d\Omega} = \frac{N_\alpha(\Omega, \Delta\Omega)}{\Delta\Omega \cdot n \cdot J}, \tag{1.13}$$

is called the *differential cross section* for channel-α. This quantity is very useful, since it does not depend on experimental details (detector size, incident flux, target thickness). It depends exclusively on the physics of the projectile and target particle interaction. The elastic scattering cross section plays a unique role, since it appears in any collision.

Frequently, one is interested in the integrated cross section (over scattering angles). As an example, the (total) elastic cross section is given as

$$\sigma_{el} = \int d\Omega \left[\frac{d\sigma_{el}(\Omega)}{d\Omega} \right]. \tag{1.14}$$

For the comparison between theory and experiment, it is necessary to have cross sections in the same reference frame. Of course, the measured cross section is obtained in the laboratory frame, where the target is at rest. From the theoretical point of view, however, it is important to take advantage of the translational invariance of the projectile-target Hamiltonian, introducing the center of mass frame (CM). Calling $\mathbf{r}_p$ and $\mathbf{r}_t$ the vector positions of the projectile and target respectively, and $\mathbf{p}_p$ and $\mathbf{p}_t$ their associated momenta, one introduces the transformations

$$\mathbf{r}_p; \mathbf{r}_t \rightarrow \mathbf{R}_{cm} = \frac{m_p \mathbf{r}_p + m_t \mathbf{r}_t}{M}; \quad \mathbf{r} = \mathbf{r}_p - \mathbf{r}_t, \tag{1.15}$$

and

$$\mathbf{p}_p; \mathbf{p}_t \rightarrow \mathbf{P}_{cm} = \mathbf{p}_p + \mathbf{p}_t; \quad \mathbf{p} = \frac{m_t \mathbf{p}_p - m_p \mathbf{p}_t}{M}, \tag{1.16}$$

where m_p and m_t are respectively the projectile and target masses and $M = m_p + m_t$. Let us discuss the consequences of this transformation on the system Hamiltonian in the simple case of potential scattering, where intrinsic degrees of freedom are not taken into account. With this transformation, the Hamiltonian,

$$H = \frac{\mathbf{p}_p^2}{2m_p} + \frac{\mathbf{p}_t^2}{2m_t} + V(\mathbf{r}_p - \mathbf{r}_t),$$

separates into a center of mass and a relative term

$$H = H(\mathbf{R}_{cm}, \mathbf{P}_{cm}) + H(\mathbf{r}, \mathbf{p}).$$

Above,

$$H(\mathbf{R}_{cm}, \mathbf{P}_{cm}) = \frac{\mathbf{P}_{cm}^2}{2M}; \qquad H(\mathbf{r}, \mathbf{p}) = \frac{\mathbf{p}^2}{2\mu} + V(\mathbf{r}), \tag{1.17}$$

where $\mu = m_p m_t / M$ is the reduced mass. In this way, the system wave function takes the product form

$$\psi(\mathbf{R}_{cm}, \mathbf{r}, t) = \psi(\mathbf{R}_{cm}, t) \cdot \psi(\mathbf{r}, t).$$

The center of mass wave function is a trivial plane wave which can be ignored in the context. The relative wave function, which contains all the relevant information, is given by the Schrödinger equation:

$$\left[-\frac{\hbar^2}{2\mu}\nabla_{\mathbf{r}}^2 + V(\mathbf{r})\right]\psi(\mathbf{r},t) = i\hbar\frac{\partial\psi(\mathbf{r},t)}{\partial t}. \tag{1.18}$$

It should be remarked that the transformation to center of mass and relative motion variables has a very important practical consequence. The number of coordinates in the Schrödinger equation is reduced from six $(\mathbf{r}_p, \mathbf{r}_t)$ to three $(\mathbf{r})$.

For the comparison of data with theoretical cross sections, it is necessary to relate the collision energies and the cross sections in the CM and in the laboratory (L) frames. The collision energies[*] in these frames are[§]

$$E_{lab} = \frac{\mathbf{P}_p^2}{2m_p}, \qquad E_{cm} = \frac{\mathbf{p}^2}{2\mu}.$$

Using Eq. (1.16) and the definition of the reduced mass, we find the relation

$$E_{lab} = \frac{M}{m_t}E_{cm}. \tag{1.19}$$

When the target is much heavier than the projectile, E_{lab} is very close to E_{cm}. Otherwise, it can be much larger.

To transform the theoretical cross section to the laboratory frame, one should consider that the count rate $N_\alpha(\Omega, \Delta\Omega)$ in the detector is independent of the reference frame. Similarly, the incident flux is also independent of the reference frame. The flux is given as the product of the beam density, the same for any frame, and the relative velocity of the projectiles. Since in the laboratory frame the incident velocity of the projectile is $\mathbf{v}_p$ and the target is at rest, the relative velocity, obtained by differentiation of Eq. (1.15) with respect to time, is $\mathbf{v} = \mathbf{v}_p$. Therefore, the flux is the same in the CM and in the laboratory frames. The desired transformation then depends exclusively on the direction angle Ω ($\Omega \equiv \theta, \varphi$) and the elementary solid angle $\Delta\Omega$. We can write

$$\frac{d\sigma_{el}(\Omega_{lab})}{d\Omega_{lab}} = \frac{d\sigma_{el}(\Omega_{cm})}{d\Omega_{cm}} \cdot \frac{d\Omega_{cm}}{d\Omega_{lab}}, \tag{1.20}$$

where $d\Omega_{cm}/d\Omega_{lab}$ is a trivial geometrical factor, which can be calculated within elementary classical mechanics. It is evaluated in the rolling text, for the case of elastic scattering.

1.6.1 Transformation between laboratory and center of mass system

The initial velocities are $\mathbf{v}_p$ and $\mathbf{v}_t = 0$, or (taking time derivatives of Eq. (1.15)) $\mathbf{V}_{cm} = m_p\mathbf{v}_p/M$ and $\mathbf{v} = \mathbf{v}_p$. In the center of mass the absolute magnitude of the target and projectile velocities are given by V_{cm} and $v_p - V_{cm} = m_t v/M$, respectively (use Eqs. 1.15 to derive this relation). The final velocities in the laboratory are $\mathbf{v}_p'$ and $\mathbf{v}_t'$. In the center of mass the projectile velocity can be expressed in terms of the final center of mass and the relative velocities as $\mathbf{v}_p' = \mathbf{V}_{cm}' + m_t\mathbf{v}'/M$ (see Figure 1.8). Momentum and energy conservation requires that $\mathbf{V}_{cm} = \mathbf{V}_{cm}'$ and $v' = v$. Since the center of mass velocity is assumed to be along the z-axis, the transformation will keep unchanged the polar coordinate, φ, so that

$$\frac{d\Omega_{cm}}{d\Omega_{lab}} = \frac{\sin\theta_{cm} \cdot d\theta_{cm}}{\sin\theta_{lab} \cdot d\theta_{lab}}. \tag{1.21}$$

---

[*]Note that E_{cm} is the energy of the relative motion, which corresponds to the total collision energy in the CM frame. It is **not** the energy of the center of mass motion in the laboratory frame, as this notation could suggest.

[§]We assume that the target is at rest in the laboratory frame. This assumption is not valid in experiments involving colliding beams.

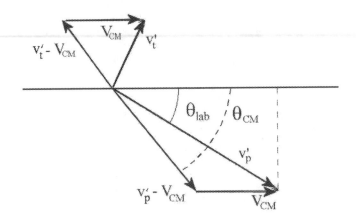

FIGURE 1.8 Transformation of center of mass velocities to the laboratory.

The angles θ_{cm} and θ_{lab} are given by

$$\cos\theta_{cm} = \frac{\mathbf{v}' \cdot \hat{\mathbf{z}}}{v},$$
(1.22)

and

$$\cos\theta_{lab} = \frac{\mathbf{v}'_p \cdot \hat{\mathbf{z}}}{v'_p}.$$
(1.23)

Replacing

$$\mathbf{v}'_p \cdot \hat{\mathbf{z}} = V_{cm} + \frac{m_t}{M}\mathbf{v}' \cdot \hat{\mathbf{z}} = V_{cm} + \frac{m_t}{M}v\cos\theta_{cm},$$

and

$$v'_p = \sqrt{V_{cm}^2 + \left(\frac{m_t}{M}\right)^2 v^2 + 2\frac{m_t}{M}V_{cm}v\cos\theta_{cm}},$$

in Eq. (1.23), we get

$$\cos\theta_{lab} = \frac{V_{cm} + \frac{m_t}{M}v\cos\theta_{cm}}{\sqrt{V_{cm}^2 + \left(\frac{m_t}{M}\right)^2 v^2 + 2\frac{m_t}{M}V_{cm}v\cos\theta_{cm}}}.$$
(1.24)

A simpler formula can be derived if we divide both the numerator and the denominator by V_{cm}. Calling

$$x = \left(\frac{m_t}{M}\frac{v}{V_{cm}}\right)^{-1} = m_p/m_t,$$

we obtain

$$\cos\theta_{lab} = \frac{1 + x^{-1}\cos\theta_{cm}}{\sqrt{1 + x^{-2} + 2x^{-1}\cos\theta_{cm}}}.$$
(1.25)

The above equation indicates that if the projectile is heavier than the target ($x > 1$) it is always deflected into the forward hemisphere, $\cos\theta_{lab} > 0$. Solving Eq. (1.25) for $\cos\theta_{cm}$, we obtain the inverse transformation

$$\cos\theta_{cm} = \cos\theta_{lab}\left[x\cos\theta_{lab} + \sqrt{1 - x^2\sin^2\theta_{lab}}\right] - x.$$
(1.26)

Differentiating both sides of Eq. (1.24) and using Eqs. (1.21) and (1.20), we obtain the desired transformation

$$\frac{d\sigma_{el}(\Omega_{lab})}{d\Omega_{lab}} = \frac{d\sigma_{el}(\Omega_{cm})}{d\Omega_{cm}} \cdot \frac{\left(1 + 2x\cos\theta_{cm} + x^2\right)^{3/2}}{|1 + x\cos\theta_{cm}|}.$$
(1.27)

For inelastic collisions, $A(a, b)B$, one can show that Eqs. (1.24)–(1.27) are still valid, if one uses

$$x = \left[\frac{m_a m_b}{m_A m_B} \frac{E_{cm}}{E_{cm} + Q} \right]^{1/2}, \tag{1.28}$$

where $E_{cm} = m_a E_{lab}/(m_a + m_A)$ is the collision energy in the center of mass.

When $x = 1$, as for the elastic scattering of particles of the same mass, Eq. (1.26) gives $\theta_{cm} = \theta_{lab}/2$, so that θ_{lab} cannot be larger than $\pi/2$. Eq. (1.27) becomes in this case

$$\frac{d\sigma_{el}(\Omega_{lab})/d\Omega_{lab}}{d\sigma_{el}(\Omega_{cm})/d\Omega_{cm}} = 4 \cos \theta_{lab}. \tag{1.29}$$

Thus, even if the angular distribution is isotropic in the center of mass, it will be proportional to $\cos \theta_{lab}$ in the laboratory. Also, it is easy to infer from Figure 1.8 that the scattered and recoil particles move at right angles at the laboratory, i.e., $\mathbf{v}'_p \cdot \mathbf{v}'_t = 0$.

Similar equations can be deduced in the relativistic case. We shall only summarize the results here. For more details, see Ref. [4]. In the laboratory, the scattering angle is given by [4]

$$\theta_{lab} = \arctan \left\{ \frac{\sin \theta_{cm}}{\gamma \left[\cos \theta_{cm} + x g(x, \mathcal{E}_1) \right]} \right\}, \tag{1.30}$$

where, $x = m_p/m_t$,

$$\mathcal{E}_1 = \frac{K_{lab}}{m_p c^2}, \tag{1.31}$$

where K_{lab} is the kinetic energy of the projectile in the laboratory frame and m_p is the projectile rest mass. Thus, $\mathcal{E}_1$ is the kinetic energy of the incident particle in units of the rest mass. The function $g(x, \mathcal{E}_1)$ and γ in Eq. (1.30) can be expressed as

$$g(x, \mathcal{E}_1) = \frac{1 + x\left(1 + \mathcal{E}_1\right)}{1 + x + \mathcal{E}_1}, \qquad \gamma = \frac{1 + x + \mathcal{E}_1}{\sqrt{(1+x)^2 + 2x\mathcal{E}_1}}. \tag{1.32}$$

The relativistic Lorentz factor $\gamma = \left(1 - V_{cm}^2/c^2\right)^{-1/2}$ pertains the motion of the center of mass system with respect to the laboratory. At non-relativistic energies, γ and g tend to unity, and Eq. (1.30) reduces to Eq. (1.25). The laboratory cross section is

$$\frac{d\sigma_{el}(\Omega_{lab})}{d\Omega_{lab}} = \frac{\left\{ \gamma^2 \left[x g(x, \mathcal{E}_1) + \cos \theta \right]^2 + \sin^2 \theta \right\}^{3/2}}{\gamma \left[1 + x g(x, \mathcal{E}_1) \cos \theta \right]} \frac{d\sigma_{el}(\Omega_{cm})}{d\Omega_{cm}}. \tag{1.33}$$

It is also easy to see that in the non-relativistic limit ($g \longrightarrow 1$, $\gamma \longrightarrow 1$), the above equation reduces to Eq. (1.27).

1.6.2 Kinematic conditions

In the reaction $a + A \longrightarrow b + B$ if we assume that b is at rest in the laboratory, the energy ϵ_i available initially in the center of mass is given by the sum of the energy of the two nuclei. With the aid of $\mathbf{V}_{cm} = m_a \mathbf{v}_a/M$, one obtains

$$\epsilon_i = \frac{1}{2} m_a v_a^2 \left(1 - \frac{m_a}{m_a + m_A} \right)^2 + \frac{1}{2} m_A v_a^2 \left(\frac{m_a}{m_a + m_A} \right)^2 = E_a \frac{\mu}{m_a}. \tag{1.34}$$

The final kinetic energy is given, naturally, by

$$\epsilon_f = \epsilon_i + Q, \tag{1.35}$$

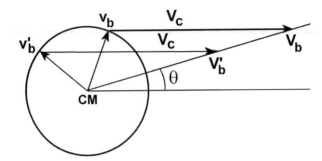

FIGURE 1.9   The diagram of velocities shows how, starting from two different velocities v_b and v_b' in the CM, one can arrive to the corresponding velocities V_b and V_b' in the L, causing that one can observe particles with different energies at the same angle θ. V_C is the speed of the L in the CM.

because Q measures exactly the energy gained with the rearrangement during the reaction. Using (1.34) in (1.35),

$$\epsilon_f = E_a \frac{\mu}{m_a} + Q. \tag{1.36}$$

If E_a is the same as the energy threshold (1.11), the final energy in the CM is written:

$$\epsilon_f = -Q \frac{m_a + m_A - Q/c^2}{m_A - Q/c^2} \frac{m_A}{m_a + m_A} + Q \cong 0, \tag{1.37}$$

where we made use of (1.7). The amount Q/c^2 was neglected in the end result of (1.37) since it is small compared with the involved masses. The result shows that, at the threshold, the particles have zero final kinetic energy at the CM. All the initial kinetic energy was consumed to supply the mass gain, i.e., the Q-value of the reaction. The final energy (1.37) is not strictly zero because the equation (1.6) and the equations derived from that were not written in a fully relativistic form. When this is done, Eq. (1.11) should be replaced by*

$$T_t = \frac{-Q(m_a + m_A + m_b + m_B)}{2m_A}, \tag{1.38}$$

which produces the exact result $\epsilon_f = 0$ when, in the passage from the L to the CM, the appropriate relativistic transformations are used.

The analysis of the collision in the CM is also convenient to understand physically the possibility of two values of the energy E_b at a given angle for a single incident energy E_a. In the CM, for a given value of E_a corresponds a single value of E_b that respects the conservation of the zero total momentum and of the energy. Particles with energy E_b are detected at all angles and this gives opportunity to the appearance of a double values for the energy in the L. The diagram of velocities in Figure 1.9 illustrates how this happens.

It is also easy to understand from the diagram why the duplicity of energy is limited to smaller angles than 90°. Also, it is not difficult to see that the phenomenon disappears at a certain energy value of the incident particle.

---

*T_t is used here as kinetic energy of a particle to comply with the usual relativistic notation where E_t is used for total energy of a particle = rest energy + kinetic energy.

1.7   Classical scattering

We first study the elastic cross section from the classical mechanics point of view (see, for example, Ref. [4]). We consider the relative motion of the projectile-target system with the Hamiltonian $H(\mathbf{r}, \mathbf{p})$ of Eq. (1.17). If the potential $V(r)$ is conservative and spherically symmetric, the energy E and the angular momentum vector $\mathbf{L}$ are conserved. Conservation of the orientation of $\mathbf{L}$ implies that the trajectory is kept on the plane determined by the initial velocity, $\mathbf{v}$, and the target. A typical situation is represented in Figure 1.10. Choosing the z-axis parallel to the incident velocity, the trajectory is fully determined by the absolute value of the initial velocity:

$$v = \sqrt{2E/\mu}, \tag{1.39}$$

and the impact parameter,

$$b = \frac{L}{\mu v}. \tag{1.40}$$

We describe the motion in terms of the coordinates r and ϕ on the collision plane, indicated in Figure 1.10. Using conservation of energy and angular momentum, we get

$$E = \frac{1}{2}\mu\left[\left(\frac{dr}{dt}\right)^2 + r^2\left(\frac{d\phi}{dt}\right)^2\right] + V(r), \tag{1.41}$$

and

$$L = \mu vb = \mu r^2\left(\frac{d\phi}{dt}\right) \quad\longrightarrow\quad \frac{d\phi}{dt} = \frac{vb}{r^2}. \tag{1.42}$$

Taking $d\phi/dt$ from Eq. (1.42) and replacing in Eq. (1.41), we get

$$\frac{dr}{dt} = \pm v\sqrt{\left(1 - \frac{b^2}{r^2} - \frac{V(r)}{E}\right)}, \tag{1.43}$$

where the minus and plus signs apply for the incoming and outgoing branches of the trajectory, respectively. We now replace $d\phi/dt = [dr/dt][d\phi/dr]$ in Eq. (1.42) and get

$$\frac{dr}{dt} = \frac{vb}{r^2}\left[\frac{d\phi}{dr}\right]^{-1}. \tag{1.44}$$

Using Eqs. (1.43) and (1.44), we obtain the differential equation

$$\frac{d\phi}{dr} = \pm\frac{b}{r^2\sqrt{1 - b^2/r^2 - V(r)/E}}.$$

Integrating over the incoming and outgoing branches of the trajectory and using the appropriate sign in each case, we get the asymptotic value of ϕ for a collision with impact parameter b:

$$\phi(b) = 2\int_{R_{ca}}^{\infty}\frac{bdr}{r^2\sqrt{1 - b^2/r^2 - V(r)/E}}.$$

The lower limit of the integral, R_{ca}, is determined by the condition

$$1 - b^2/R_{ca}^2 - \frac{V(R_{ca})}{E} = 0. \tag{1.45}$$

The deflection of the projectile is determined from $\phi(b)$ (see Figure 1.10), through the expression $\Theta(b) = \pi - \phi(b)$, or explicitly

$$\Theta(L) = \pi - 2\int_{R_{ca}}^{\infty}\frac{bdr}{r^2\sqrt{1 - b^2/r^2 - V(r)/E}}. \tag{1.46}$$

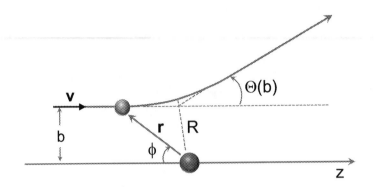

FIGURE 1.10   Representation of a collision between classical particles.

$\Theta(b)$ is called the *deflection function*. It plays a very important role in the classical and in semi-classical descriptions of the differential elastic cross section. Frequently, the deflection function is given in terms of the angular momentum L, instead of b. In this case, we have

$$\Theta(L) = \pi - \int_{R_{ca}}^{\infty} \frac{L\,dr}{r^2 \sqrt{2\mu\left[E - L^2/2\mu r^2 - V(r)\right]}}, \tag{1.47}$$

or

$$\Theta(L) = \pi - \int_{R_{ca}}^{\infty} \frac{L\,dr}{r^2 \sqrt{2\mu\left[E - V_L(r)\right]}}, \tag{1.48}$$

where $V_L(r)$ is the effective potential (containing the centrifugal term)

$$V_L(r) = V(r) + \frac{L^2}{2\mu r^2}. \tag{1.49}$$

1.8   The classical cross section

To evaluate $d\sigma(\Omega)/d\Omega$, we consider a beam of projectiles approaching a single target particle with random impact parameters, with intensity of J projectiles per unit area per unit time. If the potential is spherically symmetric, the cross section has axial symmetry and, therefore, $N(\Omega, \Delta\Omega) \to N(\theta, 2\pi \sin\theta\Delta\theta)$. In this way, $N(\Omega, \Delta\Omega)$ corresponds to the number of projectiles scattered between two cones with apertures θ and $\theta + \Delta\theta$, indicated in Figure 1.11. Using the inverse of the deflection function*, we can determine the impact parameters b and $b + \Delta b$ which lead to the deflections θ and $\theta + \Delta\theta$, respectively. The number of particles scattered within the solid angle $\Delta\Omega$ is then given by the number of incident particles on the area between the circles with radii b and $b + \Delta b$, $\Delta S = 2\pi b\Delta b$. Namely,

$$N(\Omega, \Delta\Omega) = J\Delta S = 2\pi b\Delta b J.$$

---

*Note that the inversion of the deflection function may present problems. In some situations different impact parameters result in the same deflection angle. In this case the cross section is given by the sum of contributions from these impact parameters. The second difficulty is when no impact parameter leads to the required deflection. In this case, the classical cross section vanishes for this particular angle.

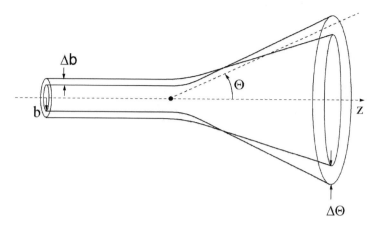

FIGURE 1.11 The elementary solid angle $\Delta\Omega$ and the associated area $2\pi b\, db$.

Inserting this result in Eq. (1.13), we get (with $n = 1$)

$$\frac{d\sigma(\Omega)}{d\Omega} = \frac{b}{\sin\theta}\left|\frac{\Delta b}{\Delta\theta}\right|, \tag{1.50}$$

where the absolute value is used to guarantee positive cross sections. Using the deflection function, we can express $\Delta\theta$ in terms of Δb. Replacing

$$\Delta\theta = \frac{d\Theta(b)}{db}\Delta b$$

in Eq. (1.50), we obtain the classical differential cross section

$$\frac{d\sigma(\Omega)}{d\Omega} = \frac{b}{\sin\theta}\left|\frac{d\Theta(b)}{db}\right|^{-1}. \tag{1.51}$$

In most situations this expression is inappropriate for comparison with experiment. However, this classical treatment gives a good insight of the problem and is helpful as a starting point for approximations. If different values of the impact parameter, $b_1, b_2, \cdots, b_n$ lead to the same angle θ, the classical cross section is generalized as

$$\frac{d\sigma(\Omega)}{d\Omega} = \sum_{i=1}^{n}\frac{b_i}{\sin\theta}\left|\frac{d\Theta(b)}{db}\right|^{-1}_{b=b_i}. \tag{1.52}$$

In fact, the quantum mechanical result is rather different. The above formula has the important shortcoming of neglecting interference between the different trajectories leading to the same scattering angle.

1.9 Rutherford scattering

Let us apply the above results to the Rutherford scattering. This is the case of the Coulomb potential between point particles[§],

$$V_C(r) = \frac{q_p q_t}{r}, \tag{1.53}$$

[§]We use here the CGS unit system.

where q_p and q_t are respectively the projectile and target charges. Because of the slow fall-off with distance, Coulomb scattering requires a special treatment.

Let us first study Coulomb scattering using classical mechanics. It is convenient to introduce the *Sommerfeld parameter*

$$\eta = \frac{q_p q_t}{\hbar v},$$ (1.54)

and the half distance of closest approach in a head-on collision

$$a = \frac{q_p q_t}{2E}.$$ (1.55)

These quantities are related as

$$\eta = ka.$$ (1.56)

Using Eqs. (1.53), (1.54), and (1.55) in the expression for the deflection function (Eq. (1.47)), we can write

$$\frac{\Theta_c(b)}{2} = \frac{\pi}{2} - \frac{\phi(b)}{2},$$ (1.57)

where $\phi(b)$ is given by the integral

$$\frac{\phi(b)}{2} = b \int_{R_{ca}}^{\infty} \frac{dr}{r\sqrt{r^2 - 2ar - b^2}}.$$ (1.58)

Before evaluating the integral, we must find the distance of closest approach, R_{ca}, as a function of b. Setting the closest approach condition: $\dot{r} = 0$ in Eq. (1.43) and using Eq. (1.55), we get

$$R_{ca} = a + \sqrt{a^2 + b^2}.$$ (1.59)

The integral of Eq. (1.58) is given in standard integral tables (see, for example [4]),

$$\frac{\phi(b)}{2} = \left[\sin^{-1}\left(-\frac{ar + b^2}{r\sqrt{a^2 + b^2}} \right) \right]_{R_{ca}}^{\infty} = \frac{\pi}{2} - \sin^{-1}\left[\frac{a}{\sqrt{a^2 + b^2}} \right].$$ (1.60)

Using Eq. (1.60) in Eq. (1.57), we get

$$\frac{\Theta_c}{2} = \sin^{-1}\left[\frac{a}{\sqrt{a^2 + b^2}} \right].$$ (1.61)

The above equation implies that:

$$\sin\left(\frac{1}{2}\Theta_c \right) = \frac{a}{\sqrt{a^2 + b^2}}, \qquad \cos\left(\frac{1}{2}\Theta_c \right) = \frac{b}{\sqrt{a^2 + b^2}}, \qquad \tan\left(\frac{1}{2}\Theta_c \right) = \frac{a}{b}.$$ (1.62)

The deflection function can be specified by any of these equations. Usually, one uses

$$\Theta_c = 2\tan^{-1}\left(\frac{a}{b} \right).$$ (1.63)

The classical cross section is given by Eq. (1.51), with $\Theta = \Theta_c$. Evaluating the derivative of Eq. (1.63),

$$\frac{d\Theta_c}{db} = -\frac{2a}{a^2 + b^2},$$

we get

$$\frac{d\sigma_c(\Omega)}{d\Omega} = \frac{b}{\sin\theta}\frac{a^2 + b^2}{2a} = \frac{a^2}{2\sin\theta}\frac{1}{\dfrac{a}{b}\left(\dfrac{a^2}{a^2 + b^2} \right)}.$$

Using Eqs. (1.62) and writing $\sin\theta = 2\sin(\theta/2)\cos(\theta/2)$, we get the *Rutherford cross section*

$$\frac{d\sigma_c(\Omega)}{d\Omega} = \frac{a^2}{4}\left[\frac{1}{\sin^4(\theta/2)}\right]. \tag{1.64}$$

Note that this expression diverges as $\theta \to 0$. This expresses the fact that the infinite range of the Coulomb interaction deflects projectiles with any impact parameter. Collisions with $b \to \infty$, contribute to $\theta = 0$ and the cross section becomes infinite at this angle.

1.10 Orbiting, rainbow, and glory scattering

Inspecting the deflection function (Eqs. (1.46) or (1.48)), one realizes that the integrand is singular at the lower integration limit. One should therefore investigate the influence of this behavior on the deflection function. For this purpose, we write the integral appearing in Eq. (1.48),

$$I(L) = \int_{R_{ca}}^{\infty} \frac{L\,dr}{r^2\sqrt{2\mu\left[E - V_L(r)\right]}},$$

as the sum $I(L) = I_1(L) + I_2(L)$, where

$$I_1(L) = \int_{R_{ca}}^{R_{ca}+\Delta} \frac{L\,dr}{r^2\sqrt{2\mu\left[E - V_L(r)\right]}}, \tag{1.65}$$

and

$$I_2(b) = \int_{R_{ca}+\Delta}^{\infty} \frac{L\,dr}{r^2\sqrt{2\mu\left[E - V_L(r)\right]}}.$$

When Δ is finite, $I_2(b)$ is bound*. In this way, only $I_1(L)$ could give rise to singularities. To investigate this integral, we change $r \to x = r - R_{ca}$, introduce the notation

$$F(x, L) = 2\mu\left[E - V_L(x + R_{ca})\right], \tag{1.66}$$

and express $I_1(L)$ as the limit

$$I_1(L) = L\lim_{\epsilon\to 0}\left[\int_{\epsilon}^{\Delta} \frac{dx}{(x + R_{ca})^2\sqrt{F(x, L)}}\right]. \tag{1.67}$$

Choosing a small Δ, we can expand

$$F(x, L) = F'(0, L)x + F''(0, L)\frac{x^2}{2} + \mathcal{O}\left[x^3\right], \tag{1.68}$$

in the full interval $[\epsilon, \Delta]$. Above, primes and double-primes stand for first and second derivatives with respect to r. If $F'(0, L) \neq 0$, we obtain

$$I_1(L) = \lim_{\epsilon\to 0}\left[\frac{2L}{R_{ca}^2\sqrt{F'(0, L)}}\sqrt{x}\left(1 + \mathcal{O}\left[x\right]\right)\right]_{\epsilon}^{\Delta} = \frac{2L}{R_{ca}^2\sqrt{F'(0, L)}}\sqrt{\Delta}. \tag{1.69}$$

We conclude that $I_1(L)$ is finite and therefore the deflection function is well defined.

---

*In any relevant situation, the potential asymptotically goes to zero, at least as fast as $1/r$ (Coulomb case). As shown in the previous sections, this integral converges, in spite of the infinite upper limit.

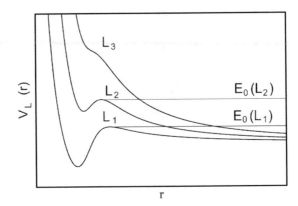

FIGURE 1.12   The effective potentials and the orbiting energy for different angular momenta.

The situation is rather different when $F'(0, L)$ vanishes, which can be expressed by the condition

$$V_L'(R_{ca}) = 0. \tag{1.70}$$

Since R_{ca} depends on both E and L, the *orbiting* condition of Eq. (1.70) is a relation between E and L. That is, it defines the orbiting energy $E_O(L)$ for a given angular momentum or, equivalently, the orbiting angular momentum for a given energy $L_O(E)$.

For repulsive potentials, $V'(r) < 0 \implies V_L'(r) < 0$ for any r and L, so that the above condition is never met. However, if the potential is attractive, at least over some r-range, this condition may possibly be satisfied. Let us consider the orbiting condition in the case of a potential $V(r)$ that is attractive everywhere. This is illustrated in Figure 1.12, which shows the effective potentials * for the angular momenta $L_1, L_2,$ and L_3. The orbiting energy in each case is determined by the condition that the classical turning point is located at the potential maximum. This procedure is indicated in the Figure for the angular momenta L_1 and L_2 (see left side of Figure 1.13). There is a critical value of the angular momentum, L_{crit}, above which the effective potential has no maximum. Above this limit no orbiting occurs. An example of this situation is the effective potential for L_3 in Figure 1.12 (see also Figure 1.13).

When the pair of variables $\{E, L\}$ satisfies the orbiting condition, the dominant term in the expansion of Eq. (1.68) is of second order and it leads to the integral

$$I_1(L) = L\sqrt{\frac{2}{F''(0, L)}} \lim_{\epsilon \to 0} \left[\int_\epsilon^\Delta \frac{dx}{x} \right] \to \infty.$$

In this case, $I(L)$ has a logarithmic divergence and we can we can write

$$\lim_{L \to L_O} \Theta(L) \to -\infty. \tag{1.71}$$

The meaning of this equation is that in a collision with energy E and angular momentum L_O the projectile is caught in an orbit with constant radius $r = R_{ca}$ and it does not emerge. In collisions with $L \simeq L_O$ the projectile goes around the target several times before coming out. In Figure 1.14 we show deflection functions presenting orbiting. In (a), the potential is attractive

---

*Note that, although $V(r) < 0$, the centrifugal terms for $L_1, L_2,$ and L_3 dominate the effective potential, so that $V_L(r) > 0$.

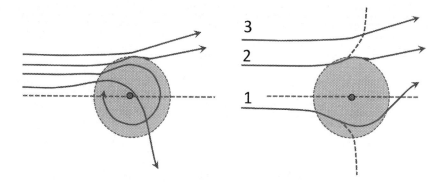

FIGURE 1.13 Classical trajectories for scattering by a potential, attractive at short distances and repulsive at large distances. On the left, from top to bottom, Rutherford, grazing, orbiting and plunging orbits are shown. On the right are shown three different trajectories leading to the same scattering angle.

everywhere, so that the deflection angle is negative for any angular momentum. In (b), the potential is the superposition of a short-range attractive part with a long-range repulsion. The orbiting angular momentum, L_O, is indicated in each case.

In the scattering from a potential which is attractive at small separations and repulsive at large distances, as the one used in Figure 1.14(b), other interesting phenomena can occur (see right side of Figure 1.13). The first one is *rainbow scattering*, arising for the angular momentum L_R (see Figure 1.14(b)). At this angular momentum, the deflection function goes through a maximum. This has two important consequences. The first one is that, according to classical mechanics, no particle of the incident beam is deflected to angles $\theta > \Theta_R$. The second consequence is that the scattering cross section at this angle diverges. This could be checked inserting $[d\Theta/db]_{b=b_R} = [d\Theta/dL]_{L=L_R} = 0$ in Eq. (1.51). In fact, the predictions of quantum mechanics, which are consistent with observation, are not the same. The cross section remains finite at $\theta = \Theta_R$ and it does not vanish at larger angles. However, The classical results are qualitatively right, in the sense that the observed cross section presents a maximum near the rainbow angle and it shows a shadow region beyond this angle, where $d\Theta/d\Omega$ decreases rapidly. This situation is analogous to the rainbow phenomenon, which occurs in the scattering of light from liquid drops contained in clouds.

Close to the rainbow angle $\Theta = \Theta_R$ we may use a parabolic approximation

$$\Theta = \Theta_R - \alpha^2 (L - L_R)^2 , \tag{1.72}$$

where the constant α is determined by the curvature of $\Theta(L)$ near $L = L_R$. From Eq. 1.51 this gives the differential cross-section near $\Theta = \Theta_R$ as

$$\frac{d\sigma}{d\Omega} = \frac{L_R}{2\alpha p^2 (\theta - \theta_R)^{1/2} \sin\theta_R} , \quad \text{for} \quad \theta < \theta_R$$

$$= 0 , \quad \text{for} \quad \theta > \theta_R .$$

This shows explicitly the divergence at the rainbow angle. In general the cross section will not be zero for $\theta > \theta_R$ because of the contributions from the negative branches of $\Theta(L)$ which lead to the orbiting singularity. However, because $d\Theta/dL$ is large near $L = L_0$, these contributions are small and decrease exponentially with increasing θ.

The classical cross section also diverges when the projectile is scattered at $\theta = 0$ or $\theta = \pi$, in a collision with finite impact parameter, since in such cases $\sin\theta = 0$. This can happen in the

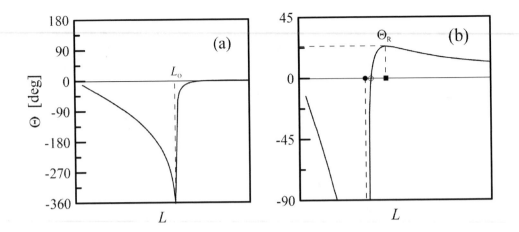

FIGURE 1.14   Deflection functions for attractive potentials. In (a), the potential is attractive everywhere. The deflection angle is then negative. In (b), the potential is attractive at small separations and repulsive at large distances. The orbiting angular momentum, L_O, is indicated in each case. In (b) are also indicated the rainbow and glory angular momenta.

scattering from potentials with attractive and repulsive regions as the one used in Figure 1.14(b). In such cases, for some particular angular momentum, a combination of attraction and repulsion along the trajectory may lead to scattering to the forward or the backward direction. In 1.14 b, this behavior occurs at the angular momentum L_G, for which the scattering angle vanishes. An analogous phenomenon, called *Glory* effect, occurs in Optics. There, a "halo" around the shadow of an object is produced by light scattering to forward angles. As in the case of the rainbow scattering, wave mechanical effects eliminate the divergence of the cross section.

1.11   Stationary scattering of a plane wave

We now consider the cross section from the quantum mechanics point of view. We first study the stationary states in the scattering of a plane wave from a short-range potential, $V(\mathbf{r})$. Although this problem does not seem to correspond to the scattering of a beam of particles, we will show in the next section that it is a good approximation to it. In this case, the wave function $\psi(\mathbf{r},t)$ must satisfy the Schrödinger Eq. (1.18). A stationary state with energy E can be written

$$\psi(\mathbf{r},t) = \psi(\mathbf{r})e^{-iEt/\hbar} \equiv \psi(\mathbf{r})e^{-i\omega t}. \tag{1.73}$$

Using Eq. (1.73), Eq. (1.18) can be put in the form

$$\left(\nabla^2 + k^2\right)\psi(\mathbf{r}) = U(r)\psi(\mathbf{r}). \tag{1.74}$$

Above,

$$\omega = \frac{E}{\hbar}, \qquad k = \frac{\sqrt{2\mu E}}{\hbar} \quad \text{and} \quad U(r) = \frac{2\mu}{\hbar^2}V(r). \tag{1.75}$$

We look for a solution of Eq. (1.74) corresponding to the distortion of the plane wave

$$\phi_{\mathbf{k}}(\mathbf{r}) = Ae^{i\mathbf{k}\cdot\mathbf{r}}, \tag{1.76}$$

where A is an arbitrary normalization constant. Treating the RHS of Eq. (1.74) as a source, we write the desired wave function in the form

$$\psi_{\mathbf{k}}^{(+)}(\mathbf{r}) = \phi_{\mathbf{k}}(\mathbf{r}) + \psi_{\mathbf{k}}^{sc}(\mathbf{r}), \tag{1.77}$$

where $\psi_{\mathbf{k}}^{sc}(\mathbf{r})$is a particular solution which sets the boundary conditions. In our case, it should be an outgoing spherical wave, produced by the scatterer. For this solution, we use the notation $\psi_{\mathbf{k}}^{(+)}(\mathbf{r})$, where (+) stands for the outgoing character of the spherical wave. Near the detector, $r \to \infty$ and $\psi_{\mathbf{k}}^{sc}(\mathbf{r})$ should have the asymptotic behavior

$$\psi_{\mathbf{k}}^{sc}(\mathbf{r}) \to f(\Omega)\frac{e^{ikr}}{r}. \tag{1.78}$$

Above, $f(\Omega) \equiv f(\theta, \varphi)$ is the *scattering amplitude*, which plays a major role in cross section calculations. It is responsible for the orientation dependence of the strength of the outgoing spherical wave. The factor $1/r$ keeps constant the flux through portions of spherical surfaces intercepting a cone with vertex on the scatterer. Using Eqs. (1.78) and (1.76) in Eq. (1.77) we get the asymptotic form of the wave function

$$\psi_{\mathbf{k}}^{(+)}(\mathbf{r}) \to A \left(e^{i\mathbf{k}\cdot\mathbf{r}} + f(\Omega)\frac{e^{ikr}}{r} \right). \tag{1.79}$$

Eq. (1.79) corresponds to the superposition of the incident plane wave with an outgoing spherical emergent wave, both propagating with speed (phase velocity)

$$v_p = \frac{\omega}{k} = \frac{1}{2}\left(\frac{\hbar k}{\mu} \right). \tag{1.80}$$

It should be pointed out that this speed, which is obtained from the condition of constant phase, $d(kr - \omega t)/dt = 0$, is half of that corresponding to a classical particle with momentum $\hbar k$. This is not a real problem as the plane wave cannot represent a localized particle, which should be given by a wave packet. The velocity of the classical particle then corresponds to the group velocity of the wave packet and there is no inconsistency.

We now turn to the calculation of the differential scattering cross section, defined in Eq. (1.13). Firstly, we must obtain the count rate $N(\Omega, \Delta\Omega)$. This quantity is given by the flux of the scattered current $\mathbf{j}_{sc}$,

$$\mathbf{j}_{sc} = \frac{\hbar}{2\mu i} \left[(\psi_{\mathbf{k}}^{sc})^* \boldsymbol{\nabla}\psi_{\mathbf{k}}^{sc} - \psi_{\mathbf{k}}^{sc}\boldsymbol{\nabla} (\psi_{\mathbf{k}}^{sc})^* \right], \tag{1.81}$$

through the detector surface

$$\Delta\mathbf{S} = \Delta S\hat{\mathbf{r}} = \Delta\Omega r^2\hat{\mathbf{r}}, \tag{1.82}$$

where $\hat{\mathbf{r}}$ is the unit vector along the radial direction. We get,

$$N(\Omega, \Delta\Omega) = \mathbf{j}_{sc} \cdot \Delta\mathbf{S} = \left(\Delta\Omega r^2\right)\hat{\mathbf{r}} \cdot \mathbf{j}_{sc}. \tag{1.83}$$

Writing the gradient in polar coordinates,

$$\boldsymbol{\nabla} = \hat{\mathbf{r}}\frac{\partial}{\partial_r} + \hat{\boldsymbol{\theta}}\frac{1}{r}\frac{\partial}{\partial_\theta} + \hat{\boldsymbol{\varphi}}\frac{1}{r\sin\theta}\frac{\partial}{\partial_\varphi}, \tag{1.84}$$

and using the asymptotic form of $\psi_{\mathbf{k}}^{(+)}$ (Eq. (1.78)) in Eq. (1.83)), we obtain

$$N(\Omega, \Delta\Omega) \simeq |f(\Omega)|^2 |A|^2 \Delta\Omega r^2 \left\{ \frac{\hbar}{2\mu i} \left[\frac{e^{-ikr}}{r}\frac{d}{dr}\left(\frac{e^{ikr}}{r}\right) - \frac{e^{ikr}}{r}\frac{d}{dr}\left(\frac{e^{-ikr}}{r}\right) \right] \right\}.$$

Carrying out the derivation with respect to r, we get

$$N(\Omega, \Delta\Omega) = v|f(\Omega)|^2 |A|^2 \Delta\Omega, \tag{1.85}$$

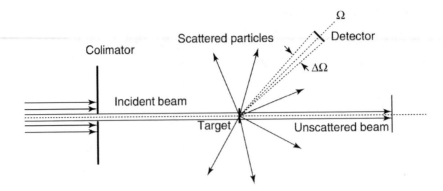

FIGURE 1.15   Schematic representation of a scattering experiment.

where $v = \hbar k/\mu$ is the velocity of a classical particle with momentum $\hbar k$. The incident flux, J, corresponds to the projection of the incident current on the beam direction, which we assume to be the z-axis. Namely,

$$J = \hat{\mathbf{z}} \cdot \mathbf{j}_{in} = \frac{\hbar}{2\mu i} \left[e^{-ikz} \frac{d}{dz} \left(e^{ikz} \right) - e^{ikz} \frac{d}{dz} \left(e^{-ikz} \right) \right] = v \left| A \right|^2 . \tag{1.86}$$

Substituting Eqs. (1.85) and (1.86) in Eq. (1.13), we obtain the differential cross section

$$\frac{d\sigma_{el}(\Omega)}{d\Omega} = \left| f(\Omega) \right|^2 . \tag{1.87}$$

It should be pointed out that the total current

$$\mathbf{j} = \frac{\hbar}{2\mu i} \left[\left(\psi_{\mathbf{k}}^{(+)} \right)^* \boldsymbol{\nabla} \psi_{\mathbf{k}}^{(+)} - \psi_{\mathbf{k}}^{(+)} \boldsymbol{\nabla} \left(\psi_{\mathbf{k}}^{(+)} \right)^* \right] = \mathbf{j}_{in} + \mathbf{j}_{sc} + \mathbf{j}_{int}, \tag{1.88}$$

contains the term

$$\mathbf{j}_{int} = \frac{\hbar}{\mu i} \, \text{Re} \left\{ \phi_{\mathbf{k}}^* \boldsymbol{\nabla} \psi_{\mathbf{k}}^{sc} + \left(\psi_{\mathbf{k}}^{sc} \right)^* \boldsymbol{\nabla} \phi_{\mathbf{k}} \right\},$$

which has been disregarded. This procedure is justified by the oscillating behavior of $\mathbf{j}_{int}$ at any $\theta \neq 0$. Since any experiment has a finite angular resolution, the average over the detector aperture leads to destructive interference. Therefore, its contribution can be neglected, except at very forward angles. For a detailed discussion of this point, we refer to [5].

Of course, the scattering of a beam of wave packets is the appropriate quantum mechanics description of a scattering experiment. The collision of one wave packet with a single target is represented in Figure 1.15. The wave packet goes through a collimator with aperture d, of the order of a few millimeters. The size of the wave packet, Δr, should have the same order of magnitude; $\Delta r \sim d \sim 1$ mm. To reach the detector, it traverses a distance D_1 – from the collimator to the target, and a distance D_2 – from the target to the detector. Typically, these distances are of the order $D \sim 10\,\text{cm} - 1\,\text{m}$.

By using the scattering formalism for wavepackets, it can be shown [5] that the condition of validity of the plane wave description of scattering reduces to the expression ($\lambda = 2\pi/k$)

$$\sqrt{\lambda D} \ll d,$$

which is met in most relevant situations.

1.12   Appendix 1.A – Systems of units

In the scattering of microscopic systems, typical values of lengths, masses, charges, energies etc. are extremely small when expressed in the Gaussian unit system – used in the present text – or any other macroscopic system of units. Therefore, it is convenient to introduce appropriate microscopic units yielding physical scales of the order of magnitude close to one. We discuss this problem below, in two situations, where different scales are used: (a) the scattering of nuclei at non-relativistic energies and, (b) the scattering of atoms, ions, and molecules.

1.12.1   Nuclear Collisions

In the scattering of nuclear systems, the typical size is the nuclear radius: $R \sim 10^{-13} - 10^{-12}$cm. One then uses as unit of length the Fermi* (fm), defined by the relation: 1 fm $= 10^{-13}$ cm. This leads naturally to the cross section unit of fm^2; 1 fm$^2 = 10^{-26}$ cm^2. However, the cross sections are usually expressed in barns; 1 b $= 10^{-24}$ cm, millibarns; 1 mb $= 10^{-3}$ b, microbarns; 1 μb $= 10^{-6}$ b, nanobarns; 1 nb $= 10^{-9}$ b or picobarns; 1 pb $= 10^{-12}$ b.

In nuclear physics, electromagnetic forces play a major role. In such systems, the natural unit of charge is the absolute value of the electronic charge e, which has the value $e = 4.803$ esu $= 1.6 \times 10^{-19}$ C. Usually, a scattering experiment uses a beam of charged projectiles accelerated by electromagnetic forces. In an electrostatic accelerator the kinetic energy of a projectile of charge q subjected to a potential ΔU is $E_{lab} = q\Delta U$. This leads to the use of the electron Volt, defined as the kinetic energy acquired by an electron through the acceleration by a potential of 1 Volt. It can be easily checked that 1 eV $= 1.602 \times 10^{-12}$ erg. For investigations of the internal structure of nuclei, the projectile's energy is usually high enough to surmount the potential barrier (resulting from the interplay between short-range attractive forces and Coulomb repulsion) and excite the intrinsic states of interest. This amounts to energies of the order of MeV (10^6 eV) or GeV (10^9 eV). Nowadays, the beam energy is frequently expressed by the ratio between the collision energy in the laboratory frame to the number of nucleons in the projectile, denoted by ϵ. That is $E_{lab} = \epsilon \cdot A_p$. It can be readily checked that $E_{cm} = \epsilon \cdot A_p A_T / (A_p + A_T)$.

Projectile's and target's masses are frequently multiplied by c^2 and given in MeV. Some examples are:

$$m_p c^2 = 938.272 \text{ MeV}$$
$$m_n c^2 = 939.565 \text{ MeV}$$
$$m_e c^2 = 510.998 \text{ keV}$$
$$\text{atomic mass unit}: m_u c^2 = 931.494 \text{ MeV}$$
$$M(^{12}\text{C})c^2 = 12\, m_u c^2,$$

where the labels p, n, e stand respectively for proton, neutron and electron, m_N is the average mass of a bound nucleon inside the nucleus and $M(^{12}\text{C})$ is the mass of the $A = 12$ carbon isotope.

The Coulomb potential energy in a collision between nuclei with atomic numbers Z_P and Z_T can be written (for the Gaussian system of units)

$$V(r) = \frac{Z_p Z_T e^2}{r}.$$

In order to express it in MeV and r in fm, we must evaluate e^2 accordingly. Writing

$$e^2 \equiv \frac{e^2}{\hbar c}\hbar c,$$

---

*This unit is also known as femtometer. For this reason, one uses the notation "fm" – instead of "Fm".

and using the explicit value of the fine structure constant

$$\alpha = \frac{e^2}{\hbar c} = \frac{1}{137.036},$$

and the constant

$$\hbar c = 197.327 \text{ MeV fm},$$

we get

$$e^2 = 1.43998 \text{ MeV fm}.$$

1.12.2   Collisions of atoms, ions, or molecules

Atoms and molecules are about 10^5 times larger than nuclei. Their radii can be measured in Angstroms, where 1 Å= 10^{-8} cm = 10^5 fm. On the other hand their appropriate energy scale is much smaller. Typical excitation energies are of the order of the eV or a fraction of it. The Coulomb energy can be expressed in terms of these units using the relations $\hbar c = 1973.27$ eV.Å and $e^2 = 14.4$ eV.Å. However, in practical applications Hartree's atomic units (a.u.) are used. The basic quantities in this system of units are discussed below.

In Hartree's atomic unit system, the unit of length is the radius of the classical orbit of the electron in a hydrogen atom. According to Bohr's atomic model, this radius is

$$1 \text{ a.u.} = a_0 = \frac{\hbar^2}{m_e e^2} = 5.29177 \times 10^{-9} \text{ cm} = 0.529177 \text{ Å}.$$

An important remark is that, despite the use of atomic units for lengths, cross sections in atomic and molecular physics are usually given in cm^2. The units of mass and charge are mass and the charge of the electron, that is, m_e and e. To complement the unit system, one adopts $\hbar$ as the unit of action and, as in the Gaussian system, sets $\varepsilon_0 = 1/4\pi$, $\mu_0 = 4\pi/c^2$. In this way, in a.u. we have: $m_e = e = a_0 = \hbar = 4\pi\varepsilon_0 = \mu_0 c^2/4\pi = 1$. The atomic units for other physical quantities are defined in terms of these basic units. We have given below a few examples: (a) The unit of velocity is that of an electron in Bohr's orbit, v_0. It can be readily checked that 1 a.u. = $v_0 = \alpha c = 2.18769 \times 10^8$ cm/s, so that $c = 137.036$ a.u. The a.u. of energy, the Hartree (H) is twice the ionization energy of the hydrogen atom. That is, $1\text{H} = e^2/a_0 = 27.2114$ eV. (b) the unit of time is 1 a.u. = $a_0/v_0 = 2.41888 \times 10^{-17}$ s.

In the case of electromagnetic radiation, the photon energy can also be specified by the frequency or by the wavelength. The spectroscopic tradition uses MHz (mega Hertz) for photon frequencies in the infrared and microwave regions and wave numbers in the visible and ultraviolet regions. With this in mind, the energy of 1 a.u. corresponds to a frequency of $\nu = 6.57968 \times 10^9$ MHz, wave length $\lambda = 455.633$ Å and wave number $k = 219475$ cm^{-1}. This energy can also be expressed as a temperature (in units of the Boltzmann constant, $k = 8.617 \times 10^{-5}$ eV/K), 1 a.u. = 3.15777×10^5 K.

1.13   Appendix 1.B – Useful constants and conversion factors

The fundamental constants presented here are recommended by the CODATA (Committee on Data for Science and Technology), USA, 2018. The values are periodically revised and can be accessed electronically at the NIST (National Institute of Standards and Technology) website.

1.13.1 Constants

Electric charge	$e = 1.602\,176\,634 \times 10^{-19}$ C
	$e^2 = 1.439\,98$ MeV fm
Planck constant	$h = 6.626\,070\,15 \times 10^{-27}$ erg s
	$= 4.135\,667\,696 \times 10^{-21}$ MeV s
	$\hbar = h/2\pi = 1.054\,571\,817 \times 10^{-27}$ erg s
	$= 6.582\,119\,569 \times 10^{-22}$ MeV s
Speed of light	$c = 299\,792\,458$ m/s
	$\hbar c = 1.973\,269\,804 \times 10^{-11}$ MeV cm $= 197.327$ MeV fm
Gravitational constant	$G = 6.674\,30 \times 10^{-11}$ m^3 kg^{-1} s^{-2}
Boltzmann constant	$k = 1.380\,649 \times 10^{-16}$ erg/K
Avogadro number	$N_A = 6.022\,140\,76 \times 10^{23}$ mol^{-1}
Molar volume	$V_m = 22.413\,969\,54$ l/mol (273.15 K; 101 325 Pa)
Faraday constant	$F = 96\,485.332\,12$ C/mol
Compton wavelength	$\lambda_e = \hbar/m_e c = 386.159\,267\,96$ fm (electron)
	$\lambda_p = \hbar/m_p c = 0.210\,308\,910\,336$ fm (proton)
Nuclear magneton	$\mu_N = 3.152\,451\,258\,44 \times 10^{-14}$ MeV/T
Bohr magneton	$\mu_B = 5.788\,381\,806 \times 10^{-11}$ MeV/T
Fine structure constant	$\alpha = e^2/\hbar c = 1/137.035\,999\,084$
Electron classical radius	$r_e = e^2/m_e c^2 = 2.817\,940\,326$ fm
Bohr radius	$a_0 = \hbar^2/m_e e^2 = 0.529\,177\,210\,903 \times 10^{-8}$ cm

1.13.2 Masses

Electron	$m_e = 9.109\,383\,701 \times 10^{-28}$ g $= 5.485\,799\,090 \times 10^{-4}$ u
	$= 0.510\,998\,950$ MeV/c^2
Muon	$m_\mu = 0.113\,428\,925$ u $= 105.658\,375$ MeV/c^2
Pions	$m_{\pi^0} = 134.976\,6$ MeV/c^2
	$m_{\pi^\pm} = 139.570\,18$ MeV/c^2
Proton	$m_p = 1.672\,621\,923 \times 10^{-24}$ g $= 1.007\,276\,466$ u
	$= 938.272\,088$ MeV/c^2
Neutron	$m_n = 1.008\,664\,915$ u $= 939.565\,420$ MeV/c^2
Hydrogen atom	$m_H = 1.007\,825\,05$ u $= 938.791$ MeV/c^2

1.13.3 Conversion factors

Length	1 fermi = 1 fm $= 10^{-15}$ m $= 10^{-13}$ cm
Area	1 barn = 1 b $= 10^{-24}$ cm^2 $= 10^2$ fm^2
Mass	1 unit of atomic mass = 1 u $= (1/12)$ m($^{12}_6$C)
	$= 1.660\,539\,066 \times 10^{-24}$ g $= 931.494\,102$ MeV/c^2
	$= 1822.872\ m_e$
Energy	1 eV $= 1.602\,177 \times 10^{-12}$ erg
	$= 1.073\,544 \times 10^{-9}$ u·c^2
	1 erg $= 10^{-7}$ J
Temperature ($k = 1$)	1 MeV$= 1.16 \times 10^{10}$ K$= 1.78 \times 10^{-30}$ kg

1.14 Exercises

1. (a) What kinetic energy must be given to a helium atom in order to increase its mass by 1%? (b) What are the mean velocity and the mean kinetic energy of a helium atom at standard temperature and pressure?

2. Alpha-particles from ^{218}Po (E$_\alpha = 6.0$ MeV) are used to bombard a gold foil. (a) How close to the gold nucleus can these particles reach? (b) What is the nuclear radius of gold according to the radius-mass relation $R = r_o A^{1/3}$ ($r_o = 1.3$ fm)? (c) Calculate the distance of closest approach for 5 MeV α-particles to a gold target.

3. In a Rutherford scattering experiment ^{2}H atoms of 150 keV are used to bombard a thin ^{59}Ni foil having a surface density of 67×10 g/cm^2. The detector subtends a solid angle of 1.12×10 sr and detects 4816 deuterons out of a total of 1.88×10 incident on the target.

Calculate (a) the differential cross-section (in barns). (b) What is the distance between the target and the solid state detector, which has a surface area of 0.2 cm^2?

4. In a Rutherford scattering on a silver foil using α-particles from a thin-walled radon tube, the following data were observed: $d\sigma/d\Omega = 22(\theta = 150°)$, $47(105°)$, $320(60°)$, $5260(30°)$, $105400(15°)$ barns per steradian. Calculate the energy of the incident α-particles.

5. What is the Q-value for the reactions: (a) $^{11}\text{B}(d,\alpha)^9\text{Be}$; (b) $^7\text{Li}(p,n)^7\text{Be}$?

6. What is the maximum velocity that a deuteron of 2 MeV can impart to a ^{16}O atom?

7. Calculate the mass of an electron accelerated through a potential of 2×10^8 V. Note that the relativistic equations relating energy E, momentum p, kinetic energy K, mass m, and velocity v, are given by

$$E^2 = p^2 c^2 + m_0^2 c^4, \qquad K = E - m_0 c^2,$$

$$p = \gamma m_0 v, \quad E = \gamma m_0 c^2, \quad \gamma = \left(1 - v^2/c^2\right)^{-1/2}, \tag{1.89}$$

and that the mass increases with the speed as

$$m = \gamma m_0. \tag{1.90}$$

8. ^{12}C atoms are used to irradiate ^{239}Pu to produce an isotope of berkelium. What is the Coulomb barrier height?

9. Measurements made on the products of the reaction $^7\text{Li}(d,\alpha)^5\text{He}$ have led to an isotopic mass of 5.0122 for the hypothetical nuclide He. Show that this nuclear configuration cannot be stable by considering the reaction $^5\text{He} \longrightarrow {}^4\text{He} + \text{n}$.

10. In an experiment, one hopes to produce the long-lived (2.6 y) ^{22}Na through a (d,2n) reaction on neon. What is (a) the Q-value, (b) the threshold energy, (c) the Coulomb barrier height, and (d) the minimum deuteron energy for the reaction? The mass excesses (in keV) are -5185 for ^{22}Na and -8027 for ^{22}Ne. Note that the mass excess is defined as $M - A$, where M is expressed in mass units of $M(^{12}_6\text{C})/12$.

11. Complete the reactions:

$$\text{p} + \qquad \longrightarrow {}^{28}\text{Si} + \text{n}$$
$$^{197}\text{Au} + {}^{12}\text{C} \longrightarrow \qquad + \gamma$$
$$^{235}\text{U} + \text{n} \longrightarrow {}^{100}\text{Mo} + \qquad + 3\text{n}.$$

12. What spin and parities can be expected in ^{20}Ne formed from the reaction $\alpha + {}^{16}\text{O}$?

13. Obtain the threshold for the production of ^{10}B in the reaction of Figure 1.6 for the scattering angles (a) $0°$ and (b) $60°$. Verify if the results are compatible with the values of the figure.

14. (a) Write a relativistic expression that relates the CM energy with the L energy for a system of two particles, the second one at rest in the L. (b) Show that, if the energy of the incident particle in the L is the value T_t given by Eq. (1.38), the total kinetic energy in the CM after the collision is zero, showing that T_t is the threshold energy of the reaction.

15. Describe what happens in the diagram of the Figure 1.9 when the energy of the incident particle is equal: (a) to the threshold energy; (b) to the energy E'_a [Eq. (1.12)]. Still using the diagram, say why the duplicity of energy does not happen in endothermic reactions.

16. Show that Eqs. (1.24)–(1.27) are still valid for inelastic collisions, $A(a,b)B$, if one uses Eq. (1.28).

17. In a scattering experiment, a beam of ^{9}Be ($Z_p = 4$) nuclei with energy $E_{lab} = 19$ MeV impinges on a thin solid target of ^{64}Zn ($Z_T = 30$). The scattered particles are measured by a set of five detectors distributed on a circumference at the angles (referred to the beam direction) $\theta_1, \cdots, \theta_5$. The yields in the detectors, $N(\theta)$, are listed in the table below. The experimental setup is such that the product $J \cdot n \cdot \Delta\Omega$ is 0.29×10^{27} cm^{-2}. Transform these angles to the CM-frame and obtain the corresponding experimental cross sections. Plot the experimental points in comparison with the Rutherford cross section

$$\left(\frac{d\sigma}{d\Omega}\right)_{Ruth} = \frac{a^2}{4}\left[\frac{1}{\sin^4(\theta/2)}\right], \quad \text{with} \quad a = \frac{Z_P Z_T e^2}{2E},$$

where e is the absolute value of the electronic charge and $v = \sqrt{2E_{lab}/M_P}$.

$\theta(\text{deg})$	30	60	90	120	150
$N(\theta)$	32983 ± 181	2296 ± 50	545 ± 10	181 ± 7	82 ± 3

18. Calculate the classical impact parameter necessary to give an orbital angular momentum of $l = 1$ for an n-p scattering event when $E_{Lab} = 10$ MeV.

19. Low-energy neutrons are scattered by protons. Let θ and ϕ be the emerging angles of neutrons and protons, respectively. (a) Show that, for a given event, $\theta + \phi = 90°$. (b) The scattering is isotropic in the center of mass and Eq. (1.29) shows that the neutron angular distribution in the laboratory system is given by $\sigma(\theta) = 4\cos\theta\,\sigma(\Theta)$. Show that for the protons there is the relationship $\sigma(\phi) = 4\cos\phi\,\sigma(\Phi)$. (c) Being $\sigma(\Theta)$ and $\sigma(\Phi)$ constants, the functions $\sigma(\theta)$ and $\sigma(\phi)$ have maxima in $0°$. How this result harmonizes with the result of the item (a)?

20. Why it is not possible for a proton at rest to scatter another proton, of low energy, if both spins have the same direction?

21. A neutron of kinetic energy E_1 is elastically scattered by a nucleus of mass M, remaining with a final kinetic energy E_2. (a) Being Θ the scattering angle in the center of mass, show that

$$\frac{E_1}{E_2} = \frac{1}{2}\left[(1+\alpha) - (1-\alpha)\cos\Theta\right],$$

where $\alpha = [(M-1)/(M+1)]^2$. (b) What is the maximum loss of kinetic energy as a function of E_1 and M? For which angle Θ it occurs? Which angle θ in the laboratory corresponds to this value of Θ?

22. Consider the scattering of a particle of mass μ and energy E from the spherically symmetric potential $V(r) = \lambda/r^2$. Show that the deflection function is

$$\Theta(b) = \pi\left[1 - \frac{b}{\sqrt{b^2 + \lambda/E}}\right],$$

where b is the impact parameter. Then prove that the classical cross section can be written (see [4]):

$$\frac{d\sigma}{d\Omega} = \frac{\pi\lambda}{E\sin(\beta\pi)}\left[\frac{1-\beta}{\beta^2(2-\beta)^2}\right],$$

with $\beta = \theta/\pi$.

23. A particle of mass m and energy E is scattering from a potential with the form $V(r) = -A/r^4$, where A is a positive constant. Find the orbiting angular momentum in this case.

24. A one-dimensional plane wave propagates freely with energy E until it reaches a thin slab of absorbing material with thickness a, represented by the imaginary potential $V(r) = -iW_0$. In this expression, W_0 is a positive constant much smaller than E. Estimate the attenuation factor $\alpha = j_{in}/j_{out}$, where j_{in} and j_{out} are respectively the incident and the emergent currents.

25. A beam of protons with energy $E_{lab} = 10$ keV goes through a collimator with radius $d = 2$ mm and impinges on a target of He atoms, located 1.5 m away from the collimator. The scattered particles are detected by a set of detectors placed around the target, at the distance of 1.0 m. Show that the approximation of a packet with constant shape is justified in this experiment.

References

1. Rutherford E 1911 *Phil. Mag.* **xxi** 669
2. Taylor J 2012 *Scattering Theory: The Quantum Theory of Nonrelativistic Collisions* (Dover)
3. Bertulani C A 2007 *Nuclear Physics in a Nutshell* (Princeton Press)
4. Goldstein H, Poole C P, Safko J L 2013 *Classical Mechanics* (Pearson)
5. Joachain C J 1975 *Quantum Collision Theory* (Oxford: North-Holland)

2

The partial-wave expansion method

2.1 The scattering wave function

As we have shown in the previous chapter, the differential cross section for elastic scattering is given in terms of the scattering amplitude $f(\Omega)$, which can be isolated in the asymptotic form of the stationary wave function. Finding this cross section then amounts to solving the Schrödinger equation for the short-range potential V, namely,

$$H\psi = E\psi, \tag{2.1}$$

where E is the collision energy and H is the Hamiltonian

$$H = K + V. \tag{2.2}$$

We are interested in the solution $\psi_{\mathbf{k}}^{(+)}(\mathbf{r})$, which has the asymptotic form of Eq. (1.79),

$$\psi_{\mathbf{k}}^{(+)}(\mathbf{r}) \to A\left(e^{i\mathbf{k}\cdot\mathbf{r}} + f(\Omega)\frac{e^{ikr}}{r}\right), \tag{2.3}$$

where A is an arbitrary normalization constant.

There are different methods to solve this problem. Here, we will see in detail the method of *partial waves* with the scattering amplitude written as a sum of contributions from different angular momenta, given in terms of phase–shifts. The phase shifts are then obtained by solving the radial equations. The method is very convenient for spherically symmetric potentials, $V(\mathbf{r}) = V(r)$, at low collision energies.

2.2   Radial equation

The kinetic energy operator in spherical coordinates is given by

$$K = -\frac{\hbar^2}{2\mu} \left[\frac{1}{r^2} \frac{\partial}{\partial r} \left(r^2 \frac{\partial}{\partial r} \right) + \frac{1}{r^2 \sin^2 \theta} \frac{\partial}{\partial \theta} \left(\sin \theta \frac{\partial}{\partial \theta} \right) + \frac{1}{r^2 \sin^2 \theta} \frac{\partial^2}{\partial \varphi^2} \right],$$

where $\{r, \theta, \varphi\}$ are the spherical coordinates corresponding to the vector $\mathbf{r}$. Since the explicit form of the orbital angular momentum operator [1, 2, 3] L^2 is

$$L^2 = -\frac{\hbar^2}{\sin^2 \theta} \left[\frac{\partial}{\partial \theta} \left(\sin \theta \frac{\partial}{\partial \theta} \right) + \frac{\partial^2}{\partial \varphi^2} \right],$$

we can write

$$K = -\frac{\hbar^2}{2\mu} \left[\frac{1}{r^2} \frac{\partial}{\partial r} \left(r^2 \frac{\partial}{\partial r} \right) - \frac{L^2}{\hbar^2 r^2} \right]. \tag{2.4}$$

Eq. (2.4), together with the spherical symmetry of the potential, guarantees that the Hamiltonian commutes with the operators L^2 and L_z. Therefore, one can express the solutions of Eq. (2.1) as linear combinations of simultaneous eigenfunctions of the operator set $\left\{ H, L^2, L_z \right\}$;

$$\psi_{klm}(\mathbf{r}) = R_l(k, r) Y_{lm}(\theta, \varphi), \tag{2.5}$$

where $Y_{lm}(\theta, \varphi)$ are the spherical harmonics and $R_l(k, r)$ are the solutions of the radial equation

$$-\frac{\hbar^2}{2\mu} \left[\frac{1}{r^2} \frac{d}{dr} \left(r^2 \frac{d}{dr} \right) - \frac{l(l+1)}{r^2} \right] R_l(k, r) + V(r) R_l(k, r) = E R_l(k, r). \tag{2.6}$$

Above, we have used the wave number $k = \sqrt{2\mu E/\hbar^2}$ to characterize the collision energy, instead of the quantum number E.

The terms of (2.5) can be understood as *partial waves*, from which the general solution Ψ can be constructed. Frequently, one uses the modified radial wave function $u_l(k, \rho)$, defined by the relation

$$R_l(k, r) = \frac{u_l(k, r)}{kr}. \tag{2.7}$$

This wave function satisfies a simpler Schrödinger equation. Upon substituting Eq. (2.7) into Eq. (2.6), we get

$$-\frac{\hbar^2}{2\mu} \left[\frac{d^2}{dr^2} - \frac{l(l+1)}{r^2} \right] u_l(k, r) + V(r) u_l(k, r) = E u_l(k, r). \tag{2.8}$$

Equation (2.8) corresponds to a one-dimensional Schrödinger equation with the effective l-dependent potential

$$V_l(r) = V(r) + V_l^{Cf}(r), \tag{2.9}$$

where $V_l^{Cf}(r)$ is the centrifugal potential

$$V_l^{Cf}(r) = \frac{\hbar^2}{2\mu} \frac{l(l+1)}{r^2}. \tag{2.10}$$

It is convenient to introduce the variable $\rho = kr$ and rewrite the above equations as

$$\left[\frac{d^2}{d\rho^2} + \frac{2}{\rho} \frac{d}{d\rho} + \left(1 - \frac{l(l+1)}{\rho^2} \right) \right] R_l(k, \rho) = U(\rho) R_l(k, \rho), \tag{2.11}$$

where

$$U(\rho) = \frac{V(\rho/k)}{E}, \tag{2.12}$$

and

$$\left[\frac{d^2}{d\rho^2} + 1 - \frac{l(l+1)}{\rho^2} \right] u_l(k, \rho) = U(\rho) u_l(k, \rho). \tag{2.13}$$

2.3   Free particle in spherical coordinates

In the case of a free particle, $V = 0$ and the RHS of Eqs. (2.11) and (2.13) vanish. Eq. (2.11) then reduces to the spherical Bessel equation. Its solutions can be expressed as linear combinations of the independent pair of real solutions $j_l(\rho)$ and $n_l(\rho)$. The former is the spherical *Bessel function*, which is regular at the origin. The latter is the spherical *Neumann function*, which diverges at $\rho = 0$. Asymptotically, at the limiting regions $\rho \to 0$ and $\rho \to \infty$, the functions exhibit the behaviors[*]

$$j_l(\rho \to 0) \sim \frac{\rho^l}{(2l+1)!!}, \qquad j_l(\rho \to \infty) \sim \frac{\sin(\rho - l\pi/2)}{\rho} \tag{2.14}$$

$$n_l(\rho \to 0) \sim -\frac{(2l-1)!!}{\rho^{l+1}}, \qquad n_l(\rho \to \infty) \sim -\frac{\cos(\rho - l\pi/2)}{\rho}.$$

These equations show that for large values of the argument these functions oscillate with decreasing amplitudes, within the envelope $\pm 1/\rho$.

The regular solutions satisfy the orthogonality relation

$$\int_0^\infty r^2 j_l(kr) j_l(k'r) dr = \frac{\pi}{2k^2} \delta(k - k'). \tag{2.15}$$

The properties of $j_l(\rho)$ and $n_l(\rho)$ can be found in many books (see, for example, Refs. [1], [5], or Appendix C of Ref. [6]) and numerical values are obtained with subroutines available in standard computer libraries (see Ref. [7], for example). For $l = 0$ and $l = 1$, they have the simple forms

$$j_0(\rho) = \frac{\sin \rho}{\rho}, \qquad j_1(\rho) = \frac{\sin \rho}{\rho^2} - \frac{\cos \rho}{\rho},$$

$$n_0(\rho) = -\frac{\cos \rho}{\rho}, \qquad n_1(\rho) = -\frac{\cos \rho}{\rho^2} - \frac{\sin \rho}{\rho}. \tag{2.16}$$

For higher l, j_l and n_l can be obtained with the help of the recursion relations [5]

$$w_l(\rho) = (2l+1)\frac{w_{l+1}(\rho)}{\rho} - w_{l-1}(\rho), \tag{2.17}$$

where w stands for j or n. They can also be obtained through successive derivations with respect to the argument,

$$j_l(\rho) = (-\rho)^l \left(\frac{1}{\rho} \frac{d}{d\rho} \right)^l j_0(\rho), \qquad n_l(\rho) = (-\rho)^l \left(\frac{1}{\rho} \frac{d}{d\rho} \right)^l n_0(\rho).$$

The solutions of the spherical Bessel equation can also be expressed in terms of the complex functions $h_l^{(+)}(\rho)$ and $h_l^{(-)}(\rho)$, called *Haenkel functions*, defined as[†]

$$h_l^{(\pm)}(\rho) = j_l(\rho) \pm i n_l(\rho). \tag{2.18}$$

Asymptotically, $h_l^{(+)}$ and $h_l^{(-)}$ behave, respectively, as an outgoing and an incoming spherical wave[‡]. Namely,

$$h_l^{(\pm)}(\rho \to \infty) \sim (\mp i) \frac{e^{\pm i(\rho - l\pi/2)}}{\rho}. \tag{2.19}$$

---

[*]To extend Eq. (2.14) to $l = 0$, we adopt the convention $(-1)!! = 1$.

[†]Usually, these functions are denoted $h_l^{(1)}$, which corresponds to $h_l^{(+)}$, and $h_l^{(2)}$, which corresponds to $h_l^{(-)}$. We choose the notation $h_l^{(\pm)}$ in order to stress the incoming $(-)$ or outgoing $(+)$ nature of these functions.

[‡]These behaviors are clear in the asymptotic phases of the wave functions generated by $h_l^{(\pm)}$: $\varphi^{(\pm)} = \pm\rho - \omega t$. Since for $h_l^{(+)}$ the condition $d\varphi/dt = 0$ leads to a positive phase velocity $(v_p = d\rho/dt = \omega)$, this spherical wave moves away from the origin. The opposite occurs for $h_l^{(-)}$.

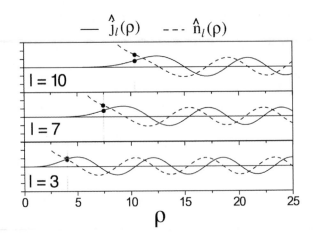

FIGURE 2.1 The functions $\widehat{j}_l(\rho)$ and $\widehat{n}_l(\rho)$ for $l = 3$, 7 and 10. The vertical dotted line represents the boundary between the classically allowed and forbidden regions.

Similarly, free particle solutions can be found for Eq. (2.13). In this case, we have the *Ricatti-Bessel equation*. The solutions corresponding to j_l and n_l are, respectively, the *Ricatti-Bessel function*, $\widehat{j}_l$, and the *Ricatti-Neumann function*, $\widehat{n}_l$, given as [8]

$$\widehat{j}_l(\rho) = \rho j_l(\rho); \qquad\qquad \widehat{n}_l(\rho) = \rho n_l(\rho). \tag{2.20}$$

These functions are illustrated in Figure 2.1, for a few partial waves. For large values of ρ, they oscillate sinusoidally with constant amplitude. At small ρ, the regular $\widehat{j}_l(\rho)$ approaches zero while $\widehat{n}_l(\rho)$ diverges. The transition between these trends occurs at the points

$$\rho_t = \sqrt{l(l+1)} \; (\simeq l, \text{ for large } l), \tag{2.21}$$

indicated by vertical dotted lines in Figure 2.1.

The meaning of ρ_t can be better understood if we get back to the variable r and consider the Riccatti-Bessel equation in the form of Eq. (2.13), with $U = 0$. In this case, it corresponds to a one-dimensional Schrödinger equation with an effective potential $V_l(r) = V_l^{Cf}(r)$. The point

$$r_t = \frac{\rho_t}{k} = \frac{\sqrt{l(l+1)}}{k}$$

is then the classical turning point, given by the condition

$$V_l(r_t) = \frac{\hbar^2}{2\mu} \frac{l(l+1)}{r_t^2} = E.$$

Therefore, the transition points are the boundaries between the classically allowed and the classically forbidden regions.

The Ricatti-Haenkel, $\widehat{h}_l^{(\pm)}$, functions are defined in a similar way,

$$\widehat{h}_l^{(\pm)}(\rho) \equiv \rho h_l^{(\pm)}(\rho) = \widehat{j}_l(\rho) \pm i \widehat{n}_l(\rho), \tag{2.22}$$

and they have the asymptotic behavior

$$\widehat{h}_l^{(\pm)}(\rho \to \infty) \sim (\mp i) \, e^{\pm i(\rho - l\pi/2)}. \tag{2.23}$$

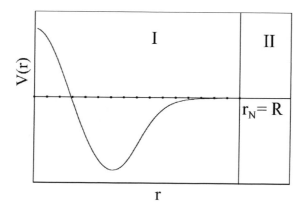

FIGURE 2.2 Schematic representation of the potential and the mesh points for a numerical calculation. The solution in region II is known analytically.

2.4 Phase shifts

The radial wave functions $u_l(k,r)$ can be obtained through numerical integration of Eq. (2.8). Let us consider a potential $V(r)$ with range R, as represented in Figure 2.2. The point $r = R$ divides the r-axis in region I, where the potential is different from zero and region II, where the particle moves freely.

In region II, $u_l(k,r)$ satisfies the Ricatti-Bessel equation and, therefore, can be put in the general form

$$u_l^{II}(k,r) = \alpha_l \left[\hat{j}_l(kr) + \beta_l \hat{n}_l(kr) \right],
\tag{2.24}$$

where α_l is a normalization constant. It will soon become clear that α_l has no impact on the scattering amplitude. The effects of the potential are contained in the constant β_l, which vanishes when $V = 0$. The logarithmic derivative at the boundary of the external region is then expressed in terms of β_l as

$$\mathcal{L}^{II} = R \frac{\left[du_l^{II}(k,r)/dr \right]_{r=R}}{u_l^{II}(k,R)} = kR \left[\frac{\hat{j}_l'(kR) + \beta_l \hat{n}_l'(kR)}{\hat{j}_l(kR) + \beta_l \hat{n}_l(kR)} \right],
\tag{2.25}$$

where primes stand for derivatives with respect to the argument[§].

To determine β_l, one should match Eq. (2.25) with the logarithmic derivative at the boundary, calculated from the internal region. For this purpose, we replace the r-axis by a mesh consisting of the points $\{r_0, r_1, \cdots, r_N\}$, with $r_0 = 0$ and $r_N = R$, and approximate the derivatives of the radial equation by finite differences. We then start integrating the differential equation from the origin. Good behavior of the wave function requires $u_l^I(k,0) \equiv y_0 = 0$. The initial value of the derivative $\left[du_l^I(k,r)/dr \right]_{r=0} \equiv y_0'$, can be chosen arbitrarily (1 for example). This choice will only affect α_l, which, as we mentioned, will not affect the scattering amplitude. With the help of a numerical procedure, like *Runge-Kutta* [7], the radial function and its first derivative can be obtained at all mesh points. The logarithmic derivative is then calculated from the results at the last mesh point, $r_N = R$,

$$\mathcal{L}^I = R \left(\frac{y_N'}{y_N} \right),
\tag{2.26}$$

[§]As defined here, the logarithmic derivatives are dimensionless. For convenience, later in this book, we will define them without the radius R factor, i.e. with dimension of inverse of distance.

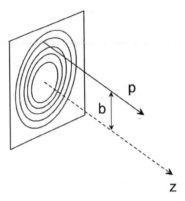

FIGURE 2.3 Classical representation of a plane wave: to each partial wave l corresponds an impact parameter b.

with the notation $y_i \equiv u_l^I(k, r_i)$ and $y_i' \equiv \left[du_l^I(k, r)/dr \right]_{r=r_i}$.

Since $\mathcal{L}^I$, $\hat{j}_l'(\bar{\rho})$, $\hat{n}_l'(\bar{\rho})$, $\hat{j}_l(\bar{\rho})$ and $\hat{n}_l(\bar{\rho})$ are known quantities at given energy, the matching condition

$$\mathcal{L}^{II} = \mathcal{L}^I \tag{2.27}$$

provides an equation for β_l. For the calculation of the scattering amplitude, it is convenient to replace β_l by a new parameter δ_l, defined by the relation

$$\beta_l = \tan \delta_l. \tag{2.28}$$

It should be remarked that Eq. (2.28) only determines δ_l modulo π. That is, it does not distinguish among δ_l, $\delta_l \pm \pi$, $\delta_l \pm 2\pi$, $\delta_l \pm 3\pi$, etc. However, this indetermination leads to no serious difficulty, since any value satisfying Eq. (2.28) yields the same scattering amplitude. Eq. (2.24) then takes the form

$$u_l^{II}(k, r) = \alpha_l \left[\hat{j}_l(kr) - \tan \delta_l \hat{n}_l(kr) \right]. \tag{2.29}$$

Using Eq. (2.28) in Eq. (2.25) and the continuity of the logarithmic derivatives (Eq. (2.27)), we obtain the explicit result

$$\tan \delta_l = -\frac{kR\hat{j}_l'(kR) - \hat{j}_l(kR)\mathcal{L}^I}{kR\hat{n}_l'(kR) - \hat{n}_l(kR)\mathcal{L}^I}. \tag{2.30}$$

The meaning of δ_l becomes clear if we look at the asymptotic behavior of $u_l(k, r)$. Taking the limits $\hat{j}_l(kr \to \infty)$ and $\hat{n}_l(kr \to \infty)$ (using Eqs. (2.14)), we get

$$\begin{aligned} u_l(k, r \to \infty) &= \alpha_l' \left[\sin(kr - l\pi/2) \cos \delta_l + \sin \delta_l \cos(kr - l\pi/2) \right] \\ &= \alpha_l' \sin(kr - l\pi/2 + \delta_l), \end{aligned} \tag{2.31}$$

where $\alpha_l' = \alpha_l/\cos \delta_l$ is a new normalization constant. In the absence of potential, the argument of the sinus in the second line of the above equation reduces to the asymptotic phase of the Ricatti-Bessel function: $kr - l\pi/2$. Therefore, the potential V leads to the shift δ_l in this phase, termed hence a *phase shift*.

Frequently, the solutions of (2.13) in region II are expressed in terms of the Ricatti-Haenkel functions, in the form

$$u_l(k, r > R) = \gamma_l \left[\hat{h}_l^{(-)}(kr) - S_l \hat{h}_l^{(+)}(kr) \right], \tag{2.32}$$

where

$$S_l = e^{2i\delta_l}. \tag{2.33}$$

The normalization of the radial wave function is irrelevant for the scattering amplitude calculation of the present section. To avoid ambiguities later, we will adopt the normalization

$$\gamma_l = \frac{i}{2},\qquad (2.34)$$

which leads to the asymptotic forms

$$u_l(k, r \to \infty) = e^{i\delta_l} \sin\left(kr - l\pi/2 + \delta_l\right) \qquad (2.35a)$$

$$\equiv \frac{1}{2i}\left[S_l e^{i(kr - l\pi/2)} - e^{-i(kr - l\pi/2)}\right]. \qquad (2.35b)$$

It is also useful to present the plane wave by an expansion in partial waves. We write $\mathbf{k} \cdot \mathbf{r} = kr \cos\theta$ and use the well known *Bauer's expansion* [1]

$$e^{ikr\cos\theta} = \sum_{lm} 4\pi Y^*_{lm}(\hat{\mathbf{r}}) Y_{lm}(\hat{\mathbf{k}}) i^l j_l(kr) = \sum_l (2l+1) P_l(\cos\theta) i^l j_l(kr). \qquad (2.36)$$

Expression (2.36) has the form of Eq. (2.5). This means that the plane wave $e^{i\mathbf{k}\cdot\mathbf{r}}$ can be understood as the sum of a set of partial waves, each one with orbital angular momentum $\sqrt{l(l+1)}\hbar$. The terms $j_l(kr) P_l(\cos\theta)$ specify the radial and angular dependence of the partial wave l, the weight of the contribution of each term being given by the amplitude $(2l+1)$ multiplied by the phase factor i^l.

Using classical arguments we can give an interpretation to this amplitude value. Let us consider a surface normal to the plane wave propagation direction, and imagine a set of circles of radius $b_l = l\lambda$, with the wavelength $\lambda = \lambda/2\pi = 1/k$, centered at the point where the z axis crosses the surface (see Figure 2.3). If the beam of particles moves along the z axis, the classical angular momentum of a particle about the origin of the coordinate system is the product of the impact parameter b and the linear momentum $p = \hbar k$. Hence, all particles that pass by a ring of internal radius b_l and external radius b_{l+1} will have orbital angular momentum between $l\lambda\hbar k = l\hbar$ and $(l+1)\lambda\hbar k = (l+1)\hbar$. In the classical limit l is large and $l+1 \cong l$. So we can say that all the particles that pass by the ring have orbital angular momentum $l\hbar$. However, still using the classical reasoning, a particle belonging to an uniform beam can have any impact parameter, and its probability to pass by one of the rings is proportional to the area A of that ring:

$$A = \pi(b^2_{l+1} - b^2_l) = \pi\lambda^2\left[(l+1)^2 - l^2\right] = \pi\lambda^2(2l+1).$$

We see that $(2l+1)$ is the relative probability that a particle in an uniform beam has an orbital angular momentum $l\hbar$, which is the classical limit for the orbital angular momentum $\sqrt{l(l+1)}$ associated to the partial wave l.

2.5   Scattering amplitude and cross sections

With the radial wave functions, we have the eigen-functions ψ_{klm}. To solve the scattering problem it is necessary to build the linear combination of these functions which has the asymptotic behavior of Eq. (2.3). Choosing the z-axis along the beam direction ($\mathbf{k} = k\hat{\mathbf{z}}$) and taking into account the axial symmetry of the problem, the expansion can only involve states with $m = 0$. Therefore, we can replace in Eq. (2.5) $Y_{lm}(\theta, \phi) \to P_l(\cos\theta)$. The required wave function can then be written

$$\psi^{(+)}_{\mathbf{k}}(\mathbf{r}) = \sum_l C_l P_l(\cos\theta) \frac{u_l(k, r)}{kr}. \qquad (2.37)$$

Near the detector, $r \to \infty$, so that we can use Eq. (2.35b) and the wave function can be put in the form

$$\psi^{(+)}_{\mathbf{k}}(\mathbf{r}) \to \frac{1}{2i} \sum_{l=0}^{\infty} C_l P_l(\cos\theta) \frac{S_l\, e^{i(kr - l\pi/2)} - e^{-i(kr - l\pi/2)}}{kr} = \frac{e^{-ikr}}{r} X_- + \frac{e^{ikr}}{r} X_+, \qquad (2.38)$$

with

$$X_- = \sum_l P_l(\cos\theta)\left[-\frac{C_l}{2ik}e^{il\pi/2}\right], \tag{2.39}$$

and

$$X_+ = \sum_l P_l(\cos\theta)\left[\frac{C_l}{2ik}e^{-il\pi/2}e^{2i\delta_l}\right]. \tag{2.40}$$

To obtain and expression for the scattering amplitude in terms of phase shifts, we must compare Eq. (2.38) with the scattering boundary condition expressed in Eq. (2.3). For this purpose, it is necessary to separate the incoming and outgoing components of the incident plane-wave (Eq. (2.3)). Using Eq. (2.36) we obtain

$$\phi_{\mathbf{k}}(\mathbf{r}) = A\sum_{lm}4\pi Y_{lm}^*(\hat{\mathbf{r}})Y_{lm}(\hat{\mathbf{k}})i^l\frac{\hat{j}_l(kr)}{kr} \tag{2.41}$$

$$= A\sum_l(2l+1)P_l(\cos\theta)i^l\frac{\hat{j}_l(kr)}{kr}. \tag{2.42}$$

Taking the asymptotic form of $\hat{j}_l(kr)$ and writing $\sin(kr - l\pi/2)$ in exponential form, we get

$$e^{ikr\cos\theta} = \frac{1}{2i}\sum_{l=0}^{\infty}(2l+1)i^l P_l(\cos\theta)\left[\frac{e^{i(kr-l\pi/2)} - e^{-i(kr-l\pi/2)}}{kr}\right], \tag{2.43}$$

that represents the asymptotic form of a plane wave. The first term inside brackets corresponds to an outgoing spherical wave and the second to an ingoing spherical wave. Thus, each partial wave in (2.43) is, at large distances from the origin, a superposition of two spherical waves, an ingoing and an outgoing. The total radial flux for the wave function $\phi_{\mathbf{k}}(\mathbf{r}) = Ae^{ikr\cos\theta}$ vanishes, since the number of free particles that enters into a region is the same that exits. This can be easily shown using (2.43) in the definition of the probability current, i.e.

$$\mathbf{j}_{\mathrm{sc}} = \frac{\hbar}{\mu}\,\mathrm{Im}\left(\Psi^*\boldsymbol{\nabla}\Psi\right)$$

[an exercise proposes this demonstration for the more general expression (2.38)].

We can rewrite (2.43) as

$$e^{ikr\cos\theta} \to \frac{e^{-ikr}}{r}\left\{\sum_l P_l(\cos\theta)\left[-\frac{2l+1}{2ik}i^l e^{il\pi/2}\right]\right\} + \frac{e^{ikr}}{r}\left\{\sum_l P_l(\cos\theta)\left[\frac{2l+1}{2ik}i^l e^{-il\pi/2}\right]\right\}.$$

Inserting this expansion in Eq. (2.3), we obtain

$$\psi_{\mathbf{k}}^{(+)} \to \frac{e^{-ikr}}{r}\bar{X}_- + \frac{e^{ikr}}{r}\bar{X}_+, \tag{2.44}$$

with

$$\bar{X}_- = \sum_l P_l(\cos\theta)\left[-A\frac{2l+1}{2ik}i^l e^{il\pi/2}\right], \tag{2.45}$$

and

$$\bar{X}_+ = A\frac{f(\theta)}{r} + A\sum_l P_l(\cos\theta)\left[\frac{2l+1}{2ik}i^l e^{-il\pi/2}\right]. \tag{2.46}$$

Since e^{-ikr}/r and e^{ikr}/r are linearly independent, the equivalence of Eqs. (2.44) and (2.38) requires that

$$\bar{X}_- = X_- \quad \text{and} \quad \bar{X}_+ = X_+. \tag{2.47}$$

Eq. (2.47) leads to the expansion coefficients

$$C_l = A\,(2l+1)\,i^l. \tag{2.48}$$

Inserting the explicit values of the coefficients C_l in Eq. (2.37), we get the partial-wave expansion of the wave function

$$\psi_{\mathbf{k}}^{(+)}(\mathbf{r}) = A \sum_l (2l+1) P_l(\cos\theta) i^l e^{i\delta_l} \frac{u_l(k,r)}{kr} \tag{2.49a}$$

$$= A \sum_{lm} 4\pi Y_{lm}^*(\hat{\mathbf{k}}) Y_{lm}(\hat{\mathbf{r}}) i^l e^{i\delta_l} \frac{u_l(k,r)}{kr}. \tag{2.49b}$$

We now evaluate the scattering amplitude. Using Eqs. (2.40), (2.45), (2.46) and (2.48) in Eq. (2.47), we obtain

$$f(\theta) = \frac{1}{2ik} \sum_l (2l+1) P_l(\cos\theta) [S_l - 1] \tag{2.50a}$$

$$\equiv \frac{1}{k} \sum_l (2l+1) P_l(\cos\theta) e^{i\delta_l} \sin\delta_l. \tag{2.50b}$$

A useful expression for $f(\theta)$ can be derived multiplying both sides of Eq. (2.50a) by $\cos\theta$ and using the recursion relations for the Legendre polynomials [5]

$$\cos\theta \, (2l+1) \, P_l(\cos\theta) = (l+1) \, P_{l+1}(\cos\theta) + l P_{l-1}(\cos\theta)$$

After some algebra, one obtains (see, e.g. Ref. [4])

$$(1 - \cos\theta) f(\theta) = \frac{1}{2ik} \sum_l P_l(\cos\theta) \left[(2l+1)S_l - lS_{l-1} - (l+1)S_{l+1} \right]. \tag{2.51}$$

The elastic cross section is then obtained from the formula

$$\frac{d\sigma}{d\Omega} = |f(\theta)|^2. \tag{2.52}$$

An expression for the total elastic cross section, σ_{el}, is obtained through angular integration of Eq. (2.52). Using Eq. (2.50b) and the orthogonality relation for the Legendre polynomials,

$$\int P_l(\cos\theta) P_{l'}(\cos\theta) d(\cos\theta) = \frac{2}{2l+1} \delta_{l,l'}, \tag{2.53}$$

we get

$$\sigma_{el} = \frac{\pi}{k^2} \sum_l (2l+1) \left[1 + |S_l|^2 - 2\,\text{Re}\,\{S_l\} \right] \tag{2.54}$$

$$= \frac{4\pi}{k^2} \sum_l (2l+1) \sin^2\delta_l. \tag{2.55}$$

From the partial-wave expansions for σ_{el} and for $f(\theta = 0)$, one can trivially derive the *Optical Theorem*. Since at $\theta = 0$, $P_l(\cos\theta) \equiv P_l(1) = 1$, the imaginary part of Eq. (2.50b) can be written

$$\text{Im}\,\{f(\theta = 0)\} = \sum_l \frac{(2l+1)}{k} \sin^2\delta_l. \tag{2.56}$$

Comparing Eqs. (2.55) and (2.56) we obtain the *Optical Theorem*,

$$\sigma_{el} = \frac{4\pi}{k} \text{Im}\,\{f(0)\}. \tag{2.57}$$

The theorem connects the total cross section with the scattering amplitude at the angle zero and this is physically understandable: the cross section measures the removal of particles from the incident beam that arises, by its turn, from the destructive interference between the incident and

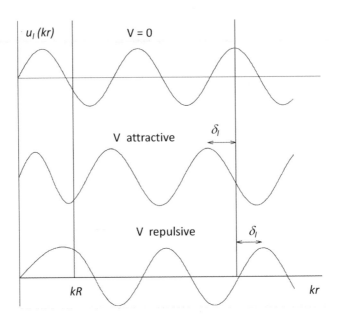

FIGURE 2.4 Radial part of the wave function for three different potentials, showing how the phase-shift sign is determined by the function behavior in the region $r < R$ where the potential acts.

scattered beams at zero angle. The optical theorem is not restricted to elastic scattering, being also valid for inelastic processes.

The phase shifts are evaluated by solving Eq. (2.8) for each l and comparing the phase of $u_l(r)$, for some large r, with the phase of $\hat{\jmath}_l(kr)$ for the same value of r. This is depicted in Figure 2.4, for a generic value of l and three different situations of the potential $V(r)$.

The first curve shows $u_l(r)$ for the case in which $V(r) = 0$ for all r. In this case, $u_l(r) = \hat{\jmath}_l(kr)$ and we do not have phase shifts for any l. The middle curve shows $u_l(r)$ when one introduces a small attractive potential acting inside a certain radius R, that is, $V(r) < 0$ for $r < R$, and $V(r) = 0$ when $r > R$. From (2.8) we see that, with this attractive potential, $|E - V(r)| > E$ in the potential region and the quantity $d^2 u_l/dr^2$ will be greater in that region than in the region of zero potential. Thus, $u_l(r)$ will oscillate rapidly for $r < R$. For $r > R$, the behavior is the same that in the case $V(r) = 0$, except that the phase is displaced. In this way we see that with a small attractive potential, $u_l(r)$ is "pulled into", which turns its phase advanced and the phase shift positive. The last curve shows $u_l(r)$ for the case of a small repulsive potential, i.e., $V(r) > 0$ when $r < R$, and $V(r) = 0$ when $r > R$. In this case, $|E - V(r)| < E$ in the potential region and the quantity $d^2 u_l/dr^2$ will be smaller in that region than in the region of zero potential. The result is that, for a repulsive potential, $u_l(r)$ is "pulled out", its phase is retarded and the phase shift is negative.

In the sum (2.55), each angular momentum contributes at most with a cross section

$$(\sigma)_{\text{max}} = \frac{4\pi}{k^2}(2l + 1), \tag{2.58}$$

a value of the same order of magnitude as the maximum classical cross section per unity $\hbar$ of angular momentum. Actually, if we use the estimate $b = l/k$ for the impact parameter, the contribution of an interval $\Delta l = 1$, or $\Delta b = 1/k$, for the total cross section will be

$$\sigma_l = 2\pi b \Delta b = 2\pi \frac{l}{k^2}. \tag{2.59}$$

For large l this agrees with $(\sigma)_{\text{max}}$, except for a factor 4. The difference is due to the unavoidable presence of diffraction effects for which the wave nature of matter is responsible.

The partial-wave analysis (2.50b), gives an exact procedure to solve the scattering problem at all energies. For a given potential $V(r)$, equation (2.8) should be solved and its asymptotic solutions (2.35a) used to find the phase shifts δ_l. The infinite number of terms of (2.50b) is not a problem in practice since

$$\lim_{l \to \infty} \delta_l = 0, \tag{2.60}$$

a result that can be verified examining (2.35a): for large l the centrifugal potential term, proportional to $l(l + 1)$, is totally dominant, turning irrelevant the phase shifts generated by $V(r)$. However, the sum (2.50b) will have, at high energies, the contribution of many terms since, in this case, $kR \gg 1$ and for all l up to $l_{\max} \simeq kR$, there will be appreciable phase shifts. The partial-wave analysis is of great utility especially when few partial waves contribute.

2.6 Integral formulae for the phase shifts

The phase shifts can also be evaluated by an integral involving radial wave functions. We start by multiplying Eq. (2.13) by $\hat{j}_l(\rho)$, the Ricatti-Bessel equation

$$\hat{j}_l''(\rho) + \left(1 - \frac{l(l+1)}{\rho^2}\right) \hat{j}_l(\rho) = 0 \tag{2.61}$$

by $u_l(\rho)$, and subtracting the resulting equations. We obtain

$$\frac{d^2 u_l(k, \rho)}{d\rho^2} \hat{j}_l(\rho) - u_l(k, \rho) \hat{j}_l''(\rho) = U(\rho) \hat{j}_l(\rho) u_l(k, \rho), \tag{2.62}$$

where we have used the notation $U(\rho) = V(\rho/k)/E$. Now we integrate both sides over ρ from zero to ∞ and carry out the integration by parts on the LHS. We get

$$\left[\frac{du_l(k, \rho)}{d\rho} \hat{j}_l(\rho) - u_l(k, \rho) \hat{j}_l'(\rho)\right]_0^\infty = \int_0^\infty d\rho\, \hat{j}_l(\rho) U(\rho) u_l(k, \rho). \tag{2.63}$$

Since both $\hat{j}_l(\rho)$ and $u_l(k, \rho)$ are regular at the origin, the LHS (left hand side) of Eq. (2.63) has only contribution from the upper limit, where $u_l(k, \rho)$ can be replaced by its asymptotic form which we adopt here as

$$u_l(k, \rho \to \infty) = \sin(\rho - l\pi/2 + \delta_l). \tag{2.64}$$

Using this result together with the asymptotic form of $\hat{j}_l(\rho)$ in Eq. (2.63), we get*,

$$\sin \delta_l = -\int d\rho\, \hat{j}_l(\rho) U(\rho) u_l(k, \rho). \tag{2.65}$$

or

$$\sin \delta_l = -\left(\frac{2\mu}{\hbar^2}\right) \int dr\, r\, j_l(kr) V(r) u_l(k, r). \tag{2.66}$$

Although for direct evaluation of the phase-shift Eq. (2.65) is not an improvement over Eq. (2.30), it is very useful for the development of approximations. If the potential is weak ($U(\rho) \ll 1$ for any ρ), the radial wave function cannot be very different from the Ricatti-Bessel function and one can approximate,

$$\sin \delta_l \simeq -\int d\rho\, U(\rho) \hat{j}_l^2(\rho) = -k\left(\frac{2\mu}{\hbar^2}\right) \int dr\, r^2\, V(r) j_l^2(kr). \tag{2.67}$$

This equation is similar to the Born approximation, which will be discussed in a forthcoming chapter. Since $j_l^2(\rho) \geq 0$, one concludes that the sign of the phase shift is determined by the sign of

---

*If a different normalization for the radial wave function is adopted, Eq. (2.65) will have a different form. If, for example, one uses $\alpha_l = 1$ in Eq. (2.29), as in Ref. [6], the LHS of Eq. (2.65) changes to $\tan \delta_l$.

the potential $V(r)$. Attractive potentials ($V(r) < 0$) give rise to positive phase shifts while repulsive potentials lead to negative phase shifts. This property played an important role in the discussion on the existence of a repulsive core in the strong interaction between two protons, which took place some decades ago [9]. The fact that the experimental S-wave phase shift in $p - p$ collisions changed from positive to negative at the energy of 250 MeV indicated that the picture of a fully attractive nuclear interaction was not consistent with the data.

2.7   Hard sphere scattering

An interesting problem which can easily be handled is the scattering from an infinitely repulsive square well potential. This potential gives an approximation to hard-cores in the nuclear interaction. The hard sphere potential is,

$$V(r) = 0, \quad \text{for } r > R,$$
$$V(r) \to +\infty, \quad \text{for } r < R. \tag{2.68}$$

In this case, the wave function vanishes in region I and its continuity at the boundary requires that* $u_l^{II}(k, R) = 0$. Using its explicit form, as given in Eq. (2.29), this condition leads to

$$\tan \delta_l = \frac{\hat{j}_l(kR)}{\hat{n}_l(kR)}. \tag{2.69}$$

This result becomes particularly simple for S-waves. Using Eqs. (2.16) and (2.20), we get

$$\tan \delta_0 = -\tan(kR) \quad \to \quad \delta_0 = -kR. \tag{2.70}$$

Using (2.55) we get for the s-wave cross section

$$\sigma_{el}^{(0)} = \frac{4\pi}{k^2} \sin^2 \delta_0 \longrightarrow 4\pi R^2, \tag{2.71}$$

where the right side is the limit when $E \longrightarrow 0$. We see that the cross section given above is a factor of 4 larger than what one would expect with a classical description. The reason is the unavoidable diffraction contributions in the scattering of a wave by an obstacle. The experimental findings corroborate the results of quantum mechanics calculations for σ_{el}.

2.8   Resonances

Although the phase shifts are usually slowly varying functions of the collision energy, there may be situations where a particular phase shift δ_l changes rapidly with E. In this case, the corresponding partial wave tends to give important contribution to the cross section. This behavior is called a *resonance*. In many important examples, the description of the resonance requires the consideration of intrinsic degrees of freedom. This is, for example, the case of the compound nucleus resonances in nuclear physics (see, for example, Ref. [10]). However, a deep attractive potential can also give rise to resonances. In this section we study such resonances, for the simple case of S-waves.

As a starting point, we write the asymptotic wave function as in Eq. (2.32). For convenience, let us change the normalization so that $\gamma_l = -i$ in Eq. (2.34). Thus,

$$u_0^{II}(k, r) = e^{-ikr} - S_0 e^{ikr}, \tag{2.72}$$

where $S_0 = e^{2i\delta_0}$. We now use this equation to calculate the logarithmic derivative

$$\mathcal{L}^{II} = R \frac{\left[du_0^{II}(k, r)/dr \right]_{r=R}}{u_0^{II}(k, R)} = -ikR \left[\frac{1 + S_0 e^{2ikR}}{1 - S_0 e^{2ikR}} \right].$$

---

*Note that for infinite potentials the continuity of the derivative $u_l'(k, r)$ is not required. In this particular case, it is discontinuous at $\rho = kR$.

Using the matching condition $\mathcal{L}^I = \mathcal{L}^{II}$, one can express S_0 in terms of the internal logarithmic derivative $\mathcal{L}^I$. Namely,

$$S_0 = e^{-2ikR} \left[\frac{\mathcal{L}^I + ikR}{\mathcal{L}^I - ikR} \right]. \tag{2.73}$$

Now we separate real and imaginary parts of $\mathcal{L}^I$

$$\mathcal{L}^I = a - bi, \tag{2.74}$$

and rewrite Eq. (2.73) as

$$S_0 = -e^{-2ikR} \left[\frac{(kR - b) - ia}{(kR + b) + ia} \right]. \tag{2.75}$$

Using Eq. (2.75) in the S-wave contribution for the scattering amplitude $f_0(\theta) \equiv f_0$,

$$f_0 = \frac{1}{2ik}(S_0 - 1), \tag{2.76}$$

we get

$$f_0 = \frac{1}{2ik} \left\{ -e^{-2ikR} \left[\frac{(kR - b) - ia}{(kR + b) + ia} \right] - 1 \right\}.$$

Adding and subtracting e^{-2ikR} within the curly brackets, f_0 takes the form

$$f_0 = f_{pot} + f_{res}, \tag{2.77}$$

with

$$f_{pot} = \frac{1}{2ik} \left(e^{-2ikR} - 1 \right), \tag{2.78}$$

and

$$f_{res} = \frac{e^{-2ikR}}{k} \left[\frac{kR}{a - i(kR + b)} \right]. \tag{2.79}$$

Eq. (2.78) gives the S-wave term of Eq. (2.50a) with the phase shift

$$\delta_{pot} = -kR.$$

Therefore, it corresponds to the scattering from a hard sphere with radius R, as given in Eq. (2.70). The resonant behavior has then to be contained in f_{res}. A resonance occurs when the real part of $\mathcal{L}^I$ vanishes. In this way, the condition

$$a(E_r) = 0 \tag{2.80}$$

is equivalent to an equation for the resonance energy, E_r. To study the cross section in the neighborhood of the resonance, we expand $a(E)$ around $E = E_r$,

$$a(E) \simeq \left[\frac{\partial a(E)}{\partial E} \right]_{E=E_r} [E - E_r] \equiv a'(E_r)[E - E_r], \tag{2.81}$$

and introduce the notation

$$\Gamma_e = -\frac{2kR}{a'(E_r)}, \qquad \Gamma_a = -\frac{2b}{a'(E_r)}, \qquad \Gamma = \Gamma_e + \Gamma_a. \tag{2.82}$$

The quantities Γ_e, Γ_a and Γ are known as the scattering width, the absorption (or reaction) width and the total width, respectively.

Using the expansion for $a(E)$, the resonant amplitude can be written[*]

$$f_{res} \simeq -\frac{e^{-2ikR}}{k} \left(\frac{\Gamma_e/2}{(E - E_r) + i\Gamma/2} \right). \tag{2.83}$$

---

[*]If the present study is extended to include complex potentials, the resonant behavior also appears in the absorption cross section. In this case, the cross section takes the form
$$\sigma_a = \left(\pi/k^2 \right) \left[1 - |S_0|^2 \right] = \left(\pi/k^2 \right) \Gamma_e \Gamma_a / \left[(E - E_{res})^2 + \Gamma^2/4 \right].$$

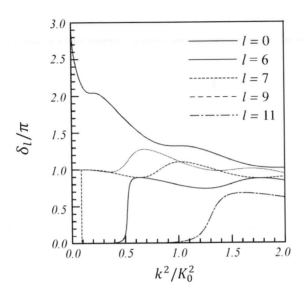

FIGURE 2.5   Phase shifts for the scattering from a square well with characteristic parameter $KR = 10$.

For a strong and isolated resonance, the contribution from f_{res} to the cross section gets so large that f_{pot} can be neglected. In this case, the integrated elastic cross section becomes

$$\sigma_{el} = \frac{\pi}{k^2} \left[\frac{\Gamma_e^2}{(E - E_r)^2 + \Gamma^2/4} \right]. \tag{2.84}$$

This result is known as the *Breit-Wigner formula*. The cross section is maximal at $E = E_r$ and has the half-width $|E - E_r| = \Gamma/2$. For real potentials, as assumed in this chapter, the logarithmic derivative is real and, therefore, $b = 0$. In this case the reaction width vanishes and $\Gamma_e = \Gamma$.

As shown in the Appendix 2.A, close to a resonance the phase shift approaches the value $\pi/2$. Thus, close to a resonance we can use the expansion

$$\delta_l \simeq \frac{\pi}{2} - (E_r - E) \frac{d\delta_l}{dE}. \tag{2.85}$$

Defining

$$\Gamma = 2 \left(\frac{d\delta_l}{dE} \right)^{-1} \tag{2.86}$$

in the above equation, we get

$$\tan \delta_l = \frac{\Gamma/2}{E_r - E}. \tag{2.87}$$

Thus,

$$\sigma_l = |f_l|^2 = \frac{\pi}{k^2} \left[\frac{\Gamma^2}{(E - E_r)^2 + \Gamma^2/4} \right],$$

which coincides with Eq. (2.84), in the absence of absorption. The relation (2.86) is useful to relate the resonance width to the derivative of the phase shift at the resonance energy.

2.9   Scattering from an attractive square-well

Let us now calculate phase shifts in the scattering from a square-well, for an arbitrary partial wave. In this case, the radial wave function in the internal region becomes

$$u_l^I(k, r) = A\hat{\jmath}_l(Kr), \qquad K = \sqrt{2m(E + V_0)}/\hbar,$$

l	n	ε
0	0	-0.92
	1	-0.68
	2	-0.29
1	0	-0.83
	1	-0.52
	2	-0.08
2	0	-0.73
	1	-0.34
3	0	-0.60
	1	-0.14
4	0	-0.46
5	0	-0.29
6	0	-0.11

TABLE 2.1   Bound states of the square-well discussed in the text.

l	ε	γ
7	0.09	0.000007
8	0.30	0.002
9	0.52	0.012
10	0.77	0.026
11	1.04	0.57

TABLE 2.2   Locations and widths of the resonances at the partial waves $l = 7, \cdots, 10$.

where V_0 is the depth of the potential.

The logarithmic derivative is

$$\mathcal{L}^I = KR \frac{\hat{j}_l'(KR)}{\hat{j}_l(KR)}. \tag{2.88}$$

Inserting Eq. (2.88) in Eq. (2.187), we get

$$\delta_l = \tan^{-1}\left[\left(\frac{k\hat{j}_l'(kR)\hat{j}_l(KR) - K\hat{j}_l(kR)\hat{j}_l'(KR)}{k\hat{n}_l'(kR)\hat{j}_l(KR) - K\hat{n}_l(kR)\hat{j}_l'(KR)}\right)\right]. \tag{2.89}$$

Applying the procedure of section Appendix 2.A, we obtain absolute phase shifts for the same potential well with $K_0R = 10$, for several partial-waves. Here, $K_0 = \sqrt{2mV_0}/\hbar$. We start from a very high energy and progressively reduce it, adding or subtracting π to eliminate the discontinuities. Note that the phase shifts depend exclusively on the characteristic parameter of the well, K_0R, and on the wave number k. The results are shown in Figure 2.5, for a few relevant partial-waves. The x-axis gives the normalized scattering energy $\varepsilon = E/V_0 = k^2/K_0^2$. The behavior of $\delta_l(\varepsilon)$ can be better understood in connection to the bound states of the potential. In Table 2.1 they are given as functions of the angular momentum in $\hbar$-units, l, and the number of nodes of the wave function, n. We see that the potential binds three states for each of the angular momentum values $l = 0, 1$, two states for $l = 2, 3$ and a single state for $l = 4, 5$, and 6. There are no bound state for higher angular momenta. According to Levinson's theorem (Appendix 2.A), the phase shifts should have the limits $\delta_l(k \to 0) = 3\pi$ for $l = 0$ and 1, $\delta_l(k \to 0) = 2\pi$ for $l = 2$ and 3, and $\delta_l(k \to 0) = \pi$ for $l = 4$, 5, and 6. At higher partial-waves the phase shifts should go to zero in the $k \to 0$ limit. This behavior is confirmed in Figure 2.5. For $l \geq 7$, the bound state occurring at $l = 6$ gives place to a resonance. It is very sharp for $l = 7$ and becomes progressively broader as the angular momentum increases. These resonances represent virtual (or almost) bound states at increasing energy, for increasing l. These virtually bound states arise from constructive interferences by the reflection of the internal waves off the wall of the potential.

Fitting with Breit-Wigner shapes, we get the resonance parameters indicated in Table 2.2. In this table, the widths Γ are normalized with respect to the potential depth, as $\gamma = \Gamma/V_0$.

2.10   Low energy scattering: scattering length

As $E \to 0$, only S-waves contribute to the scattering amplitude and Eq. (2.50b) becomes

$$f = \frac{1}{k} e^{i\delta_0} \sin \delta_0. \tag{2.90}$$

It is convenient to re-write this equation in the form

$$f = \left(\frac{ke^{-i\delta_0}}{\sin\delta_0}\right)^{-1} = \frac{1}{k\cot\delta_0 - ik}. \tag{2.91}$$

To obtain a low-energy approximation for f, we must evaluate $k\cot\delta_0$ in the $k \to 0$ limit. We start from Eq. (2.30) for $l = 0$, so that we can replace $\hat{\jmath}_l(\rho) = \sin\rho$, $\hat{\jmath}'_l(\rho) = \cos\rho$, $\hat{n}_l(\rho) = -\cos\rho$ and $\hat{n}'_l(\rho) = \sin\rho$. Using the notation $x = kR$, we get

$$\tan\delta_0 = \frac{x\cos(x) - \mathcal{L}^I\sin(x)}{x\sin(x) + \mathcal{L}^I\cos(x)}, \tag{2.92a}$$

and we can write

$$k\cot\delta_0 = \frac{x}{R}\left[\frac{x\sin(x) + \mathcal{L}^I\cos(x)}{x\cos(x) - \mathcal{L}^I\sin(x)}\right]. \tag{2.93}$$

At low energies, $x \to 0$ and, to lowest order in k, we then replace $\cos x \simeq 1$; $\sin x \simeq x$; $\mathcal{L}^I \simeq L$. Eq. (2.92a) then becomes

$$\tan\delta_0 = -kR\left[\frac{L-1}{L}\right]. \tag{2.94}$$

Introducing the *scattering length* a, defined as

$$a \equiv -\lim_{k\to 0}\left(\frac{\tan\delta_0}{k}\right) = R\left[\frac{L-1}{L}\right], \tag{2.95}$$

we get

$$\tan\delta_0 = -ka. \tag{2.96}$$

To study the behavior of the elastic cross section at low energies for any partial wave, we consider Eq. (2.8) as $E \to 0$. In this limit, the internal radial wave function $u^I_l(k,r)$ tends to some function $z_l(r)$ and its logarithmic derivative at the boundary R can be written

$$L = \lim_{k\to 0}\mathcal{L}^I = R\lim_{k\to 0}\left\{\frac{[du^I_l(k,r)/dr]_{r=R}}{u^I_l(k,R)}\right\} = R\frac{[dz_l(r)/dr]_{r=R}}{z_l(R)}. \tag{2.97}$$

Taking the $k \to 0$ limit in Eq. (2.30), the phase shifts are given by the relation

$$\tan\delta_l = \frac{kR\hat{\jmath}'_l(kR) - L\hat{\jmath}_l(kR)}{kR\hat{n}'_l(kR) - L\hat{n}_l(kR)}. \tag{2.98}$$

Since $k \to 0$, we can replace $\hat{\jmath}_l(kR)$, $\hat{n}_l(kR)$, $\hat{\jmath}'_l(kR)$ and $\hat{n}'_l(kR)$ by their small argument limits (Eq. (2.14) and their derivatives. This yields,

$$\tan\delta_l = k^{2l+1}\left[\frac{R^{2l+1}}{(2l+1)!!(2l-1)!!}\frac{(l+1-L)}{l+L}\right]. \tag{2.99}$$

The above expression becomes particularly simple for $l = 0$. This case is particularly important and will be discussed in detail below.

Since at low energies $\tan\delta_0$ is very small, Eq. (2.95) gives* $\delta_0 \simeq -ka$. Comparing this result with Eq. (2.70), we conclude that the scattering length can be interpreted as the radius of an "effective hard-sphere" which produces the same low-energy S-wave phase shift. In this limit, the cross section becomes isotropic and its integration over angle can be trivially evaluated. Replacing $\sin\delta_0 \simeq -ka$ in Eq. (2.55), we obtain

$$\sigma_{el} = 4\pi a^2. \tag{2.100}$$

---

*We should point out that the discussion below is not applicable for absolute S-wave phase shifts of an attractive potential with bound states. According to the Levinson theorem (see Appendix 2.A), $\lim_{k\to 0}\delta_0 = n\pi \neq 0$, where n is the number of bound states with $l = 0$.

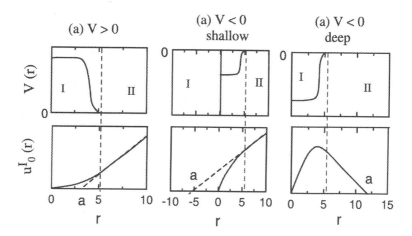

FIGURE 2.6 Emergence of the scattering length for repulsive and attractive potentials.

The scattering length is qualitatively related with the potential in a simple way. Firstly, we note that the $k \to 0$ limit implies $\tan \delta_0 \to 0$ and also allows the use of the small argument approximations (Eq. (2.14)) at $r \approx R$. Namely: $\hat{\jmath}_0(kr) \simeq kr$ and $\hat{n}_0(kr) \simeq -1$. The wave function in the external region (see Eq. (2.29)) near R can be approximated as

$$u_l^{II}(k,r) \simeq A\left(kr + \tan \delta_0\right),$$

or, using Eq. (2.95),

$$u_l^{II}(k,r) = Ak\left(r - a\right). \tag{2.101}$$

Therefore, near R the external wave function is linear and the scattering length is given by the point where the straight line intercepts the horizontal axis. This establishes a qualitative relation between the potential and a. Firstly we consider a repulsive potential, as represented in the upper part of Figure 2.6a. In the $k \to 0$ limit, region I is always classically forbidden. Therefore, the wave function $u_l^I(k,r)$ grows very rapidly (exponentially for a square well) as r approaches R. The continuity of the logarithmic derivative requires that u_l^I and u_l^{II} match as indicated in the bottom part of Figure 2.6a. Therefore, the scattering length must be positive and smaller than the matching radius, indicated by vertical dashed lines.

We now turn to attractive potentials. We first consider a shallow potential well, not deep enough to have a bound state. In this case, the phase of $u_l^I(k, R)$ is very small and the matching must be as indicated in Figure 2.6b. The scattering length is then negative. Finally, we consider a deeper potential well, such that the internal wave function is strongly bent and its phase at R is between $\pi/2$ and π. This corresponds to the condition for the well to have a single bound state. In this case, $u_l^{II}(k,r)$ matches a decreasing wave function and the result is represented in Figure 2.6c. The scattering length is then positive and larger than the matching radius. This analysis is summarized below.

$V(r) > 0$ (repulsive)	$R > a > 0$
$V(r) < 0$ (attractive) - 0 b. st.	$a < 0$
$V(r) < 0$ (attractive) - 1 b. st.	$a > R > 0$

To close this section, we evaluate the scattering amplitude to the same order in k. Using Eq. (2.96) we obtain $k \cot \delta_0 = -1/a$ and, upon inserting this result in Eq. (2.91), the scattering amplitude becomes

$$f = \frac{1}{-1/a - ik}. \tag{2.102}$$

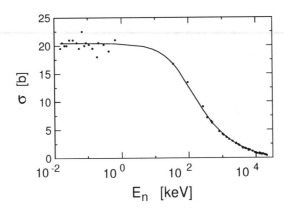

FIGURE 2.7   Neutron-proton cross section. The experimental points were obtained from Refs. [11] and [13]. The curve was calculated using Eq. (2.124).

2.11   Scattering length for nucleon-nucleon scattering

Let us make an estimate of the neutron-proton cross section. Let us use for V_0, the approximate depth of the deuteron potential, the value of 34 MeV. We have

$$K_0 = \frac{\sqrt{2mV_0}}{\hbar} = \frac{\sqrt{2mc^2V_0}}{\hbar c} = \frac{\sqrt{938.93 \times 34}}{197.33} = 0.91 \text{ fm}^{-1}, \qquad (2.103)$$

where m is the reduced mass, equal to half a nucleon mass. Using $R \cong 2.1$ fm as the deuteron radius, we have (see Eq. (2.197))

$$
\begin{aligned}
a &= -\frac{\delta_0}{k} = R\left[1 - \frac{\tan(K_0R)}{K_0R}\right] \\
  &= 2.1 \times 10^{-13}\left[1 - \frac{\tan(0.91 \times 2.1)}{0.91 \times 2.1}\right] = 5.2 \times 10^{-13} \text{ cm.} \qquad (2.104)
\end{aligned}
$$

Hence,
$$\sigma = 4\pi(5.2 \times 10^{-13})^2 \text{ cm}^2 \cong 3.5 \text{ b.} \qquad (2.105)$$

Figure 2.7 shows experimental values of the neutron-proton cross sections up to 20 MeV. In the zero energy limit the cross section has the value $\sigma = (20.43 \pm 0.02)$ b, 6 times greater than the value obtained in (2.105). The reason of this discrepancy was explained by Wigner, proposing that the nuclear force depends on the spin, being different when nucleons collide with parallel (triplet) spins or when they collide with antiparallel spins (singlet). As there are three triplet states and only one singlet, an experiment where the nucleons are not polarized will register three times more events of the first type than of the second, resulting for the cross section the combination:

$$\sigma = \frac{3}{4}\sigma_t + \frac{1}{4}\sigma_s. \qquad (2.106)$$

If then $\sigma = 20.4$ b and $\sigma_t = 3.4$ b, one gets for σ_s the value of 71 b! The explanation for this high value is found in the fact that the singlet potential is shallower than the triplet one, being in the threshold for the appearance of the first bound state. This gives place to a resonance when the incident particle has very low energy, as it happens in the present case.

There are ways to find if the singlet potential has a bound state with a very low negative energy or if the resonance occurs at a very low yet positive energy. As we have seen above, the sign of the scattering length can show us if a resonance occurs with a negative (bound) state or a positive (virtual) energy.

The combination of (2.100) and (2.106) gives

$$\sigma = \pi(3a_t^2 + a_s^2), \tag{2.107}$$

where a_t and a_s are the scattering length for the triplet and singlet potentials, with the respective cross sections σ_t and σ_s given by expressions such as (2.100). In an equation of the form (2.107) the sign of the scattering amplitude does not matter, due to an incoherent combination of singlet and triplet scattering there. The cross section is proportional to the square of the amplitudes, in the same way as the intensity of a light beam is proportional to the square of the magnitude of the electric (or magnetic) field . One form of coherent scattering is achieved when the wavelength of the incident particles is of the order of the distance between the nuclei inside a molecule. In the case of neutron incident on H_2, the distance between the protons in the molecule is about 0.8×10^{-8} cm and the coherent scattering is reached with a neutron energy near 2×10^{-3} eV. The scattering of these very slow neutrons by a hydrogen molecule produces an interference phenomenon similar to that occurring with light waves that emerge from the two slits of an Young experiment. An additional ingredient in the present case is that the H_2 molecule can be in two states, one with the spins of the protons forming a triplet (ortho-hydrogen) and the other where the spins form a singlet (para-hydrogen). When a neutron interacts with an ortho-hydrogen molecule the two scattering amplitudes are of the same type and when the interaction is with para-hydrogen they are of different types. Schwinger and Teller [14] have found expressions for the scattering cross sections of slow neutrons with the ortho- and para-hydrogen:

$$\sigma = c_1(a_t - a_s)^2 + c_2(3a_t + a_s)^2, \tag{2.108}$$

where c_1 and c_2 are numerical coefficients, different for the two types of hydrogen and that also depend upon the incident neutron energy and upon the gas temperature. This last dependence is natural since the incident neutrons have very low velocities and the thermal molecular motion, in this case, has a non negligible influence. As an example, at a temperature of 20.4 K and with 0.0045 eV neutrons the coefficients are, for the ortho-hydrogen, $c_1 = 13.762$ and $c_2 = 6.089$ and for the para-hydrogen, $c_1 = 0$ and $c_2 = 6.081$. Note that in (2.108), contrary to (2.107), the signs of a_t and a_s are important for the calculation of σ.

From (2.108) it is possible to obtain the scattering lengths a_t and a_s if the experimental cross sections are known. Measurements of these cross sections using gaseous hydrogen have been done since 1940 [12], improving former works done with liquid H_2, where the effect due to the intermolecular forces are difficult to separate. The cross sections were measured from the room temperature, where the proportion is 75% of ortho-hydrogen and 25% of para-hydrogen, down to 20 K. At the lowest temperature only the para-hydrogen exists since it has a greater binding energy than the ortho-hydrogen and can be formed by the decay of the para-hydrogen, provided that the process is accelerated by a catalyzer (a substance with paramagnetic atoms that induces the spin change of one of the protons of the H_2 molecule).

The results found for the scattering lengths in these and in other experimental works using different methods are not yet free of some systematic errors. Houk [13] recommends the values

$$a_t = (5.423 \pm 0.005) \text{ fm}, \qquad a_s = (-23.71 \pm 0.01) \text{ fm} \tag{2.109}$$

as being the most accurate for the scattering lengths. Note that these values also satisfy (2.107). The negative sign of the singlet scattering length is the answer to our question: the proton-neutron system has no other bound state than the deuteron ground state. The resonance in the low-energy scattering of neutron by protons is due to a state of the system with a small positive energy.

2.12   The effective range formula

We now look for better approximations to the scattering amplitude, evaluating $k \tan \delta_0$ to second order in k. For this purpose, we use the expansions

$$\sin x \simeq x - \frac{x^3}{6}; \quad \cos x \simeq 1 - \frac{x^2}{2}, \tag{2.110}$$

and take into account second order corrections to $\mathcal{L}^I$. Writing the differential equations for $u_0(k, r)$ and $z(r)$, i.e. for $E \neq 0$ and $E = 0$ cases,

$$u_0''(k, r) - U(r)u_0(k, r) = -k^2 u_0(k, r); \tag{2.111}$$

$$z''(r) - U(r)z(r) = 0, \tag{2.112}$$

where $U(r) = 2\mu V(r)/\hbar^2$, and carrying out the operation

$$z(r) \times \text{ Eq. } (2.111) - u_0(k, r) \times \text{ Eq. } (2.112) ,$$

we get

$$\frac{d}{dr}\left(z(r)\, u_0'(k, r) - u_0(k, r)z'(r)\right) = -k^2 u_0(k, r)z(r).$$

Integrating the above equation between $r = 0$ and $r = R$, and using the regularity of u_0 and z, we get

$$z(R)\, u_0'(k, R) - u_0(k, R)z'(R) = -k^2 \int_0^R dr u_0(k, r)z(r). \tag{2.113}$$

Now, we multiply both sides by $R/\left[z(R)u_0(k, R)\right]$. Using the definitions of $\mathcal{L}^I$ and L (see Eq. (2.97)), we obtain

$$\mathcal{L}^I - L = -k^2 \frac{R}{u_0(k, R)z(R)} \int_0^R dr u_0(k, r)z(r).$$

Since the difference $u_0(k, r) - z(r)$ leads to corrections of order k^3 or higher, we can replace $u_0 \to z$. The logarithmic derivative then becomes

$$\mathcal{L}^I \simeq L - x^2 \Delta, \tag{2.114}$$

with

$$\Delta = \frac{1}{Rz^2(R)} \int_0^R dr z^2(r). \tag{2.115}$$

Note that this result is independent of the normalization of $z(r)$.

Inserting Eqs. (2.110) and (2.114) in Eq. (2.93), and keeping terms up to x^2, we get

$$k \cot \delta_0 \simeq \frac{1}{R}\left[\frac{L + x^2\left(1 + \Delta - L/2\right)}{1 - L + x^2\left(L/6 + \Delta - 1/2\right)}\right] \simeq -\frac{1}{a} + \frac{k^2}{2}\left[2R\left(\frac{1 - L - \Delta + L^2/3}{(L - 1)^2}\right)\right]$$

or

$$k \cot \delta_0 \simeq -\frac{1}{a} + \frac{1}{2}k^2 r_{eff} \tag{2.116}$$

where r_{eff} is the *effective range*, given by[*]

$$r_{eff} = 2R\left[\frac{1 - L - \Delta + L^2/3}{(L - 1)^2}\right]. \tag{2.117}$$

The effective range can be expressed in terms of the range of the potential and the scattering length. Using Eq. (2.95) we get

$$r_{eff} = 2R\left[\left(1 - \frac{R}{a}\right)(1 - \Delta) + \frac{1}{3}\left(\frac{R}{a}\right)^2\right]. \tag{2.118}$$

---

[*]Usually, textbooks (Ref. [6], for example) give an alternative expression for r_{eff}, in terms of integrals involving $Z(r)$ and an auxiliary function which does not vanish at the origin. It can be easily shown that their expression is identical to Eq. (2.117).

We remark here some interesting properties of a and r_{eff} for very weak potentials. Let us consider the $U \to 0_\pm$ limit, where 0_+ and 0_- are used to denote cases of repulsive and attractive potentials, respectively. The discussion of Figure 2.6 leads to the conclusion

$$\lim_{U \to 0_\pm} a = 0_\pm. \tag{2.119}$$

Therefore, in this limit the scattering length approaches zero from positive values, for repulsive potentials, or negative values, for attractive potentials. Inspection of Eq. (2.118) shows that the effective range diverges in this limit. However, it can be shown that the product $a r_{eff}$ remains finite and is given by

$$a r_{eff} = \pm 2 R^2 / 3,$$

with + and - for repulsive and attractive potentials, respectively.

Eq. (2.116) leads to an improved expression for the scattering amplitude. Upon inserting this equation into Eq. (2.91), we obtain

$$f = \frac{1}{g(k^2) - ik} \tag{2.120}$$

where

$$g(k^2) = -\frac{1}{a} + \frac{1}{2} k^2 r_{eff}. \tag{2.121}$$

Eq. (2.116) yields also an improved expression for the S-wave phase shift. It can be written

$$\tan \delta_0 = -ka \left[\frac{1}{1 - \frac{1}{2} k^2 \left(a r_{eff} \right)} \right]. \tag{2.122}$$

As an application of the above discussed low-energy approximations, we consider the S-wave phase shifts in the scattering from a square well with depth V_0 and range R. In this case,

$$z(r) = A \sin (K_0 r),$$

where $K_0 = \sqrt{2\mu V_0}/\hbar$, and we get

$$L = K_0 R \cot (K_0 R)$$

$$\Delta = \frac{1}{2} \left[\frac{1}{\sin^2 (K_0 R)} - \frac{\cot (K_0 R)}{K_0 R} \right].$$

Figure 2.8 shows S-wave phase shifts in the scattering from a square well, in comparison with the first order approximation (Eq. (2.96)) and the effective range formula for $\tan \delta_0$ (Eq. (2.122)). The results are plotted against kR, in the range $1 > kR > 0$. Above $kR = 1$, these approximations are not expected to be reliable. In Figure 2.8a, we used the same deep potential of the previous section, with $K_0 R = 10$. In this case, the scattering length is positive and $\lim_{k \to 0} \delta_0 = 3\pi$. The exact phase shifts are the same as in Figure 2.13 (Appendix 2.A), except that only a very small part of the figure is shown. In the present figure, the upper limit of the x-axis ($kR = 1$) corresponds to the $k^2/K_0^2 = 0.01$. In Figure 2.8b, we used a much shallower potential, with $K_0 R = 1.5$. It has been chosen such as to be just below the limit to have a bound state ($K_0 R = \pi/2$). In this case, the scattering length is negative and very large. As a result, the phase shift changes more rapidly with k. The figure shows that the first order approximation breaks down at very low values of kR, as anticipated. On the other hand, the effective range approximation turned out to be surprisingly accurate, even at $kR = 1$.

2.13 Effective range for nucleon-nucleon scattering

Gathering Eqs. (2.54) (for $l = 0$) and (2.116), we can express the nucleon-nucleon cross section at low energies by

$$\sigma = \frac{4\pi}{k^2} \left(\frac{1}{1 + \cot^2 \delta_0} \right) = \frac{4\pi a^2}{a^2 k^2 + (1 - \frac{1}{2} a r_{eff} k^2)^2}, \tag{2.123}$$

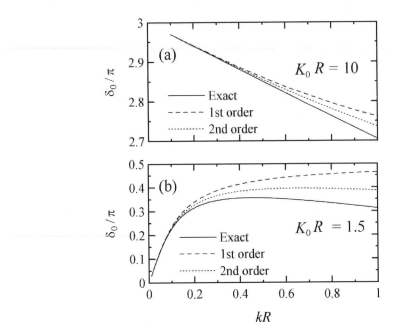

FIGURE 2.8   Exact vs. approximate low-energy S-wave phase shifts in the scattering from a square well.

where the influence of the potential is represented by two parameters, the effective range r_{eff} and the scattering length a. Thus the cross section is not affected by the details of the assumed form of the potential since with any other reasonable form it will be always possible to adjust the depth and the range in such a way to reproduce the values of a and r_{eff}. The reciprocal result is that a study of low-energy scattering does not lead to information about the form of the nucleon-nucleon potential. The theory of effective range is therefore sometimes called *form independent approximation*.

For the application of the cross section (2.123) we have to remember that there are in fact two potentials, one for the singlet and another for the triplet state and (2.123) should be, using (2.106), more appropriately written

$$\sigma = \frac{3}{4} \frac{4\pi a_t^2}{a_t^2 k^2 + (1 - \frac{1}{2} a_t r_{eff}^{(t)} k^2)^2} + \frac{1}{4} \frac{4\pi a_s^2}{a_s^2 k^2 + (1 - \frac{1}{2} a_s r_{eff}^{(s)} k^2)^2} \,, \tag{2.124}$$

implying the existence of four parameters to be determined: a_t, a_s, $r_{eff}^{(t)}$, $r_{eff}^{(s)}$, namely, the scattering lengths in the singlet and triplet states and the respective effective ranges. For the first two we have the established values (2.109). The effective range in the triplet state, $r_{eff}^{(t)}$, can be obtained from a well known experimental information, the deuteron binding energy. For that goal it is enough to see that there is no restriction to extend the above theory to negative energies, namely, to bound states. So let us use for u a decreasing exponential for the deuteron radial wave function, outside the potential

$$u_0 = e^{-\sqrt{2\mu E_B} r/\hbar}, \tag{2.125}$$

already properly normalized. $E_B = 2.2$ MeV is the deuteron binding energy and $\mu = m_N/2$ its reduced mass. Following manipulations similar as for Eq. (2.113), with ground-state and low-energy scattering wave functions, alternatively exact and asymptotic, Eqs. (2.125) and (2.64), we get

$$-\frac{\sqrt{2\mu E_B}}{\hbar} + \frac{1}{a_t} = -\frac{2\mu E_B}{\hbar^2} \frac{r_{eff}^{(t)}}{2}, \tag{2.126}$$

expression that allows to deduce $r_{eff}^{(t)}$ from a_t and E_B. The sign of the right side of (2.126) was introduced because in the case of negative energy the sign of the second term of (2.111) should be

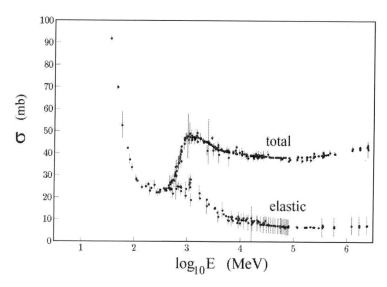

FIGURE 2.9 Total and elastic proton-proton cross sections as a function of the laboratory energy [15].

changed. Using the values of E_B and (2.109), we get the effective range for the triplet potential:

$$r^{(t)}_{eff} = 1.76 \text{ fm.} \tag{2.127}$$

The value of $r^{(s)}_{eff}$ cannot be obtained as a direct result of an experiment and is normally used in (2.124) as the parameter that best reproduces experimental values of the cross section. The value

$$r^{(s)}_{eff} = 2.56 \text{ fm.} \tag{2.128}$$

produces cross sections with (2.124) in very good agreement with experiment, as we can see in Figure 2.7.

The low-energy neutron-neutron scattering does not present any additional theoretical difficulty as compared to the neutron-proton scattering, since in both cases the nuclear force is the only agent. The problem here is of experimental character since a neutron target is not available and the study of the interaction should be done indirectly.

One of the employed methods consists to analyze the energy spectrum of the protons resulting from the reaction

$$\mathrm{n} + \mathrm{d} \rightarrow \mathrm{p} + \mathrm{n} + \mathrm{n}. \tag{2.129}$$

It is a continuum spectrum but presents a peak near the minimum energy. This indicates a resonance related to the formation of the di-neutron virtual state and the peak width can give an information about the scattering length. When one collects this and other results from reaction with two neutron formation one can extract the average values

$$a_s = (-17.6 \pm 1.5) \text{ fm,} \qquad r^{(s)}_{eff} = (3.2 \pm 1.6) \text{ fm,} \tag{2.130}$$

for the scattering length and effective range of the neutron-neutron scattering, respectively. These values are closer to the proton-proton nuclear scattering than the neutron-proton scattering. Hence, stronger than the charge independence, there is the indication of a charge symmetry of the nuclear force.

When the energy of the incident nucleon reaches some tenths of MeV new modifications in the elastic scattering treatment are necessary: waves with $l > 0$ begin to be important and the differential cross section, according to Eq. (2.52) and Eq. (2.50a), will be determined by the interference of different l values. If, for example, $l = 0$ and $l = 1$ waves are present in the scattering, then

$$\frac{d\sigma}{d\Omega} = \frac{1}{k^2} \left[\sin^2 \delta_0 + 6 \sin \delta_0 \sin \delta_1 \cos(\delta_0 - \delta_1) \cos \theta + 9 \sin^2 \delta_1 \cos^2 \theta \right]. \tag{2.131}$$

As a consequence, the interference between s and p scattered waves leads to breaking of the scattering symmetry about the angle $\theta = 90°$, that would exist if each wave scattered independently.

Up to an energy close to 280 MeV the elastic scattering is the only process to occur in a nucleon-nucleon collision. The nucleons do not have low-energy excited states and the weak interaction is very slow to manifest itself in a scattering process. Figure 2.9 sketches the proton-proton total cross section behavior as a function of the energy. The smooth decrease in energy is interrupted at 280 MeV (140 MeV in the center of mass), value that sets the threshold of pion creation. With the beginning of the contribution of these inelastic processes the total cross section separates from the elastic one.

2.14   Coulomb scattering

An essential condition for the scattering wave function to have the asymptotic behavior of Eq. (1.79) is that the potential $V(r)$ goes to zero faster than $1/r$ as $r \to \infty$. Otherwise, the incident wave function is distorted even at asymptotic distances. This is the case of the Coulomb potential between point particles (we use here the CGS unit system),

$$V_C(r) = \frac{q_p q_t}{r}, \tag{2.132}$$

where q_p and q_t are respectively the projectile and target charges. Therefore, Coulomb scattering requires a special treatment.

It is convenient to introduce the *Sommerfeld parameter*

$$\eta = \frac{q_p q_t}{\hbar v} \tag{2.133}$$

and the *half distance of closest approach* in a head-on collision

$$a = \frac{q_p q_t}{2E}. \tag{2.134}$$

These quantities are related as

$$\eta = ka. \tag{2.135}$$

One should solve the Schrödinger equation

$$-\frac{\hbar^2}{2\mu}\nabla^2 \phi_c(\mathbf{r}) + \frac{q_p q_t}{r}\phi_c(\mathbf{r}) = E\phi_c(\mathbf{r}),$$

or

$$\left[\nabla^2 + k^2 - \frac{2\eta k}{r}\right]\phi_c(\mathbf{r}) = 0. \tag{2.136}$$

Owing to the long range of the Coulomb field, the wave function does not have the asymptotic form of Eq. (1.79). Choosing the incident beam along the $z-$axis, we try the ansatz

$$\phi_c(\mathbf{r}) = Ce^{ikz}g(r-z), \tag{2.137}$$

where $g \to 1$ if the strength of the Coulomb field goes to zero ($\eta \to 0$). Inserting Eq. (2.137) in Eq. (2.136) and calling $\xi = r - z$, we get*

$$\xi g''(\xi) + (1 - ik\xi)g'(\xi) - \eta k g(\xi) = 0, \tag{2.138}$$

where primes stand for derivatives with respect to the argument. We then change variables

$$\xi \to s = ik\xi, \qquad g(\xi) \to w(s) = g(\xi) \tag{2.139}$$

---

*The details of this procedure are given in Ref. [6]. One writes the laplacian in terms of parabolic coordinates $\{\xi, \zeta, \varphi\}$, where φ is the azimuthal angle, $\xi = r - z$ and $\zeta = r + z$, and looks for a solution in product form, $\psi_c(\mathbf{r}) = g(\xi) \cdot f(\zeta)$. It is easy to show that $f(\zeta) = \exp(-ik\zeta/2)$ satisfies the Schödinger equation, if $g(\xi)$ satisfies Eq. (2.138).

and use the notation

$$d = -i\eta, \quad c = 1. \tag{2.140}$$

Eq. (2.138) takes the form

$$sw''(s) + (c - s)w'(s) - dw(s) = 0. \tag{2.141}$$

We identify above the *Kumer-Laplace equation*, which has the *confluent hypergeometric function*, $_1F_1(d; c; s)$, as regular solution. Replacing d, c, and s according to Eqs. (2.139) and (2.140) in the argument of $_1F_1$ and using Eq. (2.137), we get

$$\phi_c(r, z) = C e_1^{ikz} F_1(-I\eta; 1; ik(r - z)). \tag{2.142}$$

The function $_1F_1(-i\eta; 1; ik(r - z))$ can be written as (see, for example, appendix B of Ref. [3], or Ref. [5])

$$_1F_1(-i\eta; 1; ik(r - z)) = W_1(-i\eta; 1; ik(r - z)) + W_2(-i\eta; 1; ik(r - z)),$$

where W_1 and W_2 are integrals in the complex plane. The wave function $\phi_c(r, z)$ is then given by the sum

$$\phi_c(r, z) = \psi_i(r, z) + \psi_s(r, z), \tag{2.143}$$

where

$$\psi_i(r, z) = C e^{ikz} W_1(-i\eta; 1; ik(r - z)), \tag{2.144}$$

$$\psi_s(r, z) = C e^{ikz} W_2(-i\eta; 1; ik(r - z)). \tag{2.145}$$

Choosing the normalization constant as

$$C = \Gamma(1 + i\eta) e^{-\pi\eta/2},$$

where Γ is the gamma-function (see Ref. [5]), and using the asymptotic forms of W_1 and W_2, we obtain

$$\psi_i(r, z)_{|r - z| \to \infty} \to e^{ikz} \cdot \left[e^{-i\eta \ln(k(r - z))} \right] \tag{2.146}$$

$$\psi_s(r, z)_{|r - z| \to \infty} \to f_c(\theta) \frac{e^{ikr}}{r} \cdot \left[e^{-i\eta \ln(2kr)} \right], \tag{2.147}$$

so that

$$\phi_c(r, z)_{|r - z| \to \infty} \to e^{ikz} \cdot \left[e^{-i\eta \ln(k(r - z))} \right] + f_c(\theta) \frac{e^{ikr}}{r} \cdot \left[e^{-i\eta \ln(2kr)} \right]. \tag{2.148}$$

Above, $f_c(\theta)$ is the *Coulomb scattering amplitude*,

$$f_c(\theta) \doteq -\frac{\eta}{2k \sin^2(\theta/2)} e^{-i\eta \ln(\sin^2 \theta/2)} e^{2i\sigma_0}, \tag{2.149}$$

where σ_0 is the S-wave *Coulomb phase shift*

$$\sigma_0 = \arg \Gamma(1 + i\eta). \tag{2.150}$$

Note that Eqs. (2.146) and (2.147) are not valid at $\theta = 0$. In this case, $|r - z| = 0$ and the condition $|r - z| \to \infty$ cannot be satisfied.

Inspecting Eqs. (2.146) and (2.147), we conclude that ψ_i and ψ_s correspond respectively to the incident plane wave and a scattered wave, as in Eq. (1.79). However, due to the long range of the interaction, the waves have a logarithmic distortion given by the factors within square brackets.

Now we turn to the elastic cross section. It can easily be checked that the distortion factors in Eqs. (2.146) and (2.147) lead to additional terms in the incident and in the scattered currents. However, these terms are proportional to $1/r$, so that they vanish asymptotically. Therefore, the elastic cross section is given by

$$\frac{d\sigma_c(\Omega)}{d\Omega} = |f_c(\theta)|^2.$$

Using Eq. (2.149), we get

$$\frac{d\sigma_c(\Omega)}{d\Omega} = \frac{a^2}{4} \left[\frac{1}{\sin^4(\theta/2)} \right].$$

A very interesting feature of Coulomb scattering is that the above quantum mechanical cross section is identical to the classical cross section, of Eq. (1.64).

2.14.1   Partial-wave expansion

In the case of pure Coulomb scattering, partial-wave expansions are not useful. Firstly they are not necessary, since the scattering amplitude is known analytically. Secondly, it is not practical because of convergence problems with the partial-wave series, specially at small angles and for large η. This is a consequence of the long range of the Coulomb interaction. However, partial-wave expansions become very useful when the potential contains a short-range term, in addition to the Coulomb potential. This situation is frequently found in nuclear physics, where the nucleus-nucleus potential is the sum of the Coulomb repulsion with a short-ranged term, arising from nuclear forces. It is also the case of collisions between ionized atoms. We first consider the case of pure Coulomb scattering.

The wave function $\phi_c(\mathbf{r})$ can be formally expanded in partial waves as in Eq. (2.37),

$$\phi_c(\mathbf{r}) = \sum_{l=0}^{\infty} C_l P_l(\cos\theta) \frac{u_l(k,r)}{kr}, \tag{2.151}$$

where $u_l(k,r)$ satisfies the radial equation

$$-\frac{\hbar^2}{2\mu} \left[\frac{d^2}{dr^2} - \frac{2\eta k}{r} - \frac{l(l+1)}{r^2} \right] u_l(k,r) = E u_l(k,r).$$

Making the change of variable $r \to \rho = kr$, the radial equation takes the form

$$u_l''(\rho) + \left[1 - \frac{l(l+1)}{\rho^2} - \frac{2\eta}{\rho} \right] u_l(\rho) = 0. \tag{2.152}$$

One identifies above the Coulomb wave equation [5]. Its solutions can be expressed as linear combinations of the independent pair of real solutions $F_l(\eta,\rho)$ and $G_l(\eta,\rho)$. $F_l(\eta,\rho)$ is the regular solution and $G_l(\eta,\rho)$ diverges at the origin (except for $l=0$). The main properties of these functions are discussed in standard text books (see appendix B of Ref. [3]). They are summarized below:

(1) In the $\rho \to 0$ limit, the functions exhibit the behavior

$$F_l(\eta, \rho \to 0) \to A_l \rho^{l+1}, \qquad G_l(\eta, \rho \to 0) \to \frac{1}{(2l+1)A_l} \rho^{-l} \tag{2.153}$$

where

$$A_0 = \left(\frac{2\pi\eta}{(e^{2\pi\eta}-1)} \right)^{1/2}, \qquad A_{l>0} = \frac{A_0}{(2l+1)!!} \cdot \Pi_{s=1}^{l} \left(1 + \frac{\eta^2}{s^2} \right)^{1/2}.$$

(2) In the $\rho \to \infty$ limit, the functions exhibit the behavior

$$F_l(\eta, \rho \to \infty) \to \sin\left[\rho - \frac{l\pi}{2} - \eta \ln(2\rho) + \sigma_l \right] \tag{2.154}$$

$$G_l(\eta, \rho \to \infty) \to \cos\left[\rho - \frac{l\pi}{2} - \eta \ln(2\rho) + \sigma_l \right], \tag{2.155}$$

where

$$\sigma_l = \arg \Gamma(1 + l + i\eta) \tag{2.156}$$

are the *Coulomb phase shifts*. Usually, one evaluates $\sigma_0 = \arg \Gamma(1 + i\eta)$ and obtain Coulomb phase shifts for higher partial-waves from the series

$$\sigma_l = \sigma_0 + \sum_{s=0}^{l} \tan^{-1}\left(\frac{\eta}{s} \right).$$

(3) In the weak interaction limit: $V_c \to 0 \Rightarrow \eta \to 0$, the functions approach

$$F_l(\eta \to 0, \rho) \to \widehat{j}(\rho), \qquad G_l(\eta \to 0, \rho) \to \widehat{n}_l(\rho). \tag{2.157}$$

Eqs. (2.157) are not surprising since Eq. (2.152) reduces to the Ricatti-Bessel equation when $\eta = 0$.

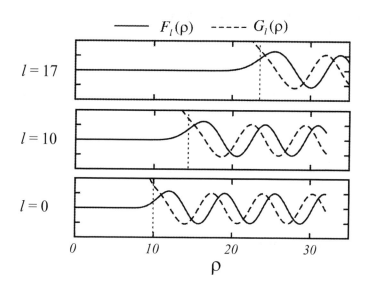

FIGURE 2.10   Coulomb wave functions for $\eta = 5$ and $l = 0$, 10 and 17.

To illustrate the behavior of the Coulomb wave functions, we present in Figure 2.10 $F_l(\eta, \rho)$ and $G_l(\eta, \rho)$ as functions of ρ, in the case $\eta = 5$, for a few partial waves. Quantitatively, they are similar to the corresponding Ricatti-Bessel functions, shown in Figure 2.1. For large values of ρ, $F_l(\eta, \rho)$ and $G_l(\eta, \rho)$ oscillate with constant amplitude, out of phase with respect to each others. For small ρ, $F_l(\eta, \rho)$ converges to zero while $G_l(\eta, \rho)$ grows continuously. The transition between these behaviors occurs at the classical turning points, $\rho_t = kR_{ca}$. Using Eq. (1.59), we can write

$$\rho_t = k\left(a + \sqrt{a^2 + b^2}\right) = \eta + \sqrt{\eta^2 + l^2}. \tag{2.158}$$

Note that the S-wave Coulomb turning point does not vanish with $l = 0$, in contrast with that for the Ricatti-Bessel function (Eq. (2.21)). This is because in the present case the S-wave effective potential contains the Coulomb repulsion.

The solutions of the Coulomb wave equation can also be expressed as linear combinations of incoming and outgoing Coulomb waves, respectively $H_l^{(-)}(\rho)$ and $H_l^{(+)}(\rho)$. They are defined as

$$H_l^{(\pm)}(\eta, \rho) = G_l(\eta, \rho) \pm i F_l(\eta, \rho). \tag{2.159}$$

Using Eqs. (2.154) and (2.155) we find their asymptotic forms

$$H_l^{(\pm)}(\eta, \rho \to \infty) \to e^{\pm i\left[\rho - \frac{l\pi}{2} - \eta \ln(2\rho) + \sigma_l\right]}. \tag{2.160}$$

As the strength of the Coulomb field goes to zero ($\eta \to 0$), $H_l^{(\pm)}$ converge to the Ricatti-Haenkel functions,

$$H_l^{(\pm)}(\eta \to 0, \rho) \to \hat{h}_l^{(\pm)}(\rho). \tag{2.161}$$

To find the partial-wave expansion of the wave function $\phi_c(\mathbf{r})$, we replace $u_l(k, r) = F_l(\eta, \rho)$ in Eq. (2.151) and determine the coefficients C_l by imposing that it has the asymptotic behavior of Eq. (2.148). Using the asymptotic form of $F_l(\eta, \rho)$, we find*

$$\phi_c(\mathbf{r}) = A \sum_{l=0}^{\infty} (2l + 1) i^l e^{i\sigma_l} P_l(\cos\theta) \frac{F_l(\eta, \rho)}{kr}. \tag{2.162}$$

---

*For a detailed proof, see Ref. [3].

In the $\eta \to 0$ limit, $\sigma_l \to 0$, $F_l(\eta, \rho) \to \hat{\jmath}_l(\rho)$ and Eq. (2.162) reduces to the Bauer's expansion for the plane wave (Eq. (2.36)), except for the normalization factor A.

It is also useful to write the partial-wave expansion in terms of incoming and outgoing Coulomb waves. Using Eq. (2.159), Eq. (2.162) becomes

$$\phi_c(\mathbf{r}) = A \left(\frac{1}{2kr} \right) \sum_{l=0}^{\infty} (2l+1) i^{l+1} P_l(\cos\theta) e^{i\sigma_l} \left[H_l^{(-)}(\eta, \rho) - H_l^{(+)}(\eta, \rho) \right]. \tag{2.163}$$

2.14.2   Coulomb and short-range potentials combined

Let us consider the scattering from the potential

$$V(r) = V_c(r) + V_N(r), \tag{2.164}$$

where $V_c(r)$ is the Coulomb potential of Eq. (2.132) and $V_N(r)$ is a short-range potential, which vanishes for $r > R$. In this case the Schrödinger equation can be put in the form

$$\left[\nabla^2 + k^2 - \frac{2\eta k}{r} \right] \psi(\mathbf{r}) = U_N(r)\psi(\mathbf{r}), \tag{2.165}$$

where

$$U_N(r) = \frac{2\mu}{\hbar^2} V_N(r). \tag{2.166}$$

Following the procedures of the previous sections, we look for a solution in the form

$$\psi^{(+)}(\mathbf{r}) = \phi_c(\mathbf{r}) + \psi_{sc}(\mathbf{r}), \tag{2.167}$$

where the Coulomb wave function, $\phi_c(\mathbf{r})$, plays the role of the "homogeneous" solution and $\psi_{sc}(\mathbf{r})$ is a spherical outgoing wave, resulting from the action of $V_N(r)$. In the present case, however, the asymptotic form of $\psi_{sc}(\mathbf{r})$ is distorted by the Coulomb potential, in the same way as the Coulomb scattered wave (Eq. (2.147)). Namely,

$$\psi_{sc}(\mathbf{r})_{|r-z|\to\infty} \to A f_N(\theta) \frac{e^{ikr}}{r} \cdot \left[e^{-i\eta \ln(2kr)} \right], \tag{2.168}$$

where $f_N(\theta)$ is the correction to the Coulomb scattering amplitude accounting for the short-range potential.

To find $f_N(\theta)$, we expand $\psi^{(+)}(\mathbf{r})$ in partial waves

$$\psi^{(+)}(\mathbf{r}) = \sum_{l=0}^{\infty} C_l P_l(\cos\theta) \frac{u_l(k,r)}{kr}, \tag{2.169}$$

where $u_l(k,r)$ are the solutions of the radial equation

$$-\frac{\hbar^2}{2\mu} \left[\frac{d^2}{dr^2} - \frac{l(l+1)}{r^2} - \frac{2\eta k}{r} \right] u_l(k,r) + V_N(r)u_l(k,r) = Eu_l(k,r). \tag{2.170}$$

For $r > R$, $V_N(r)$ vanishes and Eq. (2.170) reduces to the Coulomb wave equation. Therefore, $u_l(k,r)$ can be expressed as the linear combination

$$u_l(k,r) = \frac{i}{2} \left[H_l^{(-)}(\eta, \rho) - S_l H_l^{(+)}(\eta, \rho) \right] e^{-i\delta_l}, \tag{2.171}$$

where $S_l = e^{2i\delta_l}$ are the partial wave components of the S-matrix associated with V_N, and δ_l are the corresponding phase shifts. The normalization of u_l is such that it reduces respectively to $F_l(\eta, \rho)$ in the $V_N \to 0$ and $\eta \to 0$ limits. We now consider the asymptotic form of $\psi^{(+)}(\mathbf{r})$. Using Eq. (2.160) and inserting Eq. (2.171) in Eq. (2.169), we can write

$$\psi^{(+)}(\mathbf{r})_{|r-z|\to\infty} = \frac{e^{-i[kr-\eta \ln(2kr)]}}{r} X_- + \frac{e^{i[kr-\eta \ln(2kr)]}}{r} X_+, \tag{2.172}$$

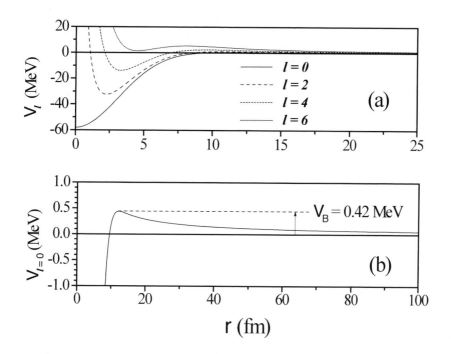

FIGURE 2.11 Effective potentials in $\alpha - \alpha$ scattering for a few partial waves. The lower panel shows details of the Coulomb barrier, for $l = 0$, $V_{l=0}(r)$.

with

$$X_{\pm} = \sum_{l=0}^{\infty} P_l(\cos\theta) \left[\pm \frac{C_l}{2ik} e^{\mp il\pi/2} e^{\pm i(\sigma_l + \delta_l)} \right] . \tag{2.173}$$

Eq. (2.172) must be consistent with the asymptotic form of Eq. (2.167). Using Eq. (2.168) and the partial-wave expansion for ϕ_c (Eq. (2.163)) with $H_l^{(\pm)}$ replaced by their asymptotic forms, we get

$$\psi^{(+)}(\mathbf{r})_{|r-z|\to\infty} = \frac{e^{-i[kr - \eta \ln(2kr)]}}{r} X_- + \frac{e^{i[kr - \eta \ln(2kr)]}}{r} X_+, \tag{2.174}$$

with

$$X_- = \sum_{l=0}^{\infty} P_l(\cos\theta) \left[-A \frac{(2l+1)i^l}{2ik} e^{il\pi/2} \right] \tag{2.175}$$

and

$$X_+ = Af_N(\theta) + \sum_{l=0}^{\infty} P_l(\cos\theta) \left[A \frac{(2l+1)}{2ik} e^{2i\sigma_l} \right] . \tag{2.176}$$

Identifying the coefficients of $e^{-i[kr - \eta \ln(2kr)]}/r$ in Eqs. (2.172) and 2.174) we get

$$C_l = A(2l+1) i^l e^{i(\sigma_l + \delta_l)}$$

and the expansion of Eq. (2.169) becomes

$$\psi^{(+)}(\mathbf{r}) = A \sum_{l=0}^{\infty} (2l+1) i^l e^{i(\sigma_l + \delta_l)} \frac{u_l(k,r)}{kr} P_l(\cos\theta). \tag{2.177}$$

As mentioned before, different normalizations are used for the the radial wave function. Also frequently used is the normalization which leads to the asymptotic form

$$u_l(k,r) = \frac{i}{2}\left[H_l^{(-)}(\eta,\rho) - S_l H_l^{(+)}(\eta,\rho)\right]. \tag{2.178}$$

In this case the scattering state is written

$$\psi^{(+)}(\mathbf{r}) = A\sum_{l=0}^{\infty}(2l+1)i^l e^{i\sigma_l}\frac{u_l(k,r)}{kr}P_l(\cos\theta). \tag{2.179}$$

Repeating the above procedure for the coefficients of $e^{i[kr-\eta\ln(2kr)]}/r$ we obtain

$$f_N(\theta) = \frac{1}{2ik}\sum_{l=0}^{\infty}(2l+1)P_l(\cos\theta)e^{2i\sigma_l}\left(S_l - 1\right). \tag{2.180}$$

To get the cross section, we consider that the wave function can be written, as in Eq. (2.148) with the replacement

$$f_c(\theta) \rightarrow f(\theta) = f_c(\theta) + f_N(\theta). \tag{2.181}$$

Therefore, the cross section is

$$\frac{d\sigma_c(\Omega)}{d\Omega} = |f_c(\theta) + f_N(\theta)|^2, \tag{2.182}$$

with $f_c(\theta)$ given by Eq. (2.149) and $f_N(\theta)$ by Eq. (2.180).

Usually one is interested in investigating the action of the short-ranged potential. For this purpose, the cross sections are usually normalized with respect to the corresponding pure Coulomb (or Rutherford) ones.

The evaluation of the phase shifts δ_l are carried out in the same way as in the absence of Coulomb integration. The boundary R divides the r-axis into an internal and an external region. The radial equation is solved numerically in the internal region and the logarithmic derivative, $\mathcal{L}^I$, is calculated. Matching it to the one calculated with the radial wave function at the external region, one finds an expression for the phase shift δ_l. This expression is identical to Eq. (2.30) with the replacement of Ricatti-Bessel and Ricatti-Neumann functions by F_l and G_l, respectively. It becomes

$$\tan\delta_l = \frac{kRF_l'(kR) - F_l(kR)\mathcal{L}^I}{kRG_l'(kR) - G_l(kR)\mathcal{L}^I}. \tag{2.183}$$

2.15   An illustration: $\alpha - \alpha$ scattering

To close this chapter we study the elastic scattering of two α-particles. The $\alpha - \alpha$ potential is approximated as

$$V(r) = V_C(r) + V_N(r), \tag{2.184}$$

where V_N is the attractive gaussian potential

$$V_N(r) = -V_0 e^{-r^2/R_0^2} \tag{2.185}$$

and

$$V_C(r) = \frac{4e^2}{r}, \quad \text{for } r > R_c \simeq R_p + R_t = 2R_\alpha$$
$$= \frac{4e^2}{2R_c}\left(3 - \frac{r^2}{R_c^2}\right), \quad \text{for } r < R_c. \tag{2.186}$$

The Coulomb potential, V_C, takes into account the finite size of the α-particles. If they do not overlap ($r > R_c$) it corresponds to the point charge interaction of Eq. (2.132). If they overlap, it is approximated by the potential of a point charge $2e$ inside a uniformly charged sphere of radius R_c

l	$V_{B,l}$ (MeV)	$E_{r,l}$ (MeV)	$\Delta_l = V_{B,l} - E_{r,l}$
4	2.30	1.60	0.70
6	5.24	4.90	0.34

TABLE 2.3   Resonance energies and the barrier heights for $l = 4$ and 6.

and total charge $2e$. For the numerical calculations, we use typical Nuclear Physics values: $V_0 = 60$ MeV, $R_c = R_0 = 4.5$ fm (4.5×10^{-15} m).

The effective potentials $V_l(r) = V(r) + \left(\hbar^2/2\mu\right) l(l + 1)/r^2$ are shown in Figure 2.11, for a few partial waves. Details of the Coulomb barrier are displayed in the inset. The barrier height is $V_B = 0.42$ MeV.

In Figure 2.12, we show the phase shifts of the short-range potential, $\delta_l^N(E)$, for the partial waves* $l = 0, 2, \cdots, 10$. We plot absolute phase shifts, as discussed in Appendix 2.A. The S-wave phase shift decreases monotonically from the value 4π, as the collision energy grows. A similar behavior is found for $l = 2$, except for its smaller low-energy limit, $\delta_2^N(0) = 3\pi$. According to Levinson's theorem, this means that the potential has four bound states with $l = 0$ and three with $l = 2$. For $l = 4$, the phase shift has the value π at $E = 0$, indicating the existence of a single bound state. As the energy increases, it grows slowly until the energy $E = 1.60$ MeV, where it exhibits a sharp resonance. It then falls to 0 as $E \to \infty$. The phase shifts for higher partial waves vanish at $E = 0$. Therefore, the potential has no bound states above $l = 4$. δ_6^N shows a broader resonance at 4.90 MeV and δ_8^N and δ_{10}^N have broad maxima before going to zero asymptotically.

In Table 2.3, we give the barrier heights, $V_{B,l}$, and the resonance energies, $E_{r,l}$, for $l = 4$ and 6. These quantities play a central role in the resonance widths. The width Γ_l is related to the lifetime τ_l through the uncertainty relation, $\Gamma_l = \hbar/\tau_l$, and the life-times are inversely proportional to the corresponding tunnelling probability, T_l. Therefore, we can write $\Gamma_l \sim T_l$. Since the tunnelling probability decreases rapidly with Δ_l, the same should occur with Γ_l. This trend is confirmed in Figure 2.12. From Table 2.3 we find that $\Delta_6 = 0.34$ MeV $< \Delta_4 = 0.70$ MeV. Accordingly, inspecting Figure 2.12, it is clear that $\Gamma_6 > \Gamma_4$.

2.16   Appendix 2.A – Absolute phase shifts and Levinson Theorem

The phase shifts are defined modulo π. This means that one cannot distinguish among δ_l, $\delta_l \pm \pi$, $\delta_l \pm 2\pi, \cdots$, etc. This ambiguity can be eliminated if we replace Eq. (2.30) by

$$\delta_l = \tan^{-1}\left[-\frac{kR\hat{\jmath}_l'(kR) - \hat{\jmath}_l(kR)\mathcal{L}^I}{kR\widehat{n}_l'(kR) - \widehat{n}_l(kR)\mathcal{L}^I}\right]. \tag{2.187}$$

Now, δ_l is uniquely defined, since it is contained in the interval $\{-\pi/2, +\pi/2\}$. However, this procedure has the serious disadvantage of leading to phase shifts which are not continuous functions of the collision energy. As the energy varies, a phase shift would have a jump of π whenever it reaches $\pm\pi/2$.

A search for an absolute definition of δ_l should begin with a study of its behavior in the limiting situations $E \to 0$ and $E \to \infty$. In the high-energy limit, the phase shift should satisfy the condition $\tan \delta_l \to 0$. If the potential is limited everywhere and has finite range it becomes negligible as compared to E in the Schrödinger equation. The dynamics then reduces to that of a free particle. On the other hand, it is straightforward to show that $\lim_{k \to 0} [\tan \delta_l] = 0$.

---

*As will be discussed in the next chapter, the $\alpha - \alpha$ scattering amplitude has only contribution from even partial waves. This is a consequence that projectile and target are identical particles with an even number of fermions.

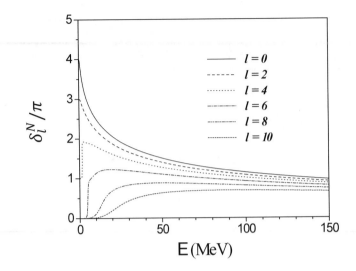

FIGURE 2.12   Phase shifts of the short-range potential in $\alpha - \alpha$ collision.

We make the reasonable choice

$$V(r) = 0 \rightarrow \delta_l = 0 \qquad \text{for any } l, \tag{2.188}$$

and look for absolute phase shifts by imposing continuity with respect to E. This requirement implies that one has to add or subtract π whenever the phase shift of Eq. (2.187) goes through a jump, to retain its continuity. Following this procedure, one could find that

$$\Delta = \delta_l(k \rightarrow 0) - \delta_l(k \rightarrow \infty) \neq 0, \tag{2.189}$$

although (for S-wave scattering from a square well, there is an anomalous situation where $\lim_{k \to 0} \tan \delta_0 \neq 0$.)

$$\tan[\delta_l(k \rightarrow 0)] - \tan[\delta_l(k \rightarrow \infty)] = 0.$$

To get a better insight of the problem, we consider the $k \rightarrow 0$ limit in S-wave scattering from the attractive square well potential

$$V(r) = 0, \qquad \text{for } r > R, \tag{2.190}$$

$$V(r) = -V_0, \quad \text{for } r < R. \tag{2.191}$$

In this case, the radial wave function inside the well is known analytically:

$$u_0^I(k,r) = A\sin(Kr), \tag{2.192}$$

where

$$K = \frac{\sqrt{2\mu(V_0 + E)}}{\hbar}. \tag{2.193}$$

The internal logarithmic derivative at the boundary R then is

$$\mathcal{L}^I \equiv R\frac{\left[du_0^I(k,r)/dr\right]_{r=R}}{u_0^I(k,R)} = KR\cot(KR). \tag{2.194}$$

Using Eq. (2.194) and the explicit forms of $\hat{\jmath}_0(kr)$, $\hat{n}_0(kr)$, $\hat{\jmath}_0'(kr)$, $\hat{n}_0'(kr)$,

$$\hat{\jmath}_0(kr) = \sin(kr); \qquad \hat{\jmath}_0'(kr) = \cos(kr),$$
$$\hat{n}_0(kr) = \cos(kr) \qquad \hat{n}_0'(kr) = -\sin(kr),$$

in Eq. (2.30) we get

$$\tan \delta_0 = \frac{k \cos(kR) - K \sin(kR) \cot(KR)}{k \sin(kR) + K \cos(kR) \cot(KR)} = \frac{\frac{k}{K} \tan(KR) - \tan(kR)}{1 + \frac{k}{K} \tan(KR) \tan(kR)}. \tag{2.195}$$

Calling

$$A = \tan^{-1}\left[\frac{k}{K} \tan(KR)\right], \quad B = kR,$$

we use the trigonometric identity

$$\frac{\tan A - \tan B}{1 + \tan A \tan B} = \tan(A - B)$$

in Eq. (2.195) and get

$$\delta_0 = \tan^{-1}\left[\frac{k}{K} \tan(KR)\right] - kR. \tag{2.196}$$

At this stage, we leave open the possibility of summing or subtracting integer multiples of π to δ_0. The need for such a procedure becomes clear as we investigate the low-energy behavior of δ_0 as a function of the potential depth, V_0.

Initially, we consider a very weak potential so that,

$$K_0 R = \left(\frac{\sqrt{2\mu V_0}}{\hbar}\right) R \ll 1.$$

In the low-energy limit, $k \to 0$ and we can approximate

$$\delta_0 \simeq kR \left(\frac{\tan(KR)}{KR} - 1\right) \simeq kR \left(\frac{\tan(K_0 R)}{K_0 R} - 1\right). \tag{2.197}$$

Therefore, the phase shift of Eq. (2.197) goes to zero proportionally to k. This is consistent with our assumption of zero phase shift for a free particle.

We now let the potential strength grow until it satisfies the condition

$$K_0 R = \left(\frac{\sqrt{2\mu V_0}}{\hbar}\right) R = \frac{\pi}{2}.$$

In this case, the approximation of Eq. (2.197) breaks down and Eq. (2.196) must be used. Replacing $K_0 R = \pi/2$ and then taking the limit $k \to 0$, we find the anomalous (in the sense that $\lim_{k \to 0} [\tan \delta_0] \neq 0$) phase shift

$$\lim_{k \to 0} \delta_0 = \frac{\pi}{2}.$$

Let us study the context of this potential in more details. Plugging the value $KR = \pi/2$ in Eq. (2.194) we find that $\mathcal{L}^I = 0$. Therefore, the internal wave function has an horizontal tangent at the matching radius. It is the limiting case where the internal wave function can be matched to a decaying exponential in the external region. It corresponds to a bound state with zero binding energy. Therefore, the minimal depth for a square well to bind an S-state is

$$V_0 = \frac{\pi^2 \hbar^2}{8\mu R^2}. \tag{2.198}$$

Let us now consider a deeper potential such that

$$\frac{3\pi}{2} > K_0 R > \frac{\pi}{2}. \tag{2.199}$$

In this case, Eq. (2.197) can again be used and δ_0 goes to zero in the $k \to 0$ limit. However, now it approaches zero from negative values. For the phase shift to grow monotonically with the potential depth, it is necessary to sum the constant phase π to Eq. (2.197). Namely,

$$\delta_0 \simeq \pi + kR \left(\frac{\tan(KR)}{KR} - 1\right). \tag{2.200}$$

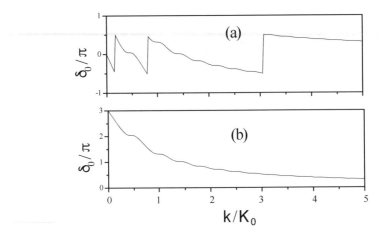

FIGURE 2.13   The phase shifts defined in the $\{-\pi/2, \pi/2\}$ interval (a) and the absolute phase shifts (b).

If we let the potential get deeper, a new anomaly will occur when $K_0 R = 3\pi/2$. In this case,

$$\lim_{k \to 0} \delta_0 = \frac{3\pi}{2}.$$

It corresponds to the situation where the potential has two bound states. The one it had in the previous case plus a second bound state with one node (besides the one at the origin) and zero energy. As the depth exceeds this value, again a constant phase π has to be added to Eq. (2.200).

Following the same procedures successively, we conclude that the anomaly occurs whenever the depth of the potential reaches the value

$$V_0 = \frac{\pi^2 \hbar^2}{2\mu R^2} \left(n - \frac{1}{2}\right)^2, \tag{2.201}$$

where $n = 1, 2, \cdots$, and it corresponds to the condition for the potential to bind n S-states. Using the same arguments as above, we get

$$\delta_0(k \to 0) = n\pi. \tag{2.202}$$

The generalization of Eq. (2.202) for an arbitrary potential and an arbitrary partial-wave is the *Levinson Theorem*[*].

A practical method to obtain absolute phase shifts is to calculate δ_l for a very high energy, where it is close to zero, and then reduce the energy gradually. One can start, for example, with $E = 100\bar{V}$, where $\bar{V}$ is the value of $V(r)$ where it is largest. As the energy is reduced, the phase shift changes monotonically until it reaches the value $\pm\pi/2$. At this point, it jumps to $\mp\pi/2$. Whenever this occurs, one eliminates the jump adding or subtracting π. This procedure leads to a continuous, uniquely defined phase shift, which reaches the value predicted by the Levinson Theorem as $k \to 0$. This is illustrated in figure 2.13, through the S-wave phase shift for a square well with $K_0 R = 10$. The phase shifts are plotted as a function of k^2/K_0^2, which is equal to the scattering energy normalized with respect to the depth of the well. In this case, the potential can bind 3 S-states ($KR = \pi/2, 3\pi/2, \text{and } 5\pi/2$). In Figure 2.13a, the phase shifts are obtained directly

---

[*] See Ref. [16]. A detailed proof of the Levinson Theorem is also presented in sect. (12-e) of Ref. [8].

from Eq. (2.187). In Figure 2.13b, we have summed π whenever needed to guarantee continuity. We note that the absolute phase shift of Figure 2.13b is a smooth function of k. In accordance with the Levinson Theorem, $\delta_0(k \to 0) = 3\pi$.

To conclude this appendix, we want to point out an interesting feature of low-energy scattering from attractive potentials. Looking at Figure 2.13b, we see that the phase shift can take the values π and 2π for finite k-values. When this occurs, $S_0 = 1$ and the contribution from $l = 0$ to the scattering amplitude vanishes. If this happens at a very low energy, the partial-wave series can be truncated at $l = 0$ and the elastic cross section becomes very small. At this energy, the target will be transparent to the beam. This phenomenon, known as the *Ramsauer-Townsend effect*, has been observed in collisions of 0.7 eV electrons in A, Kr, and Xe targets. The quantum theory gives a simple explanation of this effect, which cannot be explained by the classical theory. The atomic field of inert gases decreases faster with distance than the fields of other atoms; as a first approximation, we can replace this field by a rectangular well and use (2.55), with $l = 0$, and (2.196) to evaluate the cross section of the low-energy electrons. For an electron energy of approximately 0.7 eV, we get $\sigma \sim 0$, if we use R equal to the atomic dimensions.

2.17   Exercises

1. Show that if $S_l(k)$ has the form (2.33), with δ_l real, the wave function (2.38) describes an elastic scattering. Suggestion: show that the flux of the probability current vector throughout a sphere that involves the scattering center is zero.

2. Derive the expression (Eq. (2.51))

$$(1 - x) f(\theta) = \frac{1}{2ik} \sum_l P_l(x) \left[(2l + 1)S_l - lS_{l-1} - (l + 1)S_{l+1} \right],$$

where $x = \cos \theta$.

3. Matching $l = 0$ wave functions at the boundary $r = c + b$ of a square-well repulsive core potential leads to the transcendental equation $K \cot Kb = k \left[\cot k(c + b) + \delta \right]$. Determine the radius of the repulsive core (c) given the information that the triplet S phase shift is zero for $E_{Lab} = 350$ MeV. Use $V_0 = 73$ MeV, and $b = 1.337$ fm

4. (a) Derive an expression for the s-wave phase shift for a particle with energy E incident upon a square-well potential of strength V and radius R . Calculate the wave function for $V = E/2$ and $V = -E/2$, and $R = 6\lambda$.

   (b) What is the s-wave contribution to the total cross section? Under what circumstances will the cross section vanish for zero bombarding energy (Ramsauer-Townsend effect)?

5. Consider the scattering of a particle of mass μ and energy E, from a potential barrier centered at $r = R_0$ with width a

$$V(r) = 0, \quad \text{for } r < R_0 - \frac{a}{2}$$
$$V(r) = \frac{C}{a} \quad \text{for } R_0 - \frac{a}{2} < r < R_0 + \frac{a}{2}$$
$$V(r) = 0 \quad \text{for } r > R_0 + \frac{a}{2}.$$

   Calculate the S-wave phase shift in the $a \to 0$ limit, where the barrier becomes a delta-function.

6. Consider the "delta-shell" potential ($\gamma > 0$)

$$V(r) = \gamma \delta(r - a),$$

   and the scattering problem for the s-wave. Answer the following questions:

   a) In the limit $\gamma \longrightarrow \infty$, the regions inside ($r < a$) and outside ($r > a$) the shell decouple. What are the values of k for the "bound states" confined inside the shell?

b) Show that the phase shift is given by

$$e^{2i\delta_0} = \frac{1 + \left(2m\gamma/\hbar^2 k\right) e^{-ika} \sin\left(ka\right)}{1 + \left(2m\gamma/\hbar^2 k\right) e^{ika} \sin\left(ka\right)}.$$

c) Verify that $\gamma \longrightarrow \infty$ gives the hard sphere, while $\gamma \longrightarrow 0$ no scattering.

d) Plot the behavior of the partial wave cross section σ_0 for appropriate values of parameters. Identify peaks due to the hard sphere scattering as well as resonances.

e) Identify the location of poles in the large γ approximation up to $O(\gamma^{-2})$, and see that the real values of k correspond to those of the "bound states" in the large γ limit.

f) Work out the wave function for the complex values of k at the poles analytically (no expansion in γ), and plot its time-dependence to identify the wave flowing to infinity resulting from the decay of the quasi-bound state.

7. Derive an integral expression for the phase shifts in the scattering from a potential with range R, in terms of the radial wave function $\tilde{u}_l(k,r)$, normalized such that

$$\tilde{u}_l(k, r > R) = \hat{\jmath}_l(kr) - \tan\delta_l \hat{n}_l(kr).$$

8. Alpha particles of 8 MeV energy are scattered from a gold foil target. The back-angle cross section, $d\sigma/d\Omega(180°)$, is found to be reduced by 5% from that given by the Rutherford cross section. Assume that the reduction is caused by a short-range modification of the Coulomb potential which affects only s-waves.

a) Is the short-range potential attractive or repulsive?

b) Deduce the modification in the s-wave phase shift caused by the short-range potential.

c) Deduce the modification for the Rutherford cross section reduced by 10%.

d) At what angles would you expect corrections due to the atomic shielding to be important (that is, for the theory to break down)?

9. Assume the interaction potential between the neutron and proton to be exponential, namely $V = V_0 e^{-r/r_0}$ where V_0 and r_0 are respectively the depth and range of the nuclear potential.

a) Write down the Schrödinger equation (in the center-of-mass system) for the ground state of the deuteron for the case in which the angular momentum $l = 0$.

b) Let $x = e^{-r/r_0}$ and $\psi(r) = u(r)/r$. Show that the Schrödinger equation gives a Bessel function. Write down the general solution of this equation.

c) Applying the boundary conditions ($\psi =$ finite at $r = 0$ and ∞), determine the relation between V_0 and r_0.

10. Determine the scattering length for the attractive Yukawa potential $V(r) = -(V_0 r_0/r)e^{-r/r_0}$ with $V_0 = g\hbar^2/2mr_0^2$ by a numerical integration of the Schrödinger equation for $g = 1$.

11. If both $l = 0$ and $l = 1$ waves contribute to n-p scattering, evaluate the scattering angle dependence of the differential elastic scattering cross section in the simplifying assumption of a single δ_1 phase shift. Plot the expected cross section as a function of scattering angle assuming $\delta_0 = 45°$ and $\delta_1 = 30°$.

12. Suppose that in an elastic scattering experiment between two structureless particles the center-of-mass differential cross section may be represented by

$$\frac{d\sigma}{d\Omega} = A + BP_1(\cos\theta) + CP_2(\cos\theta) + \cdots$$

Express the coefficients A, B and C in terms of the phase shifts δ_l.

References

1. Merzbacher E 2011 *Quantum Mechanics* (Wiley)
2. Edmonds A R 1996 *Angular Momentum in Quantum Mechanics* (Princeton University Press)
3. Messiah A 2017 *Quantum Mechanics* (Dover)
4. Brink D M 1986 *Semiclassical Methods for Nucleus-Nucleus Scattering* (Cambridge University Press)
5. Abramowitz M and Stegun I 1965 *Handbood of Mathematical Functions* (Dover)
6. Joachain C J 1984 *Quantum Collision Theory* (Elsevier)
7. Press W H et al. 2007 *Numerical Recipes* (Cambridge University Press)
8. Taylor J 2012 *Scattering Theory: The Quantum Theory of Nonrelativistic Collisions* (Dover)
9. Brink D M 1965 *Nuclear Forces* (Pergamon Press)
10. Feshbach H 1993 *Theoretical Nuclear Physics - Nuclear Reactions* (Wiley)
11. Adair R K 1950 *Rev. Mod. Phys.* **22** 249
12. Squires G L and Stewart A T 1955 *Proc. Roy. Soc.* **A230** 19
13. Houk T L 1971 *Phys. Rev.* **C3** 1886
14. Schwinger J and Teller E 1937 *Phys. Rev.* **52** 286
15. Barnett R M et al. 1996 *Phys. Rev.* **D54** 1
16. Levinson N 1949 *Kgl. Danske Videnskab. Selskab. Mat.-Fys. Medd.* **25** (9)

3

Formal scattering theory

3.1 Introduction: Green's functions

wave function In this chapter we develop the main concepts of formal scattering theory, introducing the Green's functions, transition and scattering matrices. The scattering of particles with spin and of particles with internal structure is discussed later in the chapter. The important case of absorptive potentials is discussed at the end of this chapter.

Let us start with the Schrödinger equations for a free particle with wave vector $\mathbf{k}$ and energy $E_k = \hbar^2 k^2 / 2\mu$, and for the scattering of this particle from a short-range potential $V(r)$ ($V(r) = 0$ for $r > R$). Using Dirac's notation, these equations can be written as

$$(E_k - H_0)\,|\phi_{\mathbf{k}}\rangle = 0, \tag{3.1}$$

and

$$(E_k - H)\,|\psi_{\mathbf{k}}\rangle = 0. \tag{3.2}$$

Above, H_0 is the kinetic energy operator and $H = H_0 + V$. Since $H_0^\dagger = H_0$ and $H^\dagger = H$, the states $|\phi_{\mathbf{k}}\rangle$ and $|\psi_{\mathbf{k}}\rangle$ can be normalized so as to satisfy the relations

$$\langle \phi_{\mathbf{k}'} | \phi_{\mathbf{k}} \rangle = \delta\left(\mathbf{k} - \mathbf{k}'\right), \qquad \langle \psi_{\mathbf{k}'} | \psi_{\mathbf{k}} \rangle = \delta\left(\mathbf{k} - \mathbf{k}'\right), \tag{3.3}$$

and

$$\langle \psi_m | \psi_n \rangle = \delta(m, n), \qquad \langle \psi_m | \psi_{\mathbf{k}} \rangle = 0. \tag{3.4}$$

Above, m, and n stand for negative energy states in case the Hamiltonian $H = H_0 + V$ binds. $\delta(m, n) \equiv \delta_{mn}$ and $\delta(\mathbf{k} - \mathbf{k}')$ are respectively the usual Kronecker's and the Dirac's deltas. The identity operator in the Hilbert spaces spanned by $|\phi_\mathbf{k}\rangle$ and $|\psi_\mathbf{k}\rangle$ can be written as

$$\int |\phi_\mathbf{k}\rangle \, d^3\mathbf{k} \, \langle\phi_\mathbf{k}| = \mathbf{1}, \tag{3.5}$$

and

$$\int |\psi_\mathbf{k}\rangle \, d^3\mathbf{k} \, \langle\psi_\mathbf{k}| + \sum_n |\psi_n\rangle \langle\psi_n| = \mathbf{1}. \tag{3.6}$$

The free particle's, $G_0(E)$, and the full, $G(E)$, *Green's operators* are defined as,

$$G_0(E) = \frac{1}{E - H_0}, \qquad G(E) = \frac{1}{E - H}. \tag{3.7}$$

To find a relation between G and G_0, we use the operator identity

$$A^{-1} = B^{-1} + B^{-1}(B - A)A^{-1}. \tag{3.8}$$

Setting $A = E - H \equiv E - (H_0 + V)$ and $B = E - H_0$, we get

$$G(E) = G_0(E) + G_0(E)VG(E). \tag{3.9}$$

Else, setting $A = E - H_0$ and $B = E - (H_0 + V)$, we get

$$G(E) = G_0(E) + G(E)VG_0(E). \tag{3.10}$$

The state $|\psi_\mathbf{k}\rangle$ satisfies equations involving the *Green's functions*. To derive these equations, we put Eq. (3.2) in the form

$$(E - H_0) |\psi_\mathbf{k}\rangle = V |\psi_\mathbf{k}\rangle \tag{3.11}$$

and write

$$|\psi_\mathbf{k}\rangle = |\phi_\mathbf{k}\rangle + |\psi_{sc}\rangle. \tag{3.12}$$

Inserting Eq. (3.12) in Eq. (3.11) and using Eq. (3.1), we get

$$(E_k - H_0) |\psi_{sc}\rangle = V |\psi_\mathbf{k}\rangle. \tag{3.13}$$

Multiplying from the left with $G_0(E)$ and replacing $|\psi_{sc}\rangle = |\psi_\mathbf{k}\rangle - |\phi_\mathbf{k}\rangle$, we get the desired equation,

$$|\psi_\mathbf{k}\rangle = |\phi_\mathbf{k}\rangle + G_0(E_k)V |\psi_\mathbf{k}\rangle. \tag{3.14}$$

A similar equation can be obtained in terms of the full Green's function. For this purpose, we use Eq. (3.12) in the RHS of Eq. (3.13) and re-arrange the equation in the form

$$(E_k - H_0 - V) |\psi_{sc}\rangle = V |\phi_\mathbf{k}\rangle.$$

Multiplying from the left with $G(E_k)$, we get

$$|\psi_{sc}\rangle = G(E_k)V |\phi_\mathbf{k}\rangle.$$

Adding $|\phi_\mathbf{k}\rangle$ to both sides and using Eq. (3.12), we obtain

$$|\psi_\mathbf{k}\rangle = |\phi_\mathbf{k}\rangle + G(E_k)V |\phi_\mathbf{k}\rangle = [1 + G(E_k)V] |\phi_\mathbf{k}\rangle. \tag{3.15}$$

Eqs. (3.14) and (3.15) are usually called *Lippmann-Schwinger equations*.

3.2   Free particle's Green's functions

To evaluate the free particle's Green's function, we use Eq. (3.5) write its *spectral decomposition*,

$$G_0(E_k) \equiv \int d^3\mathbf{k}' \frac{|\phi_{\mathbf{k}'}\rangle \langle \phi_{\mathbf{k}'}|}{E_k - E_{k'}} = -\left(\frac{2\mu}{\hbar^2}\right) \int d^3\mathbf{k}' \frac{|\phi_{\mathbf{k}'}\rangle \langle \phi_{\mathbf{k}'}|}{k'^2 - k^2}, \tag{3.16}$$

and take matrix elements in the coordinate representation. We get

$$G_0(E_k; \mathbf{r}, \mathbf{r}') = -\left(\frac{2\mu}{\hbar^2}\right) \int d^3\mathbf{k}' \frac{\phi_{\mathbf{k}'}(\mathbf{r})\phi_{\mathbf{k}'}^*(\mathbf{r}')}{k'^2 - k^2}. \tag{3.17}$$

Next, we use the normalized free particle's wave function

$$\phi_{\mathbf{k}'}(\mathbf{r}) = \frac{1}{(2\pi)^{3/2}} e^{i\mathbf{k}'\cdot\mathbf{r}}, \tag{3.18}$$

introduce the notation $\mathbf{R} = \mathbf{r} - \mathbf{r}'$ and evaluate the integral in spherical coordinates, with the z-axis along the $\mathbf{R}$-direction. Owing to the axial symmetry, the integration over the azimuthal angle produces the trivial factor 2π and we get[*]

$$G_0(E_k; \mathbf{r}, \mathbf{r}') = -\left(\frac{2\mu}{\hbar^2}\right) \frac{1}{(2\pi)^2} \int \frac{k'^2 dk'}{k'^2 - k^2} \int_{-1}^{1} d(\cos\theta_{k'}) e^{ik'R\cos\theta_{k'}} = \left(\frac{2\mu}{\hbar^2}\right) \frac{i}{(2\pi)^2 R} (I_1 + I_2), \tag{3.19}$$

with

$$I_1 = \int_0^\infty dk' \frac{k' e^{ik'R}}{k'^2 - k^2}, \qquad I_2 = -\int_0^\infty dk' \frac{k' e^{-ik'R}}{k'^2 - k^2}. \tag{3.20}$$

Changing $k' \to -k'$ in the second integral of Eq. (3.20), one gets

$$I_1 + I_2 = \int_{-\infty}^{\infty} dk' \frac{k' e^{ik'R}}{k'^2 - k^2}. \tag{3.21}$$

Using Eq. (3.21) and the identity

$$\frac{k}{k'^2 - k^2} = \frac{1}{2}\left[\frac{1}{k' - k} + \frac{1}{k' + k}\right]$$

in Eq. (3.19), we get

$$G_0(E_k; \mathbf{r}, \mathbf{r}') = \left(\frac{2\mu}{\hbar^2}\right) \frac{i}{2(2\pi)^2 R} \int_{-\infty}^{\infty} dk' \left[\frac{1}{k' - k} + \frac{1}{k' + k}\right] e^{ik'R}. \tag{3.22}$$

The above integral is not well defined owing to the poles of the integrand at $k' = \pm k$. However, meaningful Green's functions $G_0^{(+)}(E_k; r, r')$ and $G_0^{(-)}(E_k; r, r')$ can be obtained shifting the poles onto the complex-plane, according to the prescription:

$$\pm k \to \pm\left(k + i\epsilon'\right) \quad (G_0(E_k; \mathbf{r}, \mathbf{r}') \to G_0^{(+)}(E_k; \mathbf{r}, \mathbf{r}')); \quad \text{or}$$

$$\pm k \to \pm\left(k - i\epsilon'\right) \quad (G_0(E_k; \mathbf{r}, \mathbf{r}') \to G_0^{(-)}(E_k; \mathbf{r}, \mathbf{r}')). \tag{3.23}$$

These procedures correspond to redefining the Green's functions as

$$G_0^{(\pm)}(E_k) = \lim_{\epsilon \to 0} \left(\frac{1}{E_k - H_0 \pm i\epsilon}\right), \tag{3.24}$$

---

[*]The fact that the Green's function depends on $\mathbf{r}$ and $\mathbf{r}'$ only through the distance $R = |\mathbf{r} - \mathbf{r}'|$ is a consequence of the translational and rotational invariances of H_0.

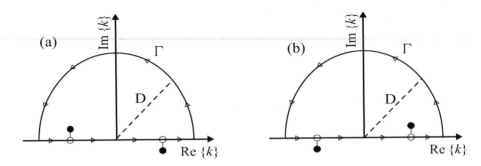

FIGURE 3.1 The closed contour on the complex plane and the shifts of the poles away from the real axis (solid circles). (a) corresponds to $G_{E_i}^{(+)}$ while (b) corresponds to $G_{E_i}^{(-)}$.

where $\epsilon = \left(\hbar^2 k/\mu\right)\epsilon'$ is an infinitesimal quantity. It is equivalent to the *analytical continuation* of the variable E_k onto the complex plane, to the energies $E_k \pm i\epsilon$. In a similar way, we get

$$G^{(\pm)}(E_k) = \lim_{\epsilon \to 0}\left(\frac{1}{E_k - H \pm i\epsilon}\right). \tag{3.25}$$

Eqs. (3.24) and (3.25), together with the hermiticity of the Hamiltonians, lead to the property

$$\left[G_0^{(\pm)}(E_k)\right]^\dagger = G_0^{(\mp)}(E_k), \qquad \left[G^{(\pm)}(E_k)\right]^\dagger = G^{(\mp)}(E_k). \tag{3.26}$$

Using the standard result (see, for example, Ref. [1])

$$\lim_{\epsilon \to 0}\left[\frac{1}{x \pm i\epsilon}\right] = \mathcal{P}\left(\frac{1}{x}\right) \mp i\pi\delta(x), \tag{3.27}$$

where $\mathcal{P}$ stands for the Cauchy's principal value, $G_0^{(\pm)}(E_k)$ is split as

$$G_0^{(\pm)}(E_k) = \mathcal{P}\left(\frac{1}{E_k - H_0}\right) \mp i\pi\delta(E_k - H_0). \tag{3.28}$$

The first term at the RHS of Eq. (3.28) corresponds to a standing wave Green's function [1] and the second term is the on-the-energy-shell part.

With the modification of the poles, Eq. (3.22) can be written

$$G_0^{(\pm)}(E_k; \mathbf{r}, \mathbf{r}') = \left(\frac{2\mu}{\hbar^2}\right)\frac{i}{2\left(2\pi\right)^2 R}I^{(\pm)}, \tag{3.29}$$

where

$$I^{(\pm)} = \int_{-\infty}^{\infty} dk'\left\{\frac{1}{[k' - (k \pm i\epsilon')]} + \frac{1}{[k' + (k \pm i\epsilon')]}\right\}e^{ik'R}. \tag{3.30}$$

These integrals can be evaluated on the complex plane, over the closed contour represented in Figure 3.1. In the $D \to \infty$ limit, it corresponds to the real axis plus a semi-circle of infinite radius, Γ. The exponential factor in the integrand of Eq. (3.30) can be written

$$e^{iR\left(\text{Re}\{k'\} + i\,\text{Im}\{k'\}\right)} = e^{iR \cdot \text{Re}\{k'\}} \times e^{-R \cdot \text{Im}\{k'\}}.$$

Considering that $R = |\mathbf{r} - \mathbf{r}'|$ is a positive definite quantity and that $\text{Im}\{k'\} > 0$ over the path Γ, the factor $e^{-R \cdot \text{Im}\{k'\}}$ goes to zero as $D \to \infty$. Therefore, the integral over Γ vanishes and the integral over the closed contour reduces to that over the real axis, as in Eq. (3.30). In this way, it is given by the residue of the pole enclosed by the contour. Namely, $k + i\epsilon$ for $I^{(+)}$ and $-(k - i\epsilon')$ for $I^{(-)}$. Taking the contour in the counterclockwise sense, we get

$$I^{(\pm)} = 2\pi i \times \text{Res}\left\{k' = \pm\left(k \pm i\epsilon'\right)\right\} = 2\pi i e^{\pm ikR}. \tag{3.31}$$

Inserting Eq. (3.31) in Eq. (3.29), we get

$$G_0^{(\pm)}(E_k; \mathbf{r}, \mathbf{r}') = -\left(\frac{2\mu}{\hbar^2}\right) \frac{1}{4\pi} \frac{e^{\pm ik|\mathbf{r} - \mathbf{r}'|}}{|\mathbf{r} - \mathbf{r}'|}. \tag{3.32}$$

A practical consequence of the above discussion is that different wave functions are generated by the Green's functions $G_0^{(+)}$ and $G_0^{(-)}$. One should distinguish between the states $|\psi_{\mathbf{k}}^{(\pm)}\rangle$, given by the Lippmann-Schwinger equations

$$|\psi_{\mathbf{k}}^{(\pm)}\rangle = |\phi_{\mathbf{k}}\rangle + G_0^{(\pm)}(E_k)V|\psi_{\mathbf{k}}^{(\pm)}\rangle. \tag{3.33}$$

Equivalent Lippmann-Schwinger equations are generated by the full Green's functions $G^{(\pm)}(E_k)$, namely

$$|\psi_{\mathbf{k}}^{(\pm)}\rangle = |\phi_{\mathbf{k}}\rangle + G^{(\pm)}(E_k)V|\phi_{\mathbf{k}}\rangle \equiv \left(1 + G^{(\pm)}(E_k)V\right)|\phi_{\mathbf{k}}\rangle. \tag{3.34}$$

Although the wave functions $\psi_{\mathbf{k}}^{(+)}$ and $\psi_{\mathbf{k}}^{(-)}$ are solutions of the same equation, with the same energy, they have however different boundary conditions. This point will be made clear in the next section, when their asymptotic forms in the coordinate representation are developed.

In the coordinate representation the Lippmann-Schwinger equations are written as

$$\psi_{\mathbf{k}}^{(\pm)}(\mathbf{r}) = \phi_{\mathbf{k}}(\mathbf{r}) + \int d^3r' G_0^{(\pm)}\left(E_k; \mathbf{r}, \mathbf{r}'\right) V\left(\mathbf{r}'\right) \psi_{\mathbf{k}}^{(\pm)}(\mathbf{r}'), \tag{3.35}$$

and

$$\psi_{\mathbf{k}}^{(\pm)}(\mathbf{r}) = \phi_{\mathbf{k}}(\mathbf{r}) + \int d^3r' G^{(\pm)}\left(E_k; \mathbf{r}, \mathbf{r}'\right) V\left(\mathbf{r}'\right) \phi_{\mathbf{k}}(\mathbf{r}'). \tag{3.36}$$

It can be easily checked that the Green's functions satisfy the equations

$$\left(E_k + \frac{\hbar^2}{2\mu}\nabla_{\mathbf{r}}^2\right) G_0^{(\pm)}\left(E_k; \mathbf{r}, \mathbf{r}'\right) = \delta\left(\mathbf{r}, \mathbf{r}'\right), \tag{3.37}$$

$$\left(E_k + \frac{\hbar^2}{2\mu}\nabla_{\mathbf{r}}^2 - V\left(r\right)\right) G^{(\pm)}\left(E_k; \mathbf{r}, \mathbf{r}'\right) = \delta\left(\mathbf{r}, \mathbf{r}'\right). \tag{3.38}$$

3.3 Scattering amplitude

Using the properties (3.26) of the Green's functions, one can show that

$$\left(\psi_{\mathbf{k}}^{(\pm)}(\mathbf{r})\right)^* = \psi_{-\mathbf{k}}^{(\mp)}(\mathbf{r}). \tag{3.39}$$

Eq. (3.39) can be used to derive the asymptotic form of $\psi_{\mathbf{k}}^{(-)}(\mathbf{r})$. Since

$$\psi_{\mathbf{k}}^{(-)}(\mathbf{r}) = \left(\psi_{-\mathbf{k}}^{(+)}(\mathbf{r})\right)^*, \tag{3.40}$$

one gets

$$\psi_{\mathbf{k}}^{(-)}(\mathbf{r}) \to A\left(e^{i\mathbf{k}\cdot\mathbf{r}} + f^*(\pi - \theta)\frac{e^{-ikr}}{r}\right). \tag{3.41}$$

Note that the transformation $\mathbf{k} \to -\mathbf{k}$ leads to the change $\theta \to \pi - \theta$ in the argument of the above scattering amplitude.

Let us now investigate the nature of the states $\left|\psi_{\mathbf{k}}^{(\pm)}\right\rangle$. Using Eq. (3.32) in Eq. (3.35), we obtain

$$\psi_{\mathbf{k}}^{(\pm)}(\mathbf{r}) = \phi_{\mathbf{k}}(\mathbf{r}) - \frac{2\mu}{4\pi\hbar^2} \int d^3r' \frac{e^{\pm ik|\mathbf{r} - \mathbf{r}'|}}{|\mathbf{r} - \mathbf{r}'|} \left\langle\mathbf{r}'\left|V\right|\psi_{\mathbf{k}}^{(\pm)}\right\rangle. \tag{3.42}$$

We now take the asymptotic limit, $|\mathbf{r}| \to \infty$. Since V vanishes when $\mathbf{r}'$ is outside the potential range ($|\mathbf{r}'| > R$), $|\mathbf{r}'| / |\mathbf{r}| \to 0$ and one can replace

$$\frac{1}{|\mathbf{r} - \mathbf{r}'|} \to \frac{1}{r}. \tag{3.43}$$

The factor $\exp\left(\pm ik\,|\mathbf{r} - \mathbf{r}'|\right)$ must be handled more carefully. It is necessary to expand to first order,

$$\lim_{|\mathbf{r}| \to \infty} \left\{ \exp\left(\pm ik\,|\mathbf{r} - \mathbf{r}'|\right) \right\} = \lim_{|\mathbf{r}| \to \infty} \left\{ \exp\left(\pm ik\sqrt{r^2 + r'^2 - 2\mathbf{r}\cdot\mathbf{r}'}\right) \right\} = \exp\left(\pm i(kr - k\hat{\mathbf{r}}\cdot\mathbf{r}')\right), \tag{3.44}$$

where $\hat{\mathbf{r}}$ is the unit vector along $\mathbf{r}$. Using Eqs. (3.43) and (3.44) in Eq. (3.42) and defining the final wave vector $\mathbf{k}' = k\hat{\mathbf{r}}$, we obtain

$$\left[\psi_{\mathbf{k}}^{(\pm)}(\mathbf{r})\right]_{|\mathbf{r}| \to \infty} \to \phi_{\mathbf{k}}(\mathbf{r}) + \frac{e^{\pm ikr}}{r}\left(-\frac{2\mu}{4\pi\hbar^2}\int d^3r'\, e^{\mp i\mathbf{k}'\cdot\mathbf{r}'}\left\langle\mathbf{r}'\left|V\right|\psi_{\mathbf{k}}^{(\pm)}\right\rangle\right).$$

Or, using Eq. (3.18),

$$\left[\psi_{\mathbf{k}}^{(\pm)}(\mathbf{r})\right]_{|\mathbf{r}| \to \infty} \to \frac{1}{(2\pi)^{3/2}}\left[e^{i\mathbf{k}\cdot\mathbf{r}} + \frac{e^{\pm ikr}}{r}\left(-2\pi^2\left(\frac{2\mu}{\hbar^2}\right)\left\langle\phi_{\pm\mathbf{k}'}\left|V\right|\psi_{\mathbf{k}}^{(\pm)}\right\rangle\right)\right]. \tag{3.45}$$

It is now clear that the Green's function $G_0^{(+)}(E_k;\mathbf{r},\mathbf{r}')$ generates a state $\psi_{\mathbf{k}}^{(+)}$ which asymptotically behaves as the sum of the incident plane wave with an outgoing spherical wave. This is the desired scattering solution. A comparison with the asymptotic Form of the scattered wave (with the normalization constant $A = (2\pi)^{-3/2}$) leads to the relation

$$f(\theta) = -2\pi^2\left(\frac{2\mu}{\hbar^2}\right)\left\langle\phi_{\mathbf{k}'}\left|V\right|\psi_{\mathbf{k}}^{(+)}\right\rangle. \tag{3.46}$$

Above, θ is the observation angle which corresponds to the angle between the initial ($\mathbf{k}$) and the final ($\mathbf{k}'$) wave vectors. The meaning of the solution generated by $G_0^{(-)}(E_k;\mathbf{r},\mathbf{r}')$ is not straightforward. However, it will prove very useful for calculations.

The scattering amplitude can also be written as

$$f(\theta) = -2\pi^2\left(\frac{2\mu}{\hbar^2}\right)\left\langle\psi_{\mathbf{k}'}^{(-)}\left|V\right|\phi_{\mathbf{k}}\right\rangle. \tag{3.47}$$

3.4   Born approximation

Assuming that the scattering potential V is so weak that one can replace $\psi_{\mathbf{k}}^{(+)}$ by $\phi_{\mathbf{k}}$ in (3.46), one obtains a useful approximation for the scattering amplitude

$$f(\theta) = -2\pi^2\left(\frac{2\mu}{\hbar^2}\right)\langle\phi_{\mathbf{k}'}\left|V\right|\phi_{\mathbf{k}}\rangle = -\frac{\mu}{2\pi\hbar^2}\int d^3r\, e^{-i\mathbf{q}\cdot\mathbf{r}}\,V(\mathbf{r}), \tag{3.48}$$

or

$$f(\theta) = -2\pi^2\left(\frac{2\mu}{\hbar^2}\right)\widetilde{V}(\mathbf{q}), \tag{3.49}$$

where

$$\widetilde{V}(\mathbf{q}) = \frac{1}{(2\pi)^3}\int d^3r\, e^{-i\mathbf{q}\cdot\mathbf{r}}\,V(\mathbf{r}), \quad \text{and} \quad \mathbf{q} = \mathbf{k}' - \mathbf{k}. \tag{3.50}$$

Eq. (3.48) is known as *Born approximation*. It is accurate only in a few situations for elastic scattering. However, it has important applications in inelastic scattering experiments, e.g., electron scattering off nuclei.

3.5 Transition and scattering matrices

Eq. (3.46) indicates that the scattering amplitude is proportional to the matrix-element of the potential between a final free state, with wave vector along the observation direction, and the full scattering state. However, these states do not belong to the same orthogonal set. It is thus convenient to express the scattering amplitude in terms of matrix-elements between initial and final free states. The set of matrix-elements of T between free states, represented as

$$T_{\mathbf{k}',\mathbf{k}} \equiv \langle \phi_{\mathbf{k}'} | T | \phi_{\mathbf{k}} \rangle = \left\langle \phi_{\mathbf{k}'} \middle| V \middle| \psi_{\mathbf{k}}^{(+)} \right\rangle, \tag{3.51}$$

is called the *T-matrix*. The T-matrix-elements on-the-*energy-shell*, i.e. $T_{\mathbf{k}',\mathbf{k}}$, with $|\mathbf{k}'| = |\mathbf{k}|$, give the probability amplitude for the transition from the incident momentum $\hbar\mathbf{k}$ to the final momentum $\hbar\mathbf{k}'$, through the action of the potential V. In the absence of potential, the transition matrix vanishes identically.

Inserting Eq. (3.51) in Eq. (3.46), the scattering amplitude becomes*

$$f(\theta) = -2\pi^2 \left(\frac{2\mu}{\hbar^2} \right) T_{\mathbf{k}',\mathbf{k}}. \tag{3.52}$$

Multiplying Eqs. (3.15) by V from the left and using Eq. (3.51), we get the Lippmann-Schwinger equations for the *transition operator*,

$$T = V + V G_0(E_k) T, \quad \text{and} \quad T = V + V G(E_k) V. \tag{3.53}$$

It should be remarked that if one writes Eqs. (3.53) in the momentum representation, both *on-the-energy-shell* and *off-the-energy-shell* matrix-elements must be considered [3]. One can show that the Lippmann-Schwinger equation for the T-matrix contains the Optical Theorem [3].

We now introduce the *scattering operator*, S. It can be defined in the form

$$S_{\mathbf{k}',\mathbf{k}} = \langle \psi_{\mathbf{k}'}^{(-)} | \psi_{\mathbf{k}}^{(+)} \rangle = \langle \phi_{\mathbf{k}'} | S | \phi_{\mathbf{k}} \rangle. \tag{3.54}$$

The S-matrix has the important property of being unitary. Namely

$$\left(S^\dagger S \right)_{\mathbf{k}',\mathbf{k}} = \left(S S^\dagger \right)_{\mathbf{k}',\mathbf{k}} = \delta \left(\mathbf{k} - \mathbf{k}' \right). \tag{3.55}$$

Because the S-operator relates physical states, S-matrix elements are defined only for initial and final states having the same energy $E_i = E_f$. Thus, the S-matrix is related with the on-the–energy-shell matrix elements of the transition operator through the expression (see problem section)

$$S_{\mathbf{k}',\mathbf{k}} = \delta \left(\mathbf{k} - \mathbf{k}' \right) - 2\pi i \delta \left(E_k - E_{k'} \right) T_{\mathbf{k}',\mathbf{k}}, \tag{3.56}$$

or in operator notation

$$S(E) = 1 - 2\pi i \delta \left(E - H_0 \right) T(E). \tag{3.57}$$

This property can proved using the Lippmann-Schwinger equations. It should be pointed out that the unitarity of the scattering operator contains the Optical Theorem.

3.6 The two-potential formula

In Section 3.2 we wrote the Hamiltonian as $H = H_0 + V$, where H_0 was the kinetic energy operator K and V the total potential. We then showed that the exact solutions $\left| \psi_{\mathbf{k}}^{(\pm)} \right\rangle$ and the plane waves

---

*Note that the constant of proportionality between the scattering amplitude and the T-matrix depends on the choice of the normalization of the plane wave. We use $A = (2\pi)^{-3/2}$ and obtain $-2\pi^2 \left(2\mu/\hbar^2 \right)$. Some authors (like Austern in Ref. [2]) use $A = 1$. In this case the proportionality constant takes the value $-\mu/2\pi\hbar^2$.

$|\phi_{\mathbf{k}}\rangle$ are related by Lippmann-Schwinger equations. In some situations, it is convenient to split the total potential in two parts,

$$V = V_0 + U, \tag{3.58}$$

include V_0 in the Hamiltonian H_0 and evaluate the T-matrix in terms of waves distorted by V_0. In this case, the Lippmann-Schwinger equations relating $\left|\psi_{\mathbf{k}}^{(\pm)}\right\rangle$ with plane waves can be written as

$$\left|\psi_{\mathbf{k}}^{(\pm)}\right\rangle = |\phi_{\mathbf{k}}\rangle + G_0^{(\pm)}(E_k)\,[V_0 + U]\left|\psi_{\mathbf{k}}^{(\pm)}\right\rangle, \tag{3.59}$$

and

$$\left|\psi_{\mathbf{k}}^{(\pm)}\right\rangle = \left\{1 + \mathcal{G}^{(\pm)}(E_k)\,[V_0 + U]\right\}|\phi_{\mathbf{k}}\rangle. \tag{3.60}$$

Above, $G_0^{(\pm)}(E_k)$ is the free particle's Green's function,

$$G_0^{(\pm)}(E) = \frac{1}{E - K \pm i\epsilon}, \tag{3.61}$$

where K stands for the kinetic energy operator, and $\mathcal{G}^{(\pm)}(E_k)$ is the full Green's function

$$\mathcal{G}^{(\pm)}(E) = \frac{1}{E - H \pm i\epsilon} \equiv \frac{1}{E - [K + V_0 + U] \pm i\epsilon}. \tag{3.62}$$

Note that in this section we use the notation $\mathcal{G}^{(\pm)}(E)$ for the full Green's function, while up to this point it has been denoted $G^{(\pm)}(E)$. The reason for this change will be clear below.

The states $|\phi_{\mathbf{k}}\rangle$ and $\left|\psi_{\mathbf{k}}^{(\pm)}\right\rangle$ can also related with waves distorted by the potential V_0, $\left|\chi_{\mathbf{k}}^{(\pm)}\right\rangle$. These waves satisfy the Schrödinger equation

$$(E - H_0)\left|\chi_{\mathbf{k}}^{(\pm)}\right\rangle = 0, \tag{3.63}$$

with

$$H_0 = K + V_0. \tag{3.64}$$

Proceeding as in Section 3.1 (setting $U = 0$ and replacing $H_0 \to K$, $H \to H_0$), we get the Lippmann-Schwinger equations

$$\left|\chi_{\mathbf{k}}^{(\pm)}\right\rangle = |\phi_{\mathbf{k}}\rangle + G_0^{(\pm)}(E_k)\,V_0\left|\chi_{\mathbf{k}}^{(\pm)}\right\rangle, \tag{3.65}$$

$$= \left[1 + G^{(\pm)}(E_k)\,V_0\right]|\phi_{\mathbf{k}}\rangle, \tag{3.66}$$

where

$$G^{(\pm)}(E) = \frac{1}{E - H_0 \pm i\epsilon}. \tag{3.67}$$

To relate the exact states with those distorted by V_0, we follow again the steps of Section 3.1. It is straightforward to derive the equations

$$\left|\psi_{\mathbf{k}}^{(\pm)}\right\rangle = \left|\chi_{\mathbf{k}}^{(\pm)}\right\rangle + G^{(\pm)}(E_k)\,U\left|\psi_{\mathbf{k}}^{(\pm)}\right\rangle \tag{3.68}$$

$$= \left[1 + \mathcal{G}^{(\pm)}(E_k)\,U\right]\left|\chi_{\mathbf{k}}^{(\pm)}\right\rangle. \tag{3.69}$$

With the help of the above equations, we can derive the important result for the on-the-energy-shell T-matrix-elements

$$T_{\mathbf{k}',\mathbf{k}} = \left\langle \chi_{\mathbf{k}'}^{(-)}\left|V_0\right|\phi_{\mathbf{k}}\right\rangle + \left\langle \chi_{\mathbf{k}'}^{(-)}\left|U\right|\psi_{\mathbf{k}}^{(+)}\right\rangle \tag{3.70}$$

$$= \left\langle \phi_{\mathbf{k}'}\left|V_0\right|\chi_{\mathbf{k}}^{(+)}\right\rangle + \left\langle \chi_{\mathbf{k}'}^{(-)}\left|U\right|\psi_{\mathbf{k}}^{(+)}\right\rangle. \tag{3.71}$$

These expressions are known as the *two-potential formulae* or the *Gell-Mann Goldberger relations*. To prove these relations, we use the property* $G_0^{(-)\dagger} = G_0^{(+)}$ in the Hermitian conjugate of Eq. (3.65) to write

$$\langle \phi_{\mathbf{k}'}| = \left\langle \chi_{\mathbf{k}'}^{(-)}\right| - \left\langle \chi_{\mathbf{k}'}^{(-)}\right| V_0 G_0^{(+)}. \tag{3.72}$$

Inserting Eq. (3.72) in the definition of the T-matrix,

$$T_{\mathbf{k}',\mathbf{k}} \equiv \left\langle \phi_{\mathbf{k}'} |V_0 + U| \psi_{\mathbf{k}}^{(+)}\right\rangle, \tag{3.73}$$

we get

$$T_{\mathbf{k}',\mathbf{k}} = \left\langle \chi_{\mathbf{k}'}^{(-)}\right| V_0 + U \left|\psi_{\mathbf{k}}^{(+)}\right\rangle - \left\langle \chi_{\mathbf{k}'}^{(-)}\right| V_0 \left\{ G_0^{(+)} [V_0 + U] \left|\psi_{\mathbf{k}}^{(+)}\right\rangle\right\}. \tag{3.74}$$

Now, we use Eq. (3.59) to write

$$G_0^{(+)} [V_0 + U] \left|\psi_{\mathbf{k}}^{(+)}\right\rangle = \left|\psi_{\mathbf{k}}^{(+)}\right\rangle - |\phi_{\mathbf{k}}\rangle. \tag{3.75}$$

Using Eq. (3.75) to replace the term within curly brackets in Eq. (3.74), we obtain

$$T_{\mathbf{k}',\mathbf{k}} = \left\langle \chi_{\mathbf{k}'}^{(-)}\right| V_0 + U \left|\psi_{\mathbf{k}}^{(+)}\right\rangle - \left\langle \chi_{\mathbf{k}'}^{(-)}\right| V_0 \left|\psi_{\mathbf{k}}^{(+)}\right\rangle + \left\langle \chi_{\mathbf{k}'}^{(-)}\right| V_0 \left|\phi_{\mathbf{k}}\right\rangle \tag{3.76}$$
$$= \left\langle \chi_{\mathbf{k}'}^{(-)}\right| V_0 \left|\phi_{\mathbf{k}}\right\rangle + \left\langle \chi_{\mathbf{k}'}^{(-)}\right| U \left|\psi_{\mathbf{k}}^{(+)}\right\rangle,$$

and Eq. (3.70) is proved. The proof of Eq. (3.71) follows immediately.

3.7 Distorted wave Born approximation

The two-potential formulae are the starting points for deriving the well known *Distorted Wave Born Approximation (DWBA)*. Using Eqs. (3.52) and (3.71) we get

$$f(\theta) = f_0(\theta) - 2\pi^2 \left(\frac{2\mu}{\hbar^2}\right) \int d^3r \, \chi_{\mathbf{k}'}^{(-)*}(\mathbf{r}) \, U(\mathbf{r}) \, \psi_{\mathbf{k}}^{(+)}(\mathbf{r}). \tag{3.77}$$

The DWBA amplitude is obtained replacing $\psi_{\mathbf{k}}^{(+)}$ by $\chi_{\mathbf{k}}^{(+)}$, i.e.,

$$f_{DWBA}(\theta) = f_0(\theta) - 2\pi^2 \left(\frac{2\mu}{\hbar^2}\right) \int d^3r \, \chi_{\mathbf{k}'}^{(-)*}(\mathbf{r}) \, U(\mathbf{r}) \, \chi_{\mathbf{k}}^{(+)}(\mathbf{r}). \tag{3.78}$$

This approximation is good if U is weak compared to V_0. It is called "Born" because it is first order in the potential U but also called "distorted wave", because instead of using the plane waves as in Eq. (3.48) we used the distorted waves $\chi^{(\pm)}$, which should be a better approximation to the exact solution.

This approximation can be generalized to inelastic scattering. In this case, V_0 (and hence f_0) is chosen to describe the elastic scattering (i.e. V_0 is an optical potential), while U is the interaction which induces the non-elastic transition. The validity of the DWBA then depends upon elastic scattering being the most important event which occurs when two nuclei collide, so that inelastic events can be treated as perturbations. The corresponding inelastic scattering amplitude for a reaction $A(a,b)B$ has the form

$$f_{\text{DWBA}}^{\text{inel}}(\theta) = -2\pi^2 \left(\frac{2\mu}{\hbar^2}\right) \int \chi_\beta^{(-)*}(\mathbf{k}_\beta, \mathbf{r}_\beta) \, \langle b, B|U_{\text{int}}|a, A\rangle \chi_\alpha^{(+)}(\mathbf{k}_\alpha, \mathbf{r}_\alpha) \, d^3 r_\alpha d^3 r_\beta. \tag{3.79}$$

We will use this result in many occasions.

---

*Since in this section we are dealing with matrix-elements on-the-energy-shell, $E_k = E_{k'}$, the energy labels become superfluous.

Here, $\chi_{\mathbf{k}}^{(\pm)}$ has been generalized to χ_α and χ_β. The function χ_α describes the elastic scattering in the $\alpha = a + A$ entrance channel, arising from an optical potential V_α, while χ_β describes the elastic scattering in the $\beta = b + B$ exit channel arising from a potential V_β. The interaction potential U_{int}, which describes the non-elastic transition depends upon the type of reaction and the model chosen to describe it.

3.8   Partial-wave expansion of the S-matrix

In this section we carry out partial-wave expansion of the Lippmann-Schwinger equations. As a starting point, we introduce the sets of states $\{|\phi_{Elm}\rangle\}$ and $\{|\psi_{Elm}\rangle\}$, which are eigenstates of the operators L^2 and L_z, with eigenvalues $\hbar^2 l(l+1)$ and $\hbar m$, respectively. They satisfy the Schrödinger equations

$$(E - H_0)\,|\phi_{Elm}\rangle = 0 \tag{3.80}$$

$$(E - H)\,|\psi_{Elm}\rangle = 0, \tag{3.81}$$

and are normalized as to satisfy the relations

$$\langle \phi_{E'l'm'} |\phi_{Elm}\rangle = \delta_{l,l'}\delta_{m,m'}\delta(E - E') \tag{3.82}$$

$$\langle \psi_{E'l'm'} |\psi_{Elm}\rangle = \delta_{l,l'}\delta_{m,m'}\delta(E - E'). \tag{3.83}$$

In the coordinate representation, the wave functions $\phi_{Elm}(\mathbf{r})$ and $\psi_{Elm}(\mathbf{r})$ are products of *spherical harmonics* $Y_{lm}(\hat{\mathbf{r}})$ with appropriate radial wave functions. The completeness relation for the states $|\phi_{Elm}\rangle$ has the form

$$\sum_{lm} \int |\phi_{Elm}\rangle\, dE\, \langle \phi_{Elm}| = \mathbf{1}. \tag{3.84}$$

The solutions of the Schrödinger equation with the full Hamiltonian, $\psi_{Elm}(\mathbf{r})$, depend on the particular potential of the problem while the free spherical waves $\phi_{Elm}(\mathbf{r})$ are completely determined (see Chapter 2), except for the normalization required by Eq. (3.82). They are given by

$$\phi_{Elm}(\mathbf{r}) = A_l \frac{\hat{j}_l(kr)}{kr} Y_{lm}(\hat{\mathbf{r}}), \tag{3.85}$$

where $k = \sqrt{2\mu E}/\hbar$. Using the orthonormality of the spherical harmonics and the radial integral of Ricatti-Bessel functions Eq. (3.82) yields

$$|A_l|^2 = \frac{2\mu k}{\pi\hbar^2} \rightarrow A_l = i^l \left(\frac{2\mu k}{\pi\hbar^2}\right)^{1/2}, \tag{3.86}$$

where the arbitrary phase of A_l was chosen to be i^l. We can use an analogous procedure for the wave functions $\psi_{Elm}(\mathbf{r})$. We write

$$\psi_{Elm}(\mathbf{r}) = A_l \frac{u_l(kr)}{kr} Y_{lm}(\hat{\mathbf{r}}), \tag{3.87}$$

were A_l are the coefficients of Eq. (3.86) and $u_l(k, r)$ are solutions of the radial equation. Eq. (3.83) requires that the radial wave functions be normalized as to satisfy the relation

$$\int_0^\infty u_l^*(k, r)u_l(k', r)dr = \frac{\pi}{2}\delta(k - k'). \tag{3.88}$$

The plane waves and the scattering states are expressed in terms of $\phi_{Elm}(\mathbf{r})$ and $\psi_{Elm}(\mathbf{r})$ respectively as (adopting the normalization $A = (2\pi)^{-3/2}$

$$\phi_{\mathbf{k}'}(\mathbf{r}) = \frac{\hbar}{\sqrt{\mu k}} \sum_{l,m} Y_{lm}^*(\hat{\mathbf{k}}')\phi_{Elm}(\mathbf{r}), \tag{3.89}$$

and

$$\psi_{\mathbf{k}}^{(+)}(\mathbf{r}) = \frac{\hbar}{\sqrt{\mu k}} \sum_{l,m} Y_{lm}^*(\hat{\mathbf{k}}) \psi_{Elm}(\mathbf{r}). \tag{3.90}$$

To derive the partial-wave projected Lippmann-Schwinger equation, we carry out the partial wave expansion of $G_0^{(\pm)}(E; \mathbf{r}, \mathbf{r}')$ as*

$$G_0^{(\pm)}(E; \mathbf{r}, \mathbf{r}') = \frac{1}{rr'} \sum_{lm} Y_{lm}^*(\hat{\mathbf{r}}) g_{0,l}^{(\pm)}(r, r') Y_{lm}(\hat{\mathbf{r}}'). \tag{3.91}$$

Since we are supposing that the potential is spherically symmetric, the Hamiltonian is rotationally invariant. Therefore, the Green's function is diagonal in the angular momentum representation. We then insert this expansion, the one for the incident plane wave and the one for the scattering state with $A = 1/(2\pi)^{3/2}$, into Lippmann-Schwinger equation (Eq. (3.35)). Using the linear independence and the orthonormality of the spherical harmonics, we get

$$u_l(k, r) = \hat{\jmath}_l(kr) + \int dr' g_{0,l}^{(+)}(r, r') V(r') u_l(k, r'). \tag{3.92}$$

Equations involving the Green's function $g_{0,l}^{(-)}$ could be obtained using the partial-wave expansion of $\psi_{\mathbf{k}}^{(-)}(\mathbf{r})$. Namely,

$$u_l^*(k, r) = \hat{\jmath}_l(kr) + \int dr' g_{0,l}^{(-)}(r, r') V(r') u_l^*(k, r'). \tag{3.93}$$

However, these equations are not linearly independent of those involving $g_{0,l}^{(+)}$, since they are simply their complex conjugates.

Equivalent equations can be obtained with the full Green's function. Making the expansion

$$G^{(\pm)}(E; \mathbf{r}, \mathbf{r}') = \frac{1}{rr'} \sum_{lm} Y_{lm}^*(\hat{\mathbf{r}}) g_l^{(\pm)}(r, r') Y_{lm}(\hat{\mathbf{r}}'), \tag{3.94}$$

and following the same procedures, we obtain

$$u_l(k, r) = \hat{\jmath}_l(kr) + \int dr' g_l^{(+)}(r, r') V(r') \hat{\jmath}_l(kr'). \tag{3.95}$$

Similar equations can be obtained in terms of $g_l^{(-)}$.

It is also convenient to carry out the partial-wave expansion of the T-matrix-elements between states on the same energy-shell. For this purpose, we start from its definition (Eq. (3.51)) and use the partial-wave expansions of $\psi_{\mathbf{k}}^{(+)}$, and of $\phi_{\mathbf{k}'}$. With the help of Eqs. (3.85-3.87), we get

$$T_{\mathbf{k}', \mathbf{k}} = \left\langle \phi_{\mathbf{k}'} \left| V \right| \psi_{\mathbf{k}}^{(+)} \right\rangle = \frac{\hbar^2}{\mu k} \sum_{lm} Y_{lm}(\hat{\mathbf{k}}') Y_{lm}^*(\hat{\mathbf{k}}) T_l(E_k), \tag{3.96}$$

where we have introduced the partial-wave component of the T-matrix between states on the same energy-shell

$$T_l(E_k) = \langle \phi_{Elm} | V | \psi_{Elm} \rangle \equiv \frac{2\mu}{\pi \hbar^2 k} \int_0^\infty dr \hat{\jmath}_l(kr) V(r) u_l(k, r). \tag{3.97}$$

Using the addition theorem for the spherical harmonics

$$\sum_m Y_{lm}(\mathbf{k}') Y_{lm}^*(\hat{\mathbf{k}}) = \frac{2l+1}{4\pi} P_l(\cos\theta),$$

---

*The use of the factor $1/rr'$ in the partial-wave expansion of the Green's functions leads to a simpler form of the Lippmann-Schwinger equations for $u_l^{(\pm)}(k, r)$.

Eq. (3.96) becomes

$$T_{\mathbf{k}',\mathbf{k}} = \sum_l (2l+1) P_l(\cos\theta) \left[\frac{\hbar^2}{4\pi\mu k} T_l(E_k) \right] . \tag{3.98}$$

The same procedure can be carried out for the S-matrix. Firstly, we write the scattering operator in the representation of the $|\phi_{Elm}\rangle$ states. With the help of the completeness relation of Eq. (3.84), we get

$$S_{\mathbf{k}',\mathbf{k}} = \int\int dE dE' \sum_{lm,l'm'} \langle \phi_{\mathbf{k}'} | \phi_{E'l'm'} \rangle \langle \phi_{E'l'm'} | S | \phi_{Elm} \rangle \langle \phi_{Elm} | \phi_{\mathbf{k}} \rangle . \tag{3.99}$$

Since we are dealing with spherically symmetric potentials, the Hamiltonians H_0 and H are both rotationally invariant. This implies that the S-operator commutes with any component of the angular momentum operator. Furthermore, $\langle \phi_{E'l'm'} | S | \phi_{Elm} \rangle$ must be independent of m, since S also commutes with the operators $J_\pm = J_x \pm iJ_y$, which change m by one unit. Considering that the S-operator also commutes with the Hamiltonian H_0, we can write

$$\langle \phi_{E'l'm'} | S | \phi_{Elm} \rangle = S_l(E) \delta_{l,l'} \delta_{m,m'} \delta(E_k - E_{k'}) . \tag{3.100}$$

On the other hand, it can be easily verified that

$$\langle \phi_{Elm} | \phi_{\mathbf{k}} \rangle = \frac{\hbar}{\sqrt{\mu k}} Y^*_{l'm'}(\hat{\mathbf{k}}) \delta(E - E_k), \tag{3.101}$$

and

$$\langle \phi_{\mathbf{k}'} | \phi_{E'l'm'} \rangle = \frac{\hbar}{\sqrt{\mu k}} Y_{lm}(\mathbf{k}') \delta(E' - E_{k'}). \tag{3.102}$$

Inserting Eqs. (3.100-3.101) into Eq. (3.99), we get

$$S_{\mathbf{k}',\mathbf{k}} = \delta(E_{k'} - E_k) \frac{\hbar^2}{\mu k} \sum_{lm} Y_{lm}(\mathbf{k}') Y^*_{lm}(\hat{\mathbf{k}}) S_l(E_k) \tag{3.103}$$

$$= \delta(E_{k'} - E_k) \frac{\hbar^2}{4\pi\mu k} \sum_l (2l+1) P_l(\cos\theta) S_l(E_k) , \tag{3.104}$$

where we used the addition theorem to go from Eq. (3.103) to Eq. (3.104).

We can now relate the partial-wave components of the T- and the S-operators. For this purpose we start from Eq. (3.56) with the partial-wave expansion for the T-matrix,

$$S_{\mathbf{k}',\mathbf{k}} = \delta(\mathbf{k}' - \mathbf{k}) - 2\pi i \delta(E_{k'} - E_k) \sum_l (2l+1) P_l(\cos\theta) \left[\frac{\hbar^2}{4\pi\mu k} T_l(E_k) \right] . \tag{3.105}$$

The next step is to express $\delta(\mathbf{k}' - \mathbf{k})$ in terms of $E_k, \Omega_{\mathbf{k}}, E_{k'}, \Omega_{\mathbf{k}'}$. We must have

$$\int d\Omega_{\mathbf{k}} dk k^2 \delta(\mathbf{k}' - \mathbf{k}) = \int d\Omega_{\mathbf{k}} dE_k \delta(E_{k'} - E_k) \delta(\Omega_{\mathbf{k}'} - \Omega_{\mathbf{k}}). \tag{3.106}$$

Writing

$$dE_k = \frac{\hbar^2}{\mu k} \left(k^2 dk \right) ,$$

we can identify

$$\delta(\mathbf{k}' - \mathbf{k}) = \frac{\hbar^2}{\mu k} \delta(E_{k'} - E_k) \delta(\Omega_{\mathbf{k}'} - \Omega_{\mathbf{k}}). \tag{3.107}$$

We now use the completeness of the spherical harmonics to write

$$\delta(\Omega_{\mathbf{k}'} - \Omega_{\mathbf{k}}) = \sum_{lm} Y^*_{lm}(\Omega_{\mathbf{k}'}) Y_{lm}(\Omega_{\mathbf{k}}) = \frac{1}{4\pi} \sum_l (2l+1) P_l(\cos\theta).$$

Using this relation in Eq. (3.107) and inserting the result in Eq. (3.105), we obtain

$$S_{\mathbf{k}',\mathbf{k}} = \delta(E_{k'} - E_k) \sum_l (2l + 1) P_l(\cos\theta) \left[\frac{\hbar^2}{4\pi\mu k} (1 - 2\pi i T_l(E)) \right]. \tag{3.108}$$

Comparing Eqs. (3.104) and Eq. (3.108), we get the desired expression

$$S_l(E) = 1 - 2\pi i T_l(E). \tag{3.109}$$

The scattering amplitude can be expressed in terms of $T_l(E)$. Inserting Eq. (3.98) in Eq. (3.52) we get

$$f(\theta) = -2\pi^2 \left(\frac{2\mu}{\hbar^2} \right) \frac{\hbar^2}{4\pi\mu k} \sum_l (2l + 1) P_l(\cos\theta) T_l(E) = -\frac{\pi}{k} \sum_l (2l + 1) P_l(\cos\theta) T_l(E),$$

or

$$f(\theta) = \frac{1}{2ik} \sum_l (2l + 1) P_l(\cos\theta) \left[-2\pi i T_l(E) \right]. \tag{3.110}$$

Using Eq. (3.109), the scattering amplitude takes the form

$$f(\theta) = \frac{1}{2ik} \sum_l (2l + 1) P_l(\cos\theta) \left[S_l(E) - 1 \right]. \tag{3.111}$$

This equation was derived in Chapter 2 from the asymptotic behavior of the wave function.

3.9   Partial-wave free particle's Green's functions

Let us now evaluate the partial-wave projected Green's function for a free particle, $g_l^{(+)}(r, r')$. We start from the spectral representation of $G_0^{(+)}(E; \mathbf{r}, \mathbf{r}')$ in the spherical basis[§],

$$G_0^{(+)}(E; \mathbf{r}, \mathbf{r}') = \sum_{lm} \int_0^\infty dE' \frac{\phi_{E'lm}^*(\mathbf{r}) \phi_{E'lm}(\mathbf{r}')}{E - E' + i\epsilon}$$

$$= \frac{1}{rr'} \sum_{lm} Y_{lm}^*(\hat{\mathbf{r}}) Y_{lm}(\hat{\mathbf{r}}') \left[\frac{2\mu}{\pi\hbar^2} \int_0^\infty \frac{dE'}{k'} \frac{\hat{j}_l(k'r)\hat{j}_l(k'r')}{E - E' + i\epsilon} \right].$$

Comparing the above equation with Eq. (3.91) we get

$$g_{0,l}^{(+)}(E; r, r') = \frac{2\mu}{\pi\hbar^2} \int_0^\infty \frac{dE'}{k'} \frac{\hat{j}_l(k'r)\hat{j}_l(k'r')}{E - E' + i\epsilon}. \tag{3.112}$$

Changing variables $E' \to \hbar^2 k'^2/2\mu$, $dE' \to \left(\hbar^2/\mu \right) k' dk'$, Eq. (3.112) becomes

$$g_{0,l}^{(+)}(E; r, r') = \frac{2}{\pi} \left(\frac{2\mu}{\hbar^2} \right) \int_0^\infty dk' \frac{\hat{j}_l(k'r)\hat{j}_l(k'r')}{k^2 - k'^2 + i\epsilon'}. \tag{3.113}$$

Considering the properties of Bessel functions (see, for example, appendix C of Ref. [3])

$$j_l(-z) = (-)^l j_l(z) \to \hat{j}_l(-z) = (-)^{l+1} \hat{j}_l(z) \tag{3.114}$$

$$n_l(-z) = (-)^{l+1} n_l(z) \to \hat{n}_l(-z) = (-)^l \hat{n}(z),$$

[§]For simplicity of notation, we do not write $\lim_{\epsilon \to 0}$. Taking such limits is implicitly assumed.

we can put Eq. (3.113) in the form

$$g_{0,l}^{(+)}(E; r, r') = \frac{1}{\pi}\left(\frac{2\mu}{\hbar^2}\right)\int_{-\infty}^{\infty} dk' \frac{\hat{j}_l(k'r)\hat{j}_l(k'r')}{k^2 - k'^2 + i\epsilon'}. \tag{3.115}$$

Writing

$$\frac{1}{k^2 - k'^2 + i\epsilon'} = -\frac{1}{2k'}\left[\frac{1}{k' - (k + i\varepsilon)} + \frac{1}{k' + (k + i\varepsilon)}\right],$$

where $\varepsilon = 2k\epsilon'$ is also an infinitesimal quantity, and using this expression in Eq. (3.115) we obtain

$$g_{0,l}^{(+)}(E; r, r') = \frac{2\mu}{\hbar^2}I^{(+)}, \tag{3.116}$$

with

$$I^{(+)} = -\frac{1}{2\pi}\int_{-\infty}^{\infty} dk' \frac{\hat{j}_l(k'r)\hat{j}_l(k'r')}{k'}\left[\frac{1}{k' - (k + i\varepsilon)} + \frac{1}{k' + (k + i\varepsilon)}\right]. \tag{3.117}$$

The integral $I^{(+)}$ can be evaluated on the complex plane. For this purpose, we consider separately the situations $r' > r$ and $r' < r$. In the former case, we write

$$\hat{j}_l(k'r') = \frac{1}{2}\left(\hat{h}_l^{(+)}(k'r') + \hat{h}_l^{(-)}(k'r')\right) \tag{3.118}$$

and split $I^{(+)}$ into the terms

$$I^{(+)} = I_1^{(+)} + I_2^{(+)},$$

where

$$I_1^{(+)} = -\frac{1}{4\pi}\int_{-\infty}^{\infty} dk' \frac{\hat{j}_l(k'r)\hat{h}_l^{(-)}(k'r')}{k'}\left[\frac{1}{k' - (k + i\varepsilon)} + \frac{1}{k' + (k + i\varepsilon)}\right]$$

$$I_2^{(+)} = -\frac{1}{4\pi}\int_{-\infty}^{\infty} dk' \frac{\hat{j}_l(k'r)\hat{h}_l^{(+)}(k'r')}{k'}\left[\frac{1}{k' - (k + i\varepsilon)} + \frac{1}{k' + (k + i\varepsilon)}\right].$$

The first integral is evaluated over the closed contour formed by the real axis and the lower half of a circle of infinite radius. On the lower half-circle we can take the asymptotic form $\hat{h}_l^{(-)}(k'r') \to i\exp\left[-i\left(\text{Re}\left(k'r'\right) - l\pi/2\right)\right] \times \exp\left[\text{Im}\left(k'r'\right)\right]$, with $\text{Im}\left(k'r'\right) < 0$. Since $|\text{Im}\left(k'r'\right)| \to \infty$, the integrand vanishes and the integral over the closed contour reduces to the contribution from the real axis. Applying Cauchy's theorem, the integral is given by the residue of the pole at $k' = -(k + i\varepsilon)$, enclosed by this contour. Therefore, we have

$$I_1^{(+)} = -2\pi i\left[-\frac{1}{4\pi}\left(\frac{\hat{j}_l(-kr)\hat{h}_l^{(-)}(-kr')}{-k}\right)\right]. \tag{3.119}$$

Note that the minus sign in the RHS of the above equation arises from the clockwise sense taken in the contour. With the help of Eq. (3.114) and the property of the Ricatti-Haenkel functions

$$\hat{h}_l^{(\pm)}(-z) = (-)^l\hat{h}^{(\mp)}(z),$$

we get

$$I_1^{(+)} = -\frac{i}{2k}\left[\hat{j}_l(kr)\hat{h}_l^{(+)}(kr')\right]. \tag{3.120}$$

The integral $I_2^{(+)}$ can be evaluated in a similar way. In this case, we close the contour with the upper half-circle, where the integrand vanishes, and take it in the counterclockwise sense. The integral over the real axis is then given by the residue of the pole at $k' = k + i\varepsilon$. Namely,

$$I_2^{(+)} = 2\pi i\left[-\frac{1}{4\pi}\left(\frac{\hat{j}_l(kr)\hat{h}_l^{(+)}(kr')}{k}\right)\right] = -\frac{i}{2k}\left[\hat{j}_l(kr)\hat{h}_l^{(+)}(kr')\right]. \tag{3.121}$$

Adding Eqs. (3.120) and (3.121) and inserting the result in Eq. (3.116) we obtain

$$g_l^{(+)}(E; r, r') = -\left(\frac{2\mu}{\hbar^2}\right)\frac{i}{k}\left[\hat{j}_l(kr)\hat{h}_l^{(+)}(kr')\right].$$ (3.122)

For $r' < r$, we proceed in the same way, except that we use Eq. (3.118) for $\hat{j}_l(kr)$, instead of $\hat{j}_l(kr')$. It is clear that we obtain the same result, exchanging r and r'. Analogously, one can also derive the Green's functions with incoming wave boundary condition. These results can be summarized in the compact form

$$g_{0,l}^{(+)}(E; r, r') = -\left(\frac{2\mu}{\hbar^2}\right)\frac{i}{k}\left[\hat{j}_l(kr_<)\hat{h}_l^{(+)}(kr_>)\right],$$ (3.123)

where $r_<$ and $r_>$ stand, respectively, for the smaller and the larger of the two radial distances r and r'. Using the properties $g_{0,l}^{(-)} = \left(g_{0,l}^{(+)}\right)^*$ and $\hat{h}_l^{(-)} = \left(\hat{h}_l^{(+)}\right)^*$, the above result can be generalized in the form

$$g_{0,l}^{(\pm)}(E; r, r') = -\left(\frac{2\mu}{\hbar^2}\right)\frac{i}{k}\left[\hat{j}_l(kr_<)\hat{h}_l^{(\pm)}(kr_>)\right].$$

Using the explicit form of $g_{0,l}^{(+)}(E; r, r')$ in Eq. (3.92) we can investigate the asymptotic form of $u_l(k, r)$. For large r, i.e. outside the range of the potential, we can write

$$u_l(k, r \to \infty) = \hat{j}_l(kr) + i\hat{h}_l^{(+)}(kr)\left[-\frac{1}{k}\left(\frac{2\mu}{\hbar^2}\right)\int_0^\infty dr'\,\hat{j}_l(kr')V\left(r'\right)u_l(k, r')\right].$$ (3.124)

According to Eq. (3.97), we can substitute the quantity within square brackets by $-\pi T_l(E)$. Thus, we get

$$u_l(k, r \to \infty) = \hat{j}_l(kr) - i\hat{h}_l^{(+)}(kr)\pi T_l(E).$$

Using Eq. (3.118) we can write

$$u_l(k, r \to \infty) = \frac{1}{2}\left[\hat{h}_l^{(-)}(kr) + \hat{h}_l^{(+)}(kr)\right] - i\hat{h}_l^{(+)}(kr)\pi T_l(E)$$
$$= \frac{1}{2}\left[\hat{h}_l^{(-)}(k'r') - (1 - 2i\pi T_l(E))\,\hat{h}_l^{(+)}(kr)\right],$$

or, with the help of Eq. (3.109),

$$u_l(k, r \to \infty) \propto \hat{h}_l^{(-)}(k'r') - S_l(E)\hat{h}_l^{(+)}(kr).$$ (3.125)

This is the correct asymptotic form.

3.10 Collision of particles with spin

In the preceding sections we have dealt with spherically symmetric potentials and spinless particles, so that the only particular space orientation was that of the incident beam. In such cases, the partial-wave expansion of the wave function contains only $m = 0$ states and the scattering amplitude is axially symmetric. The situation is more complicated when the projectile or/and the target has/have internal degrees of freedom and thus the intrinsic state does not behave as a scalar with respect to rotations. This occurs when the system has intrinsic spin or any other kind of angular momentum. An example is a deformed molecule with rotational states. Also in such cases we loosely use the term 'spin' for the intrinsic angular momentum. For simplicity, we consider collisions of a spinless target with a projectile with spin s. The incident beam is polarized with z-component equal to ν_0 and we assume that the experimental setup is capable of distinguishing among the possible final values of the spin component along the z-axis, ν. The intrinsic states, $|s\nu\rangle$, have the properties

$$S^2|s\nu\rangle = \hbar^2 s(s+1)|s\nu\rangle, \qquad S_z|s\nu\rangle = \hbar\nu|s\nu\rangle.$$ (3.126)

In the time-independent approach, the incident projectile is described by the free wave function

$$\phi_{\mathbf{k}_0\nu_0}(\mathbf{r}) = \frac{1}{(2\pi)^{3/2}}e^{i\mathbf{k}_0\cdot\mathbf{r}}|s\nu_0\rangle$$ (3.127)

and the scattering solution becomes

$$\psi_{\mathbf{k}_0\nu_0}^{(+)}(\mathbf{r}) = \phi_{\mathbf{k}_0\nu_0}(\mathbf{r}) + \psi_{\mathbf{k}_0\nu_0}^{sc}(\mathbf{r}),\tag{3.128}$$

where

$$\psi_{\mathbf{k}_0\nu_0}^{sc}(\mathbf{r}) = \sum_{\nu=-s}^{+s}\psi_{\mathbf{k}_0\nu_0}^{sc}(\mathbf{r},\nu)\,|s\nu\rangle.\tag{3.129}$$

At asymptotic distances, $\psi_{\mathbf{k}_0\nu_0}^{sc}(\mathbf{r})$ takes the form

$$\psi_{\mathbf{k}_0\nu_0}^{sc}(\mathbf{r}) \to \frac{1}{(2\pi)^{3/2}}\sum_{\nu=-s}^{+s}f_{\nu\nu_0}(\Omega)\frac{e^{ikr}}{r}\,|s\nu\rangle,\tag{3.130}$$

and the elastic cross section for a final state with spin component along the z-axis with value ν is

$$\frac{d\sigma_\nu(\Omega)}{d\Omega} = |f_{\nu\nu_0}(\Omega)|^2.\tag{3.131}$$

The distribution of the cross section over ν is determined by the spin-dependence of the interaction. Clearly, when the potential V is independent of spin, we have

$$f_{\nu\nu_0}(\Omega) = f_{\nu_0\nu_0}(\Omega)\delta_{\nu,\nu_0}.$$

In this case, owing to the rotational invariance of the Hamiltonian, the scattering amplitude is independent of ν_0. Namely,

$$f_{\nu_0\nu_0}(\Omega) \equiv f(\Omega).$$

As a more involved illustration, we consider the case where the potential is the sum of spherical and spin-orbit terms, in the form

$$V(r) = V_0(r) + V_{SO}(r)\mathbf{L}\cdot\mathbf{S} \equiv V_0(r) + \frac{1}{2}V_{SO}(r)\left[\mathbf{J}^2 - \mathbf{L}^2 - \mathbf{S}^2\right].\tag{3.132}$$

Above, $\mathbf{J}$ is the total angular momentum operator

$$\mathbf{J} = \mathbf{L} + \mathbf{S}.\tag{3.133}$$

To carry out partial-wave expansions, we look for a set of simultaneous eigenfunctions of the Hamiltonian and angular momentum operators. These states are $\psi_{Elm}(\mathbf{r}) = Y_{lm}(\hat{\mathbf{r}})u_l(k,r)/r$, which are eigenfunctions of H, $\mathbf{L}^2$ and L_z. Here this set is no longer suitable since L_z does not commute with the spin-orbit potential contained in H. Now the appropriate set of operators is $\{H,\mathbf{L}^2,\mathbf{S}^2,\mathbf{J}^2,J_z\}$ and the eigenfunctions are[*]

$$\psi_{ElJM}(\mathbf{r}) = \mathcal{Y}_{lJM}(\hat{\mathbf{r}})\frac{u_l^J(k,r)}{r},\tag{3.134}$$

where

$$\mathcal{Y}_{lJM}(\hat{\mathbf{r}}) = \sum_{m\nu}\langle lms\nu\,|JM\rangle\,Y_{lm}(\hat{\mathbf{r}})\,|s\nu\rangle,\tag{3.135}$$

and $u_l^J(k,r)$ are the solutions of the radial equation

$$-\frac{\hbar^2}{2\mu}\left[\frac{d^2}{dr^2} - \frac{l(l+1)}{r^2}\right]u_l^J(k,r) + V_0(r)u_l^J(k,r)\tag{3.136}$$

$$+\left[\frac{1}{2}V_{SO}(r)\left(J(J+1) - l(l+1) - s(s+1)\right)\right]u_l^J(k,r) = Eu_l^J(k,r).$$

---

[*]Since the spin quantum number is fixed at the value s, we do not include it in the state labels, although the eigen-functions generally depend on s.

Above, $\langle lms\nu | JM \rangle$ are the usual Clebsh-Gordan coefficients of angular momentum coupling. Note that in the present case the radial equation depends on the total angular momentum quantum number, J. At large radial distances, the radial wave function can be written

$$u_l^J(k,r) \to \frac{i}{2} \left[e^{-i(kr-l\pi/2)} - S_l^J e^{i(kr-l\pi/2)} \right] \tag{3.137}$$

$$\equiv \frac{i}{2} \left[e^{-i(kr-l\pi/2)} - e^{2i\delta_l^J} e^{i(kr-l\pi/2)} \right]. \tag{3.138}$$

As one would expect, the phase-shifts or the components of the S-matrix are now J-dependent. The radial equation can be solved numerically, following the steps described in Chapter 2. The phase-shifts are then extracted from the asymptotic values of the logarithmic derivatives of the radial wave functions.

The scattering wave function can be expanded in the set of states ψ_{ElJM}, as

$$\psi_{\mathbf{k}_0\nu_0}^{(+)}(\mathbf{r}) = \sum_{lJM} A_{lJM} \mathcal{Y}_{lJM}(\hat{\mathbf{r}}) \frac{u_l^J(k,r)}{r}. \tag{3.139}$$

In the $r \to \infty$ limit, it takes the form

$$\psi_{\mathbf{k}_0\nu_0}^{(+)}(\mathbf{r}) \longrightarrow \frac{e^{-ikr}}{r} X_- + \frac{e^{ikr}}{r} X_+, \tag{3.140}$$

with

$$X_- = \sum_{lJM} \mathcal{Y}_{lJM}(\hat{\mathbf{r}}) \left[-\frac{A_{lJM}}{2i} e^{il\pi/2} \right], \tag{3.141}$$

$$X_+ = \sum_{lJM} \mathcal{Y}_{lJM}(\hat{\mathbf{r}}) \left[\frac{A_{lJM}}{2i} e^{-il\pi/2} S_l^J \right]. \tag{3.142}$$

To arrive at the scattering amplitude, we must compare Eq. (3.140) with the scattering boundary condition expressed in Eqs. (3.128–3.130). Using the Bauer expansion for the plane-wave and the asymptotic form of the Bessel function, the scattering boundary condition can be re-written in the form

$$\psi_{\mathbf{k}_0\nu_0}^{(+)}(\mathbf{r}) \longrightarrow \frac{e^{-ikr}}{r} X_- + \frac{e^{ikr}}{r} X_+, \tag{3.143}$$

with

$$X_- = \sum_{\nu lm} |s\nu\rangle\, Y_{lm}(\hat{\mathbf{r}}) \left[-\frac{\sqrt{\pi(2l+1)}}{ik(2\pi)^{3/2}} (-1)^l \delta_{m,0} \delta_{\nu_0,\nu} \right]; \tag{3.144}$$

and

$$X_+ = \sum_{\nu} |s\nu\rangle \left[\frac{f_{\nu\nu_0}(\Omega)}{(2\pi)^{3/2}} + \sum_{lm} Y_{lm}(\hat{\mathbf{r}}) \frac{\sqrt{\pi(2l+1)}}{ik(2\pi)^{3/2}} \delta_{\nu,\nu_0} \delta_{m,0} \right]. \tag{3.145}$$

Above, we have used the identity

$$P_l(\cos\theta) = \sqrt{4\pi/(2l+1)}\, Y_{l0}(\hat{r}) = \sum_m \sqrt{4\pi/(2l+1)}\, Y_{lm}(\hat{\mathbf{r}}) \delta_{m,0}. \tag{3.146}$$

We use the equivalence of Eqs. (3.140) and (3.143) to determine the expansion coefficients A_{lJM}. Using the inverse transformation of Eq. (3.135), namely

$$Y_{lm}(\hat{\mathbf{r}})\, |s\nu_0\rangle = \sum_{JM} \langle JM | lms\nu_0 \rangle\, \mathcal{Y}_{lJM}(\hat{\mathbf{r}}),$$

we can rewrite X_- as

$$X_- = \sum_{lJM} \mathcal{Y}_{lJM}(\hat{\mathbf{r}}) \left[-\langle J\nu_0 | l0s\nu_0 \rangle \frac{\sqrt{\pi(2l+1)}}{ik(2\pi)^{3/2}} i^l e^{il\pi/2} \delta_{M,\nu_0} \right]. \tag{3.147}$$

Equaling Eqs. (3.147) and (3.141) we obtain

$$A_{lJM} = \langle J\nu_0 | l0s\nu_0 \rangle \frac{2\sqrt{\pi(2l+1)}}{k\,(2\pi)^{3/2}} i^l \delta_{M,\nu_0}. \tag{3.148}$$

We now determine the components of the scattering amplitude, $f_{\nu\nu_0}(\Omega)$. Firstly, it is necessary to modify Eq. (3.142), replacing $Y_{lJM}(\hat{\mathbf{r}})$ by its expansion in the set of decoupled angular momentum states, $Y_{lm}(\hat{\mathbf{r}})|s\nu_0\rangle$ (Eq. (3.135)). Following this procedure and using the explicit values of the coefficients A_{lJM}, one obtains

$$X_+ = \sum_{\nu} |s\nu\rangle \left[\sum_{l,m} Y_{lm}(\hat{\mathbf{r}}) \frac{\sqrt{\pi(2l+1)}}{ik\,(2\pi)^{3/2}} \sum_J \langle lms\nu | J\nu_0 \rangle S_l^J \langle J\nu_0 | l0s\nu_0 \rangle \right]. \tag{3.149}$$

Comparing Eqs. (3.145) and (3.149), we finally obtain the partial-wave expansion for the scattering amplitudes

$$f_{\nu\nu_0}(\Omega) = \frac{1}{ik} \sum_{lm} \sqrt{\pi(2l+1)} Y_{lm}(\hat{\mathbf{r}}) \left[\bar{S}_{lm\nu} - \delta_{\nu,\nu_0}\delta_{m,0} \right], \tag{3.150}$$

where

$$\bar{S}_{lm\nu} = \sum_J \langle lms\nu | J\nu_0 \rangle S_l^J \langle J\nu_0 | l0s\nu_0 \rangle. \tag{3.151}$$

In practical situations, we must solve numerically the radial equation for $l = 0, 1, \cdots, l_{\max}$. For the partial-wave expansion to converge, $l_{\max}$ should satisfy the usual condition $l_{\max} \gg kR$, where R is the range of the potential. For each partial wave, the equation must be solved $2s+1$ times, one for each J-value. Therefore, the procedure to obtain the scattering amplitude is very similar to the one described in Chapter 2, except that, for the same potential range, the number of times the radial wave equation has to be solved is multiplied by $2s+1$.

In a more general situation, the spin-dependence of the potential could be more complicated than a spin-orbit term and it may happen that the operators L^2 and S^2 do not separately commute with the Hamiltonian. The algebra becomes then considerably more difficult. For problems of this kind, we refer to the textbook of Satchler [4].

For unpolarized beams and inclusive experiments, where the final component of the spin along the z-axis cannot be distinguished, the cross section should be averaged over the initial spin orientation and summed over the final spin state. Namely

$$\frac{d\sigma(\Omega)}{d\Omega} = \frac{1}{2s+1} \sum_{\nu\nu_0} |f_{\nu\nu_0}(\Omega)|^2 = \frac{1}{2s+1} \mathrm{Tr}(f^\dagger f).$$

3.11   Collisions of identical particles

If the projectile and the target are identical particles, quantum mechanics imposes a few changes in the results of the previous chapters. In first place, when a particle reaches a detector at the orientation $\hat{\mathbf{r}}$ with respect to the CM-frame, the experiment cannot tell if this particle is the projectile or the target. On the other hand, momentum conservation guarantees that whenever one of the particles emerges at the orientation $\hat{\mathbf{r}} \equiv \theta, \varphi$, the other particle emerges at the opposite orientation, i.e. $-\hat{\mathbf{r}} \equiv \pi - \theta, \varphi + \pi$. Therefore, the cross sections for these orientations will be mixed in some way. From the point of view of classical mechanics, the cross section, which we denote σ_{inc}, is an incoherent sum of the cross sections for the corresponding scattering of discernible particles at these orientations*. Namely[†]

$$\sigma_{inc}(\theta, \varphi) = \sigma(\theta, \varphi) + \sigma(\pi - \theta, \varphi + \pi). \qquad (3.152)$$

Note that the total elastic scattering cross section obtained by angular integration of the above equation is twice that for the collision of discernible particle under the same conditions. This result is not surprising since σ_{inc} contains equal contributions from the projectile and from the target particles. The quantum mechanical description of such collisions is more complicated, as discussed below.

In quantum mechanics, the total wave function for pairs of identical particles with integer spins, i.e. two bosons, must be symmetric with respect to the exchange of these particles, while in the case of particles with half-integer spin, i.e. two fermions, it must be anti-symmetric.

The wave function of a particle with spin can be expressed in terms of the set of coordinates $q \equiv \mathbf{r}, \mu$, where $\mathbf{r}$ is its vector position and μ is the spin component along the z-axis. In collisions of identical particles, the wave function must be an eigenstate of the exchange operator P_{pt}, given by

$$P_{pt} = P_{pt}^{\mathbf{r}} P_{pt}^{s},$$

where $P_{pt}^{\mathbf{r}}$ exchanges $\mathbf{r}_p \leftrightarrows \mathbf{r}_t$ and P_{pt}^{s} exchanges $\mu_p \leftrightarrows \mu_t$. Namely

$$P_{pt} \psi_\lambda(q_p, q_t) = \psi_\lambda(q_t, q_p) = \lambda \psi_\lambda(q_p, q_t),$$

with $\lambda = 1$ for boson-boson collisions and $\lambda = -1$ for fermion-fermion collisions. Symmetric $(\psi_+(q_p, q_t))$ or anti-symmetric $(\psi_-(q_p, q_t))$ wave functions can be generated with the help of the projector $\mathcal{A}_\lambda$, given by

$$\mathcal{A}_\lambda = \frac{1}{2} [1 + \lambda P_{pt}]. \qquad (3.153)$$

Namely,

$$\psi_+(q_p, q_t) = N\mathcal{A}_+ \psi(q_p, q_t) = \frac{1}{\sqrt{2}} [\psi(q_p, q_t) + \psi(q_t, q_p)] \qquad (3.154)$$

$$\psi_-(q_p, q_t) = N\mathcal{A}_- \psi(q_p, q_t) = \frac{1}{\sqrt{2}} [\psi(q_p, q_t) - \psi(q_t, q_p)], \qquad (3.155)$$

where $N = \sqrt{2}$ is an appropriate normalization factor.

We consider here the simple situation were the wave function in the spin space is symmetric with respect to projectile-target exchange. This condition is trivially satisfied when the identical particles have spin $s = 0$. The $s \neq 0$ case is usually more complicated. However, a simple situation is also encountered when both the projectile and the target are polarized with their spins aligned. The total spin is then $S = 2s$, and the spin wave function is fully symmetric with respect to projectile-target exchange. The spin degree of freedom can then be ignored and the eigenvalue λ is given by the exchange symmetry in the position space. The wave function should then be written as

$$\psi_\pm(\mathbf{r}_p, \mathbf{r}_t) = \frac{1}{\sqrt{2}} [\psi(\mathbf{r}_p, \mathbf{r}_t) \pm \psi(\mathbf{r}_t, \mathbf{r}_p)], \qquad (3.156)$$

with "+" for bosons and "-" for fermions.

Since the usual treatment of the scattering problem replaces the position vectors $\mathbf{r}_p, \mathbf{r}_t$ by those of the center of mass and projectile-target separation, $\mathbf{R}_{cm}, \mathbf{r}$, given by the transformations,

$$\mathbf{r}_p; \mathbf{r}_t \to \mathbf{R}_{cm} = \frac{\mathbf{r}_p + \mathbf{r}_t}{2}; \quad \mathbf{r} = \mathbf{r}_p - \mathbf{r}_t, \qquad (3.157)$$

---

*The cross-section $d\sigma_{inc}(\theta, \varphi)/d\Omega$ is only classical in the sense that the contributions from the target and the projectile to the scattering amplitudes are summed incoherently. The cross sections $d\sigma(\theta, \varphi)/d\Omega$ and $d\sigma(\pi - \theta, \pi + \varphi)/d\Omega$ are evaluated by quantum mechanics.

[†]Henceforth, we adopt the simplified notation: $d\sigma(\theta)/d\Omega \to \sigma(\theta)$.

we must discuss the effects of exchange on these variables. Inspection of Eq. (3.157) indicates that they transform as

$$P_{pt}^{\dagger} \mathbf{R}_{cm} P_{pt} = \mathbf{R}_{cm}, \qquad P_{pt}^{\dagger} \mathbf{r} P_{pt} = -\mathbf{r}. \tag{3.158}$$

In this way, we have

$$P_{pt} = \Pi,$$

where Π is the parity operator, which carries out space inversion of the vector $\mathbf{r}$. Eq. (3.156) then yields

$$\psi_{\pm}(\mathbf{r}) = \frac{1}{\sqrt{2}} \left[\psi(\mathbf{r}) \pm \psi(-\mathbf{r}) \right]. \tag{3.159}$$

Therefore, the exchange symmetry requires that the wave function has even parity, $\pi = +$, in the case of bosons, or $\pi = -$, in the case of fermions.

Equation (3.159) leads to a modification of the scattering boundary condition. Writing

$$\psi_{\pm}^{(+)}(\mathbf{r}) = \phi_{\pm}(\mathbf{r}) + \psi_{\pm}^{sc}(\mathbf{r}),$$

the asymptotic forms of the incident plane wave and the scattered wave are respectively[‡]

$$\phi_{\pm}(\mathbf{r}) \rightarrow \frac{1}{(2\pi)^{3/2}} \left[\frac{e^{i\mathbf{k}_0 \cdot \mathbf{r}} \pm e^{-i\mathbf{k}_0 \cdot \mathbf{r}}}{\sqrt{2}} \right] \tag{3.160}$$

and

$$\psi_{\pm}^{sc}(\mathbf{r}) \rightarrow \frac{1}{(2\pi)^{3/2}} \left[\frac{f(\theta) \pm f(\pi - \theta)}{\sqrt{2}} \right] \frac{e^{ikr}}{r}. \tag{3.161}$$

Above, we have assumed that the potential is spherically symmetric so that the wave function has axial symmetry and we can write $f(\hat{\mathbf{r}}) \equiv f(\theta, \varphi) = f(\theta)$ and $f(-\hat{\mathbf{r}}) = f(\pi - \theta)$. Next, we must consider the implications of these modifications on the cross section. Inspecting Eq. (3.160), we see that now there are two incident currents. One approaching from the left, arising from $\exp[i\mathbf{k}_0 \cdot \mathbf{r}]$, and one approaching from the right, arising from $\exp[-i\mathbf{k}_0 \cdot \mathbf{r}]$. The incident flux should then have contributions from these two currents. Since the opposite signs of the currents are compensated by the opposite orientation of the normals, the two contributions are identical. In this way one gets a factor 2 that cancels the factor $1/2$ arising from the extra $1/\sqrt{2}$ in the normalization of $\phi_{\pm}(\mathbf{r})$. Therefore, the incident flux is unchanged with respect to the value obtained in Chapter 1. We now consider the flux scattered onto the detector. Comparing to section (1.11), one should make the replacement

$$|f(\theta)|^2 \rightarrow \left| \frac{f(\theta) \pm f(\pi - \theta)}{\sqrt{2}} \right|^2 = \frac{1}{2} |f(\theta) \pm f(\pi - \theta)|^2,$$

where f is the scattering amplitude for discernible particles under the same conditions. Since the factor $1/2$ is cancelled by the factor 2 arising from the indiscernibility of projectile and target, discussed in the beginning of the present section, we get the cross section

$$\sigma_{\pm}(\theta) = |f_{\pm}(\theta)|^2 = |f(\theta) \pm f(\pi - \theta)|^2. \tag{3.162}$$

Above, we have introduced the notation

$$f_{\pm}(\theta) = f(\theta) \pm f(\pi - \theta). \tag{3.163}$$

Evaluating Eq. (3.162) and comparing with Eq. (3.152), we can write

$$\sigma_{\pm}(\theta) = \sigma_{inc}(\theta) + \sigma_{int}(\theta), \tag{3.164}$$

where $\sigma_{int}(\theta)$ is the interference term

$$\sigma_{int}(\theta) = \pm 2 \operatorname{Re} \left\{ f^*(\theta) f(\pi - \theta) \right\}. \tag{3.165}$$

[‡]Henceforth, we are adopting the normalization constant $A = (2\pi)^{-3/2}$.

This term, which has no classical analog, plays a very important role. While $\sigma_{inc}(\theta)$ is independent of the spin, $\sigma_{int}(\theta)$ is strongly spin-dependent.

Direct inspection of Eqs. (3.152), (3.165), and (3.164), indicates that the cross sections $\sigma_{inc}(\theta)$, $\sigma_{int}(\theta)$ and $\sigma_{\pm}(\theta)$ are symmetric with respect to $\theta = \pi/2$. This feature is connected with an important property of the partial wave expansion of $f_{\pm}(\theta)$. For simplicity, we discuss the simple case of scattering from a short-range spherically symmetric potential. Using the property of Legendre Polynomials

$$P_l(\cos(\pi - \theta)) = (-)^l P_l(\cos\theta),$$

we get

$$f_{\pm}(\theta) = \frac{1}{2ik} \sum_l [(2l + 1)P_l(\cos\theta)(S_l - 1)] \times \left(1 \pm (-)^l\right). \tag{3.166}$$

This result is analogous to the usual partial-wave expansion, except for the factor $\left(1 \pm (-)^l\right)$ within the partial-wave summation. In boson-boson (fermion-fermion) collisions, this factor doubles the contribution from even-waves (odd-waves) and eliminates those from odd-waves (even-waves). This is the justification for excluding odd-waves in the calculations for identical spinless bosons.

3.12 Scattering of clusters of identical fermions

In typical atomic and nuclear collisions the projectile and/or the target are composite particles. Usually, they are clusters of fermions*. In this way, the Pauli principle requires that the wave function be anti-symmetric with respect to the exchange of any two identical fermions, either in the same cluster or in different ones. If the projectile contains N_p fermions with coordinates $r_1, r_2, \cdots, r_{N_p}$ and the target contains N_t particles with position vectors $r_{N_p+1}, r_{N_p+2}, \cdots, r_{N_p+N_t}$, the system's wave function can be written as

$$\Psi(\mathbf{r}_1, \cdots, \mathbf{r}_{N_p}; \mathbf{r}_{N_p+1}, \cdots, \mathbf{r}_N) = \mathcal{A}\left\{\Phi_p(\mathbf{r}_1, \cdots, \mathbf{r}_{N_p})\Phi_t\left(\mathbf{r}_{N_p+1}, \cdots, \mathbf{r}_N\right)\right\}. \tag{3.167}$$

Above, Φ_p and Φ_t are respectively the anti-symmetrized intrinsic wave functions of the projectile and the target, and $N = N_p + N_t$. $\mathcal{A}$ is the many-body anti-symmetrization operator, which accounts for the $N_p!N_t!/N!$ different ways of permuting fermions between the two clusters. It can written as

$$\mathcal{A} = \sum (-)^{p_\alpha} P_\alpha, \tag{3.168}$$

where the operator P_α exchanges fermion coordinates until the permutation α is reached. The sign $(-)^{p_\alpha}$ is positive for permutations obtained through an even number of fermion pair exchange and negative otherwise.

Solving the Schrödinger equation for the many-body scattering problem is, in most cases, a hopeless task. However, the effects of Pauli principle can be included in approximations like Wheeler's *Resonating Group Method* (RGM) ([5] and references therein). The RGM avoids the complications of a coupled-channel problem, approximating the intrinsic wave functions of the projectile and the target by the Slater determinants corresponding to their ground states when they are far apart. Furthermore, these intrinsic states are kept frozen along the collision. To derive the RGM equation, one introduces the coordinates of the projectile's and target's centers of mass,

$$\mathbf{r}_p = \frac{1}{N_p}\left(\mathbf{r}_1 + \cdots + \mathbf{r}_{N_p}\right); \quad \mathbf{r}_t = \frac{1}{N_t}\left(\mathbf{r}_{N_p+1} + \cdots + \mathbf{r}_N\right), \tag{3.169}$$

---

*In the case of nuclei, there are two different fermions: protons and neutrons. However, with the isotopic spin formalism they are treated as different states of the same particle – the nucleon.

and the intrinsic coordinates[*]

$$\xi_i = \mathbf{r}_i - \mathbf{r}_p; \quad i = 1, \cdots, N_p$$
$$\xi_j = \mathbf{r}_j - \mathbf{r}_t; \quad j = N_p + 1, \cdots, N, \tag{3.170}$$

and then adopts the ansatz

$$\Psi(\mathbf{r}_1, \cdots, \mathbf{r}_N) = \Psi_{CM}(\mathbf{R}_{cm}) \cdot \mathcal{A}\left\{\psi(\mathbf{r})\Phi_p(\xi_1, \cdots, \xi_{N_p})\Phi_t\left(\xi_{N_p+1}, \cdots, \xi_N\right)\right\}. \tag{3.171}$$

It is convenient to re-write this ansatz in the form

$$\Psi = \int d^3\mathbf{r}'' \psi(\mathbf{r}'')\Phi_{\mathbf{r}''}, \tag{3.172}$$

with

$$\Phi_{\mathbf{r}''} = \Psi_{CM}(\mathbf{R}_{cm}) \cdot \mathcal{A}\left\{\delta(\mathbf{r} - \mathbf{r}'')\Phi_p(\xi_1, \cdots, \xi_{N_p})\Phi_t\left(\xi_{N_p+1}, \cdots, \xi_N\right)\right\}. \tag{3.173}$$

In Eq. (3.171), $\mathbf{R}_{cm}$ is the CM position vector and $\mathbf{r}$ is the projectile-target separation, related to $\mathbf{r}_p$ and $\mathbf{r}_t$ through Eq. (3.157).

The relative wave function ψ, which contains the relevant information for the scattering cross section, is then determined by the variational condition

$$\delta\left[\frac{\langle\Psi|\,H\,|\Psi\rangle}{\langle\Psi\,|\Psi\rangle}\right] = 0, \tag{3.174}$$

where H is the many-body collision Hamiltonian

$$H = \left[\sum_{i=1}^N -\frac{\hbar^2}{2\mu}\nabla_{\mathbf{r}_i}^2\right] + \left[\sum_{i>j=1}^N v(\mathbf{r}_i - \mathbf{r}_j)\right] - T_{CM}. \tag{3.175}$$

Note that the kinetic energy of the center of mass has been extracted from the collision Hamiltonian. In Eq. (3.171), the variations are taken with respect to the relative wave function, $\delta \equiv \delta\left[\psi(\mathbf{r})\right]$. The RGM equation is obtained inserting the wave function of Eq. (3.172) in the variational principle. Using the property $\mathcal{A}^2 = \mathcal{A}$, the variational principle leads to the RGM equation,

$$\left[-\frac{\hbar^2}{2\mu}\nabla_{\mathbf{r}}^2 + V_D(\mathbf{r})\right]\psi(\mathbf{r}) + \int d^3\mathbf{r}' V_E(\mathbf{r}, \mathbf{r}')\psi(\mathbf{r}') = E\psi(\mathbf{r}). \tag{3.176}$$

Above, $V_D(r)$ is the double-folding potential

$$V_D(\mathbf{r})\delta(\mathbf{r} - \mathbf{r}') = \int d\xi_p d\xi_t d^3\mathbf{r}''\left\{\Phi_p(\xi_p)\Phi_t(\xi_t)\,\delta(\mathbf{r} - \mathbf{r}'')\right\}\mathcal{V}(\mathbf{r}'', \xi_p, \xi_t)\left\{\Phi_p(\xi_p)\Phi_t(\xi_t)\,\delta(\mathbf{r}' - \mathbf{r}'')\right\},$$

where

$$\mathcal{V}(\mathbf{r}, \xi_p, \xi_t) = \sum_{i=1}^{N_p}\sum_{j=N_p+1}^N v(\mathbf{r}_i - \mathbf{r}_j),$$

and $V_E(r, r')$ is the non-local energy-dependent potential

$$V_E(\mathbf{r}, \mathbf{r}') = \int d\xi_p d\xi_t d^3\mathbf{r}''\left\{\Phi_p(\xi_p)\Phi_t(\xi_t)\,\delta(\mathbf{r} - \mathbf{r}'')\right\}(H - E)(\mathcal{A} - 1)\left\{\Phi_p(\xi_p)\Phi_t(\xi_t)\,\delta(\mathbf{r}' - \mathbf{r}'')\right\}, \tag{3.177}$$

---

[*]Note that these intrinsic coordinates are not linearly independent as they are related by the constraints $\sum_P \xi_i = \sum_T \xi_j = 0$. This feature must be taken into account in the evaluation of the RGM potentials which will be introduced later on.

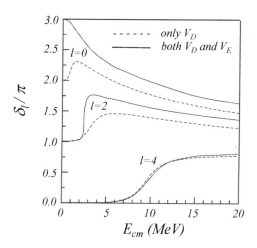

FIGURE 3.2 Phase-shift in $\alpha - \alpha$ scattering, treated as a collision of identical clusters of fermions (solid lines) and as a collision of identical structureless bosons (dashed lines).

which arises from exchange. Note that we have used the short-hand notations $\xi_p \equiv \{\xi_1, \cdots \xi_{N_p}\}$ and $\xi_t \equiv \{\xi_{N_p+1}, \cdots, \xi_N\}$.

Although the direct potential can be calculated without major difficulties, the exchange part is very complicated. The difficulty arises from the fact that as $\mathcal{A}$ exchange particles from the projectile with those from the target, the relative coordinate in the delta functions becomes different combinations of the single particle coordinates. To handle this problem, specific techniques have been developed (see, for example, Ref. [5] and references therein). In typical situations, it has short range, and so does the non-locality.

When the projectile and the target are identical clusters of fermions, the exchange potential contains a local term. It arises from the permutation P_{Tot}, where all projectile's fermions are exchanged with those of the target. In this case, it is convenient to include this term in $V_D(r)$, replacing in Eq. (3.177) $(\mathcal{A}-1) \rightarrow (\mathcal{A}-1 - P_{Tot})$. This leads to the replacement

$$\psi(\mathbf{r}) \rightarrow \frac{1}{\sqrt{2}} \left(\psi(\mathbf{r}) + \lambda\psi(-\mathbf{r})\right),$$

where λ gives the symmetry for projectile-target exchange ($\lambda = +1$ for even N_p and $\lambda = -1$ for odd N_p). The relevance of $V_E(r, r')$ is illustrated below, in the scattering of two alpha particles. We compare the results of the RGM, which includes both the local and the non-local potentials, with those obtained approximating the alpha particles by identical bosons. In the later case, only the projectile-target exchange – corresponding to the permutation P_{Tot} – is taken into account. Phase-shifts for the partial waves $l = 0, 2$ and 4 and collision energies in the range $0 - 20$ MeV are shown in Figure 3.2.

The RGM results are for a *Volkov's V1* nucleon-nucleon interaction [6] and harmonic oscillators orbitals. The Volkov potential is a two-body nucleon-nucleon potential of the form,

$$V = \sum_{i \leq j}^{4} \left(1 - M + M P_{ij}^x\right) V_{ij},$$

$$V_{ij}(r) = V_\alpha \exp\left(-r^2/\alpha^2\right) + V_\beta \exp\left(-r^2/\beta^2\right). \qquad (3.178)$$

In this equation M, α, β, V_α and V_β are parameters adjusted to reproduce the scattering lengths of nucleon-nucleon scattering, and the binding energies of light nuclei. P_{ij}^x is an exchange operator, exchanging the spatial coordinates of the nucleons i and j. The Volkov V1 interaction uses the parameters [6]

$$V_\alpha = -83.3 \text{ MeV}, \quad \alpha = 1.6 \text{ fm}, \quad V_\beta = 144.86 \text{ MeV}, \quad \alpha = 0.82 \text{ fm}.$$

One notices from Figure 3.2 that the non-local exchange potential can be neglected for $l = 4$ (or higher partial-waves), but it is very important for $l = 0$ and 2. In the latter cases, the collision is dominated by small projectile-target separations, where single particle orbitals in the two clusters becomes very similar. In this way, the Pauli Principle is violated, unless the wave function if fully anti-symmetrized. Since the exchange potential, which takes care of this effect, is neglected in the boson-boson approximation, the agreement with the RGM calculation becomes very poor. This point is discussed in detail in Ref. [5].

3.13   Imaginary potentials: absorption cross section

So far we have discussed scattering of structureless particles whose interaction potential is real. In most applications in the physical world involving nuclear, molecular, atomic and other composite system collisions, the process of scattering is a very complicated one and requires the introduction of new concepts which will be expanded upon in later chapters. A first step in this direction is the generalization of the interaction potential into a complex one which will enable us to describe elastic scattering of the composite particles. The imaginary part of this interaction mocks up the loss of flux into non-elastic channels. A fuller description of the complex potential (usually called the *optical potential*) using a microscopic treatment of the collision process will be developed afterwards. An extension to inelastic reactions and application to compound nuclear reactions will be done in the next chapter.

In this section we simply take the interaction to be complex and seek the consequences in so far as the elastic scattering observables are concerned.

Since the Hamiltonian is usually Hermitian, the total current

$$\mathbf{j} = \frac{\hbar}{2\mu i} \left[\psi^* \left(\mathbf{r} \right) \boldsymbol{\nabla} \psi \left(\mathbf{r} \right) - \psi \left(\mathbf{r} \right) \boldsymbol{\nabla} \psi^* \left(\mathbf{r} \right) \right] \tag{3.179}$$

satisfies the continuity equation, which for a stationary state is

$$\boldsymbol{\nabla} \cdot \mathbf{j} = 0.$$

However, this is not valid for complex potentials. In this case, the Hamiltonian has the form

$$H = -\frac{\hbar^2}{2\mu} \nabla^2 + V(r) - iW(r), \tag{3.180}$$

where $W(r)$ is a short-range positive function of r. The term $-iW$ takes care of the flux lost to excited channels, in an approximate way. In this case, the continuity equation must be modified. For this purpose, we write the Schrödinger equation and its complex conjugate,

$$\left[-\frac{\hbar^2}{2\mu} \nabla^2 + V(r) - iW(r) \right] \psi \left(\mathbf{r} \right) = E \psi \left(\mathbf{r} \right) \tag{3.181}$$

$$\left[-\frac{\hbar^2}{2\mu} \nabla^2 + V(r) + iW(r) \right] \psi^* \left(\mathbf{r} \right) = E \psi^* \left(\mathbf{r} \right), \tag{3.182}$$

and evaluate $\psi^* \left(\mathbf{r} \right) \times$ Eq. (3.181) minus $\psi \left(\mathbf{r} \right) \times$ Eq. (3.182). The resulting equation can be written

$$\frac{\hbar}{2\mu i} \left[\psi^* \left(\mathbf{r} \right) \nabla^2 \psi \left(\mathbf{r} \right) - \psi \left(\mathbf{r} \right) \nabla^2 \psi^* \left(\mathbf{r} \right) \right] = -\frac{2}{\hbar} W(r) \left| \psi \left(\mathbf{r} \right) \right|^2, \tag{3.183}$$

or, identifying the LHS of Eq. (3.183) as the divergence of the total current,

$$\boldsymbol{\nabla} \cdot \mathbf{j} = -\frac{2}{\hbar} W(r) \left| \psi \left(\mathbf{r} \right) \right|^2. \tag{3.184}$$

Integrating Eq. (3.184) over the volume of a large sphere with radius r and using the divergence theorem, we obtain,

$$\int d^3 \mathbf{r} \boldsymbol{\nabla} \cdot \mathbf{j} = \int d\mathbf{s} \cdot \mathbf{j} = -N_a = -\frac{2}{\hbar} \int d^3 \mathbf{r} W(r) \left| \psi \left(\mathbf{r} \right) \right|^2, \tag{3.185}$$

where N_a is the number of absorbed particles per unit time (i.e., particles removed from the elastic channel). The *absorption (or reaction) cross section* is obtained dividing N_a by the incident flux $J = |A|^2 v$. Adopting the normalization where the wave function has the asymptotic behavior with $A = 1$, we get

$$\sigma_a = \frac{2}{\hbar v} \int d^3 \mathbf{r} W(r) |\psi(\mathbf{r})|^2, \tag{3.186}$$

or, writing v in terms of E and k and using a more compact notation,

$$\sigma_a = \frac{k}{E} \langle \psi |W| \psi \rangle. \tag{3.187}$$

It should be stressed that Eqs. (3.186) and (3.187) have been derived under the assumption that ψ is normalized such that its asymptotic form has $A = 1$. Therefore, if a different normalization is used, a compensating factor must be used in Eq. (3.187). It should also be stressed that the wave function ψ contains the full effect of W. In practical applications the wave function in Eq. (3.187) should have the appropriate outgoing wave boundary condition. Namely,

$$\sigma_a = \frac{k}{E_k} \left\langle \psi_{\mathbf{k}}^{(+)} \left| W \right| \psi_{\mathbf{k}}^{(+)} \right\rangle. \tag{3.188}$$

The absorption (or reaction) cross section is a very inclusive piece of information. It gives the total flux removed from the elastic channel, without specifying how it is distributed over the open channels. Eq. (3.187) becomes particularly useful when the absorptive potential is a sum of terms, $W = W_1 + W_2 + \cdots$, which can be associated to the coupling with specific channels. In this case, the cross section for some channel i can be estimated by Eq. (3.187), with the same wave function but with the replacement $W \to W_i$.

For practical purposes, σ_a can be expressed in terms of radial integrals. The incident plane wave was written asymptotically [equation (2.38)] as a sum of ingoing and outgoing spherical waves. The expression (2.38) shows that the wave function is modified by the presence of a scattering potential $V(r)$, responsible for the appearance of a phase in the outgoing part of the wave. We saw, however, that the elastic scattering is just one of the channels for the which the reaction can be processed and we denominate it *elastic channel*. The modification of the wave function due to absorption is called the *absorption (or reaction) channel*.

The occurrence of absorption leads to a modification of the outgoing part of (2.38), now not only by a phase factor, but also by a factor that changes its magnitude, indicating that there is a loss of particles in the elastic channel. This can be expressed by

$$\Psi \sim \frac{1}{2i} \sum_{l=0}^{\infty} (2l+1) i^l P_l(\cos\theta) \frac{S_l e^{i(kr - l\pi/2)} - e^{-i(kr - l\pi/2)}}{kr}, \tag{3.189}$$

where the complex coefficient S_l is the factor mentioned above. To calculate the elastic cross section we should place Ψ in the asymptotic form. This procedure yields

$$f(\theta) = \frac{1}{2k} \sum_{l=0}^{\infty} (2l+1) i (1 - S_l) P_l(\cos\theta), \tag{3.190}$$

what results in the differential scattering cross section

$$\frac{d\sigma_{el}}{d\Omega} = |f(\theta)|^2 = \frac{1}{4k^2} \left| \sum_{l=0}^{\infty} (2l+1)(1 - S_l) P_l(\cos\theta) \right|^2. \tag{3.191}$$

The total scattering cross section is calculated using the orthogonality of the Legendre polynomials, what results in

$$\sigma_{el} = \frac{\pi}{k^2} \sum_{l=0}^{\infty} (2l+1) |1 - S_l|^2. \tag{3.192}$$

To calculate the absorption cross section, σ_a, we use the partial-wave expansion of the wave function (with $A = 1$) together with the orthogonality relation for the Legendre polynomials in Eq. (3.188). We obtain the result

$$\sigma_a = \frac{\pi}{k^2} \sum_l (2l + 1)\mathcal{T}_l, \tag{3.193}$$

where $\mathcal{T}_l$ is the transmission coefficient

$$\mathcal{T}_l = \frac{4k}{E_k} \int_0^\infty dr |u_l(k, r)|^2 W(r). \tag{3.194}$$

It can be easily shown that the transmission coefficients are related to the l-components of the S-matrix as

$$\mathcal{T}_l = 1 - |S_l|^2. \tag{3.195}$$

For this purpose, we evaluate (see section (1.11))

$$\sigma_a \equiv \frac{N_a}{J} = -\frac{1}{v} \int d\mathbf{s} \cdot \mathbf{j} = -\frac{r^2}{v} \int d\Omega j_r, \tag{3.196}$$

where j_r is the radial component of the total current,

$$j_r = \frac{\hbar}{2\mu i} \left[\left(\psi_{\mathbf{k}_i}^{(+)} \right)^* \frac{\partial}{\partial r} \left(\psi_{\mathbf{k}_i}^{(+)} \right) - \psi_{\mathbf{k}_i}^{(+)} \frac{\partial}{\partial r} \left(\psi_{\mathbf{k}_i}^{(+)} \right)^* \right] = \frac{\hbar}{\mu} \text{Im} \left\{ \left(\psi_{\mathbf{k}_i}^{(+)} \right)^* \frac{\partial}{\partial r} \left(\psi_{\mathbf{k}_i}^{(+)} \right) \right\}. \tag{3.197}$$

To calculate j_r, we recall that

$$\psi_{\mathbf{k}}^{(+)} = X_- \frac{e^{-ikr}}{r} + X_+ \frac{e^{ikr}}{r}. \tag{2.39}$$

Using the expressions for $X_\pm$ with the coefficients C_l, we can write

$$X_- = -\sum_l \frac{2l+1}{2ik} (-)^l P_l(\cos\theta) \tag{3.198}$$

$$X_+ = \sum_l \frac{2l+1}{2ik} S_l P_l(\cos\theta). \tag{3.199}$$

From Eq. (3.197) we obtain

$$j_r = \frac{v}{r^2} \left(|X_+|^2 - |X_-|^2 \right).$$

Using this result in Eq. (3.196), we get

$$\sigma_a = \int d\Omega |X_-|^2 - \int d\Omega |X_+|^2.$$

Above, the first integral represents the flux of the incident spherical wave while the second integral gives the corresponding emergent flux. The explicit form of $X_\pm$, Eqs. (3.198) and (3.199), together with the orthogonality relation for the Legendre polynomials lead to the result

$$\sigma_a = \frac{\pi}{k^2} \sum_l (2l + 1) \left[1 - |S_l|^2 \right]. \tag{3.200}$$

Comparing Eqs. (3.200) and (3.193), we have the proof of Eq. (3.195).

Note that for real potentials $|S_l| = 1$ and σ_a vanishes and we have pure scattering. The contrary, however, cannot happen, as the vanishing of σ_e also implies in the vanishing of σ_a. In general there is a region of allowed values of S_l for which the two cross sections can coexist. Such region is located below the curve shown in Figure 3.3.

The maximum of σ_a happens for $S_l = 0$, what corresponds total absorption. Let us suppose that the absorption potential is limited to the surface of a nucleus with radius $R \gg \lambda = 1/k$, that

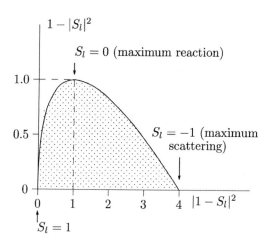

FIGURE 3.3 Area (below the curve) of allowed normalization factors for partial scattering and reaction cross sections as magnitude and phase of partial scattering matrix S_l changes.

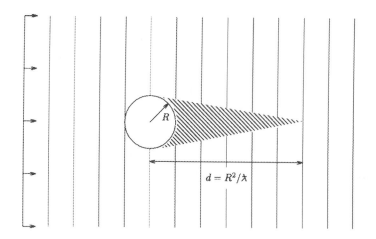

FIGURE 3.4 Scattering of waves around an obstacle showing the diffraction effect.

is, that all the particles with impact parameter smaller than the radius R are absorbed. That is equivalent to saying that all particles are absorbed for $l \leq R/\lambda$. In this case

$$\sigma_a = \pi\lambda^2 \sum_{l=0}^{R/\lambda}(2l+1) = \pi(R+\lambda)^2. \qquad (3.201)$$

This is the value that would be intuitively adequate for the total cross section, i.e., equal to the geometric cross section (the part λ can be understood as an uncertainty in the transverse position of the incident particle). But, we saw above that the presence of elastic scattering is always obligatory. As $S_l = 0$, the scattering cross section is identical to the absorption one, producing a total cross section

$$\sigma = \sigma_a + \sigma_{el} = \pi(R+\lambda)^2 + \pi(R+\lambda)^2 = 2\pi(R+\lambda)^2 \qquad (3.202)$$

that is twice the geometric cross section!

The presence of the scattering part, that turns the result (3.202) apparently strange, can be interpreted as the effect of diffraction of the plane waves at the nuclear surface (Figure 3.4). This

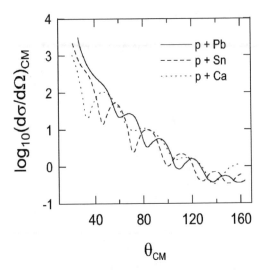

FIGURE 3.5   Elastic scattering angular distribution of protons of 30 MeV on Ca, Sn, and Pb. The theoretically derived curves are adjusted to the experimental data of Ridley and Turner [1].

effect leads to "shadow" behind the nucleus decreasing its apparent diameter so that, at a certain distance, the perturbation caused by the presence of the nucleus disappears and the plane wave is reconstructed. In this situation we can say that the part of the beam which is diffracted has to be the same as the part that is absorbed, justifying the equality of σ_a and σ_e. We can also calculate the extension of the shadow formed behind the nucleus using the uncertainty principle: the limitation of the absorption to a circle with radius R creates an uncertainty in the transverse momentum $\Delta p = \hbar/R$ and the distance d shown in Figure 3.4 is the result of the ratio $R/d = \Delta p/p$, using $p/\hbar = k = 1/\lambda$. This is the same result obtained in the study of *Fraunhofer diffraction* in the absorption of a light beam by a dark object. Only that, in this case, the distances d are large for objects of common use in optics, while in the nuclear case d can be of the order of some nuclear diameters, if the energy of the incident particles is sufficiently small. The diffraction phenomenon appears clearly in the of elastic scattering or inelastic angular distribution (differential cross section as function of the scattering angle). Figure 3.5 exhibits angular distributions for the elastic scattering of 30 MeV protons on ^{40}Ca, ^{120}Sn and ^{208}Pb. The oscillations in the cross sections are characteristic of a Fraunhofer diffraction figure, similar to light scattering by an opaque disk. The angular distances $\Delta\theta$ between the diffraction minima obey in a reasonable way the expression $\Delta\theta = \hbar/pR$ that is a consequence of the considerations above.

An interesting limit is the *Born approximation.* Here the wave function ψ is replaced by a plane wave and thus the absorption cross section of Eq. (3.187) becomes simply

$$\sigma_a = \frac{k}{E_k} \int d^3\mathbf{r} W(r). \tag{3.203}$$

If we further take $W(r)$ to be a step function, namely $W(r) = W_0\Theta(r - R)$, where R is the absorption range, we get the simple expression

$$\sigma_a = \frac{4\pi k}{3E} W_0 R^3 = \left(\frac{8\pi W_0 R^3}{3\hbar}\right)\frac{1}{v}. \tag{3.204}$$

Thus, the above approximation gives an absorption cross section which goes inversely with the velocity of the absorbed particle. This behavior is in fact observed in very low-energy neutron scattering from nuclei as well as in atoms where the phenomenon is referred to as *Wigner's law.* We should warn the reader that in these systems the Born approximation is not valid and the above result should be considered as a mere coincidence.

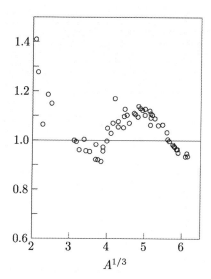

FIGURE 3.6   Dependence of the ratio $\sigma/[2\pi(r_0 A^{1/3})^2]$ constructed from the total cross section σ for neutrons of 14 MeV incident on several nuclei, showing how it fluctuates around the unitary value predicted by (3.202).

We now show the generalization of the optical theorem for complex potentials. The essential difference is that in this case $|S_l| < 1$. We get

$$\frac{4\pi}{k}\,\mathrm{Im}\,\{f(0)\} = \frac{2\pi}{k^2}\sum_l (2l+1)\,[1 - \mathrm{Re}\,\{S_l\}]. \tag{3.205}$$

It is straightforward to check that the same result is obtained adding Eqs. (3.200) and (2.54). Therefore, the Optical Theorem for complex potentials becomes

$$\sigma_{el} + \sigma_a = \frac{4\pi}{k}\,\mathrm{Im}\,\{f(0)\}. \tag{3.206}$$

Special care is required when use is made of Eq. (3.206) for charged particle scattering. In this case the Coulomb interaction renders σ_{el} infinite. In such situations, one relies on the two-potential formula for the scattering amplitude (Eq. 3.58),

$$f(\theta) = f_C(\theta) + f_N(\theta), \tag{3.207}$$

where $f_C(\theta)$ is the Coulomb scattering amplitude and $f_N(\theta)$ is the correction owing to the short-range complex potential. Carrying out partial-wave expansion for $f_N(\theta)$, we get

$$f_N(\theta) = \frac{1}{2ik}\sum_l (2l+1)e^{2i\sigma_l}\left(e^{2i\delta_l} - 1\right). \tag{3.208}$$

Above, σ_l is the Coulomb phase-shift and δ_l is the complex phase-shift arising from the short-range complex potential. The cross section $d\sigma/d\Omega$ can be related to the Coulomb cross section $d\sigma_C/d\Omega$ as

$$\frac{d\sigma}{d\Omega} - \frac{d\sigma_C}{d\Omega} = |f_N(\theta)|^2 + 2\,\mathrm{Re}\,\{f_C^*(\theta)f_N(\theta)\}. \tag{3.209}$$

Following Holdeman and Thaler [7], a modified optical theorem can be obtained by integrating the above equation in the range $\pi > \theta > \theta_0$, where θ_0 is chosen to be very small. This procedure,

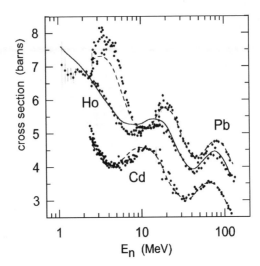

FIGURE 3.7 Total cross section for neutrons incident on cadmium, holmium and lead, showing an oscillatory behavior in its decrease with energy. The curves for cadmium and lead are results of calculations with the optical model [8].

known as the *Sum-of-Differences Method* (SOD), results in an expression which is convenient for the extraction of the σ_a. The SOD form of the optical theorem can be written as

$$\sigma_a + 2\pi \int_{\theta_0}^{\pi} d\theta \sin\theta \left(\frac{d\sigma}{d\Omega} - \frac{d\sigma_C}{d\Omega} \right) = \frac{4\pi}{k} |f_N(\theta_0)| + \sin\left[\arg\left(f_N(\theta_0)\right) - 2\sigma_0 + 2\eta \ln\left(\sin\theta_0/2\right)\right].$$

$$(3.210)$$

Equation (3.210) is the appropriate form of the optical theorem used in the analysis of the elastic scattering of charged particles in the presence of absorption.

The expression (3.202) shows that if a particle were always absorbed when it reached the nucleus, the total cross section would fall monotonically with the energy and grow linearly with $A^{2/3}$. Already with the available experimental data at the end of the decade of 1940, it could be noticed that one had perceptible deviations from these tendencies. In Figures 3.6 and 3.7, we see curves of total cross section of neutron scattering showing an undulatory behavior for its energy dependence as well as for the mass dependence of the target. We shall study in the next chapter a model that explains this behavior, introducing the idea that, when penetrating a nucleus, a nucleon has a probability smaller than 1 of being absorbed, i.e., the nucleus has a certain transparency for the incident beam. This is harmonized with the independent particle model, where a nucleon has an appreciable average free path in the nuclear matter and cannot, in this way, escape the nucleus after a few interactions.

3.14   Appendix 3.A – Analytical properties of the S-matrix

3.14.1   The Jost function

We begin this appendix by introducing the *Jost function*, which plays a central role in the analytical continuation of the S-matrix. We have seen that different normalizations of the radial equation can be used, since the relevant information for the scattering cross section are contained in its asymptotic form, rather than in its normalization. In most cases, the radial wave function is normalized so that outside the potential range it has the form

$$u_l(k, r > R) = -\frac{1}{2}\left[\hat{h}_l^{(-)}(kr) - S_l(k)\hat{h}_l^{(+)}(kr)\right],$$

and they satisfy the orthonormality relation

$$\int_0^\infty u_l^*(k,r)u_l(k',r)dr = \frac{\pi}{2}\delta(k-k').$$

To stress the energy-dependence of the S-matrix, we have used the notation $S_l(k)$, where $k = \sqrt{2\mu E/\hbar^2}$. One notices above an incoming spherical wave plus an outgoing spherical wave. The amplitude of the incoming wave is fixed, while that of the outgoing wave is given by $S_l(k)$. For real potentials the strengths of the two waves are the same and the multiplication by $S_l(k)$ only shifts the phase of the outgoing wave.

As it will become clear below, for some purposes it is convenient to normalize the radial wave function so that it reduces to the Ricatti-Bessel function in the $r \to 0$ limit. Namely,

$$\bar{u}_l(k, r \to 0) = \hat{j}_l(kr). \tag{3.211}$$

Note that we use the notation $\bar{u}_l$ for this solution. Outside the potential range, it must be a linear combination of $\hat{h}_l^{(-)}$ and $\hat{h}_l^{(+)}$. Assuming that the potential is real, these coefficients must have the same modulus so that we can write

$$\bar{u}_l(k, r > R) = -\frac{1}{2}\left[\mathcal{J}_l(k)\hat{h}_l^{(-)}(kr) - \mathcal{J}_l^*(k)\hat{h}_l^{(+)}(kr) \right]. \tag{3.212}$$

The coefficient $\mathcal{J}_l(k)$, is called the *Jost function* and it is uniquely determined by the condition of Eq. (3.211). To find an expression for $\mathcal{J}_l(k)$, we first look for an integral equation for $\bar{u}_l(k,r)$. It can be checked by direct substitution in the radial equation that $\bar{u}_l(k,r)$ satisfies the *Volterra type* integral equation

$$\bar{u}_l(k,r) = \hat{j}_l(kr) + \int_0^r dr' \bar{g}_l(r,r')V(r')\bar{u}_l(k,r'), \tag{3.213}$$

with the Green's function

$$\begin{aligned}
\bar{g}_l(r,r') &= \left(\frac{2\mu}{\hbar^2}\right)\frac{i}{k}\left[\hat{j}_l(kr)\hat{n}_l(kr') - \hat{j}_l(kr')\hat{n}_l(kr)\right] \\
&= \left(\frac{2\mu}{\hbar^2}\right)\frac{1}{2k}\left[\hat{h}_l^{(-)}(kr)\hat{h}_l^{(+)}(kr') - \hat{h}_l^{(+)}(kr)\hat{h}_l^{(-)}(kr')\right].
\end{aligned} \tag{3.214}$$

Taking the $r \to \infty$ limit in Eq. (3.213), replacing $\hat{j}_l = (1/2)\left[\hat{h}_l^{(-)} + \hat{h}_l^{(+)}\right]$ and comparing with Eq. (3.212), we get

$$\mathcal{J}_l(k) = 1 + \int_0^\infty dr \hat{h}_l^{(+)}(kr)V(r)\bar{u}_l(k,r). \tag{3.215}$$

It can be shown, that for physical values of k, i.e., real and non-negative, this integral always converges. Since in most relevant cases the effects of the potential for very large r-values are negligible (with the exception of the Coulomb potential), one can approximate $V(r > R) = 0$. In this case, the integrand is always bound and the proof of convergence is immediate. Throughout this section we will assume that this approximation holds so that we can restrict our discussion to finite-range potentials.

3.14.2 Analytical continuation in the complex k- and E-planes

Although only real and non-negative values of k are meaningful to the physical scattering problem, the radial equation can be mathematically generalized as to include complex values of the wave number. The usefulness of such solutions will be discussed in the present section.

We begin by introducing some basic concepts and properties of functions of a complex variable. The first of such concepts is that of an *analytical function*. A function of a complex variable $f(z)$ is said to be analytical at some point z_0, if it has a uniquely defined derivative at this point and also at all points in some neighborhood of z_0. The function is analytical on a domain D if it is analytical at all points on this domain. If the analyticity domain is the full complex z-plane, $f(z)$ is called an *entire function*.

The first property which will be useful for our purposes is that if a function $f(z)$ is analytical on the domain D, the function $f^*(z^*)$ is analytical on the domain D^*. The analyticity of $f(z)$ guarantees the existence of the derivative $df/dz = a + bi$, where a and b are real numbers. If z' is some point on D^*, we have

$$\frac{df^*(z'^*)}{dz'} = \left[\frac{df(z'^*)}{dz'^*}\right]^*. \tag{3.216}$$

Calling in Eq. (3.216) $z'^* = z$, z is a point on D so that we can write

$$\frac{df^*(z'^*)}{dz'} = \left[\frac{df(z)}{dz}\right]^* = (a + bi)^* = a - bi.$$

Therefore, the derivative $df^*(z'^*)/dz'$ does exist and $f^*(z'^*)$ is an analytical function on the domain D^*. Owing to this property it is convenient to rewrite Eq. (3.212) in the form

$$\bar{u}_l(k, r > \bar{R}) = \frac{i}{2}\left[\mathcal{J}_l(k)\hat{h}_l^{(-)}(kr) - \mathcal{J}_l^*(k^*)\hat{h}_l^{(+)}(kr)\right]. \tag{3.217}$$

On the physical region, k is real and Eq. (3.217) is identical to Eq. (3.212). However, the former has an important advantage for complex k-values. When $\mathcal{J}_l(k)$ is an entire function of k, as is the case for real finite-range potentials, the same holds for $\mathcal{J}_l(k^*)$.

We now turn to an important theorem involving complex functions: If a function $f(z)$ is analytical on a domain D and vanishes on a continuous line C contained in this domain, then it vanishes on the whole domain. This theorem has two important consequences. The first one is that the values of an analytical function on a curve C determine the function on any domain which contains this curve. To prove this statement, one makes the hypothesis that two functions $f_1(z)$ and $f_2(z)$ are analytical on a domain D containing a curve C and coincide on this curve. If one define $f(z) = f_1(z) - f_2(z)$, this function is analytical on D and vanishes on C. Therefore, $f(z)$ vanishes on the whole domain and $f_1(z) = f_2(z)$. This property ensures that $\mathcal{J}_l(k)$ has a unique analytical continuation on the complex plane and so does the radial function $\bar{u}_l(k, r)$. The second important consequence of the theorem is the *Schwartz reflection principle*. It states that if an analytical function $f(z)$ is real on the real axis then $f(z) = f^*(z^*)$. In this case, the function $g(z) = f(z) - f^*(z^*)$ vanishes over the real axis. Therefore it vanishes on the whole analyticity domain and we can write $f(z) = f^*(z^*)$. To apply this relation to the Jost function, we use the property that $\mathcal{J}_l(k)$ is real on the imaginary axis*. Next, we introduce a new variable $\kappa = -ik$, which is real for imaginary values of k. Replacing $k = i\kappa$ in the Jost function we get a new function $\mathcal{I}_l(\kappa)$, such that $\mathcal{I}_l(\kappa) = \mathcal{J}_l(k)$. The function $\mathcal{I}_l(\kappa)$ now satisfies the condition for application of the *Schwartz reflection principle* and we can write

$$\mathcal{I}_l(\kappa) = \mathcal{I}_l^*(\kappa^*).$$

Changing back to the variable k, we get back the function $\mathcal{J}_l$ and the above equation becomes

$$\mathcal{J}_l(-k) = \mathcal{J}_l^*(k^*). \tag{3.218}$$

We are interested in the analytical continuation in the complex k-plane of the radial wave functions $\hat{h}_l^{(+)}(kr)$, $\hat{h}_l^{(-)}(kr)$ and $\bar{u}_l(k, r)$, and also of the Jost function $\mathcal{J}_l(k)$. It can easily be shown that these radial wave functions are all entire functions. This can be done through the series expansion solutions of the radial equations (see, e.g. [9]). However, the analyticity of the Jost function must be considered with more care. For this purpose we use its integral form given in Eq. (3.215). The convergence of the integral for complex k-values depends on the radial dependence of the potential. For complex k-values the Riccatti-Haenkel functions have the asymptotic values

$$\hat{h}_l^{(+)}(kr \to \infty) = e^{-r\,\mathrm{Im}\,k}e^{ir\,\mathrm{Re}\,k}, \tag{3.219}$$

$$\hat{h}_l^{(-)}(kr \to \infty) = e^{r\,\mathrm{Im}\,k}e^{-ir\,\mathrm{Re}\,k}. \tag{3.220}$$

---

*This property is proved in textbooks such as [9].

Considering the asymptotic form of $\bar{u}_l$, given in Eq. (3.212), one concludes that the integrand of Eq. (3.215) at large r-values contains terms with orders of magnitude $V(r)e^{-2r\,\mathrm{Im}\,k}$ and $V(r)$. At the lower half of the complex k-plane, $\mathrm{Im}\,k < 0$ and $V(r)e^{-2r\,\mathrm{Im}\,k}$ diverges, if the asymptotic fall off of the potential is not fast enough. In this case, $\mathcal{J}_l(k)$ is not an analytical function of k at the lower half plane. This is the case of potentials falling off as $1/r^n$. In the case of an exponential fall off, namely $V(r \to \infty) = V_0 \exp(-\gamma r)$, the analytical continuation of $\mathcal{J}_l(k)$ can be performed in the portion of the complex plane with $\mathrm{Im}\,k > -\gamma/2$. The analytical continuation can be extended to the whole complex k-plane, if the potential has a finite range. Since in any relevant case, with the exception of the Coulomb potential, the potential can be neglected beyond some value $\bar{R}$, $\mathcal{J}_l(k)$ can be approximated by an entire function.

Using Eq. (3.218), Eq. (3.217) can be put in the form

$$\bar{u}_l(k, r > \bar{R}) = \frac{i}{2}\left[\mathcal{J}_l(k)\hat{h}_l^{(-)}(kr) - \mathcal{J}_l(-k)\hat{h}_l^{(+)}(kr)\right]. \tag{3.221}$$

Comparing Eqs. (2.33) and (3.221), we obtain the relations

$$u_l(k,r) = \frac{\bar{u}_l(k,r)}{\mathcal{J}_l(k)}, \qquad S_l(k) = \frac{\mathcal{J}_l(-k)}{\mathcal{J}_l(k)}, \tag{3.222}$$

and, recalling that $S_l(k) = \exp\left(2i\delta_l(k)\right)$, we get

$$\delta_l(k) = -\arg\left(\mathcal{J}_l(k)\right). \tag{3.223}$$

Since we are dealing with finite-range potentials, $S_l(k)$ is an analytical function of k on the whole complex k-plane, except at the zeroes of the Jost function, where $S_l(k)$ has poles. The poles of the S-matrix play a very important role, as will be discussed in the next subsections.

3.14.3 Bound states

For a bound state with angular momentum l, the radial equation has negative energy, which corresponds to an imaginary wave number. In this case, the wave function outside the potential range has the asymptotic form

$$\varphi_l(k, r > R) = Ce^{-\kappa r}, \tag{3.224}$$

where $\kappa \equiv \mathrm{Im}\,k = \sqrt{-2\mu E/\hbar^2}$. Taking the asymptotic form of Eq. (3.221) and considering Eqs. (3.219) and (3.220) for an imaginary wave number, we see that the wave function $\bar{u}_l$ will have the same asymptotic form as that of φ_l (Eq. (3.224)) when the coefficient $\mathcal{J}_l(k)$ vanishes. Therefore, bound states of the projectile-target system correspond to zeroes of the Jost function on the upper part of the imaginary axis on the complex k-plane. Eq. (3.222) indicates that these zeroes give rise to poles in the S-matrix. In this way, bound states correspond to poles of $S_l(k)$ on the imaginary k-axis.

One frequently discusses the location of poles of the S-matrix on the complex E-plane. If $k = \sqrt{2\mu/\hbar^2}E^{1/2}$ is considered as a function of a complex variable E, it has two branches. Consequently, the domain of this function is a Riemann surface with two sheets. The first and the second sheets lead to k-values on the upper and on the lower halves of the complex k-plane, respectively. The bound state associated with a pole of the S-matrix at $k = k_0 = i\kappa$ corresponds to a pole on the negative part of the real E-axis, at the location $E_0 = \left(\hbar^2/2\mu\right)k_0^2 = -\hbar^2\kappa^2/2\mu$.

An interesting situation occurs when the Jost function has a pole at the origin. For $l \neq 0$ the resulting wave function can be properly normalized so that it corresponds to a physically acceptable state. However, the opposite occurs for $l = 0$. Such unphysical states are called virtual states.

3.14.4 Resonances

We now consider poles of the S-matrix on the lower half of the complex k-plane. It can be shown (see Ref. [10], section (4.4e)) that all zeroes of the Jost function on the upper half of the complex k-plane are located on the imaginary axis. As we have seen, these zeroes correspond to the bound states of the projectile-target system. However, the Jost function may have zeroes on the lower half of the complex k-plane and accordingly the S-matrix will have poles at these points. Eq. (3.218) implies that these zeroes must occur in pairs $k_\pm = \pm a - bi$, where a and b are positive real numbers. Although the zeroes at $k_- = -a - bi$ are irrelevant for scattering theory, those at $k_+ = a - bi$ will have a very important role when they are close to the real k-axis. To see this point, we expand $\mathcal{J}_l(k)$ around k_+ and take the expansion to first order. We get

$$\mathcal{J}_l(k) \simeq \left[\frac{d\mathcal{J}_l(k)}{dk}\right]_{k=a-bi} [(k-a) + bi].$$ (3.225)

If k_+ is close to the real-axis, we can use Eq. (3.225) in Eq. (3.223) for k close to a and get

$$\delta_l(k) = -\arg\{\mathcal{J}_l(k)\} \simeq -\arg\left\{\left[\frac{d\mathcal{J}_l(k)}{dk}\right]_{k=a-bi}\right\} - \tan^{-1}\left(\frac{b}{k-a}\right).$$ (3.226)

Eq. (3.226) shows that the phase-shift is the sum of a constant "background" term

$$\delta_{bg} = -\arg\left\{\left[\frac{d\mathcal{J}_l(k)}{dk}\right]_{k=a-bi}\right\}$$ (3.227)

and a resonant term

$$\delta_{res} = -\tan^{-1}\left(\frac{b}{k-a}\right).$$

The resonant term goes through $\pi/2$ as the wave number reaches the value of the real part of the pole, a. As k moves away from the resonance, the first order expansion no longer holds. It should be pointed out that the typical resonant behavior discussed in section 2.8 is observed when the background phase-shift is very small. This always happen when the resonance energy is much lower than the barrier in the effective potential, resulting from the sum of the potential with the centrifugal term. In this case, the first order expansion of the resonant phase-shift leads to a maximum in the cross section, as represented by the Breit-Wigner shape. However, when the background part is relevant, the pole only gives rise to a rapid change of the cross section as a function of the collision energy. This change may even be a sharp decrease, coming back to the original value after the resonance.

The discussion of resonances is usually carried out in terms of the collision energy. In this case, the poles of the S-matrix on the lower half of the complex k-plane are mapped on the second sheet of the Riemann surface of the complex E-plane. In particular, a pole close to the physical axis ($\mathrm{Re}\,k > 0$) corresponds to a pole on the lower half of the complex E-plane with positive real part and negative imaginary part. It is convenient to write it in the form

$$\bar{E} = E_r - i\frac{\Gamma}{2}.$$ (3.228)

In this way, near the resonance energy, the Jost function can be approximated by the lowest order expansion

$$\mathcal{J}_l(E) \simeq C\left[(E-E_r) + i\frac{\Gamma}{2}\right].$$ (3.229)

Proceeding similarly to the expansion on the complex k-plane, we get the phase-shift

$$\delta_{res} = -\tan^{-1}\left(\frac{\Gamma/2}{E-E_r}\right).$$ (3.230)

If the background phase-shifts for all partial-waves are negligible and the resonance is isolated, the integrated elastic cross section, given by Eq. (2.55), reduces to

$$\sigma_{el} \simeq \frac{4\pi}{k^2}(2l+1)\sin^2\delta_{res}.$$ (3.231)

From Eq. (3.230), we have

$$\sin\delta_{res} = \frac{\tan\delta_{res}}{\sqrt{1+\tan^2\delta_{res}}} = \frac{-\Gamma/2}{\sqrt{(E-E_r)^2 + \Gamma^2/4}},$$

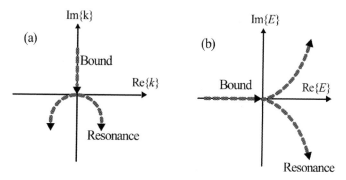

FIGURE 3.8 Poles of the S-matrix on the complex momentum and energy plane as a function of the angular momentum potential depth. The potential is the square well discussed in the text.

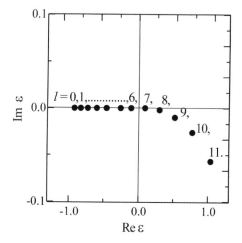

FIGURE 3.9 Poles of the S-matrix on the complex E-plane for $l = 0, 1, \cdots, 11$.

and Eq. (3.231) takes the form

$$\sigma_{el} \simeq \frac{\pi}{k^2}(2l+1)\left[\frac{\Gamma^2}{(E-E_r)^2 + \Gamma^2/4}\right]. \tag{3.232}$$

This is the *Breit-Wigner formula* in the case of S-waves in real potentials.

It is interesting to study the behavior of a pole of the S-matrix in the scattering from an attractive potential as a function of its strength V_0. Let us assume that the potential is strong enough to bind a state with energy $E = -E_0(V_0)$, which corresponds to a pole of the S-matrix on the upper part of the imaginary k-axis, or, equivalently, on the negative part of the real-axis of the first sheet of the Riemann surface on the complex E-plane. As V_0 decreases, so does the binding energy. At some critical value, the bound state becomes a sharp resonance, and the trajectory of the pole goes onto the lower half of the complex k-plane. At this point, it is split in two branches, symmetrically located with respect to the imaginary k-axis. On the complex E-plane, at this critical potential depth, the pole moves to the second sheet of the Riemann surface and the trajectory is also split in two parts. The situation is illustrated in Figure 3.8a, on the k-plane, and in Figure 3.8b for the complex E-plane. In the latter case, the two sheets of the Riemann surface are superimposed. As discussed above, the poles on the left half of the complex k-plane, which correspond to the ones on the upper half of the E-plane are not relevant for the scattering cross section.

The above discussion is of academic interest since in a real collision the depth of the attractive potential is fixed. However, the effective potential, resulting from the sum of a fixed attractive

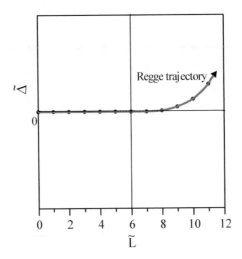

FIGURE 3.10 Illustration of the Regge trajectory of the lowest energy state. The potential is the square-well of the previous section.

potential with the centrifugal term, has its depth decreased as the angular momentum increases. Let us consider a square well deep enough to bind at least one s-state. It should satisfy the condition

$$K_0 R > \frac{\pi}{2} \quad \Rightarrow V_0 > \frac{\pi^2 \hbar^2}{8\mu R^2}.$$

We take the square-well, with $K_0 R = 10$. Figure 3.9 shows the poles of the S-matrix on the complex E-plane associated with the lowest energy state of this well, as a function of the angular momentum. The energy scales are normalized to V_0. For $l = 0, 1, \cdots, 6$ the states are bound, so that the poles are located on the real-axis, at the energy eigenvalues. Above $l = 6$, the effective potentials have no bound states and the poles correspond to resonances. They are located at $\varepsilon = \varepsilon_r - i\gamma/2$, where $\varepsilon_r = E_r/V_0$ and $\gamma = \Gamma/V_0$. The calculated bound state energies are given in Table 2.1. The resonance energies and the corresponding widths are given in Table 2.2.

3.14.5   Analytical continuation in the complex l-plane

Inspecting the differential equation for the radial wave function, it becomes clear that there are no mathematical difficulties to extend it to complex l-values. It is true that radial solutions with complex, or even non-integer, angular momenta would lead to meaningless scattering states, owing to the angular part of the wave function. Nevertheless, such solutions are a useful tool for the calculation of the scattering amplitude. There is, however, one restriction on the values of the complex variable l. For $\text{Re}\, l < -1/2$, the usual boundary condition for a regular solution may be satisfied by solutions which becomes divergent for positive l-values. Therefore, the domain of l must be the semi-plane with $\text{Re}\, l > -1/2$. For a finite range potential, the Jost function, now denoted $J(l, E)$, is an analytical function of l on this semi-plane and of E on the whole complex-plane.

Regge poles and Regge trajectories

One can now consider zeroes of $J(l, E)$ on the complex l-plane, for real values of the collision energy. These zeroes give rise to poles of the S-matrix, $S(l, E)$, on the complex l-plane, which are known as Regge poles. The position of these poles appear as a function of E. As the energy is continuously varied, each Regge pole describe a continuous line on the complex l-plane. These lines are called Regge trajectories. If the potential has bound states with angular momentum $l_1, l_2, \cdots, l_n$ (real and integer values) with energies* $E_1, E_2, \cdots, E_n$ (real and negative values), $l_1, l_2, \cdots, l_n$ are clearly points on the same Regge trajectory. As the energy becomes positive, the Regge trajectory moves away from the real axis and is given by a function $\tilde{l}(E) = \tilde{L}(E) + i\tilde{\Delta}(E)$. It is easy to show that the Regge trajectory has the following properties (see, for example, Ref. [3]):

(1) $\tilde{\Delta}(E)$ is always positive

(2) If for a given energy E_m, $\tilde{L}(E_m) = m$, where m is an integer, and $\tilde{\Delta}(E)$ is small, the phase-shift δ_m has a resonance at this energy.

As an example, we consider the attractive square-well potential of the previous section. According to the above discussion, it leads to Regge poles at the real and integer values $l = 0, 1, \cdots$. and 6, at the negative energies given in Table 2.1. The Regge trajectory passes also on the points $\tilde{l}(E) = 7 + i\tilde{\Delta}(\varepsilon_7), 8 + i\tilde{\Delta}(\varepsilon_8), \cdots, 11 + i\tilde{\Delta}(\varepsilon_{11})$, where $\varepsilon_7, \cdots, \varepsilon_{11}$ are the resonance energies given in Table 2.2. This situation is represented in Figure 3.10.

3.15 Exercises

1. Show that, if $H_0^\dagger = H_0$ and $H^\dagger = H$, the respective continuum eigenstates $|\phi_\mathbf{k}\rangle$ and $|\psi_\mathbf{k}\rangle$ can be normalized so as to satisfy the relations

$$\langle \phi_{\mathbf{k}'} | \phi_\mathbf{k} \rangle = \delta\left(\mathbf{k} - \mathbf{k}'\right), \qquad \langle \psi_{\mathbf{k}'} | \psi_\mathbf{k} \rangle = \delta\left(\mathbf{k} - \mathbf{k}'\right),$$

and that the bound-states can be chosen to satisfy

$$\langle \psi_m | \psi_n \rangle = \delta(m, n), \qquad \langle \psi_m | \psi_\mathbf{k} \rangle = 0.$$

2. Prove the Equation (3.8).

3. Show that the functions given in Eqs. (3.35) and (3.36) satisfy the Schrödinger equation, if the respective Green's function are given by Eqs. (3.37) and (3.38), respectively.

4. Using the properties of the Green's functions show that

$$\left(\psi_\mathbf{k}^{(\pm)}(\mathbf{r})\right)^* = \psi_{-\mathbf{k}}^{(\mp)}(\mathbf{r}).$$

5. Using the definition of the S-matrix, Eq. (3.54), prove its unitary relations, Eqs. (3.55).

6. Starting from the orthonormality of plane waves; $\langle \phi_{\mathbf{k}'} | \phi_\mathbf{k} \rangle = \delta\left(\mathbf{k} - \mathbf{k}'\right)$, show that the Lippmann-Schwinger equations lead to the relations

$$\left\langle \psi_{\mathbf{k}'}^{(\pm)} \middle| \psi_\mathbf{k}^{(\pm)} \right\rangle = \delta\left(\mathbf{k} - \mathbf{k}'\right).$$

7. Prove from their definitions, that S and T matrices are related by

$$S_{\mathbf{k}',\mathbf{k}} = \delta\left(\mathbf{k} - \mathbf{k}'\right) - 2\pi i \delta\left(E_k - E_{k'}\right) T_{\mathbf{k}',\mathbf{k}}.$$

The proof is very similar to those of orthogonality and unitarity given in text and previous problems.

---

*In this example, we are considering the Regge trajectory for the lowest energy states. If there are two or more bound states for a partial wave l, the state E_l referred to in the text is that with lowest energy.

8. The Schrödinger equation with a non-local potential has the form

$$-\frac{\hbar^2}{2\mu}\nabla^2\psi(\mathbf{r}) + \int d^3r' V(\mathbf{r},\mathbf{r}')\,\psi(\mathbf{r}') = E\psi(\mathbf{r}). \qquad (3.233)$$

A special case of a nonlocal is the separable potential

$$V(\mathbf{r},\mathbf{r}') = -\frac{1}{2\pi}\lambda v(r)v(r'). \qquad (3.234)$$

(a) Show that only s-waves are affected by this potential.

(b) What form must the separable potential have to affect all l's?

(c) Establish the integral equation equivalent to Eq. (3.233) including outgoing, scattered wave boundary conditions.

(d) Because of the special form of Eq. (3.234), this integral equation can be reduced to an algebraic equation which can be solved directly. Show that the scattering amplitude for momentum k is:

$$f_E = \frac{4\pi\,|g(k)|^2}{1-2\lambda J(k)/\pi}, \qquad J(k) = \int d^3q\,\frac{g^2(q)}{k^2-q^2+i\epsilon},$$

$$g(k) = \int_0^\infty dr\,r^2\,\frac{\sin kr}{kr}\,v(r).$$

9. Consider the delta-function separable potential

$$V(\mathbf{r},\mathbf{r}') = -4\pi\lambda\delta(r'-a)\,\delta(r-a).$$

(a) What is $V_l(r,r')$ for this potential?

(b) What is $T_l(E)$ for this potential?

(c) Show that the T-matrix considered as a function of the complex energy E, is discontinuous along the real positive axis (the *unitarity* or *right-hand cut*).

10. Consider the separable potential

$$V(\mathbf{r},\mathbf{r}') = \lambda v(r)v(r').$$

(a) Derive (or verify) that the off-energy-shell T matrix for this potential has the form

$$\langle\mathbf{k}'\,|T(E)|\,\mathbf{k}\rangle = \frac{\lambda g(k')g(k)}{1-\lambda J(E)},$$

where $E = k_0^2/2m$ is the energy.

(b) Determine the integral expressions for $g(k)$ and $J(E)$.

(c) Relate the value of the zero-momentum form factor $g(k=0)$ to the scattering length.

(d) Express the bound-state condition for this potential in terms of the integral $J(E)$.

(e) Deduce the number of bound states that occurs in this potential.

(f) Express the condition for this potential to produce a Breit-Wigner resonance at energy $E = k_r^2/2m$.

(g) Show that the width of this resonance is

$$\Gamma = Bg^2(k_r),$$

where B is a constant you should determine.

11. Prove that a **p**-space potential which depends on only the momentum transfer $\mathbf{k}'-\mathbf{k}$ must be local in **r**-space.

12. Show that the **r**-space representation of the potential

$$\left\langle \mathbf{k}' \left| V \right| \mathbf{k} \right\rangle = \left| a + b \left(\mathbf{k}'-\mathbf{k} \right)^2 \right| \widetilde{\rho} \left(\mathbf{k}'-\mathbf{k} \right)$$

is the local potential

$$\left\langle \mathbf{k}' \left| V \right| \mathbf{k} \right\rangle = C\delta \left(\mathbf{r}'-\mathbf{r} \right) \left[a\rho \left(r \right) - b\nabla^2 \rho \left(r \right) \right].$$

13. As seen in the text, solving the Schrödinger equation for a spin-0 particle interacting with a spin-$\frac{1}{2}$ one, via a spin-orbit interaction, is simplified considerably once appropriate spin-angle eigenfunctions are known.

(a) Derive the spin-angle functions $\mathcal{Y}^{(\pm)} \equiv \mathcal{Y}^{l,s=1/2}_{j=l\pm1/2,m=1/2}$.

(b) Show that these functions have the properties:

$$\mathbf{l} \cdot \mathbf{s} \mathcal{Y}^{(\pm)} = \left(\begin{array}{c} \frac{1}{2}l \\ -\frac{1}{2}(l+1) \end{array} \right) \mathcal{Y}^{(\pm)},$$

$$\mathbf{j} \cdot \mathbf{j} \mathcal{Y}^{(\pm)} = \left(l \pm \frac{1}{2} \right) \left(l \pm \frac{1}{2} + 1 \right) \mathcal{Y}^{(\pm)}. \tag{3.235}$$

(c) Use the properties of the spin-angle functions to show that the Schrödinger equation with a spin-orbit potential reduces to two uncoupled equations, both with the same l, but different j's. Derive these equations.

References

1. Ridley B W and Turner J F 1964 *Nucl. Phys.* **58** 497
2. Austern N 1970 *Direct Nuclear Reaction Theory* (John Wiley)
3. Joachain C J 1984 *Quantum Collision Theory* (Elsevier)
4. Satchler G R 1983 *Direct Nuclear Reactions* (Oxford University Press)
5. Friedrich H 1981 *Phys. Rep.* **74** 210
6. Volkov A B 1965 *Nucl. Phys.* **A74** 33
7. Holdeman J T and Thaler R M 1965 Phys. Rev. Lett. **14** 81; *Phys. Rev.* **B139** 1186
8. Marshak H, Langsford A, Wong C Y and Tamura T 1968 *Phys. Rev. Lett.* **20** 554
9. Taylor J 2012 *Scattering Theory: The Quantum Theory of Nonrelativistic Collisions* (Dover)
10. Newton R 2013 *Scattering Theory of Waves and Particles* (Dover)

4

Compound nucleus reactions

4.1 Introduction

The compound-nucleus model is a description of atomic nuclei proposed in 1936 by Niels Bohr [1] to explain nuclear reactions as a two-stage process comprising the formation of a relatively long-lived intermediate nucleus and its subsequent decay. First, a bombarding particle loses all its energy to the target nucleus and becomes an integral part of a new, highly excited, unstable nucleus, called a *compound nucleus*. The formation stage takes a period of time approximately equal to the time interval for the bombarding particle to travel across the diameter of the target nucleus:

$$\Delta t \sim \frac{R}{c} \sim 10^{-21} \text{ s.}$$

Second, after a relatively long period of time (typically from 10^{-19} to 10^{-15} second) and independent of the properties of the reactants, the compound nucleus disintegrates, usually into an ejected small particle and a product nucleus. For the calculation of properties for the decay of this system by *particle evaporation* one may thus borrow from the techniques of statistical mechanics. The energy distribution of the evaporated nucleons have sharp resonances, whose width is much smaller than those known from potential scattering. Qualitatively, but physically correct, this may be understood through the time-energy uncertainty relation: the relaxation to equilibrium and the subsequent "evaporation" of a nucleon takes a long time, much longer than the one it needs for a nucleon to move across the average potential. In other words, the emission of the nucleon with energy equal to that of the incident one is a concentration of the excitation energy on a single particle through a complicated process which needs a long time. Hence the associated energy width

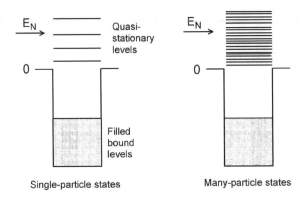

FIGURE 4.1   Energy levels in a single-particle and compound system.

will be quite narrow. For very slow neutrons the widths of these resonances are in the range of a few eV!

The basic idea is shown in Figure 4.1. First we assume that the incident projectile moves in an attractive real potential so that the quasi-stationary levels are the single-particle levels in the potential with positive energy, and a resonance will occur whenever the incident energy corresponds to the energy of one of these single-particle levels. For a simple square-well with depth V_0 and range R, adapted to reproduce the main characteristics of the bound nuclear states, the spacing of these quasi-stationary levels is about 10 MeV. It is evident that this model contains none of the features associated with the fine resonances appearing in the compound nucleus reactions. If the fine resonances are also to be associated with quasi-stationary levels, the levels in question must be closely spaced levels of a many particle system (right-side of Figure 4.1). That is, we must abandon the single-particle picture and assume that many nucleons participate in the formation of the compound system.

In a real example, the compound nucleus ^{28}Si is formed by bombarding ^{27}Al with protons. This compound nucleus is excited in a high-energy state, and may decay into (a) ^{24}Mg and ^{4}He (an alpha particle), (b) ^{27}Si and a proton, (c) a more stable form of ^{28}Si and a γ-ray, or (d) ^{24}Na plus three protons and one neutron.

The compound-nucleus model is very successful in explaining nuclear reactions induced by relatively low-energy bombarding particles (that is, projectiles with energies below about 50 MeV). In the next sections we show how microscopic, microscopic, formal, and empirical concepts are used to explain the rich physics behind the compound nucleus formation and decay.

4.2   The nucleon-nucleon interaction

We consider a group of experimental facts that indicate that the nuclear force is independent of the charge of the nucleons. This means that the force between a neutron and a proton has the same form as the force between two neutrons and between two protons, if we subtract the Coulomb part. It also means that there is a physical quantity involved for which there is a conservation law. Such quantity is the *isospin T*. In terms of the component T_z of that quantity we can express three possibilities to build a system of two nucleons: the di-neutron with $T_z = 1$, the di-proton with $T_z = -1$, and the deuteron with $T_z = 0$. T_z denotes the sum of the component-z of the isospin of each nucleon. Since we only have two nucleons, T cannot be larger than 1. Therefore, for both systems, the di-neutron and di-proton, T has to be equal to 1. For the deuteron with $T_z = 0$, T can be 0 or 1.

The wave function of a system of two nucleons can be written as the product of a space function,

Isospin wave function	T	T_Z	Symmetry by isospin exchange
$\phi_1^1 = \pi(1)\pi(2)$	1	-1	Triplet (symmetric)
$\phi_1^0 = \frac{1}{2}[\pi(1)\nu(2) + \pi(2)\nu(1)]$	1	0	
$\phi_1^{-1} = \nu(1)\nu(2)$	1	+1	
$\phi_0^0 = \frac{1}{2}[\pi(1)\nu(2) - \pi(2)\nu(1)]$	0	0	Singlet (antisymmetric)

TABLE 4.1   Isospin wave function of the two-nucleon system.

a spin function and one of isospin:

$$\Psi = \psi_{\text{spa}}\chi_s^{m_s}\phi_T^{T_z}. \tag{4.1}$$

We denote $\pi\,(T_Z = -1/2)$ as a state of the proton and $\nu(T_Z = +1/2)$ as a state of the neutron, so that $\pi(1)\,\nu(2)$ means that the first nucleon is a proton and the second is a neutron. We can build the isospin part $\varphi_{T_z}^T$ of the wave function of the two-nucleon system similarly to the case of the spin, as it is indicated in the Table 4.1.

The states ϕ_1^1 (di-neutron) and ϕ_1^{-1} (di-protron) constitute, together with ϕ_1^0, a triplet in the isospin space. Now we want to know if some member of that triplet can be part of a bound state of the two particles. To show that this is not possible, let us examine the ground state of the deuteron. This state has $J = 1$, $S = 1$ and $l = 0$. The last value indicates that the space part is symmetrical and $S = 1$ also corresponds to a symmetrical spin part. As Ψ should be antisymmetric $\phi_T^{T_z}$ should also be antisymmetric for the ground state of the deuteron and the isospin wave function of that state can only be ϕ_0^0. The function ϕ_1^0 is, therefore, the isospin wave function of an excited state of the deuteron. But we know experimentally that this state is not bound. As the nuclear force does not depend on the charge, the absence of a bound state for ϕ_1^0 should be extended to ϕ_1^1 and ϕ_1^{-1}. This last result exhibits that the bound proton-proton or neutron-neutron system does not exist. That agrees with the experimental observations.

But, how the states with $T = 1$ and $T = 0$ can correspond to different energies if the nuclear forces are independent of the charge (isospin)? That is due to the dependence of the nuclear force on the spin. To each of the groups of isospin states is associated a different orientation for the spins, so that to each group correspond different energies. The dependence of the nuclear force with the spin has a connection with the fact that there is no other state for the deuteron than the ground state (triplet spin). The force between the proton and the neutron when they have antiparallel spins (singlet) is smaller than when they have parallel spins (triplet), not being strong enough to form a bound state.

We shall analyze the possible form of a potential dependent on the spin. This potential will be written as function of the operator σ, whose components are the Pauli matrices

$$\sigma_x = \begin{pmatrix} 0 & 1 \\ 1 & 0 \end{pmatrix}, \qquad \sigma_y = \begin{pmatrix} 0 & -i \\ i & 0 \end{pmatrix}, \qquad \sigma_z = \begin{pmatrix} 1 & 0 \\ 0 & -1 \end{pmatrix}, \tag{4.2}$$

that satisfy the relations $\sigma_x^2 = \sigma_y^2 = \sigma_z^2 = \begin{pmatrix} 1 & 0 \\ 0 & 1 \end{pmatrix}$, and that are related in a simple way to the spin operator, $\mathbf{S} = \hbar\sigma/2$. To analyze the possible forms for the potential we should have in mind that it should be invariant for rotations and reflections of the system of coordinates, that is, it should be a scalar. Starting from the operators σ_1 and σ_2 of each nucleon, we can build the scalar function

$$V_\sigma(r).(A + B\boldsymbol{\sigma}_1 \cdot \boldsymbol{\sigma}_2), \tag{4.3}$$

which, in the particular case that we shall examine, $A = B = \frac{1}{2}$, assumes the form

$$V_\sigma(r) \cdot \frac{1}{2}(1 + \boldsymbol{\sigma}_1 \cdot \boldsymbol{\sigma}_2) \equiv V_\sigma(r)P_\sigma, \tag{4.4}$$

where $V_\sigma(r)$ describes the radial dependence and the operator $P_\sigma = \frac{1}{2}(1 + \boldsymbol{\sigma}_1 \cdot \boldsymbol{\sigma}_2)$ has the expected value $+1$ for the triplet state and -1 for the singlet state. This can be shown starting from the vector $\mathbf{S} = \hbar(\boldsymbol{\sigma}_1 + \boldsymbol{\sigma}_2)/2$. Since $S^2 = \hbar^2(\sigma_1^2 + \sigma_2^2 + 2\boldsymbol{\sigma}_1 \cdot \boldsymbol{\sigma}_2)/4$, then

$$\boldsymbol{\sigma}_1 \cdot \boldsymbol{\sigma}_2 = \frac{1}{2}(-\sigma_1^2 - \sigma_2^2 + 4S^2/\hbar^2), \tag{4.5}$$

the eigenvalues $\hbar^2 S(S+1)$ of S^2 are $+2\hbar^2$ for the triplet state ($S=1$) and 0 for the singlet state. The eigenvalues of $\sigma^2 = \sigma_x^2 + \sigma_y^2 + \sigma_z^2$ are equal to 3, so that the eigenvalues of $\boldsymbol{\sigma}_1 \cdot \boldsymbol{\sigma}_2$ are equal to $+1$ for the triplet state and -3 for the singlet state, resulting in the expected values of P_σ predicted above.

P_σ is known as *Bartlett potential* or spin exchange potential since, if we use for the spin the functions similar to those given in Table 4.1, we shall obtain that the operation of spin exchange is equivalent to the multiplication by a factor $+1$ for the triplet state and a factor -1 for the singlet state.

Besides the spin, it is common to consider the inclusion of other quantities in the nuclear potential. Thus, for example, we can assume that the nuclear force depends on the parity of the wave function that describes the two particles. A way of expressing that dependence is the addition of the term

$$V_r(r)P_r, \tag{4.6}$$

denominated *Majorana potential*, that contains the operator P_r which exchanges the space coordinates of the two particles. The eigenvalues of P_r are $+1$ and -1, if the wave function is even or odd, respectively.

The charge independence of the nuclear force is related to the invariance of the Hamiltonian under rotation in the isospin space. Based on that, we can also add to the potential a term created from the isospins t_1 and t_2. In a similar way to (4.6), we define the quantity

$$V_t(r) \cdot \frac{1}{2}(1 + \mathbf{t}_1 \cdot \mathbf{t}_2) \equiv V_t(r)P_t, \tag{4.7}$$

where the operator P_t changes the isospin of the two particles. The antisymmetry of the total wave function implies that P_t is not independent of P_σ and P_r, existing the relation

$$P_t = -P_\sigma P_r, \tag{4.8}$$

that can be verified easily by the application of both sides to (4.1). The operator $P_\sigma P_r$ is known as *Heisenberg potential*.

Gathering the terms presented up to now, we can write the expression that represents the central part of the nucleon-nucleon potential:

$$V_c(r) = V_W(r) + V_r(r)P_r + V_\sigma(r)P_\sigma + V_t(r)P_t, \tag{4.9}$$

where it was included the portion V_W dependent only of r, usually referred to as *Wigner potential*.

The expression (4.9) cannot, however, be the final form of the nucleon-nucleon interaction. The existence of other terms becomes necessary to explain certain experimental results. One of them is the absence of a well defined value for l, represented by the disturbance of a state $l=2$ caused in the ground state $l=0$ of the deuteron. That forces us to think in a non-central nucleon-nucleon interaction potential, $V(r)$, since a central potential $V(r)$ conserves angular momentum and has l as a good quantum number.

It is common to describe the action of those non-central forces by a function of the angles between the spin vectors of the neutron and of the proton and of the radial vector $\mathbf{r}$ that separates them. Such potential is known as *tensor potential*. The candidate functions to represent the tensor potential should have as first requirement to be a scalar. Thus, being $\mathbf{u}_r$ the unitary vector in the direction $\mathbf{r}$, products of the type $\boldsymbol{\sigma}_1 \cdot \mathbf{u}_r$, $\boldsymbol{\sigma}_2 \cdot \mathbf{u}_r$ and $(\boldsymbol{\sigma}_1 \times \boldsymbol{\sigma}_2) \cdot \mathbf{u}_r$ must be rejected for being pseudo-scalars, that is, they change sign in a reflection of the system of coordinates. Powers of those expressions are useless since we have, for example, $(\boldsymbol{\sigma} \cdot \mathbf{u}_r)^2 = 1$ and $(\boldsymbol{\sigma} \cdot \mathbf{u}_r)^3 = \boldsymbol{\sigma} \cdot \mathbf{u}_r$. In this situation, the simplest form of scalars we are looking for is $(\boldsymbol{\sigma}_1 \cdot \mathbf{u}_r)(\boldsymbol{\sigma}_2 \cdot \mathbf{u}_r)$ (the alternative $(\boldsymbol{\sigma}_1 \times \mathbf{u}_r)(\boldsymbol{\sigma}_2 \times \mathbf{u}_r)$ is not relevant, for being a linear combination of the previous and $\boldsymbol{\sigma}_1 \cdot \boldsymbol{\sigma}_2$). This expression is usually modified to satisfy the condition that the average about all directions is zero. Since we know that the average of $(\mathbf{A} \cdot \mathbf{u}_r)(\mathbf{B} \cdot \mathbf{u}_r)$ is $\frac{1}{3}\mathbf{A} \cdot \mathbf{B}$, we define the potential tensor as

$$V_{12}(r)S_{12} \quad \text{with} \quad S_{12} = 3\left(\boldsymbol{\sigma}_1 \cdot \mathbf{u}_r\right)\left(\boldsymbol{\sigma}_2 \cdot \mathbf{u}_r\right) - \left(\boldsymbol{\sigma}_1 \cdot \boldsymbol{\sigma}_2\right). \tag{4.10}$$

For the singlet state, $\boldsymbol{\sigma}_1 = -\boldsymbol{\sigma}_2$; where it follows that $(\boldsymbol{\sigma}_1 \cdot \boldsymbol{\sigma}_2) = -\sigma_1^2 = -3$, and that $(\boldsymbol{\sigma}_1 \cdot \mathbf{u}_r)(\boldsymbol{\sigma}_2 \cdot \mathbf{u}_r) = -(\boldsymbol{\sigma}_1 \cdot \mathbf{u}_r)^2 = -1$. Thus, for the singlet state, $S_{12} = 0$, that is, the tensor force is zero. That is an expected result, since there is no preferential direction for the singlet state.

The terms proposed up to now are all characteristic of a local potential, an expression that denotes a potential that it is perfectly defined at each point **r** of the space. Potentials dependent on momentum are, on the other hand, example of potentials that do not only depend on one point and that are called non-local. Among these, it is common to include in the nuclear potential a term of the form

$$V_{LS}(r)\mathbf{l} \cdot \mathbf{s} = V_{LS}(r)\frac{\hbar}{2}(\mathbf{r} \times \mathbf{p}) \cdot (\boldsymbol{\sigma}_1 + \boldsymbol{\sigma}_2), \tag{4.11}$$

linear in **p**, known as *spin-orbit interaction*. This interaction can be observed, for example, in the scattering of polarized protons by a spinless target nucleus (see Figure 4.9). Depending in which direction the proton travels, the spin **s** and the angular momentum **l** can be parallel or antiparallel. Therefore, the term $V_{LS}(r)\mathbf{l} \cdot \mathbf{s}$ in the potential has the scalar product some times positive, other times negative. That leads to an asymmetry in the scattering cross section (Figure 4.9).

Gathering the terms for the nuclear potential proposed here, we have

$$V(r) = V_c(r) + V_{12}(r)S_{12} + V_{LS}(r)\mathbf{l} \cdot \mathbf{s}. \tag{4.12}$$

To establish the form of the unknown functions contained in (4.12) we can adopt the approach that this potential describes correctly the experimental observations on nucleon-nucleon scattering, or the properties of certain nuclei, as the deuteron. The values of these functions should be adjusted in a way that they satisfy the approach above; we shall get a *phenomenological potential*. Phenomenological potentials are broadly employed, not only in the construction of nucleon-nucleon forces, but also in the interaction of complex nuclei, where the participation of the individual nucleons becomes extremely difficult to describe.

Another way to attack the nuclear force problem is analyzing the meson exchange processes directly. The simplest exchange potential is due to the exchange of just one pion [2]. But, only the long distance part of the potential can be explained in that way. Since the pion has spin zero, its wave function should be described by the Klein-Gordon equation

$$\left(\nabla^2 - \frac{m_\pi^2 c^2}{\hbar^2}\right)\Phi = \frac{1}{c^2}\frac{\partial^2\Phi}{\partial t^2}. \tag{4.13}$$

Performing a separation of variables, we obtain a time independent wave equation when the total energy of the pion is equal to 0 (binding energy equal to the rest mass),

$$(\nabla^2 - \mu^2)\phi = 0, \tag{4.14}$$

where

$$\mu = \frac{m_\pi c}{\hbar}. \tag{4.15}$$

An acceptable solution for (4.14) is

$$\phi = g\frac{e^{-\mu r}}{r}, \tag{4.16}$$

in which g is a constant that has the same role as the charge in the case of electrostatics, where the potential that results from the interaction between two equal charges is $qV = q^2/r$. That interaction is due to the continuous exchange of virtual photons between the charges.

We can assume that the potential between two nucleons is proportional to the wave function of the pion, that is, to the probability amplitude that the emitted pion finds itself close to the other nucleon. We thus find the Yukawa potential

$$V = g^2\frac{e^{-\mu r}}{r}, \tag{4.17}$$

where we used the factor g^2, in analogy with the electrostatics. The potential above decays exponentially, and its range can be estimated by

$$R \cong \frac{1}{\mu} = \frac{\hbar}{m_\pi c} \cong 0.7 \text{ fm}. \tag{4.18}$$

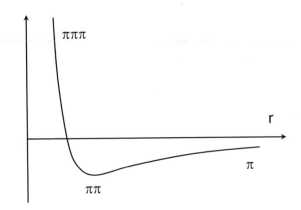

FIGURE 4.2   Schematic representation of the nucleon-nucleon interaction.

The force field between two protons, or two neutrons, can only be produced by the exchange of neutral pions. Between a proton and a neutron the exchange can be done by means of charged pions.

It is a well-established experimental fact that the nuclear force is strongly repulsive at very short distances and the form of the central part of the nuclear potential should be given, schematically, by Figure 4.2. The potential well, that is, its attractive part of medium range, can be described by the exchange of two pions. It is interesting to observe that this part of the potential is created in a similar way as the Van der Waals force between two molecules (Figure 4.3).

The molecular binding can be described by the exchange of two photons. The first photon, emitted by the molecule 1, induces an electric dipole in the molecule 2 and this dipole emits a virtual photon that induces another electric dipole in the molecule 1. The interaction between the two dipoles gives rise to the Van der Waals force. The pions take the place of the photons in the case of the nuclear forces. The production of an electric dipole is similar to the excitation of a nucleon to a $\triangle$-resonance. In this way, the nuclei are bound by a type of Van der Waals force (Figure 4.4).

Since the pion carries isospin and as the dependence of the nucleon-nucleon interaction is accompanied by a dependence in the spin, it can be shown [2] that the interaction due to the exchange of one pion should be of the form

$$V^{\mathrm{OPEP}} = \frac{g^2}{3\hbar c} m_\pi c^2 \frac{e^{-\mu r}}{\mu r} \left(\mathbf{t}^{(1)} \cdot \mathbf{t}^{(2)} \right) \left[\boldsymbol{\sigma}^{(1)} \cdot \boldsymbol{\sigma}^{(2)} + \left(1 + \frac{3}{\mu r} + \frac{3}{\mu^2 r^2} \right) S_{12} \right], \qquad (4.19)$$

where the acronym OPEP arises from "*one-pion-exchange potential*". This potential describes well the nucleon-nucleon scattering for angular momenta $l \geq 6$. The high value of this limit shows that the OPEP potential describes the nuclear force in a reasonable way at great distances ($r \geq 2$ fm).

The short range part of the nucleon-nucleon potential represented in Figure 4.2 is due to the exchange of three pions, or more. The essential part of this process can be described by the effective exchange of a resonance of three pions, known by ω-meson with spin 1 and $m_\omega = 783.8$ MeV. The ω exchange is important for two properties of the nuclear force: the repulsive part of the potential and the spin-orbit interaction.

Both properties also have analogies with the electromagnetic case. In the electromagnetism, the exchange of a photon also gives rise to the repulsive force between charges of same sign. In the nuclear force case, due to the ω-meson big mass, the repulsive force is of short range. Starting from this argument we can also conclude that the strongly repulsive potential becomes a strongly attractive potential for a nucleon-antinucleon pair at short distances.

At intermediate distances the nucleon-nucleon potential is, as we already emphasized, adequately described in terms of the exchange of two pions. Another way to describe this part of the potential is with the exchange of a single particle, the ρ-meson, of mass (768.1 ± 0.5) MeV/c^2. It

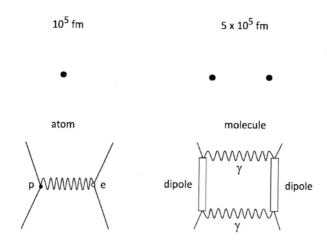

FIGURE 4.3 The Coulomb force between protons and electrons in an atom can be described in quantum electrodynamics by the exchange of one photon, and the Van der Waals force by the exchange of more than one photon.

is believed that this meson is built up of two pions, thus the equivalence with the exchange of two pions follows.

The potentials derived from the hypothesis of $\pi, \rho,$ and ω exchange consist of combinations of central, tensorial, spin-orbit parts, and of terms of higher order. The radial functions that accompany these terms have a total of up to 50 parameters, that are adjusted to the experimental data of the deuteron and to the nucleon-nucleon scattering.

Some authors have used purely phenomenological parameterizations for the nuclear potential, following the potential form due to the exchange of mesons. Those potentials possess attractive and repulsive components. At great distances they are reduced to the OPEP potential, while at small distances they possess an extremely repulsive part. That repulsive part is usually called by "hard core", with $V(r) \to \infty$ for $r < r_c \cong 0.4$ fm. Other authors use a repulsive potential that goes to infinity only for $r \to 0$. These potentials are known as "soft core" potentials. The most popular of these potentials is the *Reid soft-core potential* [3]. It has the form

$$V = V_c(\mu r) + V_{12}(\mu r)S_{12} + V_{LS}(\mu r)\mathbf{l} \cdot \mathbf{s}, \tag{4.20}$$

where

$$V_c(x) = \sum_{n=1}^{\infty} a_n \frac{e^{-nx}}{x} \; ; \qquad V_{LS}(x) = \sum_{n=1}^{\infty} c_n \frac{e^{-nx}}{x}, \tag{4.21}$$

and

$$V_{12}(x) = \frac{b_1}{x} \left[\left(\frac{1}{3} + \frac{1}{x} + \frac{1}{x^2} \right) e^{-x} - \left(\frac{b_0}{x} + \frac{1}{x^2} \right) e^{-b_0 x} \right] + \sum_{n=2}^{\infty} b_n \frac{e^{-nx}}{x}. \tag{4.22}$$

The constants are different for all values of T, S and L. Only $a_1, b_1,$ and c_1 are given in order to reproduce the OPEP potential at great distances. For $l > 2$, the Reid potential is replaced by the OPEP potential. The Reid potential is quite realistic and describes well, within its range of validity, the properties of a system of two nucleons.

4.3   The nucleus as a strongly absorbing medium

The problem we face in describing nuclear scattering with strong absorption is similar to the one we discussed in the previous chapters for nucleon-nucleon collisions, using interactions like the ones

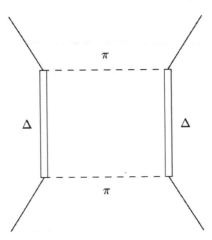

FIGURE 4.4 Nuclear force due to the exchange of two pions. The Δ-particle is a kind of polarized nucleon in the pion field generated by the other nucleon.

in the previous section. Extending this method to collisions of composite particles (e.g., neutron-nucleus collisions), one might at first just assume that the collision of a nucleon N with a nucleus $\mathcal{N}$ can be described as *potential scattering*. One only would have to find an appropriate form for this potential "$V(r)$". Indeed, one has been able to solve the latter task to great satisfaction. Later on we will discuss possible, realistic forms of "$V(r)$". However, it turns out that a reaction of N with $\mathcal{N}$ is a more complicated process. In experiments this is seen by looking at the total cross section, for instance. For pure potential scattering the latter would have to be identical to the elastic one σ_{el}. Truth is that a good deal of the total reaction cross section

$$\sigma_{\text{tot}} = \sigma_{\text{inel}} + \sigma_{\text{el}}$$

goes into *inelastic* and *absorptive processes*. There are situations in which the inelastic cross section is of the same order as the elastic one. This feature reminds one of the scattering at a *black sphere*.

The physical reason for this behavior is not difficult to understand. When a N hits a $\mathcal{N}$ the sub-structure of the latter becomes relevant practically at all energies. This is an essential, qualitative difference to the NN scattering where the sub-N degrees of freedom, which actually consist of different type of particles (*quarks and gluons*), are seen only at large energies. It is understood, of course, that such effects will eventually become relevant here as well. However, one must not forget that the impinging nucleon "sees" partners of its own type, with which it may perform individual NN collisions before it may leave the nucleus again. In such processes it is likely that it looses some of its original energy. In this sense the projectile particle is "lost" in the particular "channel" associated to its original energy or wave number. Moreover, it might get stuck in the nucleus altogether after having given all of its initial energy to the other nucleons, some of which may then be "evaporated" afterwards. Hence, with respect to the beam particle, one may speak of "absorption" either in literal or in a more general sense. As explained in Chapter 2 such processes may still be described by a partial-wave analysis if one replaces (δ_l real)

$$e^{2i\delta_l} \qquad \text{by} \qquad S_l(E) = \eta_l(E)\, e^{2i\delta_l},$$

where $\eta_l(E)$ has to lie between 0 and 1, as discussed in section 3.13.

We have already discussed the *strong absorption model* for the total cross section in Section 3.13. One can also apply this model for the differential elastic cross section. This allows us to understand the main features of the cross sections displayed in Figures 5 and 7 of Chapter 3. In analogy to the case of the black sphere, we want to define this model by assuming that outgoing partial waves exist only for angular momentum l above a certain value L. The latter shall be approximated, however, in a slightly different way. Rather than speaking of a sphere of some given radius we define the latter by the classical turning point for a particle of momentum $p = \hbar k$. Discarding the Coulomb interaction the turning point may be estimated from the centrifugal barrier as $\hbar^2 k^2 / 2\mu = \hbar^2 l(l+1)/2\mu R^2 \approx \hbar^2 (l+1/2)^2 / 2\mu R^2$. Thus, we will use

$$S_l = \begin{cases} 1 & \text{for} \quad l > L = kR - 1/2 \\ 0 & \text{for} \quad l \leq L = kR - 1/2 \end{cases}, \tag{4.23}$$

assuming that the energy is sufficiently high so that the cross section is dominated by partial waves of large l. The scattering amplitude becomes

$$f(\theta) = \frac{i}{2k} \sum_{l=0}^{L} (2l+1) P_l(\cos\theta). \tag{4.24}$$

This expression can be evaluated on a computer. But, to gain insight we now exploit the following asymptotic expressions for the Legendre polynomials [4]

$$P_l(\theta) = 1 - \frac{l(l+1)\theta^2}{2} \qquad \text{for} \qquad l\theta \ll 1,$$

$$P_l(\theta) = \frac{2\sin\left[(l+1/2)\theta + \pi/4\right]}{\sqrt{(2l+1)\pi\sin\theta}} \qquad \text{for} \qquad l\theta \gg 1. \tag{4.25}$$

For *forward angles* one readily verifies that the differential cross section becomes

$$\frac{d\sigma_{\text{el}}}{d\Omega} \approx \frac{k^2 R^4}{2} \left(1 - \frac{k^2 R^2 \theta^2}{4}\right)^2 \qquad \text{for} \qquad kR\theta \ll 1.$$

Notice that the condition given here is meant as a restriction for the angle; otherwise the form found in this estimate is only valid for $L \simeq kR \gg 1$. For larger angles the lower part of Eq. (4.25) can be exploited. In this case the summation over l cannot be performed analytically. One may, however, benefit again from the fact that the expression (4.24) will be governed by large l values and replace the sum over discrete values by an integral over a continuous variable. In this way the amplitude can be written as

$$f(\theta) \approx \frac{i}{k\sqrt{\pi\sin\theta}} \int_0^{kR-1/2} dl \sqrt{2l+1} \sin\left[(l+\frac{1}{2})\theta + \frac{\pi}{4}\right] \approx \frac{i\sqrt{2}}{k\sqrt{\pi\sin\theta}} \int_0^{kR} dx \sqrt{x} \sin\left[x\theta + \frac{\pi}{4}\right]$$

$$\approx \frac{i\sqrt{2kR}}{k\theta\sqrt{\pi\sin\theta}} \cos\left[kR\theta + \frac{\pi}{4}\right].$$

The trick to calculate the integral is as follows. The factor $\sqrt{x}$ implies that it will be dominated by contributions at the upper end. But for large x the function $\sin\left(x\theta + \frac{\pi}{4}\right)$ is strongly oscillating. On this scale $\sqrt{x}$ may be considered smooth such that it may be taken out of the integral.

For the cross section one thus gets

$$\frac{d\sigma_{\text{el}}}{d\Omega} \approx \frac{2R}{\pi k^2 \theta^2 \sin\theta} \cos^2\left(kR\theta + \frac{\pi}{4}\right) \qquad \text{for} \qquad kR\theta \ll 1. \tag{4.26}$$

Inspecting the two forms of the cross section the following feature becomes apparent. It has a maximum at zero degree and, on the general trend, it becomes smaller with larger θ. Superimposed to this behavior are oscillations. Indeed, as seen from Eq. (4.26) the cross section may become zero at certain angles. This formula suggests that the first one at which this happens is given by

$$\theta_{\text{min}} \approx \frac{5\pi}{4kR}. \tag{4.27}$$

(It is the first angle for which the cosine becomes zero and where $kR\theta$ is larger than unity). One recognizes from Eq. (4.27) that the scattering is more forward peaked the larger the nucleus. For neutrons of about 100 MeV, typical values range from about 20 to 30 degrees for lighter nuclei and about 10 to 20 degrees for heavier. This feature may be understood as a kind of "macroscopic limit": $\theta_{\min} \to 0$ for $kR \to \infty$. Indeed, one may convince one selves that the following limit holds true

$$\lim_{L\to\infty} \sum_{l=0}^{L} (2l+1) P_l(\cos\theta) = \delta(\theta). \tag{4.28}$$

As one may imagine looking at these formulas, they may actually be used to get an estimate for nuclear radii. The phenomena discussed here goes by the name of *Fraunhofer diffraction* (also discussed in Section 3.13). As a matter of fact, it is not only this example for which there exists a similar in nuclear physics. In heavy ion collisions one also observes *Fresnel diffraction*, "rainbow angles" and other exotic features (see Section 1.10).

As we see from Figure 5 of Chapter 3, the elastic cross sections decrease much faster than Eq. (4.26). This can be understood in terms of the diffuseness of the nuclear surface. For simplicity, let us take the Born approximation. We see that the scattering amplitude is proportional to the Fourier transform of the nuclear potential. To make matters even simpler and more transparent, we take the one-dimensional case. For a potential of the form $V(x) = \left(x^2 + R^2\right)^{-1}$, we have ($q = 2k \sin(\theta/2)$ is the momentum transfer)

$$\int_{-\infty}^{\infty} dx e^{iqx} \frac{1}{x^2 + R^2} = \frac{\pi}{R} e^{-qR}. \tag{4.29}$$

This certainly gives an exponential falloff, but we have lost the oscillations. What is needed is a form of the potential that preserves the size but smooths only the edge. Such a simple form is

$$V(x) = \frac{1}{1 + \exp\left[(x-R)/a\right]}. \tag{4.30}$$

This is the *Fermi* or *Woods-Saxon* form corresponding to a size R and an edge diffusivity of thickness a. For an analogy with the case of medium and heavy nuclei, we take $R \gg a$. The Fourier integral of the form (4.30) cannot be evaluated in closed form. However, for $R \gg a$ and $qa \gg 1$, the integral is dominated over a wide range of q by the closest singularity of $V(x)$. If we close the integration contour in the upper half of the plane, this is a simple pole at $x = R + i\pi a \equiv x_0$. The residue at the pole is $-a$. Taking this pole's contribution only, we have (see Ref. [5], Section 2)

$$\int_{-\infty}^{\infty} dx e^{iqx} \frac{1}{1 + \exp\left[(x-R)/a\right]} \cong -a \int_{-\infty}^{\infty} dx e^{iqx} \frac{1}{x + x_0} = -2\pi i a e^{iqx_0}.$$

Adding a similar contribution from the pole at $x_0' = R - i\pi a$, yields a scattering amplitude of the form

$$f \cong 4\pi a \sin(qR) e^{-\pi qa}. \tag{4.31}$$

In three-dimension these developments get more complicated. But the message is the same: the oscillations in the scattering cross sections are due to the size R of the nuclear system, while the exponential falloff is set by the skin's thickness a. The one-dimensional analogy allows us to understand the principal features of nuclear diffraction scattering. Looking at the Figure 3.5 we see that the minima of the cross sections are somehow filled. This minimum filling in arises from a combination of the Coulomb force, the real part of the underlying potential projectile-nucleon amplitude, spin effects, etc. Only this filling-in-effect depends on the details of the underlying dynamics. The rest, including the magnitude of the cross section, is determined largely by the nuclear geometry.

4.4   Mean free path of a nucleon in nuclei

The strong absorption model is an idealization. In a real experiment it will never be such that the scattering amplitude will be determined by the step function, Eq. (4.23) for S_l. Fortunately, there is a very natural way for improvements, namely the introduction of imaginary potentials. Because of the analogies to optical phenomena it may be no surprise that they are called "optical potentials" and that one speaks of "optical models".

Let us for the moment not worry how and why such a generalization might be justified from a microscopic point of view. Instead, we just take for granted to be given a non-hermitian Hamiltonian. For simplicity let us take a one-dimensional case and let us assume further that V and W are just constants. Then a solution of the time-dependent Schrödinger equation can be written as

$$\psi(x,t) = \exp\left[(-iV - W)\,t/\hbar\right] \exp\left[-i\frac{p^2}{2m_N}t\right] \psi(x,0). \tag{4.32}$$

For $W > 0$ the amplitude of this wave function decays exponentially, as it should for an absorptive process. The stationary solutions obey the equation

$$\left[-\frac{\hbar^2}{2m_N}\frac{\partial^2}{\partial x^2} + V - iW\right]\phi(x) = E\phi(x), \qquad (\text{with} \quad E = E^*). \tag{4.33}$$

This equation is satisfied by the ansatz $\phi \sim \exp(iKx) = \exp\left[(ik - \kappa)x\right]$. The two quantities k and κ, the real and imaginary parts of K, can be determined by

$$-\frac{\hbar^2}{2m_N}\left(k^2 - \kappa^2\right) = E - V, \qquad \text{and} \qquad \frac{\hbar^2}{2m_N}k\kappa = W. \tag{4.34}$$

It is seen that κ gets positive for positive W. This means that a wave traveling in the x-direction experiences an attenuation according to which the density behaves like

$$|\phi(x)|^2 \sim \exp\left[-2\kappa x\right]. \tag{4.35}$$

One may thus define a mean free path

$$\lambda = \frac{1}{2\kappa} = \frac{\hbar^2}{2m_N}\frac{k}{W}. \tag{4.36}$$

This has an analogy with electromagnetic waves travelling in optical media. Such waves get attenuated as soon as the index n of refraction is complex, which in most of the cases is due to a complex dielectric constant. One may introduce an absorption coefficient α being proportional to the imaginary part of n (see for instance Ref. [6]). In analogy to Eq. (4.35) the connection from this α to our κ would be $\alpha = 2\kappa$. How big the attenuation actually is depends on the frequency of the electromagnetic wave. Likewise, in our case it depends on the energy (or on the real part k of our wave number K). Moreover, as we will see soon, the W itself will vary with energy.

For interactions of neutrons and nuclei, we expect that the imaginary part increases with energy in the interior of the nucleus. This feature is not difficult to understand on account of our intuitive reasoning on the physical nature of absorption. The larger the energy of the incoming nucleon the easier it will be to excite the nucleus as a whole. The magnitude of this potential, on the other hand, is much more difficult to understand.

Let us look at the mean free part, which for the interior of the nucleus may be estimated on the basis of our considerations above. Using a typical value of $W = 10$ MeV in Eq. (4.36) one finds the nucleonic mean free path to be of the order of the size of a heavier nucleus. For the example of a neutron of 40 MeV one gets $\lambda \simeq 5$ fm $\cong R$. This is an astonishingly large number on the account of the large value of the angle integrated cross section σ for the NN scattering. For a relative kinetic energy of the order of 40 MeV one would get a value of about $\sigma \simeq 17$ fm^2. Estimating the mean free path like one would do if the nucleons would behave like a system of classical particles one would find

$$\lambda_{\text{class}} \simeq \frac{1}{\rho\sigma} \approx 0.4 \text{ fm} \ll \lambda. \tag{4.37}$$

In such a case the nucleons would behave like the molecules or atoms of a classical gas or liquid, or like an ensemble of billiard balls. Eq. (4.37) assumes the subsequent collisions of the nucleons to be independent of each other, which actually is not the case as we shall argue below. Within such a picture, formula (4.37) is understood immediately if one realizes that λ_{class} defines that volume in which on average a moving nucleon meets one scattering partner, for which reason one must have $\rho\sigma\lambda_{\text{class}} \simeq 1$. To get the number shown in Eq. (4.37) the density has been taken to be 0.17 nucleons per fm^3.

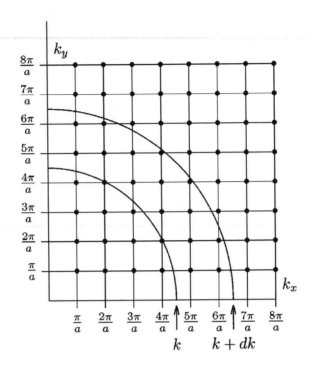

FIGURE 4.5   Allowed states in the part of the momentum space contained in the plane $k_x k_y$. Each state is represented by a point in the lattice.

There are various lessons one may learn from this discussion. First of all, it becomes apparent that the nucleons inside a nucleus do not at all behave like the constituents of a free classical gas. This is not such a big surprise on behalf of our expectation that we are dealing with a quantum many body system. Therefore the nucleons inside the nucleus must be described by a many-body wave function $\Psi(\mathbf{r}_1, \cdots, \mathbf{r}_A)$. The first important feature to be considered then is the *Pauli principle*. This cannot be all one needs to account for; otherwise, one would be able to strictly justify the picture of independent particles. But for the moment let us just look at the exclusion principle. There are two phenomena which diminish its importance and which are relevant in the present context. First of all, its effects decrease with decreasing density. This fact offers an explanation why, for the lower energies, the $W(r)$ gets much larger in the nuclear surface. There the density is smaller than in the interior, implying that the scattering of the nucleons with each other resembles more the one in free space. Secondly, the impact of the Pauli principle becomes weaker and weaker when the nucleus as a whole gets heated up, which may be known from quantum statistical mechanics.

4.5   Fermi gas model

This model, quite simple in its structure, is based on the fact that the nucleons move almost freely inside the nucleus due to the Pauli principle. Since two of them cannot occupy the same energy state, they do not scatter, as all possible final states that could be scattered to are already occupied by other nucleons. But when a nucleon approaches the surface and tries to fly off the nucleus, it suffers an attraction force by the nucleons that are left behind, what forces it to come back to the interior. Inside the nucleus it feels the attraction forces of all the nucleons that are around it, what results in a net force approximately equal to zero. We can imagine the nucleus as being a balloon, inside of which the nucleons move freely, but occupying states of different energy.

The nucleons in the Fermi gas model obey the Schrödinger equation for a free particle,

$$-\frac{\hbar^2}{2m}\nabla^2\Psi = E\Psi, \tag{4.38}$$

where m is the nucleon mass and E its energy. To simplify let us assume that instead of a sphere, the region to which the nucleons are limited to is the interior of a cube. The final results of our calculation will be independent of this hypothesis. In this way, Ψ will have to satisfy the boundary conditions

$$\Psi(x, y, z) = 0, \tag{4.39}$$

for

$$x = 0, y = 0, z = 0, \qquad \text{and} \qquad x = s, y = s, z = s,$$

where s is the side of the cube. The solution of (4.38) and (4.39) is given by

$$\Psi(x, y, z) = A\,\sin(k_x x)\,\sin(k_y y)\,\sin(k_z z), \tag{4.40}$$

where

$$k_x s = n_x \pi, \qquad k_y s = n_y \pi, \quad \text{and} \quad k_z s = n_z \pi, \tag{4.41}$$

and $n_x, n_y,$ and n_z are positive integers and A is a normalization constant.

For each group (n_x, n_y, n_z) we have an energy

$$E(n_x, n_y, n_z) = \frac{\hbar^2 k^2}{2m} = \frac{\hbar^2}{2m}(k_x^2 + k_y^2 + k_z^2) = \frac{\hbar^2 \pi^2}{2ms^2}n^2, \tag{4.42}$$

where $n^2 = n_x^2 + n_y^2 + n_z^2$. Equations (4.41) and (4.42) represent the quantization of a particle in a box, where $k \equiv (k_x, k_y, k_z)$ is the momentum (divided by $\hbar$) of the particle in the box. Due to the Pauli principle, a given momentum can only be occupied by at most four nucleons: two protons with opposite spins and two neutrons with opposite spins. Consider the space of vectors k: by virtue of (4.42), for each cube of side length π/s exists, in this space, only one point that represents a possible solution in the form (4.40). The possible number of solutions (see Figure 4.5) $n(k)$ with the magnitude of k between k and $k + dk$ is given by the ratio between the volume of the spherical slice displayed in the Figure and the volume $(\pi/s)^3$ for each allowed solution in the k-space:

$$dn(k) = \frac{1}{8}4\pi k^2 dk\,\frac{1}{(\pi/s)^3}, \tag{4.43}$$

where $4\pi k^2 dk$ is the volume of a spherical box in the k-space with radius between k and $k + dk$. Only $1/8$ of the shell is considered, since only positive values of k_x, k_y and k_z are necessary for counting all the states with eigenfunctions defined by (4.40). With the aid of (4.42) we can make the energy appear explicitly in (4.43):

$$dn(E) = \frac{\sqrt{2}m^{3/2}s^3}{2\pi^2\hbar^3}E^{1/2}dE. \tag{4.44}$$

The total number of possible states of the nucleus is obtained by integrating (4.44) from 0 to the minimum value needed to include all the nucleons. This value, E_F, is called the Fermi energy. Thus, we obtain

$$n(E_F) = \frac{\sqrt{2}m^{3/2}s^3}{3\pi^2\hbar^3}E_F^{3/2} = \frac{A}{4}, \tag{4.45}$$

where the last equality is due to the mentioned fact that a given state can be occupied by four nucleons. Inverting (4.45) we obtain

$$E_F = \frac{\hbar^2}{2m}\left(\frac{3\pi^2\rho}{2}\right)^{2/3}, \tag{4.46}$$

where $\rho = A/s^3$. We assume that the maximum energy is the same for both nucleons, what means equal values for protons and neutrons. If that case is not true, the Fermi energy for protons and

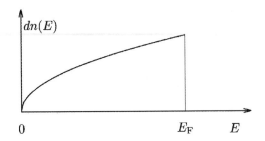

FIGURE 4.6   Fermi distribution for $T = 0$.

neutrons will be different. If $\rho_p = Z/s^3$ $\rho_n = N/s^3$ are the respective proton and neutron densities, we will have

$$E_F(\mathrm{p}) = \frac{\hbar^2}{2m}(3\pi^2\rho_p)^{2/3}, \tag{4.47}$$

and

$$E_F(\mathrm{n}) = \frac{\hbar^2}{2m}(3\pi^2\rho_n)^{2/3}, \tag{4.48}$$

for the corresponding Fermi energies.

The number of nucleons with energy between E and $E + dE$, given by (4.44), is plotted in Figure (4.6) as a function of E. This distribution of particles is denominated Fermi distribution for $T = 0$, which is characterized by the absence of any particle with $E > E_F$. That corresponds to the ground state of the nucleus. An excited state $(T > 0)$ can be obtained by the passage of a nucleon to a state above the Fermi level, leaving a vacancy (hole) in its energy state occupied previously.

If we use $\rho = 1.72 \times 10^{38}$ nucleons/cm^3 = 0.172 nucleons/fm^3, which is the approximate density of all nuclei with $A \gtrsim 12$, we obtain

$$k_F = \frac{\sqrt{2mE_F}}{\hbar} = 1.36 \text{ fm}^{-1}, \tag{4.49}$$

which corresponds to

$$E_F = 37 \text{ MeV}. \tag{4.50}$$

We know that the separation energy of a nucleon is of the order of 8 MeV. Thus, the nucleons are not inside a well with infinite walls as we supposed, but in a well with depth $V_0 \cong (37 + 8)$ MeV = 45 MeV.

The Fermi gas model is useful in numerous situations in nuclear physics. For example, we are now in conditions of explaining the origin of the asymmetry term, of the liquid drop model for the nucleus (Chapter 5). Let us imagine the nucleus as a mixture of a proton gas with Fermi energy $E_F(\mathrm{p})$ and a neutron gas with Fermi energy $E_F(\mathrm{n})$. Taking C as a constant, we can write

$$E_F(\mathrm{p}) = C(Z/A)^{2/3}, \qquad\qquad E_F(\mathrm{n}) = C(N/A)^{2/3}. \tag{4.51}$$

If dn is the number of particles with energy between E and $E + dE$, the total energy E_T of the gas is written as

$$E_T = \int_0^{E_F} E\, dn = \frac{3}{5}ZE_F(\mathrm{p}) \qquad \text{for the protons}$$

$$= \frac{3}{5}NE_F(\mathrm{n}) \qquad \text{for the neutrons}, \tag{4.52}$$

where we used Eq. (4.44). Defining C' as a new constant, we can write the total energy

$$E(Z, A) = C'A^{-2/3}(Z^{5/3} + N^{5/3}), \tag{4.53}$$

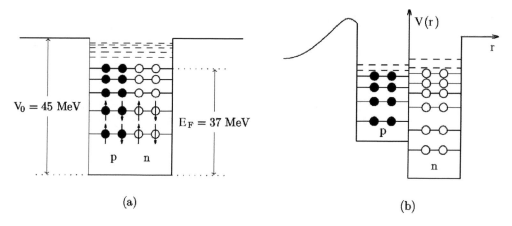

FIGURE 4.7 (a) Potential well and states of a Fermi gas for $T = 0$. In two of the levels we show the spins associated to each nucleon. (b) When one takes into account the Coulomb force the potentials for protons and neutrons are different and we can imagine a well for each nucleon type.

with the constraint $Z + N = A$. The minimum of the energy (4.53) happens when $Z = N = A/2$. Calling $E(Z, A)_{\min}$ that value, let us calculate

$$\Delta E = E(Z, A) - E(Z, A)_{\min} = C' A^{-2/3} \left[N^{5/3} + Z^{5/3} - 2 \left(A/2 \right)^{5/3} \right]. \tag{4.54}$$

Let us define $D = (N - Z)/2 = N - A/2 = A/2 - Z$; we have

$$\Delta E = C' A^{-2/3} \left[(A/2 + D)^{5/3} + (A/2 - D)^{5/3} - 2 \left(A/2 \right)^{5/3} \right]. \tag{4.55}$$

Expanding $(D + A/2)^{5/3}$ and $(-D + A/2)^{5/3}$ in a Taylor series

$$f(x + a) = f(a) + x f'(a) + \frac{x^2}{2} f''(a) + \cdots, \tag{4.56}$$

we obtain

$$\Delta E = \frac{10}{9} C' \frac{(Z - A/2)^2}{A} + \cdots. \tag{4.57}$$

We see that the unbalance between the proton and the neutron number increases the energy of the system (decreasing the binding energy) by the amount specified in (4.57). That justifies the existence of the asymmetry term

$$B_4 = -a_A \frac{(Z - A/2)^2}{A} \tag{4.58}$$

in the liquid drop model.

Despite its simplicity, the Fermi gas model is able to explain many of the nuclear properties exposed in this and the next chapters. At the beginning, the occupation of the states indicates that, in a light nucleus, $Z \cong N$ since in this way the energy is lowered. For heavy nuclei the Coulomb force makes the proton well more shallow than the one for neutrons; as a consequence, the proton number is smaller than that of neutrons, what happens in practice (see Figure 4.7). Another explained characteristic is the verified abundance of even-even nuclei contrasted to the almost non-existence of stable odd-odd nuclei and it is easy to see why that happens: when we have a nucleon isolated in a level, the lower possible state of energy for a next nucleon is in that same level. In other words: in an odd-odd nucleus we have one isolated proton and one isolated neutron, each one in their potential well. But, between those states there is, generally, a difference of energy, creating the possibility of passage of one of the nucleons to the well of the other through a β-emission and, thus, the nucleus stops being stable.

4.6 Formal theory of the optical potential

Although the many-body problem is intractable, the use of a formal theory can provide a connection between particular phenomena and the general properties of the many-body system. The construction of a theory of nuclear reactions has been approached in several different ways, and we follow here the theory due to Feshbach [7].

We consider the problem of a nucleon scattered by a nucleus, and assume that the incident energy is such that only the elastic channel is open. We have to solve the Schrödinger equation

$$(E - H)\Psi = 0, \tag{4.59}$$

with the Hamiltonian H given by

$$H = H_0 + V + H'(\xi), \tag{4.60}$$

where $H_0 = T$ is just the kinetic energy operator for the incident nucleon, V is the interaction potential, and $H'(\xi)$ is the internal Hamiltonian of the target nucleus which has coordinates ξ. The eigenstates of $H'(\xi)$ are $\Phi_\alpha(\xi)$ with energy ε_α, and for convenience we put the ground state energy ε_0 equal to zero.

The total wave function Ψ may be written as an expansion in the complete set of nuclear states $\Phi_\alpha(\xi)$

$$\Psi = \sum_\alpha \psi_\alpha \Phi_\alpha(\xi). \tag{4.61}$$

It is assumed here the incident nucleon is distinguishable, and the problem of antisymmetrization between the incident nucleon and the target nucleon will not be considered.

Because of our initial assumptions, the open channel part of Ψ is just $\psi_0 \Psi_0$, and in order to define the optical potential for this problem we must find the equation for ψ_0. In Feshbach's formalism this is done by defining projection operators P and Q which project on and off the open channels. Thus,

$$P\Psi = \psi_0 \Phi_0, \qquad Q\Psi = (1 - P)\Psi. \tag{4.62}$$

Also,

$$P^2\Psi = P\Psi, \qquad Q^2\Psi = Q\Psi, \qquad PQ\Psi = QP\Psi = 0. \tag{4.63}$$

The construction of these projection operators depends on the problem under consideration, and in the present case it will be satisfactory to take

$$P = |\Phi_0)(\Phi_0|, \tag{4.64}$$

where the round brackets imply summations and integration over the target co-ordinates. The Schrödinger equation is now

$$(E - H)(P + Q)\Psi = 0, \tag{4.65}$$

and multiplication on the left by P yields

$$(E - PHP)P\Psi = (PHQ)Q\Psi, \tag{4.66}$$

while multiplication on the left by Q yields

$$(E - QHQ)Q\Psi = (QHP)P\Psi, \tag{4.67}$$

where we have used equations (4.63). Equation (4.67) may be inverted to give

$$Q\Psi = \frac{1}{E - H_{QQ}} H_{QP} P\Psi, \tag{4.68}$$

where $H_{QQ} = QHQ$, $H_{PQ} = PHQ$, etc. (If there are open non-elastic channels, it is necessary to replace $E - H_{QQ}$ in Eq. (4.68) by $E - H_{QQ} + i\epsilon$ in order to ensure that $Q\Psi$ has only outgoing waves in these channels.) The equation for $P\Psi$ can be obtained by combining equations (4.68) and (4.66) to give

$$\left(E - H_{PP} - H_{PQ} \frac{1}{E - H_{QQ}} H_{QP} \right) P\Psi = 0. \tag{4.69}$$

If we now use the form for the projection operator P given in equation (4.64), we have

$$H_{PP} = |\Phi_0) \left[H_0 + (\Phi_0| V |\Phi_0) \right] (\Phi_0|, \qquad H_{PQ} = |\Phi_0) (\Phi_0| VQ,$$

and Eq. (4.69) becomes

$$\left[E - H_0 - (\Phi_0| V |\Phi_0) - \left(\Phi_0 \left| VQ \frac{1}{E - H_{QQ}} QV \right| \Phi_0 \right) \right] \psi_0 = 0. \tag{4.70}$$

Examination of Eq. (4.70) leads to the definition of the generalized optical potential as

$$V_{opt} = (\Phi_0| V |\Phi_0) + \left(\Phi_0 \left| VQ \frac{1}{E - H_{QQ}} QV \right| \Phi_0 \right). \tag{4.71}$$

The second term in Eq. (4.70) will vary rapidly with energy whenever E is in the vicinity of an eigenvalue of H_{QQ} and will therefore give rise to a marked fluctuation in the cross section in the energy region. This suggests that the occurrence of an eigenvalue of H_{QQ} should be associated with resonance scattering. If the eigenstates of H_{QQ} are given by

$$(E_t - H_{QQ}) \Psi_t = 0,$$

the propagator can be written, using Eq. (3.17), as

$$\frac{1}{E - H_{QQ}} = \sum_t \frac{|\Psi_t\rangle \langle \Psi_t|}{E - E_t},$$

where the angular bracket indicates integration and summation over all coordinates. For scattering in the vicinity of an isolated resonance at $E = E_s$ the generalized optical potential can be rewritten in the form

$$V_{\mathrm{opt}} = (\Phi_0| V |\Phi_0) + \sum_{t \neq s} \frac{(\Phi_0|VQ |\Psi_t) \langle \Psi_t| QV|\Phi_0)}{E - E_t} + \frac{(\Phi_0|VQ |\Psi_s) \langle \Psi_s| QV|\Phi_0)}{E - E_s}$$

$$= U + \frac{(\Phi_0|VQ |\Psi_s) \langle \Psi_s| QV|\Phi_0)}{E - E_s}. \tag{4.72}$$

The term defined as U will vary slowly with energy and includes the effects of distant resonances. In this way the distant compound resonances combine to give the same effect as a single-particle resonance.

We may define scattering solutions $\chi^\pm$ which satisfy the equation

$$(E - H_0 - U) \chi^\pm = 0, \tag{4.73}$$

and, using Eq. (3.68), the wave function ψ_0 is given by

$$\psi_0^+ = \chi_0^+ + \frac{1}{E - H_0 - U + i\epsilon} \frac{(\Phi_0|VQ |\Psi_s) \langle \Psi_s| QV|\Phi_0)}{E - E_s} \psi_0^+. \tag{4.74}$$

The transition matrix element is given by

$$T_{fi} = \left\langle \phi |U| \chi_0^+ \right\rangle + \frac{(\Phi_0|VQ |\Psi_s) \langle \Psi_s| QV|\Phi_0)}{E - E_s} \psi_0^+, \tag{4.75}$$

where the first term is the transition matrix element for elastic potential scattering. We now manipulate Eq. (4.74) in an algebraic manner by multiplying on the left with $(E - E_s) \langle \Psi_s| QV|\Phi_0)$ and integrating over all coordinates. This gives

$$\left[E - E_s - \langle \Psi_s| QV|\Phi_0) \frac{1}{E - H_0 - U + i\epsilon} (\Phi_0|VQ |\Psi_s) \right] \left\langle \Psi_s |QV| \psi_0^+ \Phi_0 \right\rangle$$

$$= (E - E_s) \left\langle \Psi_s |QV| \Phi_0 \chi_0^+ \right\rangle. \tag{4.76}$$

The coefficient of $\left\langle \Psi_s \left| QV \right| \psi_0^+ \Phi_0 \right\rangle$ in this expression may be rewritten using the relation

$$\lim_{\epsilon \longrightarrow +0} \frac{1}{Z - Z_0 + i\epsilon} = \frac{\mathcal{P}}{Z - Z_0} - i\pi\delta\left(Z - Z_0\right), \tag{4.77}$$

where $\mathcal{P}$ indicates the principal value, and defining the quantities Δ_s and Γ_s as

$$\Delta_s = \left\langle \Psi_s \right| QV \left| \Phi_0 \right\rangle \frac{\mathcal{P}}{E - H_0 - U} \left(\Phi_0 \left| VQ \right| \Psi_s \right) \tag{4.78}$$

$$\Gamma_s = 2\pi \left\langle \Psi_s \left| QV\delta\left(E - H_0 - U\right) VQ \right| \Psi_s \right\rangle, \tag{4.79}$$

so that Eq. (4.76) becomes

$$\left[E - E_s - \Delta_s + \frac{1}{2} i\Gamma_s \right] \left\langle \Psi_s \left| QV \right| \psi_0^+ \Phi_0 \right\rangle = \left(E - E_s \right) \left\langle \Psi_s \left| QV \right| \Phi_0 \chi_0^+ \right\rangle,$$

and finally, substituting for $\left\langle \Psi_s \left| QV \right| \psi_0^+ \Phi_0 \right\rangle$ into Eq. (4.75) we obtain

$$T_{fi} = \left\langle \phi \left| U \right| \chi_0^+ \right\rangle + \frac{\left(\chi_0^- \Phi_0 \left| VQ \right| \Psi_s \right) \left\langle \Psi_s \right| QV \left| \Phi_0 \chi_0^+ \right\rangle}{E - E_s - \Delta_s + \frac{1}{2} i\Gamma_s} \psi_0^+. \tag{4.80}$$

This is just the Breit-Wigner one-level formula that we have first seen in Section 2.8, and hence the existence of an eigenvalue of H_{QQ} at E_s gives rise to a resonance at $E_s + \Delta_s$ with width Γ_s. The energy shift Δ_s and the width Γ_s arise as a result of the coupling of the compound state Ψ_s with the product state for the entrance channel.

If the ground state of target nucleus represented by Φ_0 is a closed shell, the product state in the entrance channel, represented by $\chi\Phi_0$ is a one-particle, no-hole state. In contrast the compound state consists of linear combinations of shell model states in which at least one nucleon has been excited. In the framework of the shell model we associate QV with the *residual interaction* and if this is a two-body interaction the only components of Ψ_s which will connect with the entrance channel are those that differ by a single *particle-hole interaction*, i.e. the components with two-particle one-hole character. The compound nucleus occurs through a succession of collisions of the incident nucleon with the target nucleons. In the first collision a two-particle one-hole state is formed through the action of the residual interactions; such a state is called a *doorway state* since it alone is connected with the incident channel and it is coupled to more complicated particle-hole configurations by successive collisions.

4.7   Empirical optical potential

In the previous section we have developed a formal theory of the optical potential. In practice, however, the complications involved in the inclusion of all relevant reaction channels is avoided by the use of empirical optical potentials. In this model the interaction between the nuclei in a reaction is described by a potential $U(r)$, being r the distance between the center of mass of the two nuclei. This idea is similar to the one of the shell model. It replaces the complicated interaction that a nucleon has with the rest of the nucleus with a potential that acts on the nucleon. The potential $U(r)$ includes a complex part that, as we shall see, takes into account the absorption effects, i.e., the inelastic scattering. The nuclear scattering is treated in similar form as the scattering of light by a transparent glass sphere and the name of the model derives from this analogy. In the case of the light, the absorption is included by using a complex refraction index.

In its most commonly used form, the optical potential is written as the sum

$$U(r) = U_R(r) + U_I(r) + U_D(r) + U_S(r) + U_C(r), \tag{4.81}$$

which contains parameters that can vary with the energy and the masses of the nuclei and that should be chosen by an adjustment to the experimental data. Obviously, the optical potential $U(r)$ will only make sense if these variations are small for close masses or neighboring energies.

The first part of (4.81),

$$U_R(r) = -Vf(r, R, a), \tag{4.82}$$

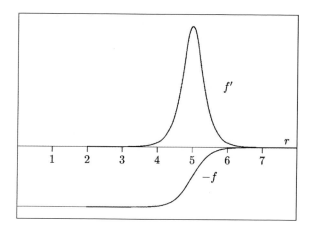

FIGURE 4.8 Schematic representation of the Woods-Saxon factor (4.83) and of its derivative.

is real and represents a nuclear well with depth V, being multiplied by a Woods-Saxon form factor

$$f(r, R, a) = \{1 + \exp[(r - R)/a]\}^{-1}, \tag{4.83}$$

where R is the radius of the nucleus and a measures the diffuseness of the potential, i.e., the width of the region where the function f is sensibly different from 0 or 1 (see Figure 4.8). This produces a well with round borders, closer to the reality than a square well. In (4.82) and (4.83), V, R and a are treated as adjustable parameters.

The absorption effect or, in other words, the disappearance of particles from the elastic channel, is taken into account including the two following imaginary parts:

$$U_I(r) = -iW f(r, R_I, a_I), \tag{4.84}$$

and

$$U_D(r) = 4ia_I W_D \frac{d}{dr} f(r, R_I, a_I). \tag{4.85}$$

An imaginary part produces absorption. It is easy to see this for the scattering problem of the square well: if an imaginary part is added to the well,

$$
\begin{aligned}
U(r) &= V_0 - iW_0 && (r < R) \\
&= 0 && (r > R),
\end{aligned} \tag{4.86}
$$

it appears in the value of $K = [2m(E + V_0 + iW_0)]^{1/2}/\hbar$. This will produce a term of decreasing exponential type in the internal wave function, as in Eq. (4.35). Thus, it corresponds to an absorption of particles from the incident beam.

The expression (4.84) is responsible for the absorption in the whole volume of the nucleus, but (4.85), built from the derivative of the function f, acts specifically in the region close to the nuclear surface, where the form factor f suffers its largest variation (see Figure 4.8). These two parts have complementary goals: at low energies there are no available unoccupied states for nucleons inside the nucleus and the interactions are essentially at the surface. In this situation $U_D(r)$ is important and $U_I(r)$ can be ignored. On the other hand, at high energies the incident particle has larger penetration and in this case the function $U_I(r)$ is important.

As with the shell model potential, a *spin-orbit interaction* term is added to the optical potential (see appendix A for the discussion of bound states in a real potential). This term, which is the fourth part of (4.81), is usually written in the form

$$U_S(r) = \mathbf{s} \cdot \mathbf{1} \left(\frac{\hbar}{m_\pi c^2}\right)^2 V_s \frac{1}{r} \frac{d}{dr} f(r, R_S, a_S), \tag{4.87}$$

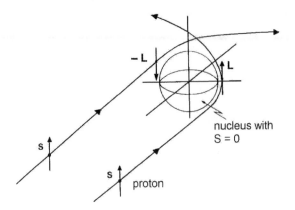

FIGURE 4.9   Scattering of polarized protons off nuclei.

incorporating a normalization factor that contains the mass of the pion m_π. **s** is the spin operator and **l** the angular orbital momentum operator. As with $U_D(r)$, the part $U_S(r)$ is also only important at the surface of the nucleus since it contains the derivative of the form factor (4.83). The values of V_S, R_S and a_S must be adjusted by the experiments. The spin orbit interaction leads to asymmetric scattering due to the different signs of the product **s** $\cdot$ **l** as the projectile passes by one or the other side of the nucleus (see Figure 4.9).

The presence of the term (4.87) is necessary to describe the effect of *polarization*. Through experiments of double scattering it can be verified that proton or neutron beams suffer strong polarization at certain angles. This means that the quantity

$$P = \frac{N_u - N_d}{N_u + N_d}, \tag{4.88}$$

where N_u is the number of nucleons in the beam with spin upward and N_d with spin downward, has a value significantly different from zero at these angles. With the inclusion of the term (4.87), the optical model is able to reproduce in many cases the experimental values of the polarization (4.88).

Finally, a term corresponding to the Coulomb potential is added to (4.81) whenever the scattering involves charged particles. It has the form

$$U_C(r) = \frac{Z_1 Z_2 e^2}{2R_c}\left(3 - \frac{r^2}{R_c^2}\right) \qquad (r \leq R_c)$$

$$= \frac{Z_1 Z_2 e^2}{r} \qquad (r > R_c), \tag{4.89}$$

where it is assumed that the nucleus is a homogeneously charged sphere of radius equal to the *Coulomb barrier radius* R_c, which defines the region of predominance of each one of the forces - nuclear or Coulomb.

Figures 4.10 and 4.11 exhibit the result of the application of Eq. (4.81) to the elastic scattering of 17 MeV protons on several light nuclei. The angular distribution viewed in Figure 4.10 is very well reproduced by the model, which also reproduces correctly (Figure 4.11) the polarization (4.88) for copper as a function of the scattering angle.

Note that in the case of inclusion of a spin-orbit potential the elastic cross section is calculated as

$$\frac{d\sigma}{d\Omega} = |F(k,\theta)|^2 + |G(k,\theta)|^2, \tag{4.90}$$

where k and θ are the magnitude of the momentum and the scattering angle, respectively. The

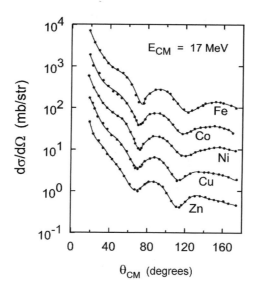

FIGURE 4.10 Angular distribution of the elastic scattering of 17 MeV protons on nuclei in the region $Z = 26 - 30$.

so-called spin-no-flip and spin-flip amplitudes F and G can be expanded in partial waves as

$$F(k, \theta) = \sum_{l \geq 0} \left[(l + 1) f_{l+}(k) + l f_{l-}(k) \right] P_l (\cos \theta),$$

$$G(k, \theta) = \sum_{l \geq 1} \left[f_{l+}(k) - f_{l-}(k) \right] \sin \theta \frac{d}{d \cos \theta} P_l (\cos \theta), \qquad (4.91)$$

where P_l is a Legendre polynomial and the partial-wave amplitudes $f_{l\pm}$ are related to the phase-shifts by

$$f_{l\pm}(k) = \frac{1}{2ik} \left[\exp(2i\delta_{l\pm}) - 1 \right]. \qquad (4.92)$$

$\delta_{l\pm}$ are the nuclear phase-shifts for $j = l \pm \frac{1}{2}$.

The optical model has a limited group of adjustable parameters and is not capable to describe abrupt variations in the cross sections, as it happens for isolated resonances. However, it can do a good description of the cross sections in the presence of the oscillations of large width in the continuous region, as it treats these as an undulatory phenomenon. Such is the case of Figure 3.7 of Chapter 3, that shows the total cross section for neutrons incident on targets of cadmium and lead.

For the scattering of low-energy nucleons off heavy nuclei, the surface difuseness parameters of the optical potentials given above is usually taken as

$$R = r_0 A^{1/3}, \qquad \text{with} \qquad r_0 = 1.25 \text{ fm},$$

$$a = 0.65 \text{ fm}, \qquad \text{and} \qquad a_I = a_S = 0.47 \text{ fm}. \qquad (4.93)$$

For nuclei heavier than oxygen, i.e. $A \geq 16$ the parameters V, W, U_I and U_D are practically independent of particle number. However, they depend on the nucleon's energy E. For neutrons

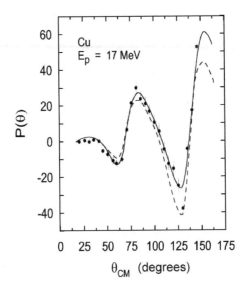

FIGURE 4.11   The polarization value (4.88) in the scattering of 9.4 MeV protons on copper. The curves are, in both cases, obtained with fits from the optical model [8].

with energy $E < 100$ MeV one finds

$$V \simeq -50 \text{ MeV} - 48 \text{ MeV} \left(\frac{N-Z}{A}\right) \widehat{t_z} + 0.3 \left(E - U_C(r)\right)$$

$$W(E) \simeq \max\left\{0.22E - 2 \text{ MeV}, \ 0\right\}$$

$$W_D \simeq \max\left\{12 \text{ MeV} - 0.25E + 24 \text{ MeV} \left(\frac{N-Z}{A}\right)\widehat{t_z}, \ 0\right\}$$

$$V_S \simeq 30 \text{ MeV fm}^2. \qquad (4.94)$$

A few remarks are in order here on these potentials and on the values of the parameters. First of all it must be stressed that we are dealing with a *phenomenological ansatz*. As always such an ansatz is largely governed by the desire for simplicity. Already from this fact it becomes evident that we must not expect the parameters to be known precisely. However, the uncertainties turn out to be not larger than about 10 to 15%, which is quite small, in particular with respect to our present purpose of learning about gross features. Indeed, they allow us to understand many basic physical aspects of nuclear physics.

The optical potential is an average over the several complicated interactions which occur in the process of forming the compound nucleus. The terms proportional to $N - Z$ are due to isospin effects. Since the nucleon-nucleon interactions contain a term of the form $\mathbf{t}_i \cdot \mathbf{t}_j$, the averaging process implied in the definition of the optical potential will give rise to a term of the form $\mathbf{t} \cdot \mathbf{T}$. The total isospin of the system can take the values $T \pm \frac{1}{2}$ where T is the isospin of the target, so that the eigenvalues of $\mathbf{t} \cdot \mathbf{T}$ are $+\frac{1}{2}T$ and $-\frac{1}{2}(T+1)$. Now for the incident neutrons the total spin can take the value $T + \frac{1}{2}$ (since $t_z = +\frac{1}{2}$), so that the real part of the potential can be written as

$$V_n = V_0 + \frac{(\mathbf{t} \cdot \mathbf{T})_N}{A} V_1 = V_0 + \frac{T}{2A} V_1,$$

where V_1 is the strength of the isospin dependent term and the factor $1/A$ is included for convenience. For protons, both total isospins values can occur and the coupling coefficients for the two cases yield the weighting factors $(2T + 1)^{-1}$ and $2T(2T + 1)^{-1}$. Thus, the mean potential for the protons is given by

$$V_p = \frac{1}{2T+1}\left[V_0 + \frac{T}{2A}V_1\right] + \frac{2T}{2T+1}\left[V_0 - \frac{(T+1)}{2A}V_1\right] = V_0 - \frac{T}{2A}V_1.$$

Since the isospin can be replaced by the neutron excess $(N - Z)$ through the relation $T = (N - Z)/2$ the real part of the optical potential can be written as

$$V = V_0 \pm \frac{1}{4} \left(\frac{N - Z}{A} \right) V_1, \tag{4.95}$$

where the plus sign refers to neutrons and the minus sign to protons. The constant V_1 is found to be approximately 100 MeV and opposite in sign to V_0, and this means that the real potential is stronger for protons than for neutrons. The second term in Eq. (4.95) is called the *symmetry term*.

For a precise determination of the parameters of the optical potential for a given energy and target a complete range of experimental data is required, e.g. differential cross-section and polarization over a wide range of angles and the absorption cross-section. Very often such a wide range of data is not available, and then certain ambiguities arise. The most important of these is the VR^n-ambiguity, where V and R are the depth and radius, respectively, of the real part of the potential and n is a constant which is about 2; if V and R are varied in such a way that VR^n remains constant the calculated differential cross-section is insensitive to the variation.

The formal theory of the optical potential shows that the optical potential is in general non-local. If V is a local interaction and exchange effects are neglected, the term $(\Phi_0| V |\Phi_0)$ in Eq. (4.71) is a single-particle potential which is local in configuration space (i.e. it has the form $V(\mathbf{r})$) and represents the interaction between the incident nucleon and the target nucleus in its ground state (potential scattering). The second term represents scattering which proceeds through an intermediate excited state of the nucleus, the operator $(E - H_{QQ})^{-1}$ being the propagator for the nucleon within the excited nucleus (see section 3.9). The presence of the operator $(E - H_{QQ})^{-1}$ also causes the potential to be non-local in configuration space. This means that the representation of V_{opt} in configuration space has the form

$$\langle \mathbf{r} | V_{\text{opt}} | \mathbf{r} \rangle = V(\mathbf{r}) \delta (\mathbf{r} - \mathbf{r}') + K (\mathbf{r}, \mathbf{r}'),$$

so that Eq. (4.70) becomes

$$[E - H_0 - V(\mathbf{r})] \psi(\mathbf{r}) = \int K(\mathbf{r}, \mathbf{r}') \psi(\mathbf{r}') d\mathbf{r}'.$$

In this special case the optical potential is real, but in general the presence of the operator $(E - H_{QQ} + i\epsilon)^{-1}$ will cause the potential to be complex.

In calculations with a phenomenological potential the effects of this non-locality are studied through the introduction of the range of the non-locality β as a parameter. It is customary to assume that the non-locality occurs only in the spin-independent term and can be described by the same form factors for the real and imaginary parts, although these assumptions are not entirely justified. In an extensive study of neutron elastic scattering, Perey and Buck [9] have shown that data covering the energy range 0.4-24 MeV can be fitted with an energy-independent non-local potential of the form

$$V(\mathbf{r}, \mathbf{r}') = U \left(\frac{|\mathbf{r} + \mathbf{r}'|}{2} \right) H \left(\frac{|\mathbf{r} - \mathbf{r}'|}{\rho} \right),$$

where U and H are taken to be of Woods-Saxon and Gaussian form respectively. They have also shown that it is possible to find an equivalent local, but energy-dependent, potential. Thus it appears that a part, at least, of the observed energy dependence of the local optical potential is due to neglect of the non-locality. The local optical potentials for complex projectiles are also energy dependent.

Many attempts have been made to calculate optical potentials from the fundamental theory. A realistic calculation of this type is an exercise in many-body theory since the formalism demands a knowledge of the particle-hole configurations involved and also the blocking effect of the Pauli principle must be taken into account at low and medium energies. These complications particularly affect the calculation of the imaginary part of the potential. For the real part, some success has been achieved using the simple formula

$$V_{opt} = \int v_{eff}(s) \, d^3 r_1 d^3 r_2 \rho_A(\mathbf{r}_1) \rho_B(\mathbf{r}_2),$$

where $\rho_i(\mathbf{r}')$, with $i = A, B$, are the nuclear matter distributions of the nuclei A and B with their center of mass at a distance $\mathbf{r}$, and $\mathbf{s} = \mathbf{r} + \mathbf{r}_2 - \mathbf{r}_1$. This expression is obtained by writing the interaction potential as a sum of two-body interactions between the projectile and the target nucleons, and has been applied to the scattering of nucleons and composite particles from nuclei at medium energies. One example is the so-called M3Y interaction [10, 11] which is given by two direct terms with different ranges, and an exchange term represented by a delta interaction:

$$v_{eff}(s) = A\frac{e^{-\beta_1 s}}{\beta_1 s} + B\frac{e^{-\beta_2 s}}{\beta_2 s} + C\delta(\mathbf{s}), \tag{4.96}$$

where $A = 7999$ MeV, $B = -2134$ MeV, $C = -276$ MeV fm^3, $\beta_1 = 4\ fm^{-1}$, and $\beta_2 = 2.5\ fm^{-1}$. Other set of parameters are found in Ref. [10].

4.8 Compound nucleus formation

It is time to analyze the mechanisms that play a role during a nuclear collision leading to the formation of a compound nucleus. The action of these mechanisms depends on the collision energy in a very pronounced way and results of the study of a given nuclear reaction only have validity for certain range of energy.

We shall assume initially that the incident particle is a neutron of low energy (< 50 MeV). When such a neutron enters the field of nuclear forces it can be scattered or begin a series of collisions with the nucleons. The products of these collisions, including the incident particle, will continue in their course, leading to new collisions and new changes of energy. During this process one or more particles can be emitted and they form with the residual nucleus the products of a reaction that is denominated by *pre-equilibrium*. But, at low energies, the largest probability is the continuation of the process so that the initial energy is distributed to all the nucleons, with no emitted particle. The final nucleus with $A + 1$ nucleons has an excitation energy equal to the kinetic energy of the incident neutron plus the binding energy the neutron has in the new, highly unstable, nucleus. It can, among other processes, emit a neutron with the same or smaller energy to the one absorbed. The de-excitation processes are not necessarily immediate and the excited nucleus can live a relatively long time. In this situation we say that there is the formation of a *compound nucleus* as intermediary stage of the reaction. In the final stage the compound nucleus can evaporate one or more particles, fission, etc. In our notation, for the most common process in which two final products are formed (the evaporated particle plus the residual nucleus or two fission fragments, etc.) we write:

$$a + A \rightarrow C^* \rightarrow B + b, \tag{4.97}$$

the asterisk indicating that the compound nucleus C is in an excited state.

The compound nucleus lives enough time to "forget" the way it was formed and the de-excitation to the final products b and B only depends on the energy, angular momentum and parity of the quantum state of the compound nucleus. An interesting experimental verification was accomplished by S. N. Ghoshal in 1950 [12]. He studied two reactions that take to the same compound nucleus, ^{64}Zn*, and he measured the cross sections of three different forms of decay, as shown in the scheme below:

$$
\begin{array}{ccc}
\text{p} + \ ^{63}\text{Cu} \searrow & \nearrow & ^{63}\text{Zn} + \text{n} \\
^{64}\text{Zn}^* & \rightarrow & ^{62}\text{Cu} + \text{n} + \text{p} \\
\alpha + \ ^{60}\text{Ni} \nearrow & \searrow & ^{62}\text{Zn} + 2\text{n}
\end{array}
\tag{4.98}
$$

The ^{64}Zn can be formed by the two reactions and decay through the three ways indicated in (4.98). If the idea of the compound nucleus is valid and if we choose the energy of the proton and of the incident α-particle to produce the same excitation energy, then the cross section for each one of the three exit channels should be independent of the way the compound nucleus was formed, that is, the properties of the compound nucleus do not have any relationship with the nuclei that formed it. This is confirmed in Figure 4.12, where one sees clearly that the cross sections depend practically

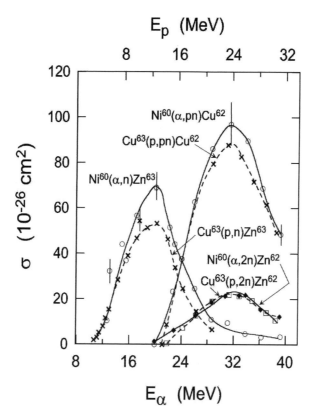

FIGURE 4.12 Cross sections for the reactions shown in (4.98). The scales of the upper axes (energy of the protons) and lower axes (energy of the α-particle) were adjusted to correspond to the same excitation energy of the compound nucleus [12].

only on the exit channels. This is called by the *independence hypothesis* for the compound nucleus formation.

Another particularity that should happen in reactions in which there is formation of a compound nucleus refers to the angular distribution of the fragments, or evaporated particles: it should be isotropic in the center of mass, and this is verified experimentally. We know, however, that the total angular momentum is conserved and cannot be "forgotten". Reactions with large transfer of angular momentum, as the ones that happen when heavy ions are used as projectiles, can show a non-isotropic angular distribution in the center of mass system.

The occurrence of a nuclear reaction in two stages allows that the cross section for a reaction A(a,b)B be written as the product,

$$\sigma(a,b) = \sigma_{CN}(a,A)P(b), \tag{4.99}$$

where $\sigma_{CN}(a,A)$ is the cross section of formation of the compound nucleus starting from the projectile a and the target A and $P(b)$ is the probability of the compound nucleus to emit a particle b leaving a residual nucleus B. If not only the particles but the quantum numbers of entrance and exit channels are well specified, i.e., if the reaction begins at an entrance channel α ends at an exit channel β, (4.99) can be written as

$$\sigma(\alpha, \beta) = \sigma_{CN}(\alpha)P(\beta). \tag{4.100}$$

We can associate the probability $P(\beta)$ to the width Γ_β of the channel β and write:

$$P(\beta) = \frac{\Gamma(\beta)}{\Gamma}, \tag{4.101}$$

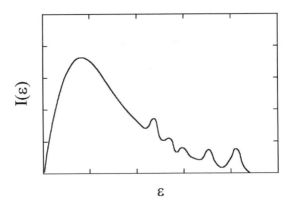

FIGURE 4.13   Energy spectrum of neutrons evaporated by a compound nucleus.

where Γ is the total width, that is, $\tau = \hbar/\Gamma$ is the half-life of disintegration of the compound nucleus. $\Gamma(\beta)$ is the partial width for the decay through channel β, and

$$\Gamma = \sum_{\beta} \Gamma(\beta).$$

In the competition between the several channels β, the nucleons have clear preference over the γ-radiation whenever there is available energy for their emission and among the nucleons the neutrons have preference as they do not have the Coulomb barrier as an obstacle. Thus, in a reaction where there is no restriction for neutron emission we can say that

$$\Gamma \cong \Gamma_n, \tag{4.102}$$

where Γ_n includes the width for the emission of one or more neutrons.

In some cases it is also useful to define the *reduced width*

$$\gamma_\beta^2 = \frac{\Gamma_\beta}{2P(\beta)},$$

where the penetrability $P(\beta)$ is the imaginary part of the logarithmic derivative of the outgoing wave in the channel β evaluated at the surface $r = R$ and multiplied by the channel radius R. For s-wave neutrons the outgoing wave is just e^{ikr} multiplied by the appropriate S-matrix element and hence $P = kR$. Thus the reduced width determines the probability that the components specified by β will appear at the surface so that the compound system can decay through this mode.

The study of the function $P(\beta)$ is done in an evaporation model that leads to results in many aspects similar to the evaporation of molecules of a liquid, with the energy of the emitted neutrons approaching the form of a Maxwell distribution

$$I(E) \propto E \exp\left(-\frac{E}{\theta}\right) dE, \tag{4.103}$$

with I measuring the amount of neutrons emitted with energy between E and $E+dE$. The quantity θ, with dimension of energy, has the role of a *nuclear temperature*. It is related to the density of levels ω of the daughter nucleus B by

$$\frac{1}{\theta} = \frac{dS}{dE}, \tag{4.104}$$

with

$$S = \ln\omega(E), \tag{4.105}$$

where, to be used in (4.103), dS/dE should be calculated for the daughter nucleus B at the maximum excitation energy that it can have after the emission of a neutron, that is, in the limit of emission of a neutron with zero kinetic energy.

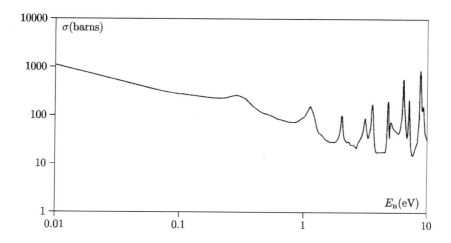

FIGURE 4.14 Total cross section for neutrons of low energy hitting ^{235}U.

The *level density* $\omega(E)$ is a measure of the number of states of available energy for the decay of the compound nucleus in the interval dE around the energy E. In this sense, the relationship (4.105) is, neglecting the absence of the Boltzmann constant, identical to the thermodynamic relationship between the entropy S and the number of states available for the transformation of a system. The entropy S defined in this way has no dimension, and (4.104) is the well-known relation between the entropy and the temperature. The last one has in the present case the dimension of energy.

The distribution of energy of neutrons emitted by a compound nucleus has the aspect of the curve shown in Figure 4.13 (see also Figure s 4.19 and 4.20). Notice that, only the low-energy part obeys (4.103) and the reason is simple: the emission of a low-energy neutron leaves the residual nucleus with a large excitation energy, a situation where the density of levels is very high. That large density of final states available turns the problem tractable with the statistical model that leads to (4.103). In the inverse situation are the contributing states of low energy of the residual nucleus. These isolated states appear as prominent peaks in the tail of the distribution. When the emission is of a proton or of another charged particle, the form of Figure 4.13 is distorted, the part of low energy of the spectrum being suppressed partially by the Coulomb barrier.

The cross section for the formation of a compound nucleus $\sigma_{CN}(\alpha)$ can be determined in a simple way if some additional hypotheses about the reaction can be done. We shall first suppose that there are only elastic scattering or formation of compound nucleus. If the projectiles are neutrons this is true for neutron energies $E_n < 50$ MeV. Then we shall assume that the elastic scattering is purely potential, without resonant elastic scattering, that is, there is no re-emission of neutrons with the same energy as the incident energy. This is equivalent to say that the probability for the exit channel to be the same as the entrance channel is very low. We shall see that this is not true for certain special values of the energy of the incident particle.

We can still write the wave function inside the nucleus as just an incoming wave

$$u_0 \cong \exp(-iKr) \qquad (r < R), \tag{4.106}$$

where $K = \sqrt{2m(E - V_0)}/\hbar$ is the wavenumber inside the nucleus, and it is assumed that the neutron with total energy E is subject to a negative potential V_0. The expression (4.106) is clearly a crude simplification for a situation where the incident neutron interacts in a complicated way with the other nucleons in the nucleus. It allows, however, to explain the average behavior of the cross sections at low energies. Starting from (4.106) one determines the value of $\mathcal{L}^I = R \ [du_0/dr]_{r=R} /u_0$:

$$\mathcal{L}^I = -iKR. \tag{4.107}$$

This means that $a = 0$ and $b = KR$ and we get the expression

$$\sigma_{\text{CN}} = \frac{\pi}{k^2} \frac{4kK}{(k+K)^2} \tag{4.108}$$

for the cross section of compound nucleus formation for neutrons with $l = 0$. At low energies, $E << |V_0|$, thus $k << K$. Under these conditions, $\sigma_{\text{CN}} = 4\pi/kK$. Thus, σ_{CN} varies with $1/k$, that is,

$$\sigma_{\text{CN}} \propto \frac{1}{v}, \tag{4.109}$$

where v is the velocity of the incident neutron. That is the well-known $1/v$ *law* that governs the behavior of the capture cross section of low-energy neutrons (Wigner's law). Figure 4.14 exhibits the *excitation function* (cross section as function of the energy) for the reaction n+^{235}U. The cross section decays with $1/v$ up to 0.3 eV, where a series of resonances start to appear. This abrupt behavior of the cross sections does not belong to the theory of the compound nucleus and the resonances appear exactly when is not possible to sustain the hypothesis that there is no return to the entrance channel, a hypothesis that was used in Eq. (4.106).

To understand these resonances, we shall use again the simple model of a single particle subject to a square-well potential. We know that inside the well the Schrödinger equation only admits solution for a discrete group of values of energy, $E_1, E_2, \cdots E_n$. A qualitative way of understanding why this happens is to have in mind that a particle is confined to the interior of the well by reflections that it has at the surface of the well.

In these reflections the wave that represents the particle should be in phase before and after the reflection and this only happens for a finite group of energies. Outside the well the Schrödinger equation does not impose restrictions and the energy can have any value. But we know from the study of the passage of a beam of particles through a potential step, that the discontinuity of the potential at the step provokes reflection even when the total energy of the particles is larger than the step, a situation where classically there would not be any difficulty for the passage of the particles. This reflection is partial and it becomes larger the closer the total energy is from the potential energy. We can say that a particle with energy slightly positive is almost as confined as a particle inside the well. From this fact results the existence of almost bound states of positive energy called by *quasi-stationary states* or *resonances*. These resonances appear as prominent peaks in the excitation function, a peak at a given energy meaning that energy coincides with a given resonance of the nucleus.

The existence of resonances can also be inferred from the properties of the wave function. In section 2.8 we have examined the simplified case where there is only elastic scattering, with the other channels closed. Let us observe the Figure 4.15. The internal and external wave functions are both sine functions, the first with wavenumber $K = \sqrt{2m(E + V_0)}/\hbar$ and the second with $k = \sqrt{2mE}/\hbar$. If E is small and V_0 is about 35 MeV, we have $K >> k$. The internal and external part should join at $r = R$ with continuous function and derivatives. As the internal frequency is much larger than the external one, the internal amplitude is quite reduced. Only at the proximity of the situation in which the derivative is zero there is a perfect matching between both and the internal amplitude is identical to the external one. The energy for which this happens is exactly the energy of resonance.

In Chapter 2 we have arrived at an expression for the cross section that describes a resonance. That formula describes elastic scattering in a resonant situation. In the case of compound nuclei we shall call it *compound elastic* cross section. A similar expression can be obtained for the reaction, or absorption cross section. Let us briefly review the concepts.

We will study resonance processes involving s-wave neutron scattering and assume that the nucleus has a well-defined surface. The nucleon does not interact with the nucleus at separation distances larger than the *channel radius* R. The logarithmic derivative is $\mathcal{L}^I = R \left[du_0/dr \right]_{r=R} / u_0$ and for a neutron wave function of the form

$$u_0 = \frac{i}{2k} \left[e^{-ikr} - \eta_0 e^{ikr} \right], \qquad r \geq R, \tag{4.110}$$

we have

$$\eta_0 = \frac{\mathcal{L}^I + ikR}{\mathcal{L}^I - ikR} e^{-i2kR}. \tag{4.111}$$

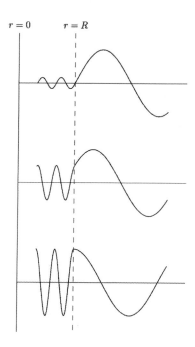

FIGURE 4.15 Connection the internal and external wave functions of the nucleus, showing as a large amplitude of the internal function is obtained at a resonance.

If $\mathcal{L}^I$ is real, $|\eta_0|$ is unity and there is no eaction, but if $\mathrm{Im}\mathcal{L}^I < 0$ then $|\eta_0| < 1$.

The scattering amplitudes and the total and reaction widths in the present case represents the scattering arising from re-emission of the absorbed neutron by the compound nucleus, or *compound-elastic scattering*. The total width is given by

$$\Gamma = -\frac{2\,(b+kR)}{a'(E_r)}, \tag{4.112}$$

so that the *reaction* (or absorption) *width* is

$$\Gamma_r = \sum_{\beta \neq \alpha} \Gamma_\beta = \Gamma - \Gamma_\alpha = -\frac{2b}{a'(E_r)}. \tag{4.113}$$

The compound elastic scattering cross section is given in our present notation,

$$\sigma_{\mathrm{ce},\alpha} = \frac{\pi}{k^2} \frac{\Gamma_\alpha^2}{(E-E_r)^2 + \Gamma^2/4}, \tag{4.114}$$

and the absorption cross section is

$$\sigma_{\mathrm{abs}} = \frac{\pi}{k^2} \frac{\Gamma_r \Gamma_\alpha}{(E-E_r)^2 + \Gamma^2/4}. \tag{4.115}$$

The cross section for compound nucleus formation is obtained by adding the cross sections for those processes which involve formation of the compound nucleus through channel α, i.e.,

$$\sigma_{\mathrm{CN}} = \sigma_{\mathrm{abs}} + \sigma_{\mathrm{ce},\alpha} = \frac{\pi}{k^2} \frac{\Gamma \Gamma_\alpha}{(E-E_r)^2 + \Gamma^2/4}, \tag{4.116}$$

and from Eqs. (4.100-4.101) the cross section for the process $\alpha \longrightarrow \beta$ is

$$\sigma_{\mathrm{abs}} = \frac{\pi}{k^2} \frac{\Gamma_\beta \Gamma_\alpha}{(E-E_r)^2 + \Gamma^2/4}. \tag{4.117}$$

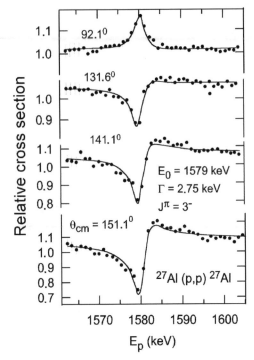

FIGURE 4.16 Differential cross section at four angles for the elastic scattering of protons off ^{27}Al, in units of the Rutherford cross section, in the neighborhood of the resonance of 1579 keV. For the larger angles we have typical interference figures between the resonant scattering and the potential scattering.

If the spins of the incident and target particles are J_a and J_A, respectively, and the incident beam is described by a single partial wave, the cross section (4.116) should be multiplied by the statistical factor (J is the total angular momentum)

$$g_J = \frac{2J+1}{(2J_a+1)(2J_A+1)}, \tag{4.118}$$

which is reduced, naturally, to the value $g = 1$ in the case of zero intrinsic and orbital angular momentum.

If the exit channel is the same as the entrance channel α, the cross section should be obtained from (2.77) and its dependence in energy is more complicated because in addition to the resonant scattering there is the potential scattering, and the cross section will contain, beyond of those two we already have, an interference term between both. The presence of these three terms results in a peculiar aspect of the scattering cross section, which differs from the simple form (4.116) for the compound nucleus cross section (see Eq. 4.131 below). This is seen in Figure 4.16 that shows the form that a resonance can take in the scattering cross section.

The region of energy where resonances show up can extend to 10 MeV in light nuclei but it ends well before this in heavy nuclei. Starting from this upper limit the increase in the density of levels with the energy implies that the average distance between the levels is smaller than the width of the levels and individual resonances cannot be observed experimentally. They form a continuum and that region is denominated the *continuum region*. But, in the continuum region, in spite that there is no characteristic narrow peaks of the individual resonances, the cross section does not vary monotonically; peaks of large width and very spaced can be observed, as we show in the example of Figure 4.17. Their presence is mainly due to interference phenomena between the part of the incident beam that passes through the nucleus and the part that passes around it [13].

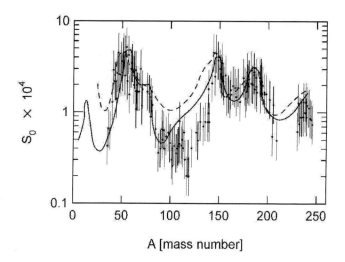

FIGURE 4.17 S-wave neutron strength function as a function of the mass number A. The solid line is the calculation of Ref. [14], while the dashed line is the calculation following Ref. [15] using deformed optical potentials.

For the decay of the compound nucleus through the channel β we can write Eq. (4.108) as

$$\sigma_{\mathrm{CN}} = \frac{\pi}{k^2} \frac{4kK}{(k+K)^2} \frac{\Gamma_\beta}{\Gamma}. \tag{4.119}$$

The quantity $4kK/(k+K)^2$ is called the *transmission coefficient* T_0 for s-wave neutrons. If $\langle \Gamma_\alpha \rangle$ is the mean width for resonances due to particles in channel α and D is the mean spacing of levels within an energy interval I we may write

$$\sigma_{\mathrm{CN}}(\alpha) = \frac{1}{I} \int_{E-I/2}^{E+I/2} \frac{\pi}{k^2} \sum_r \frac{\Gamma^r \Gamma_\alpha^r}{(E-E_r)^2 + (\Gamma^r)^2/4} dE = \frac{\pi}{k^2} \frac{2\pi}{I} \sum_r \Gamma_\alpha^r = \frac{\pi}{k^2} 2\pi \frac{\langle \Gamma_\alpha \rangle}{D}, \tag{4.120}$$

where the energy interval is chosen so that the variation of k^2 can be neglected.

Combining with Eq. (4.108) we have

$$\frac{\langle \Gamma_\alpha \rangle}{D} = \frac{1}{2\pi} \frac{4kK}{(k+K)^2} \approx \frac{2k}{\pi K}. \tag{4.121}$$

This quantity is called the s-*wave strength*. To study of the A dependence of this quantity, the comparison must be made at a fixed energy and since the strength function is measured for the different nuclei at different incident energies, one has to transform the valus thus measured to those corresponding to a fixed energy conventionally chosen to be $E_0 = 1$ eV. From Eq. (4.121) we see that the energy dependence at low energies is

$$\frac{\langle \Gamma_\alpha \rangle}{D} \propto \sqrt{E}.$$

The strength function at the conventional energy $E_0 = 1$ eV is thus related to the measured at an energy E, by the relation

$$\frac{\langle \Gamma_{0,\alpha} \rangle}{D} = \left(\frac{E_0}{E} \right)^{1/2} \frac{\langle \Gamma_\alpha \rangle}{D}.$$

In Figure 4.17 we show the neutron s-wave strength function $\langle \Gamma_{0,\alpha} \rangle /D$ as a function of A. The optical model reaction cross-section, and thus Γ_0, must have for a fixed A as a function of E (and also for a fixed E as a function of A) a resonant behavior with broad maxima whenever the energy of the neutron in the potential well coincides with that of a single neutron state. This resonant behavior, at an incident energy of $E \approx 1$ eV, is clearly evident in the figure. If one evaluates the energy of the single-particle states in a potential like Eq. (4.81) (with the imaginary part set to zero) as a function of A, one finds that the s-states have energy approximately equal to that of the potential well depth, that is practically equal to that of a neutron energy of 1 eV in the continuum, for $A \approx 13, 58, 160$. Thus the calculation made using a spherical Woods-Saxon potential shows, for $A \geq 20$, two broad maxima centered at $A \approx 58$ and 160. This calculation reproduces only the most prominent features of the experimental data. A more accurate reproduction of the data is given by a calculation with a deformed complex potential for $A \approx 140 - 200$ [15].

4.9   R-matrix

So far we have used the black-disc model to examine the s-wave phase-shifts. A more realistic S-matrix involves a description of the nuclear system in the interior of the interaction region [16]. Let us now assume that the eigenstates of the nuclear Hamiltonian in the interior region are denoted by X_λ with energy E_λ. These states are required to satisfy the boundary condition

$$r\frac{dX_\lambda}{dr} + bX_\lambda = 0 \tag{4.122}$$

at the channel radius $r = R$, where the constant b is a real number. The true nuclear wave function Ψ for the compound system is not stationary, but since the X_λ form a complete set it is possible to expand Ψ in terms of X_λ, i.e.

$$\Psi = \sum_\lambda A_\lambda X_\lambda, \tag{4.123}$$

where

$$A_\lambda = \int_0^R X_\lambda \Psi dr. \tag{4.124}$$

The differential equations for Ψ and X_λ are (for s-wave neutrons)

$$-\frac{\hbar^2}{2m}\frac{d^2\Psi}{dr^2} + V\Psi = E\Psi \tag{4.125}$$

$$-\frac{\hbar^2}{2m}\frac{d^2X_\lambda}{dr^2} + VX_\lambda = E_\lambda X_\lambda , \qquad r \leq R. \tag{4.126}$$

Multiplying Eq. (4.125) by Ψ and Eq. (4.126) by X_λ, subtracting and integrating, we have

$$\frac{\hbar^2}{2m}\left(\Psi\frac{dX_\lambda}{dr} - X_\lambda\frac{d\Psi}{dr}\right)\bigg|_{r=R} = (E - E_\lambda)\int_0^R \Psi X_\lambda dr.$$

Using Eqs. (4.122) and (4.124) we have

$$A_\lambda = (E - E_\lambda)^{-1}\frac{\hbar^2}{2mR}X_\lambda(R)\left[R\Psi'(R) + b\Psi(R)\right],$$

where the prime indicates the differentiation with respect to r. Substitutting in Eq. (4.123) gives

$$\Psi(R) = \mathcal{R}\left[R\Psi'(R) + b\Psi(R)\right], \tag{4.127}$$

where the function $\mathcal{R}$ relates the value of $\Psi(R)$ at the surface to its derivative at the surface:

$$\mathcal{R} = \frac{\hbar^2}{2mR}\sum_\lambda \frac{X_\lambda(R)X_\lambda(R)}{E_\lambda - E}. \tag{4.128}$$

Rearranging Eq. (4.127) we have

$$R\frac{\Psi'(R)}{\Psi(R)} = \frac{1 - b\mathcal{R}}{\mathcal{R}}, \tag{4.129}$$

which is just the logarithmic derivative $\mathcal{L}^I$ which can be inserted into Eq. (4.111) to determine the S-matrix element η_0 in terms of the $\mathcal{R}$ function. This gives

$$S_0 = \eta_0 = \left[1 + \frac{2ikR\mathcal{R}}{1 - (b + ikR)\mathcal{R}} \right] e^{-2ikR}.$$

Finally, we assume that E is near to a particular E_λ, say E_α, neglect all terms $\lambda \neq \alpha$ in Eq. (4.128), and define

$$\Gamma_\alpha = \hbar^2 k X_\alpha^2 (R) / m, \qquad\qquad \Delta_\alpha = -b\Gamma_\alpha / 2kR,$$

so that the S-matrix element becomes

$$S_0 = \left[1 + \frac{i\Gamma_\alpha}{(E_\alpha + \Delta_\alpha - E) - i\Gamma_\alpha/2} \right] e^{-2ikR}, \tag{4.130}$$

and the scattering cross-section is

$$\sigma_{\text{sc}} = \frac{\pi}{k^2} \left| e^{2ikR} - 1 + \frac{i\Gamma_\alpha}{(E_\alpha + \Delta_\alpha - E) - i\Gamma_\alpha/2} \right|^2. \tag{4.131}$$

From this analysis we see that the procedure of imposing the boundary conditions at the channel radius leads to isolated s-wave resonances of Breit-Wigner form (see section 2.8). If the constant b is non-zero the position of the maximum in the cross-section is shifted. The level shift does not appear in the simple form of the Breit-Wigner formula because $E_\alpha + \Delta_\alpha$ is defined as the resonance energy. In general, a nucleus can decay through many channels and when the formalism is extended to take this into account the $\mathcal{R}$-function becomes a matrix. In this $\mathcal{R}$-matrix theory the constant b is real and $X_\lambda (R)$ and E_λ can be chosen to be real so that the eigenvalue problem is Hermitian [17].

The R-matrix theory has proven over the course of time to be very useful in nuclear and atomic physics, both for the fitting of experimental data and as a tool for theoretical calculations [16]. It can be easily generalized to account for higher partial waves and spin-channels. If we define the reduced width by

$$\gamma_\lambda^2 = \frac{\hbar^2 X_\lambda^2 (R)}{2mR}, \tag{4.132}$$

which is a property of a particular state and not dependent of the scattering energy E of the scattering system, we can write

$$\mathcal{R}_{\alpha\alpha'} = \sum_\lambda \frac{\gamma_{\lambda\alpha'} \gamma_{\lambda\alpha}}{E_\lambda - E},$$

where α is the channel label. $\gamma_{\lambda\alpha}$, E_λ, and b are treated as parameters in fitting the experimental data. If we write the wave function for any channel as

$$\Psi \sim I + S_\alpha O, \tag{4.133}$$

where I and O are incoming and outgoing waves, Eq. (4.129) means

$$R \frac{I'(R) + S_\alpha O'(R)}{I(R) + S_\alpha O(R)} = \frac{1 - b\mathcal{R}}{\mathcal{R}}.$$

Thus, as in Eq. (4.130), the S-matrix is related to the R-matrix and from the above relation we obtain that,

$$S_\alpha = \frac{I(R)}{O(R)} \left[\frac{1 - \left(\mathcal{L}^I \right)^* \mathcal{R}}{1 - \mathcal{L}^I \mathcal{R}} \right]. \tag{4.134}$$

The total cross sections for states with angular momenta and spins given by l, s and J is

$$\sigma_{\alpha\alpha'} = \frac{\pi}{k_\alpha^2} \sum_{JJ'll'ss'} g_J \left| S_{\alpha Jls, \alpha'J'l's'} \right|^2, \qquad \alpha \neq \alpha', \tag{4.135}$$

with g_J given by Eq. (4.118).

In the statistical model, it can be argued that because the S-matrix elements vary rapidly with energy the statistical assumption implies that there is a random phase relation between the different components of the S-matrix. The process of energy averaging then eliminates the cross terms and gives

$$\sigma_{\text{abs}} = \sum_{\alpha' \neq \alpha, J'l's'} \sigma_{\alpha\alpha'} = \frac{\pi}{k_\alpha^2} \sum_{Jls} g_J \left[1 - |S_{\alpha Jls}|^2 \right] = \frac{\pi}{k_\alpha^2} \sum_{Jls} g_J T_{ls}^J (\alpha), \qquad (4.136)$$

where we have used the symmetry properties of the S-matrix in the form

$$\sum_{\alpha' \neq \alpha, J'l's'} S_{\alpha Jls, \alpha' J'l's'} S_{\alpha Jls, \alpha' J'l's'}^* = 1,$$

and we have introduced the general definition of the transmission coefficient

$$T_{ls}^J (\alpha) = 1 - |S_{\alpha Jls}|^2. \qquad (4.137)$$

4.10 Average of the cross sections

The theory of the empirical optical potential cannot reproduce the detailed resonance behavior of the many-body particle system, and in order to understand exactly what the cross-sections calculated from the optical potential represent we consider what is behind the average of the cross sections. We assume that the cross sections can be averaged over an energy interval I such that $D \ll I \ll E$ where D is the mean spacing of the resonances, and denote such an energy averaged quantity by $\langle \rangle$. This gives

$$\langle \sigma_{\text{scat}} \rangle = \frac{\pi}{k^2} \sum_l (2l + 1) \left\langle |1 - S_l|^2 \right\rangle$$

$$= \frac{\pi}{k^2} \sum_l (2l + 1) |1 - \langle S_l \rangle|^2 + \frac{\pi}{k^2} \sum_l (2l + 1) \left[\left\langle |S_l^2| \right\rangle - |\langle S_l \rangle|^2 \right], \qquad (4.138)$$

and

$$\langle \sigma_{\text{abs}} \rangle = \frac{\pi}{k^2} \sum_l (2l + 1) \left\langle 1 - |S_l|^2 \right\rangle = \frac{\pi}{k^2} \sum_l (2l + 1) \left(1 - \left\langle |S_l|^2 \right\rangle \right)$$

$$= \frac{\pi}{k^2} \sum_l (2l + 1) (1 - \langle S_l \rangle)^2 - \frac{\pi}{k^2} \sum_l (2l + 1) \left[\left\langle |S_l^2| \right\rangle - |\langle S_l \rangle|^2 \right]. \qquad (4.139)$$

The cross section for elastic scattering is the sum of the contribution from shape-elastic and compound elastic scattering, while the cross section for absorption is the difference between the cross-section for compound nucleus formation and compound-elastic scattering. In the continuum region as the widths become very large the spacing and the excitation functions vary smoothly with energy. Thus, we expect that

$$\left\langle |S_l^2| \right\rangle - |\langle S_l \rangle|^2 \longrightarrow 0, \qquad \sigma_{\text{ce},\alpha} \longrightarrow 0, \qquad \text{and} \qquad \sigma_{\text{CN}} = \sigma_{\text{abs}}. \qquad (4.140)$$

Thus we see that the elastic cross-section calculated from the optical potential is the shape-elastic cross section

$$\sigma_{\text{scat},\alpha} = \frac{\pi}{k^2} \sum_l (2l + 1) |1 - \langle S_l \rangle|^2, \qquad (4.141)$$

and the non-elastic cross section calculated from the optical potential is the cross section for the compound nucleus formation

$$\sigma_{\text{CN}} = \frac{\pi}{k^2} \sum_l (2l + 1) (1 - \langle S_l \rangle)^2. \qquad (4.142)$$

The validity of these results are better for high bombarding energies: for energies larger than about 10 MeV for nucleon scattering from medium and heavy nuclei and at about 20-30 MeV for light nuclei. Similarly, the differential cross section calculated from the optical potential is that for shape-elastic scattering and can be compared directly with experiment only if the compound-elastic cross section is negligible.

The single particle description of the optical potential model is evidently different than the many-particle desctiption of the compound nucleus (see Figure 4.1). The reason for the success of the optical potential model is the low-energy resolution of the experiments. In an independent-particle picture the single particle states are far apart. However, the residual interactions between the nucleons cause the nuclear states to have a more complicated structure. Since it is always possible to expand the true nuclear state in terms of a complete set of single-particle states, a given single particle state may contribute to several nuclear states, i.e. the single-particle strength is distributed over several nuclear states. If the residual interaction is weak these states are spread around the single-particle energy and the amount of mixing (overlap) of different single-particle states in the same compound state is small. Thus the low resolution of the experiments will only reveal the single-particle structure of the system.

4.11   Level densities in nuclei

A nucleus may be excited into many higher levels (or states) whose density increases rapidly with increasing excitation energy. These levels can be resolved only at low energies because as the energy increases their spacing soon becomes less than the experimental energy resolution, and also because their widths increase and eventually overlap. Since the cross-sections of many reactions depend on the level densities, it is important to know the level density as a function of excitation energy E and spin J, at energies where the levels can no longer be resolved.

For the purpose of the following discussion one must take into account that each level of spin J comprises $2J + 1$ degenerate states with different projections of J on the z-axis, so the relation between state density $\rho(A, E, J)$ and level density $\omega(A, J, E)$ is

$$\rho(A, E, J) = (2J + 1)\omega(A, E, J), \tag{4.143}$$

where the nuclear structure dependence is given through the mass number A.

These densities refer to states or levels excited in a fully equilibrated nucleus and include all possible combinations of particles and holes with the same excitation energy. If however the nucleus is not fully equilibrated, as can happen in some reactions, the nuclear states may be specified by the numbers of excited particles and holes.

Assuming that E and J are independent variables, which is only approximately true, the level density $\omega(A, E, J)$ may be factorized into the product of an energy dependent function $\rho(A, E, J)$ and a J dependent function $F(J)$. The dependence on the nuclear angular momentum J may be estimated by assuming that all the projections of the excited nucleon angular momenta on an arbitrary axis are equiprobable, and this gives

$$F(J) = \frac{2J + 1}{2\sqrt{2\pi}\sigma^3} \exp\left[-J(J + 1)/2\sigma^2\right]. \tag{4.144}$$

Comparison with measured spin distributions gives for the spin cut-off factor σ^2 the value

$$\sigma^2 \cong 0.114 A^{2/3}\sqrt{aE}, \tag{4.145}$$

and thus $F(J)$ also depends on the energy but much less sensitively than $\rho(A, E)$. (see below for a definition of the parameter a). The factor $2J + 1$ arises simply because a level of spin J actually constitutes $2J+1$ degenerate magnetic substates. The exponential cut-off reflects the fact that, as J increases, more of the excitation eenrgy E is taken up in the rotational motion and less is available for thermal motion so that the effective temperature is reduced. The more detailed considerations also lead to some additional dependence of $\rho(A, E, J)$ on E over the exponential in Eq. 4.144 as well as taking into account odd-even differences and the effects of shell structure [18].

The energy and structure dependence of the level density may be experimentally determined at low excitation energies ($E \leq 5 - 6$ MeV) by counting the cumulative number of levels $N(A, E)$ excited in many different reactions up to the energy E. This is related to the level density by

$$N(A, E) = \sum_J \int_0^E \omega(A, E', J) dE'. \tag{4.146}$$

The thermodynamical argument used to estimate the state densities is based on the relation between the entropy S and the state density ρ

$$S(A, E) = k \ln \rho(A, E). \tag{4.147}$$

If all the energy given to a nucleus becomes excitation energy without a substantial increase of the nuclear volume one has $dS = dE/\theta$ where θ is the temperature. Hence, defining a nuclear temperature $T = k\theta$ assumed to be constant in the considered energy interval one gets

$$\rho(A, E) = \rho_0 \exp(E/T). \tag{4.148}$$

A more formal derivation of these formulas is as follows [19]. We can define the *thermodynamic free energy* $F(\beta)$ of the nucleus at temperature $T = 1/\beta$ by

$$e^{-\beta F} = \int dE' \rho(E') e^{-\beta E'}. \tag{4.149}$$

We can invert the Laplace trsnform to obtain

$$\rho(E') = \frac{1}{2\pi i} \int_{\gamma - i\infty}^{\gamma + i\infty} d\beta e^{-\beta(E' - F)}. \tag{4.150}$$

The above integral can be calculated using the method of steepest descent (*saddle-point method*). The integrand has a saddle-point at $\beta = \beta(E)$ given by

$$\frac{d}{d\beta}(\beta F) = E,$$

where E has been treated as an independent variable. We note from Eq. (4.149) that

$$\frac{d}{dE}[\beta(E - F)] = \frac{d}{dE} S(E) = \beta = \frac{1}{T},$$

where

$$S(E) = \beta(E - F) = \int_0^E dE' \beta(E').$$

Equation (4.150) corresponds to the usual definition of temperature in thermodynamics. After integrating, Eq. (4.150) gives

$$\rho(E) = \frac{1}{\sqrt{-2\pi dE/d\beta}} e^S,$$

so that

$$\frac{d}{dE} \ln \rho(E) = \frac{dS}{dE} - \frac{1}{2} \frac{d}{dE} \ln\left(-\frac{dE}{d\beta}\right) = \frac{1}{T} - \frac{1}{2} \frac{d}{dE} \ln\left(T^2 \frac{dE}{dT}\right),$$

which defines the statistical nuclear temperatures. The last term in equation above is negligible so that we obtain

$$\frac{d}{dE} \ln \rho(E) \simeq \frac{dS(E)}{dE} = \frac{1}{T}, \tag{4.151}$$

which is the same as Eq. (4.147).

A more accurate expression for the energy dependence of the state density is provided by the equidistant spacing model which assumes that the one-particle states are equally spaced with

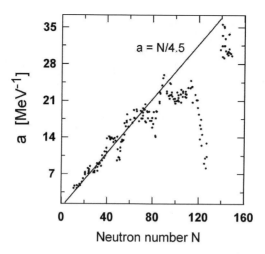

FIGURE 4.18 Values of the level density parameter a as a function of the neutron number. Far from magic regions, $a \approx N/4.5$ MeV^{-1} which approximately corresponds to $a = A/7.5$ MeV^{-1} [21].

spacing d, and that the total energy of the nucleus is simply obtained by adding the energies of the constituent nucleons. The solution of this problem can be obtained from statistical mechanics[*]

$$\rho(A, E) = \frac{1}{\sqrt{48E}} \exp\left(2\sqrt{aE}\right),$$ (4.152)

where

$$a = \frac{\pi^2}{6d}$$ (4.153)

is the level density parameter.

It is also easy to see why these formulas are reasonable. If we treat the nucleons in a nucleus as an ideal Fermi gas, similar to electrons in metal, we know that the thermal heat capacity $c = dE/dT$ is proportional to the temperature and therefore the energy E is proportional to T^2. Alternatively, if we consider the expansion of $E(T)$ at $T = 0$, then since

$$\left(\frac{dE}{dT}\right)_{T=0} = 0,$$

(because of the *Nernst heat theorem* that the specific heat vanishes at $T = 0$), the first term in the expansion of E will be proportional to T^2. We may therefore write

$$E = aT^2,$$ (4.154)

where a is a constant.

From Eqs. (4.151) and (4.154) we have

$$\frac{d}{dE} \ln \rho(E) = \sqrt{\frac{a}{E}},$$

or

$$\rho(E) \sim \exp\left(2\sqrt{aE}\right),$$

which explains Eq. (4.152).

---

[*]For a derivation, see Ref. [20].

Extensive analyzes of experimental data show that for nuclei far from the region of the magic nuclei a varies linearly with A (or with N and Z), as shown in Figure 4.18:

$$a \cong \frac{A}{k} \text{ MeV}^{-1}. \tag{4.155}$$

It is found that $k \cong 7.5 - 8$.

The Fermi gas model can be used to calculate the density of states, assuming that most of these states is distributed among many nucleons whose energy does not greatly exceed the Fermi energy. To calculate the expression for the density of states one may thus assume that the single-particle states are equidistant with spacing approximately given by the Fermi gas model at the Fermi energy. Using for (4.153) the Fermi gas model expression (4.44) for the density of nucleon states with energy approximately equal to the Fermi energy (i.e., we divide $n(E_F)/A$, given by Eq. (4.45) by $dn(E)/dE|_{E=E_F}$), one obtains

$$d = \frac{2E_F}{3A}. \tag{4.156}$$

Using this formula we find that the experimental value of k corresponds to a value of $E_F \cong 20$ MeV.

Figure 4.18 shows large deviations from the simple Fermi gas model estimate for nuclei in near magic regions (Z or $N = 50, 82, 126$). The lower values of a for the magic nuclei are due to the larger average spacing of single nucleon states around the Fermi energy as predicted by the shell model. The shell effects tend to disappear with increasing excitation energy and their average spacing approaches again that predicted by the Fermi gas model.

4.12 Compound nucleus decay: the Weisskopf-Ewing theory

At low incident energies the compound nucleus states are excited individually and each produces a resonance in the cross-section that may be described by the Breit-Wigner theory (Eq. (4.116)). As the incident energy increases compound nucleus states of higher energy are excited and these are closer together and of increasing width. Eventually they overlap and it is no longer possible to identify the individual resonances. The cross-section then fluctuates.

This fluctuating behavior is due to the interference of the reaction amplitudes corresponding to the excitation of each of the overlapping states which vanish in the energy average of the cross-section since these amplitudes are complex functions with random modulus and phase. The energy average of the cross-sections thus shows a weak energy dependence and it is predictable by the theory. To develop such a theory we consider a reaction that proceeds from the initial channel c through the compound nucleus to the final channel c'. If we forget, for the moment, that the compound nucleus may be created in states of different angular momentum J, the hypothesis of the independence of formation and decay of the compound nucleus, according to Eqs. (4.100 and 4.101) then gives for the cross-section

$$\sigma_{cc'} \sim \sigma_{CN}(c)\frac{\Gamma_{c'}}{\Gamma}, \tag{4.157}$$

where $\sigma_{CN}(c)$ is the cross-section for formation of the compound nucleus and $\Gamma_{c'}$ and Γ are, respectively, the energy-averaged width for the decay of the compound nucleus in channel c' and the energy-averaged total width.

We now use the *reciprocity theorem*, derived in the next section, that relates the cross-section $\sigma_{cc'}$ to the cross-section for the time-reversed process $c' \to c$:

$$g_c k_c^2 \sigma_{cc'} = g_{c'} k_{c'}^2 \sigma_{c'c}, \tag{4.158}$$

where $g_c = 2I_c + 1$ and $g_{c'} = 2I_{c'} + 1$ are the statistical weights of the initial and final channels, I_c and $I_{c'}$ are the spin of the projectile and the ejectile, and k_c and $k_{c'}$ their wave numbers. This gives

$$g_c k_c^2 \sigma_{CN}(c)\Gamma_{c'} = g_{c'} k_{c'}^2 \sigma_{CN}(c')\Gamma_c \tag{4.159}$$

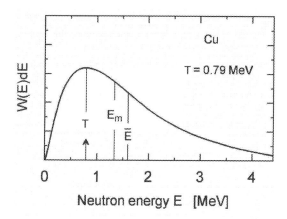

FIGURE 4.19 Spectrum of neutrons evaporated in the (n,n') reaction on Cu at 7 MeV incident energy. The residual nucleus temperature T is about 0.79 MeV, $\overline{E} = 2$ T is the mean energy of the evaporated neutrons, and $E_m = 1.7$ T is the spectrum median energy [22].

or, equivalently

$$\frac{\Gamma_c}{g_c k_c^2 \sigma_{CN}\left(c\right)} = \frac{\Gamma_{c'}}{g_{c'} k_{c'}^2 \sigma_{CN}\left(c'\right)}. \tag{4.160}$$

Since the channels c and c' are chosen arbitrarily, this relation holds for all possible channels. So,

$$\Gamma_c \propto g_c k_c^2 \sigma_{CN}\left(c\right). \tag{4.161}$$

Since the total width is obtained by summing the Γ'_cs over all open channels

$$\Gamma = \sum_c \Gamma_c, \tag{4.162}$$

the cross-section (4.157) becomes

$$\sigma_{cc'} = \sigma_{CN}\left(c\right) \frac{g_{c'} k_{c'}^2 \sigma_{CN}\left(c'\right)}{\sum_c g_c k_c^2 \sigma_{CN}\left(c\right)}. \tag{4.163}$$

Ejectiles with energy in the range $E_{c'}$ to $E_{c'} + dE_{c'}$ leave the residual nucleus with energy in the range $U_{c'}$ to $U_{c'} + dU_{c'}$ where

$$U_{c'} = E_{CN} - B_{c'} - E_{c'}, \tag{4.164}$$

and E_{CN} and $B_{c'}$ are respectively the compound nucleus energy and the binding energy of the ejectile in the compound nucleus. Introducing the density of levels of the residual nucleus $\omega(U_{c'})$, Equation (4.163) becomes

$$\sigma_{cc'} dE_{c'} = \sigma_{CN}\left(c\right) \frac{g_{c'} k_{c'}^2 \sigma_{CN}\left(c'\right) \omega\left(U_{c'}\right) dU_{c'}}{\sum_c g_c k_c^2 \sigma_{CN}\left(c\right) \omega\left(U_c\right) dU_c}, \tag{4.165}$$

or, since $k^2 = 2\mu E$,

$$\sigma_{cc'}\left(E_{c'}\right) dE_{c'} = \sigma_{CN}\left(c\right) \frac{\left(2I_{c'}+1\right)\mu_{c'} E_{c'} \sigma_{CN}\left(c'\right) \omega(U_{c'}) dU_{c'}}{\sum_c \int_0^{E_c^{\max}} \left(2I_c+1\right)\mu_c E_c \sigma_{CN}\left(c\right) \omega(U_c) dU_c}, \tag{4.166}$$

where μ_c is the reduced mass of the ejectile c. This is the *Weisskopf-Ewing formula* for the angle-integrated cross-sections.

To a good approximation, the level density $\omega(U) \propto \exp(U/T)$, so the ejectile spectrum given by the Weisskopf-Ewing theory is Maxwellian. It rises rapidly above the threshold energy, attains a maximum and then falls exponentially as shown in Figure 4.19 (see also Figure 4.13).

Since (4.161) is a proportionality relation it does not allow one to evaluate the absolute values of the decay widths. This may be obtained by use of the *detailed balance principle* which in addition to the invariance for time reversal leading to the reciprocity theorem (4.158) implies the existence of a long-lived compound nucleus state. This principle states that two systems a and b, with state densities ρ_a and ρ_b, are in statistical equilibrium when the depletion of the states of system a by transitions to b equals their increase by the time-reversed process $b \longrightarrow a$.

If $W_{ab} = \Gamma_{ab}/\hbar$ is the decay rate (probability per unit time) for transitions from a to b and $W_{ba} = \Gamma_{ba}/\hbar$ is the decay rate for the inverse process, this equality occurs when

$$\rho_a \Gamma_{ab} = \rho_b \Gamma_{ba}. \tag{4.167}$$

In the case we are interested in, a is the compound nucleus with energy E_{CN} and, if one neglects the spin dependence, its state density $\rho_{CN}(E_{CN})$ coincides with its density of levels $\omega_{CN}(E_{CN})$ (this is not true for a system with spin J since $(2J+1)$ states correspond to each level). Γ_{ab} is the width $\Gamma_{c'}$ for decay in channel c'. System b is constituted by the ejectile c' with energy from $E_{c'}$ to $E_{c'} + dE_{c'}$ and a residual nucleus with excitation energy from $U_{c'}$ to $U_{c'} + dU_{c'}$. Thus ρ_b is the product of the density of continuum states of c' and the density of levels of the residual nucleus:

$$\rho_b = \rho_{c'}(E_{c'})\omega(U_{c'}). \tag{4.168}$$

The density of the continuum states of c' is given by the Fermi gas model expression (4.44), i.e.,

$$\rho_{c'}(E_{c'}) = \frac{\mu_{c'} E_{c'} V}{\pi^2 \hbar^3 v_{c'}} g_{c'} dE_{c'}, \tag{4.169}$$

where $\mu_{c'}$ and $v_{c'}$ are, respectively, the reduced mass and the velocity of c' and V is the space volume. The decay rate for the inverse process is

$$W_{c'c} = \frac{v_{c'} \sigma_{c'}(E_{c'})}{V}, \tag{4.170}$$

where $\sigma_{c'}(E_{c'})$ is the cross-section for the inverse process (formation of the compound nucleus with energy E_{CN} from channel c') which may be evaluated with the optical potential model.

From relations (4.167) - (4.170) one finally gets

$$\Gamma_{c'} = \frac{1}{\omega_{CN}(E_{CN})} \frac{(2I_{c'}+1)\mu_{c'} E_{c'}}{\pi^2 \hbar^2} \sigma_{c'}(E_{c'})\omega(U_{c'}) dE_{c'}. \tag{4.171}$$

The Weisskopf-Ewing theory provides a simple way of estimating the energy variation, at low incident energies, of the cross-sections of all available final channels in a particular reaction, and an example of such a calculation is shown in Figure 4.20.

4.13   Reciprocity theorem

Extending the partial-wave expansion theory to include specific channels, the cross section for the reaction $\alpha \longrightarrow \beta$ is, as usual, given by the ratio of the outgoing flux in channel β and the incident flux in channel α. This gives for the angle-integrated cross section of a reaction $\alpha \longrightarrow \beta$ $(\alpha \neq \beta)$

$$\sigma_{\alpha\beta} = \frac{\pi}{k^2} \sum_l (2l+1) \left| S_l^{\alpha\beta} \right|^2, \tag{4.172}$$

and for the angle-integrated elastic cross section

$$\sigma_{\alpha\alpha} = \frac{\pi}{k^2} \sum_l (2l+1) |1 - S_l^{\alpha\alpha}|^2. \tag{4.173}$$

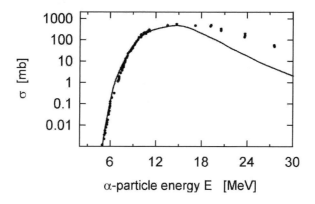

FIGURE 4.20 Excitation function for the ^{58}Ni(α,p) reaction compared with Weisskopf-Ewing calculations. The disagreement between experimental data and theoretical calculation at the higher energies, is due to the emission of *pre-equilibrium* protons.

The two expressions can be unified by writing

$$\sigma_{\alpha\beta} = \frac{\pi}{k^2} \sum_l (2l+1) \left| \delta_{\alpha\beta} - S_l^{\alpha\beta} \right|^2. \tag{4.174}$$

The scattering amplitude can be written as

$$f_{\alpha\beta}(\theta) = \frac{1}{2ik} \sum_l (2l+1) P_l(\cos\theta) \left| \delta_{\alpha\beta} - S_l^{\alpha\beta} \right|. \tag{4.175}$$

In terms of its partial-wave components, the unitarity of the S-matrix can be written as

$$\sum_\alpha S_l^{\beta\alpha} S_l^{\alpha\gamma\dagger} = \sum_\alpha S_l^{\beta\alpha} S_l^{\gamma\alpha*} = \delta_{\gamma\beta}, \tag{4.176}$$

and so

$$S_l^{\gamma\alpha} = S_l^{\alpha\gamma}. \tag{4.177}$$

From Eqs. (4.172) and (4.177) one immediately gets

$$\frac{\sigma_{\alpha\beta}}{\lambda_\alpha^2} = \frac{\sigma_{\beta\alpha}}{\lambda_\beta^2}, \qquad \text{or} \qquad k_\alpha^2 \sigma_{\alpha\beta} = k_\beta^2 \sigma_{\beta\alpha}, \tag{4.178}$$

which is known as the reciprocity theorem.

If we consider the scattering of spin particles, the reciprocity theorem has to account for the statistical weights of the channels α and β. The relation (4.178) becomes then

$$g_\alpha k_\alpha^2 \sigma_{\alpha\beta} = g_\beta k_\beta^2 \sigma_{\beta\alpha}, \tag{4.179}$$

where g_α and g_β are the total number of spin states in channels α and β, respectively, i.e., $g_\alpha = (2I_a + 1)(2I_A + 1)$ and $g_\alpha = (2I_b + 1)(2I_B + 1)$.

4.14 The Hauser-Feshbach theory

The Weisskopf-Ewing theory depends only on the nuclear level density and on the compound nucleus formation cross-section which may be easily obtained from optical model potentials. It is thus simple to use, but it has the disadvantage that it does not explicitly consider the conservation of angular momentum and does not give the angular distribution of the emitted particles. This is provided by the *Hauser-Feshbach theory*. This theory takes into account the formation of the compound nucleus in states of different J and parity π. Let us consider the case of a reaction leading from the initial channel c to a final channel c'. Since J and parity are good quantum numbers, the cross-section $\sigma_{cc'}$ is a sum of terms each corresponding to a given J^π

$$\sigma_{cc'} = \sum_{J\pi} \sigma_{cc'}^{J\pi}. \tag{4.180}$$

When one averages over the energy one further assumes that the independence hypothesis holds for each $\sigma_{cc'}^{J\pi}$. Thus,

$$\sigma_{cc'}^{J\pi} = \sigma_{CN}^{J\pi}(c) \frac{\Gamma_{c'}^{J\pi}}{\Gamma^{J\pi}}. \tag{4.181}$$

Repeating the procedure used to obtain the Weisskopf-Ewing expression gives

$$\Gamma_c^{J\pi} \propto g_c k_c^2 \sigma_{CN}^{J\pi}(c), \tag{4.182}$$

where now $g_c = (2i_c + 1)(2I_c + 1)$ and i_c and I_c are, respectively, the projectile and target spin in channel c.

If there is no pre-equilibrium emission, one may identify the compound nucleus formation cross-section

$$\sigma_{CN} = \sum_{J,\pi} \sigma_{CN}^{J\pi}, \tag{4.183}$$

with the optical model reaction cross-section

$$\sigma_R = \frac{\pi}{k^2} \sum_l (2l + 1) T_l, \tag{4.184}$$

which, if the transmission coefficients $T_l = 1 - |\langle S_l \rangle|^2$ do not depend on J, may also be written as[‡]

$$\sigma_{CN} = \frac{\pi}{k^2} \sum_{l=0}^{\infty} \sum_{s=|I-i|_{\min}}^{I+i} \sum_{J=|l-s|_{\min}}^{l+s} \frac{(2J + 1)}{(2i + 1)(2I + 1)} T_l, \tag{4.185}$$

where $\mathbf{s}$ is the channel spin $\mathbf{s} = \mathbf{i} + \mathbf{I}$. Inverting the order of summation over the various indexes, a change which does not affect the result of the sum, one gets

$$\sigma_{CN} = \frac{\pi}{k^2} \sum_J \frac{(2J + 1)}{(2i + 1)(2I + 1)} \sum_{s=|I-i|_{\min}}^{I+i} \sum_{l=|J-s|_{\min}}^{J+s} T_l. \tag{4.186}$$

Thus, comparing (4.186) with (4.183), one may write

$$\sigma_{CN} = \frac{\pi}{k^2} \frac{(2J + 1)}{(2i + 1)(2I + 1)} \sum_{s,l} T_l. \tag{4.187}$$

Using relations (4.182) and (4.187) one easily gets

$$\frac{\Gamma_c^{J\pi}}{\Gamma^{J\pi}} = \frac{\Gamma_c^{J\pi}}{\sum_c \Gamma_c^{J\pi}} = \frac{\sum_{s,l} T_l(c)}{\sum_c \sum_{s,l} T_l(c)}. \tag{4.188}$$

Thus the cross-section (4.180) for transition from the initial channel c to the final channel c' may be written

$$\sigma_{cc'} = \frac{\pi}{k^2} \sum_J \frac{(2J + 1)}{(2i_c + 1)(2I_c + 1)} \frac{\sum_{s,l} T_l(c) \sum_{s',l'} T_{l'}(c')}{\sum_c \sum_{s,l} T_l(c)}. \tag{4.189}$$

[‡] $\langle S_l \rangle$ is the average value of the scattering amplitude over several overlapping resonances.

This is the *Hauser-Feshbach expression* for the energy-averaged angle-integrated cross-section of statistical reactions. The compound nucleus states may be both of positive and negative parity. Since parity is conserved, in evaluating (4.189), one must take into account that the parity of compound nucleus states and the parity of the residual nucleus states may impose restrictions to the values of the emitted particle angular momentum. Thus, positive parity compound nucleus states decay to positive parity states of the residual only by even angular momenta and to negative parity residual nucleus states by odd angular momenta.

4.15   Appendix 4.A – The shell model

The exact Hamiltonian for a problem of A bodies can be written has

$$H = \sum_i^A T_i(\mathbf{r}_i) + V(\mathbf{r}_1, \cdots, \mathbf{r}_A), \tag{4.190}$$

where T is the kinetic energy operator and V the potential function.

If we restrict ourselves to two body interactions (i.e., nucleon-nucleon interaction), Eq. (4.190) takes the form

$$H = \sum_i^A T_i(\mathbf{r}_i) + \frac{1}{2} \sum_{ji} V_{ij}(\mathbf{r}_i, \mathbf{r}_j). \tag{4.191}$$

In the essence of the model, the nucleon i does not feel the potential $\sum_j V_{ij}$, but a central potential $U(r_i)$, that only depends on the coordinates of nucleon i. This potential can be introduced in (4.191), with the result

$$H = \sum_i^A T_i(\mathbf{r}_i) + \sum_i^A U(r_i) + H_{\text{res}}, \tag{4.192}$$

$$H_{\text{res}} = \frac{1}{2} \sum_{ji} V_{ij}(\mathbf{r}_i, \mathbf{r}_j) - \sum_i U(r_i). \tag{4.193}$$

H_{res} refers to the residual interactions, that is, to the part of potential V not embraced by the central potential U. The hope of the shell model is that the contribution of H_{res} is small or, in another way, that the shell model Hamiltonian,

$$H_0 = \sum_{i=1}^A \left[T_i(r_i) + U(r_i) \right], \tag{4.194}$$

represents a good approximation for the exact expression of H. Later we shall see that part of the lost accuracy when we pass from (4.191) to (4.194) can be recovered by an approximated treatment of the effect of the residual interaction H_{res}.

The solutions $\Psi_1(r_1), \Psi_2(r_2), \cdots$, of equation

$$H_0 \Psi = E \Psi, \tag{4.195}$$

with respective eigenvalues $E_1, E_2, \cdots$, are called orbits or orbitals. In the shell model prescription the A nucleons fill the orbitals of lower energy in a compatible way with the Pauli principle. Thus, if the sub-index 1 of Ψ_1, which represents the group of quantum numbers of the orbital 1, includes spin and isospin, we can say that a first nucleon is described by $\Psi_1(r_1), \cdots$, the A-th by $\Psi_A(r_A)$. The wave function

$$\Psi = \Psi_1(\mathbf{r}_1)\Psi_2(\mathbf{r}_2) \cdots \Psi_A(\mathbf{r}_A), \tag{4.196}$$

is solution of Eq. (4.195) with eigenvalues

$$E = E_1 + E_2 + \cdots + E_A, \tag{4.197}$$

and it would be, in principle, the wave function of the nucleus, with energy E, given by the shell model. We should have in mind, however, that we are treating a fermion system and that the total wave function should be antisymmetric for an exchange of coordinates of two nucleons. Such wave function is obtained from (4.196) for the construction of the Slater determinant

$$\Psi = \frac{1}{\sqrt{A!}} \begin{vmatrix} \Psi_1(\mathbf{r}_1) & \Psi_1(\mathbf{r}_2) & \cdots & \Psi_1(\mathbf{r}_A) \\ \Psi_2(\mathbf{r}_1) & \Psi_2(\mathbf{r}_2) & \cdots & \Psi_2(\mathbf{r}_A) \\ \vdots & \vdots & \ddots & \vdots \\ \Psi_A(\mathbf{r}_1) & \Psi_A(\mathbf{r}_2) & \cdots & \Psi_A(\mathbf{r}_A) \end{vmatrix}, \tag{4.198}$$

where the change of coordinates (or of the quantum numbers) of two nucleons changes the sign of the determinant.

An inconvenience of the construction (4.198) is that the function Ψ, by mixing well defined angular momenta J and isotopic spin T, is not more an eigenfunction of these operators. The solution for that difficulty involves the construction of linear combinations of Slater determinants that are eigenfunctions of J and T. That problem has a well-known solution but it involves a great amount of calculations. It is important to make it clear, however, that many of the properties of the nuclear states can be extracted from the shell model without the knowledge of the wave function, as we will see next.

We will analyze what is obtained when one starts with potentials $U(r)$ (Eq. 4.194) with well-known solutions. Let us examine the simple harmonic oscillator initially. Being a potential that always grows with the distance, at first it would not be adapted to represent the nuclear potential, which goes to zero when the nucleon is at a larger distance than the radius of the nucleus. It is expected, however, that this is not very important when we just analyze the bound states of the nucleus. The oscillator potential has the form

$$V(r) = \frac{1}{2}m\omega^2 r^2, \tag{4.199}$$

where the frequency ω should be adapted to the mass number A . We will seek solutions of (4.199) of the type

$$\Psi(\mathbf{r}) = \frac{u(r)}{r} Y_l^m(\theta, \phi), \tag{4.200}$$

where the substitution of Ψ in the Schrödinger equation for a particle reduces the solution of (4.200) to the solution of an equation for u:

$$\frac{d^2 u}{dr^2} + \left\{ \frac{2m}{\hbar^2}[E - V(r)] - \frac{l(l+1)}{r^2} \right\} u = 0. \tag{4.201}$$

The solution of (4.201) with the potential (4.199) is

$$u_{nl}(r) = N_{nl}\exp(-\frac{1}{2}\nu r^2)r^{l+1}\mathcal{V}_{nl}(r), \tag{4.202}$$

where $\nu = m\omega/\hbar$ and $\mathcal{V}_{nl}(r)$ is the associated Laguerre polynomial

$$\mathcal{V}_{nl}(r) = L_{n+l-\frac{1}{2}}^{l+\frac{1}{2}}(\nu r^2) = \sum_{k=o}^{n-1}(-1)^k 2^k \binom{n-1}{k}\frac{(2l+1)!!}{(2l+2k+1)!!}(\nu r^2)^k, \tag{4.203}$$

where $L_k^\alpha(t)$ are solutions of the equation

$$t\frac{d^2 L}{dt^2} + (\alpha + 1 - t)\frac{dL}{dt} + kL = 0. \tag{4.204}$$

From the normalization condition

$$\int_0^\infty u_{nl}^2(r)dr = 1, \tag{4.205}$$

we obtain that

$$N_{nl}^2 = \frac{2^{l-n+3}(2l+2n-1)!!}{\sqrt{\pi}(n-1)![(2l+1)!!]^2}\nu^{l+\frac{3}{2}}. \tag{4.206}$$

The energy eigenvalues corresponding to the wave function $\Psi_{nlm}(r)$ are

$$E_{nl} = \hbar\omega\left(2n+l-\frac{1}{2}\right) = \hbar\omega\left(\Lambda+\frac{3}{2}\right) = E_\Lambda, \tag{4.207}$$

where

$$n = 1, 2, 3, \cdots\,; \qquad l = 0, 1, 2, \cdots \qquad \text{and} \qquad \Lambda = 2n+l-2. \tag{4.208}$$

For each value of l exist $2(2l+1)$ states with the same energy (degenerate states). The factor 2 is due to two spin states. However, the eigenvalues that correspond to the same value of $2n+l$ (same value of Λ) are also degenerate. As $2n = \Lambda - l + 2 =$ even, a given value of Λ corresponds to the degenerated eigenstates

$$(n,l) = \left(\frac{\Lambda+2}{2},0\right)\left(\frac{\Lambda}{2},2\right), \cdots, (2, \Lambda-2), (1, \Lambda), \tag{4.209}$$

for Λ even and

$$(n,l) = \left(\frac{\Lambda+1}{2},1\right)\left(\frac{\Lambda-1}{2},3\right), \cdots, (2, \Lambda-2), (1, \Lambda), \tag{4.210}$$

for Λ odd.

We obtain then that the neutron or proton numbers with eigenvalues E_Λ are given by (we will use $l = 2k$ or $2k+1$, in the case that Λ is even or odd)

$$N_\Lambda = \sum_{k=0}^{\Lambda/2} 2[2(2k)+1], \qquad \text{for } \Lambda \text{ even} \tag{4.211}$$

$$N_\Lambda = \sum_{k=0}^{\Lambda-1/2} 2[2(2k+1)+1], \qquad \text{for } \Lambda \text{ odd.} \tag{4.212}$$

In both cases, the result is

$$N_\Lambda = (\Lambda+1)(\Lambda+2). \tag{4.213}$$

The quantum number Λ defines a shell and each shell can accommodate N_Λ protons and N_Λ neutrons.

The accumulated number of particles for all the levels up to Λ is

$$\sum_\Lambda N_\Lambda = \frac{1}{3}(\Lambda+1)(\Lambda+2)(\Lambda+3). \tag{4.214}$$

The levels predicted by the harmonic oscillator are given in Table 4.2. We can observe that the closed shells appear in 2, 8 and 20, being in agreement with the experimental facts, since the nucleus should close their shells (of protons and of neutrons) with a magic number. However, the same does not happen in the closed shells for nucleon number larger than 20, what is in disagreement with the experience.

For an infinite square well we have the same situation approximately. The solutions for that potential obey the equation

$$\frac{d^2u}{dr^2} + \left[\frac{2m}{\hbar^2}E - \frac{l(l+1)}{r^2}\right]u = 0, \tag{4.215}$$

whose solutions

$$u = Arj_l(kr), \tag{4.216}$$

$\Lambda =$ $2n+l-2$	$E/\hbar\omega$	l	States	$N_\Lambda =$ number of neutrons (protons)	Total
0	3/2	0	1s	2	2
1	5/2	1	1p	6	8
2	7/2	0,2	2s,1d	12	20
3	9/2	1,3	2p,1f	20	40
4	11/2	0,2,4	3s,2d,1g	30	70
5	13/2	1,3,5	3p,2f,1h	42	112
6	15/2	0,2,4,6	4s,3d,2g,1i	56	168

TABLE 4.2   Nucleon distribution for the first shells of a simple harmonic oscillator. The last column indicates the total number of neutrons (or protons) accumulated up to that shell.

orbit: nl	kR	$2(2l+1)$	Total
1s	3.142	2	2
1p	4.493	6	8
1d	5.763	10	18
2s	6.283	2	20
1f	6.988	14	34
2p	7.725	6	40
1g	8.183	18	58
2d	9.095	10	68
3s	9.425	2	70

TABLE 4.3   Proton (or neutron) distribution for the first shells of an infinite square well. The principal quantum number n indicates the order in that a zero appears for a given l in (4.217). Notice that here there is no more the degeneracy in l. The third column gives the proton and neutron numbers that can be accommodated in each orbit.

involve spherical Bessel functions that obey the boundary condition

$$j_l(kR) = 0, \tag{4.217}$$

where R is the radius of the nucleus and $k = \sqrt{2mE}/\hbar$. From (4.216) and (4.217) we build the allowed states for that potential. The Table 4.3 show these states, the nucleon numbers admitted in each one of them and the nucleon numbers that close the shells. Again here the magic numbers are just reproduced in the initial shells.

The nuclear potential should have, actually, an intermediary form between the harmonic oscillator and the square well, not being as smoothly as the first nor as abrupt as the second. It is common to use the Woods-Saxon form $V = V_0/\{1 + \exp[(r - R)/a]\}$, where V_0, r and a are adjustable parameters. A numeric solution of the Schrödinger equation with a such potential does not supply us, however, with the expected results.

A considerable improvement was obtained by Maria Mayer [23] and, independently, by Haxel, Jensen and Suess [24], in 1949, with the introduction of a term of spin-orbit interaction in the form

$$f(r)\mathbf{l} \cdot \mathbf{s}, \tag{4.218}$$

where $f(r)$ is a radial function that should be obtained by comparison with experiments. However, we will see soon that its form is not important for the effect that we want.

A spin-orbit term already appears in atomic physics as a result of the interaction between the magnetic moment of the electrons and the magnetic field created by its orbital motion. In nuclear physics this term has a different nature and it is related to the quantum field properties of an assembly of nucleons.

We will see that the addition of a term as that to the potential (4.199) alters the energy values. The new values are given to first order by:

$$E = \int \Psi^* H \Psi = \left(n + \frac{3}{2}\right)\hbar\omega + \alpha \int \Psi^* f(r)\mathbf{l} \cdot \mathbf{s}\Psi, \tag{4.219}$$

where α is a proportionality constant. If we suppose now that the spin-orbit term is small and that it can be treated as a small perturbation, the wave functions in (4.219) are basically the ones of a central potential. Recalling that $\mathbf{l} \cdot \mathbf{s} = (j^2 - l^2 - s^2)/2$ we have:

$$\int \Psi^* \mathbf{l} \cdot \mathbf{s} \Psi = \frac{l}{2} \qquad \text{for} \quad j = l + \frac{1}{2},$$

$$\int \Psi^* \mathbf{l} \cdot \mathbf{s} \ \Psi = -\frac{1}{2}(l+1) \qquad \text{for} \quad j = l - \frac{1}{2}. \tag{4.220}$$

Thus, the spin-orbit interaction removes the degeneracy in j and, anticipating that the best experimental result will be obtained if the orbitals for larger j have the energy lowered, we admit a negative value for α, what allows us to write for the energy increment:

$$\Delta E|_{j=l+1/2} = -\mid \alpha \mid \langle f(r) \rangle \frac{l}{2}, \tag{4.221}$$

$$\Delta E|_{j=l-1/2} = +\mid \alpha \mid \langle f(r) \rangle \frac{1}{2}(l+1). \tag{4.222}$$

Figure 4.21 exhibits the level scheme of a central potential with the introduction of the spin-orbit interaction. It is easy to see the effect of (4.221) and (4.222) in the energy distribution of the levels. Denominating now a shell as a group of levels of closed energy, not necessarily associated to only one principal quantum number of the oscillator, we obtain a perfect description of all the magic numbers.

The hypothesis of the shell model is, as we saw, to suppose a small effect of the residual interaction and, in that way, to assume H_{res} in (4.192) as a small perturbation. The energy correction E, as in Eq. (4.219), is given in first order perturbation theory by

$$\Delta E = \int \Psi^* H_{\text{res}} \Psi \, dv, \tag{4.223}$$

that is, ΔE is obtained by the expectation value of the residual interaction, calculated with the non-perturbed wave functions.

Let us see as example the case of two particles above a closed core. We can for this case write the residual Hamiltonian (4.193) as

$$H_{\text{res}} = \sum_{i=3}^{A} \sum_{j=3}^{A} V_{i,j}(\mathbf{r}_i, \mathbf{r}_j) - \sum_{i=3}^{A} U(r_i) + \sum_{i=3}^{A} V_{1,i} - U(r_1) + \sum_{i=3}^{A} V_{2,i} - U(r_2) + V_{1,2}. \tag{4.224}$$

The first two parts refer to the closed core and, if we admit that it is inert, they can be ignored. For the four following parts we can suppose to first approximation that the idea of the model is valid, that is, the interaction of particles 1 and 2 with the core is given by an average potential U. Thus, these parts cancel out and we remain with

$$H_{\text{res}} = V_{1,2} \tag{4.225}$$

for the use in (4.223).

The interaction potential $V_{1,2}$ of valence nucleons would be, in principle, the nucleon-nucleon potential that we studied in the beginning of this chapter. That potential is not, however, of immediate application when the particles are bound to the nucleus because the Pauli principle modifies in a drastic way the effect of the interaction for bound particles. To find a nucleon-nucleon potential that works well inside the nucleus two approaches are usual. In the first one tries to obtain a potential for bound nucleons from the nucleon potential outside the nucleus. This microscopic treatment is very complicated and the practical results are not very satisfactory. In the second case, a phenomenological treatment, a parameterized form is proposed for the interaction and the parameters are determined by a comparison with experimental data. Other conditions of importance for the choice of an effective potential are the simplicity of its use and the elements of the physics of the problem that it incorporates. An example is the surface delta interaction (SDI),

$$V(1,2) = A\delta(\mathbf{r}_1 - \mathbf{r}_2)\delta(|\mathbf{r}_1| - R_0), \tag{4.226}$$

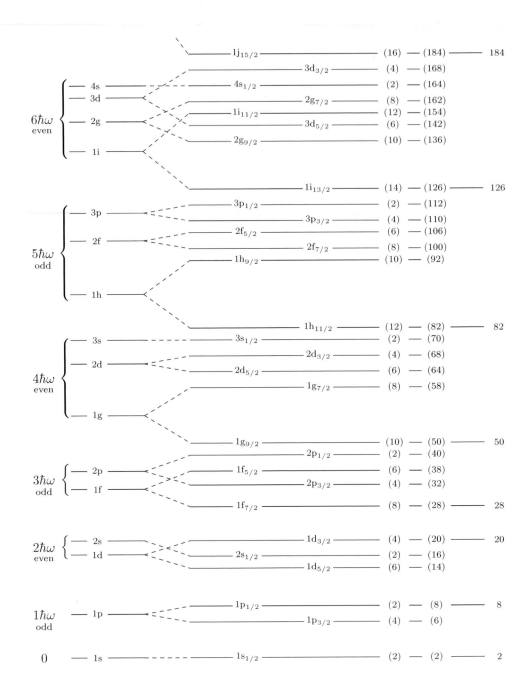

FIGURE 4.21 Level scheme of the shell model showing the break of the degeneracy in j caused by the spin-orbit interaction term and the emergence of the magic numbers in the shell closing. The values inside parenthesis indicate the number of nucleons of each type that the level admits and the values inside brackets the total number of nucleons of each type up to that level. The ordering of the levels is not rigid, and one could obtain level inversions when changes the form of the potential.

that tries to incorporate the facts that: 1) the interaction is of short range and 2) inside the nucleus the nucleons are practically free due to the action of the Pauli principle and the interaction should be located at the nuclear surface of radius R_0. The presence of the delta function turns the calculation of the matrix elements (4.223) quite simplified.

4.16   Exercises

1. A particle with spin 1 moves in a central potential of the form

$$V(r) = V_1(r) + \mathbf{S} \cdot \mathbf{L} V_2(r) + (\mathbf{S} \cdot \mathbf{L})^2 V_3(r).$$

What are the values of $V(r)$ in the states with $J = L + 1$, L, and $L - 1$?

2. Suppose that the meson π^- (spin 0 and negative parity) is captured from the orbit P in a pionic atom, giving rise to the reaction

$$\pi^- + d \longrightarrow 2n.$$

Show that the two neutrons should be in a singlet state.

3. Consider the operator S_{12} defined in (4.10). Show that, for the spin singlet and triplet state of the two particles, the following relations are valid:

$$S_{12}\chi_{\text{singlet}} = 0, \quad (S_{12} - 2)(S_{12} + 4)\chi_{\text{triplet}} = 0.$$

4. Let s_1 and s_2 be the spin operators of two particles and r the radius vector that connects them. Show that any positive integer power of the following operators

$$\mathbf{s_1} \cdot \mathbf{s_2} \quad \text{and} \quad \frac{3(\mathbf{s_1} \cdot \mathbf{r})(\mathbf{s_2} \cdot \mathbf{r})}{r^2} - (\mathbf{s_1} \cdot \mathbf{s_2})$$

can be written as a linear combination of these operators and the unit matrix.

5. Prove the relations

$$\mathbf{s_1} \times \mathbf{s_2} = \frac{2i}{\hbar}(\mathbf{s_1} \cdot \mathbf{s_2})\mathbf{s_1} - \frac{i\hbar}{2}\mathbf{s_2}$$

$$(\mathbf{s_1} \times \mathbf{r}) \cdot (\mathbf{s_2} \times \mathbf{r}) = R^2(\mathbf{s_1} \cdot \mathbf{s_2}) - (\mathbf{s_1} \cdot \mathbf{r})(\mathbf{s_2} \cdot \mathbf{r}).$$

6. Show that the tensorial force S_{12} has a zero angular average, that is, show that $\int S_{12} d\Omega = 0$.

7. Verify how much the curves of Figure 4.10 resemble the case of the diffraction of light by an opaque object, calculating the quantities $\Delta\theta = \pi/kR$ that, in a Fraunhofer diffraction, measures, for the first maxima, the angular distance between two sequential maxima.

8. Show that the average kinetic energy of the nucleons in the interior of a nucleus given by the Fermi gas model is about 23 MeV.

9. Show that the total kinetic energy of a nucleus with $N = Z$ in the Fermi gas model is given by

$$E_T = 2C_3 A^{-2/3} \left(\frac{A}{2}\right)^{5/3},$$

and find the value of C_3. Repeat the calculation for $N \neq Z$ and show that, in this case,

$$E_T' = C_3 A^{-2/3} (N^{5/3} + Z^{5/3}).$$

10. In the reactions (4.98), a) which is the excitation energy of ^{64}Zn* when it is formed by the collision of protons of 13 MeV with ^{63}Cu? b) What energy should have an α-particle colliding with ^{60}Ni to produce the same excitation energy? Compare the result with those of Figure 4.12. c) What percent of the kinetic energy of the incident particles contribute to the excitation energy of Zn?

11. When the reaction p + ^{65}Cu $\to ^{65}$Zn + n is produced by protons of 15 MeV, a continuous energy distribution is observed for the emerging neutrons, in which a sharp peak at the energy $E_n = 5.57$ MeV is superposed, associated to an excitation energy of 7.3 MeV of ^{65}Zn. What models can be used to justify this result?

12. Show that in the capture of slow neutrons the total cross section at the resonance energy has the value $\sigma = \lambda \sqrt{g\sigma_e/\pi}$, where σ_e is the scattering cross section at the same energy and g the statistical factor of the Breit-Wigner formula.

    ^{109}Ag has a neutron resonance absorption cross section at 5.1 eV with a peak value of 7600 barn and a width of 0.19 eV. Evaluate the expected value of the compound elastic cross section at the resonance.

13. The Breit-Wigner formula for the single-level resonant cross section in nuclear reactions is

$$\sigma_{ab} = \frac{\pi}{k^2} g \frac{\Gamma_a \Gamma_b}{(E - E_0)^2 + \Gamma^2/4},$$

    where

$$g = \frac{2J + 1}{(2I_A + 1)(2I_B + 1)},$$

    J being the spin of the level, I_A the spin of the incident particle and I_B the spin of the target.

    At neutron energies below 0.5 eV the cross section in ^{235}U ($I_B = 7/2$) is dominated by one resonance ($J = 3$) at a kinetic energy of 0.29 eV with a width of 0.135 eV. There are three channels, which allow the compound state to decay by neutron emission, by photon emission, or by fission. At resonance the contributions to the neutron cross-sections are: i) elastic scattering (resonant) ($\ll 1$ barn); ii) radiative capture (70 barns); iii) fission (200 barns). (a) Calculate the partial widths for the three channels. (b) How many fissions per second will there be in a sheet of ^{235}U, of thickness 1 mg/cm^{-2}, hit normally by a neutron beam of 10^5 per second with a kinetic energy of 0.29 eV? (c) How do you expect the neutron partial width to vary with the energy?

14. Evaluate the A dependence of the neutron s-wave strength function for a square well potential of depth 43 MeV and width 1.3A$^{1/3}$ fm.

15. From the relations

$$\frac{1}{T} = \frac{dS}{dE}, \qquad \text{and} \quad \rho(E) = \frac{\exp[S(U)]}{(2\pi T^2 \, dE/dT)^{1/2}},$$

    and assuming that $U = aT^n$, show that

$$\rho(E) \propto \frac{E^{-1}}{(1 + 1/n)} \exp\left[\frac{n}{n-1} a^{1/n} E^{1-1/n}\right].$$

    (Hint: Assume that E can be represented as a power series in T and $dE/dT \longrightarrow 0$, as $T \longrightarrow 0$, then if follows that the expansion must start with a term at best quadratic in T.).

16. From the statistical model show that

$$\frac{\Gamma_p}{\Gamma_n} = \frac{\int E_p \sigma(E_p) w(U_p) \, dE_p}{\int E_n \sigma(E_n) w(U_n) \, dE_n}.$$

17. Show that in a compound nuclear reaction, the angular distribution of emitted particles with respect to the direction of incidence is symmetric about 90°.

18. Show that the average spacing between energy levels according to the statistical model of the nucleus is given by

$$D\left(U\right) = \frac{1}{\rho\left(U\right)} = T\left(2\pi\frac{dU}{dT}\right)^{1/2}e^{-S},$$

where $\rho(U)$ is the number of energy levels in range dU (or dE), U is the average energy of excitation of the nucleus, T is the "nuclear temperature" in units of energy, and S is the "nuclear entropy" [25].

19. There are several methods to associate a temperature with a nucleus in an ideal gas law statistically, such as

$$E = aT^2, \qquad \text{and} \quad E = \frac{1}{11}AT^2 - T + \frac{1}{8}A^{2/3}T^{7/3},$$

where A is the mass number of the nucleus [19]. Compare these using $A = 25$.

20. Prove that in the black-nucleus approximation the inverse cross-section for emission of a charged particle is given by $\sigma \approx \pi R^2\left(1 - V_C/E\right)$, where V_C is the particle Coulomb barrier. Say whether E is the laboratory particle energy or its center of mass energy (the energy of the particle relative to the nucleus). Give an expression for the maximum value of the impact parameter and the compound nucleus angular momentum.

21. Evaluate the average angular momentum carried out by neutrons, protons and α-particles evaporated by ^{135}Tb at 22 MeV excitation energy. In this and in the following problems, when necessary, assume the nuclear temperatures to be approximately given by $T \approx \sqrt{8E/A}$ MeV and the radii for evaluating the Coulomb barriers approximately given by $R \approx 1.5\left(A_R^{1/3} + A_p^{1/3}\right)$ where A_R and A_p are the residual and the emitted particle mass.

22. Evaluate the mean energy of neutrons, protons and α-particles evaporated by ^{55}V, ^{140}Cs and ^{227}Ra at 18 MeV excitation energy.

23. Assuming that the emission of neutrons, protons and α-particles are the dominant decay modes of the compound nuclei of the previous exercise, evaluate the compound nucleus total and partial decay widths.

References

1. Bohr N 1936 *Nature* **137** 344
2. Aitchison I J R and Hey A J G 2012 *Gauge Theories in Particle Physics: A Practical Introduction* (CRC Press)
3. Reid R V 1968 *Ann. Phys.* **50** 411
4. Newton R 1982 *Scattering Theory of Waves and Particles* (Berlin: Springer Verlag, Berlin)
5. Amado R D, Dedonder J P and Lenz F 1980 *Phys. Rev.* **C21** 647
6. Jackson J D 1999 *Classical Electrodynamics* (New York: Wiley)
7. Feshbach H 1958 *Ann. Phys.* **5** 357 (1958); 1962 *Ann. Phys.* **19** 287; Lemmer R H 1966 *Rep. Prog. Phys.* **29** 131
8. Perey F G 1963 *Phys. Rev.* **131** 745
9. Perey F G and Buck B 1962 *Nucl. Phys.* **32** 353
10. Bertsch G, Borysowicz J, McManus H and Love W G 1977 *Nucl. Phys.* **A284** 399
11. Kobos A M, Brown B A, Lindsay R and Satchler G R 1984 *Nucl. Phys.* **A425** 205
12. Ghoshal S N 1950 *Phys. Rev.* **80** 939
13. McVoy K W 1967 *Ann. Phy.* **43** 91
14. Chase D M, Wilets L, and Edmonds A R 1958 *Phys. Rev.* **110** 1080
15. Buck B and Perey F G 1962 *Phys. Rev. Lett.* **8** 444
16. Lane A M and Thomas R G 1958 *Rev. Mod. Phys.* **30** 257
17. Wigner E P and Eisenbud L 1947 *Phys. Rev.* **72** 29

18. Lynn J E 1968 *The theory of neutron resonance reactions* (Oxford: Oxford Press)
19. Lang J M B and Le Coutuer K J 1954 *Proc. Phys. Soc.* **67A** 586
20. Feshbach H 1992 *Theoretical Nuclear Physics – Nuclear Reactions* (New York: John Wiley)
21. Facchini U and Saetta-Menichella 1968 E *Energ. Nucl.* **15** 54
22. Lefort M 1968 *Nuclear Chemistry* (Amsterdam: Van Nostrand)
23. Mayer MG 1949 *Phys. Rev.* **75** 1969; 1950 *Phys. Rev.* **78** 16
24. Haxel O, Jensen J H D and Suess H E 1949 *Phys. Rev.* **75** 1766
25. Bethe H A 1937 *Rev. Mod. Phys.* **9** 69

5

Fusion and fission

5.1   Introduction

The reactions in that the participant nuclei have high mass present own characteristics that deserve a study as a special topic of the theory of nuclear reactions. It is common to consider an ion with $A > 4$ as a *heavy ion*, but the properties that we enumerate below are more evident in reactions involving nuclei heavier than this limit.

1) A heavy projectile, due to its large charge, feels a Coulomb repulsion from the target nucleus. To produce a nuclear reaction it should have enough energy to overcome the Coulomb barrier. For a very heavy target as ^{238}U, for example, is necessary about 5 MeV per nucleon of the projectile to surpass the *Coulomb barrier*. The wavelength associated to the projectile becomes small compared with the dimensions of the nuclei involved and classical and semi-classical methods become useful in the description of the reaction.

2) The projectile carries a large amount of angular momentum and a good part can be transferred to the target in the reaction. Rotational bands with several dozens of units of angular momentum can be created in this process and the reactions with heavy ions are the most adapted to feed high spin levels.

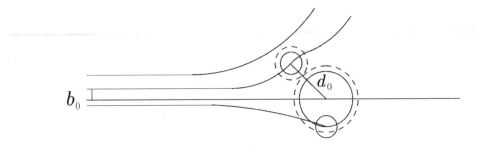

FIGURE 5.1 Projectiles with larger impact parameters than b_0 only have Coulomb interaction with the target. For $b < b_0$, projectile and target enter the radius of action of the nuclear forces. The grazing collisions have b close to b_0.

3) Direct reactions (to be studied later) and formation of compound nuclei are also common processes in reactions with heavy ions. Some peculiarities of these processes are not found in reactions in which the projectile is a nucleon. One of these processes can be understood as intermediate between a direct reaction and the formation of a compound nucleus. Fusion does not occur but projectile and target pass a relatively long time under the mutual action of the nuclear forces. Nuclear matter is exchanged between both and there is a strong heating of the two nuclei, with a large transfer of kinetic energy to the internal degrees of freedom. These are the *deep inelastic collisions* (DIC).

The reaction type that will happen depends on the smallest distance d that projectile and target will reach each other. If this distance is sufficiently large only the long range Coulomb interaction acts and, for a classical hyperbolic trajectory, d is related to the impact parameter b and to the energy E of the projectile by [1]

$$d = \frac{a}{2} + \left[\left(\frac{a}{2} \right)^2 + b^2 \right]^{1/2}, \tag{5.1}$$

where a is the smallest distance from the target that the projectile can reach in a frontal collision, and this is related to E by

$$a = \frac{Z_1 Z_2 e^2}{E}. \tag{5.2}$$

Experimentally, the variable under control is the energy E of the projectile and, for E sufficiently large, d can be small enough for the nuclear forces of short range begin to act. Collisions in the proximity of this limit are called *grazing* collisions and they are characterized by values of b_0 and d_0 shown in Figure 5.1.

Assuming that there is always reaction when $b < b_0$, the reaction cross section σ_r can be determined geometrically by

$$\sigma_r = \pi b_0^2. \tag{5.3}$$

The experimental determination of σ_r allows to establish the value of d_0 as a function of the sum $A_1^{1/3} + A_2^{1/3}$ and the result is seen in Figure 5.2. J. Wilczynski [2] verified that the straight line $d_0[\text{fm}] = 0.5 + 1.36(A_1^{1/3} + A_2^{1/3})$ adjusts well to the experimental values, what shows that the distance of grazing collision is somewhat larger than the one deduced from two touching spheres ($1.36 \text{ fm} > r_0 = 1.2 \text{ fm}$).

When the impact parameter is close to b_0 one expects nuclear reactions of short duration, without the contribution of the compound nucleus formation. Such reactions are elastic and inelastic scattering and transfer of few nucleons. When the incident energy is sufficiently large, small values of b can lead to the projectile penetrating the target. Depending on the energy and on the involved masses, the reaction can end in one of the processes below:

a) *Fusion* – is the preferred process when one has light nuclei and low energy. There is the formation of a highly excited compound nucleus that decays by evaporation of particles and γ-radiation emission, leading to a cold residual nucleus. If the energy in the CMS is close to the

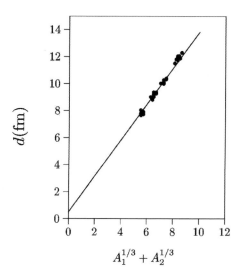

FIGURE 5.2   Smallest distance between the nuclei in a grazing collision as a function of $A_1^{1/3} + A_2^{1/3}$. The experimental points are from references [3, 4, 5].

Coulomb barrier energy the cross section of compound nucleus formation starting from two light ions is practically equal to the reaction cross section.

b) *Fission* – When the compound nucleus is heavy the fission process competes strongly with the decaying of particles in each stage of the evaporation process. A very heavy compound nucleus with a large excitation energy has a very small probability of arriving to a cold residual nucleus without fission at some stage of the de-excitation. The role of the angular momentum l transmitted to the target nucleus is also essential. The fission barrier decreases with the increase of l and for a critical value l_{crit} the barrier ceases to exist. l_{crit} depends on A and this dependence is sketched in Figure 5.3. A nucleus with angular momentum greater than l_{crit} suffers immediate fission and we shall see ahead that this is also a limiting factor in the production of superheavy elements.

c) *Deep inelastic collision (DIC)* – is a phenomenon characteristic of reactions involving very heavy nuclei ($A > 40$) and with an incident energy of 1 MeV to 3 MeV above the Coulomb barrier. In DIC the projectile and the target spend some time under mutual action, exchanging masses and energy but without arriving to the formation of a compound nucleus. The projectile escapes after transferring part of its energy and angular momentum to its internal degrees of freedom and to the target, with those values reaching 100 MeV and $50\hbar$. This dissipative property implies that the DIC are also denominated by many authors as *dissipative collisions*.

One of the most interesting aspects of DIC is the correlation between the energy dissipated in the collision and the scattering angle in the center of mass. Let us look at Figure 5.4. The trajectory 1 exhibits the projectile with an impact parameter that places it out of the range of the nuclear forces. The Coulomb scattering angle will become larger as the impact parameter decreases. In a graph as the one of Figure 5.5, where one plots the final energy against the scattering angle, trajectories of the type 1 are located in the upper branch, where there is no dissipation and the initial kinetic energy stays unaffected. This upper branch has a maximum value for the scattering angle; with a smaller value for the impact parameter that produces this maximum, the nuclear attractive force begins to act and the effects of dissipation of DIC show up. A given Coulomb scattering angle θ can also be reached by the combination of the nuclear and Coulomb forces. Only that now there is loss of energy and the events are located in the branch 2 of Figure 5.5. There still is no univocal correspondence between angle and energy because an infinite number of trajectories with absorption can lead to the same angle θ. The branch 2 should be understood as a line of maxima in a topographical representation (denominated *Wilczynski diagram*) where the perpendicular axis to the paper is proportional to the cross section $d^2\sigma/dEd\theta$.

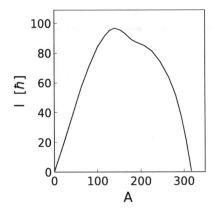

FIGURE 5.3   Critical angular momentum l_{crit} as a function of the mass number A[6].

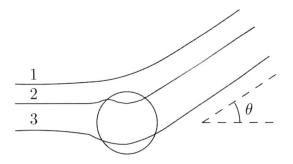

FIGURE 5.4   Reactions with different impact parameters leading to the same scattering angle θ.

5.2   The liquid drop model

This model is used in the study of fusion and fission reactions and is based on the hypothesis that the nucleus has an identical behavior as a liquid, mainly due to the saturation of the nucleon-nucleon force. This idea is the starting point to obtain an equation for the binding energy of the nucleus. In its simplest form this equation contains five contributions:

1) The main part of the binding energy is called *volume energy*. It is based on the experimental fact that the binding energy per nucleon is approximately constant (see Figure 5.6), thus the total binding energy is proportional to A:

$$B_1 = a_V A. \tag{5.4}$$

If the nucleon-nucleon interaction were the same for all the possible nucleon pairs, the total binding energy should be proportional to the total number of pairs, which is equal to $A(A-1)/2 \cong A^2/2$. Therefore, the binding energy per nucleon would be proportional to A. The fact that this energy is constant is due to the short range of the nuclear force, leading to the interaction of a nucleon with its neighbors. This property implies the saturation of the nuclear forces.

2) The surface nucleons contribute less to the binding energy since they only feel the nuclear force from the inner side of the nucleus. The number of nucleons in the surface should be proportional to the surface area, $4\pi R^2 = 4\pi r_0^2 A^{2/3}$. We should, therefore, correct equation (5.4) by adding the *surface energy*

$$B_2 = -a_S A^{2/3}. \tag{5.5}$$

3) The binding energy should also be smaller due to the Coulomb repulsion between the protons. The Coulomb energy of a charged sphere with homogeneous distribution, and total charge Ze, is

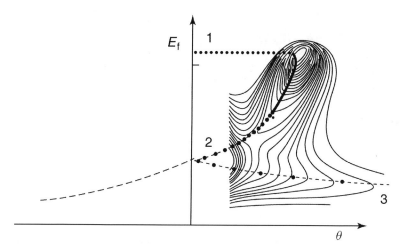

FIGURE 5.5   Lines of maxima in topographical diagrams of the final energy against the scattering angle (Wilczynski diagram).

given by $(3/5)(Ze)^2/R = (3/5)(e^2/r_0^2)(Z^2/A^{1/3})$. Thus, the *Coulomb energy* contributes negatively to the binding energy with a part given by

$$B_3 = -a_C Z^2 A^{-1/3}. \tag{5.6}$$

4) If the nucleus has a different number of protons and neutrons, its binding energy is smaller than for a symmetric nucleus. The reason for this term is clear from the Fermi gas model described in Chapter 4. This *asymmetry term* also contributes negatively and it is given by

$$B_4 = -a_A \frac{(Z - A/2)^2}{A}. \tag{5.7}$$

5) The binding energy is larger when the proton and neutron numbers are even (even-even nuclei) and it is smaller when one of the numbers are odd (odd nuclei) and even more when both are odd (odd-odd nucleus). Thus, we introduce a *pairing term*

$$B_5 = \begin{cases} +\delta & \text{for even-even nuclei} \\ 0 & \text{for odd nuclei} \\ -\delta & \text{for odd-odd nuclei.} \end{cases} \tag{5.8}$$

Empirically, we find that

$$\delta \cong a_P A^{-1/2}. \tag{5.9}$$

Gathering the terms, we obtain

$$B(Z, A) = a_V A - a_S A^{2/3} - a_C Z^2 A^{-1/3} - a_A \frac{(Z - A/2)^2}{A} + \frac{(-1)^Z + (-1)^N}{2} a_P A^{-1/2}. \tag{5.10}$$

We thus obtain an equation for the mass of a nucleus

$$m(Z, A) = Z m_p + (A - Z)m_n - B(Z, A) = Z m_p + (A - Z)m_n - a_V A + a_S A^{2/3} +$$

$$a_C Z^2 A^{-1/3} + a_A \frac{(Z - A/2)^2}{A} - \frac{(-1)^Z + (-1)^N}{2} a_P A^{-1/2}. \tag{5.11}$$

The expression (5.11) is known as *semi-empiric mass formula* or Weizsäcker formula [7]. The constants appearing in (5.11) are determined empirically, that is, from the data analysis. A good adjustment is obtained using [8]

$$a_V = 15.85 \text{ MeV/c}^2, \quad a_S = 18.34 \text{ MeV/c}^2, \quad a_C = 0.71 \text{ MeV/c}^2,$$

$$a_A = 92.86 \text{ MeV/c}^2 \quad \text{and} \quad a_P = 11.46 \text{ MeV/c}^2. \tag{5.12}$$

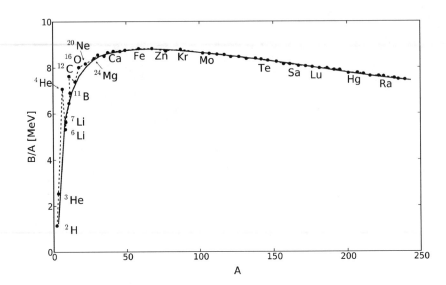

FIGURE 5.6 Average experimental values of B/A for several nuclei, and the corresponding curve calculated by Eqs. (5.10) and (5.14).

However, other good groups of parameters can also be found. Observe that a small variation in a_V or a_S leads to a great variation in the other parameters. That is due to a larger relevance of the corresponding terms in the mass formula.

Figure 5.6 exhibits a comparison of Eq. (5.10) to experimental data for many nuclei. For $A < 20$ there is not a good agreement with the experience. This is expected, since light nuclei are not so similar to a liquid drop.

Equation (5.11) allows us to deduce important properties. Observe that this equation is quadratic in Z. For odd nuclei with the same mass one has a parabola, as viewed in Figure 5.7.

For A even, we obtain two parabolas due to the pairing energy $\pm\delta$. Nuclei with a given Z can decay into neighbors by β^+ (positron) or β^- (electron) particle emission. In Figure 5.7 we see that for a nucleus with odd A there is only one stable isobar, while for even A many stable isobars are possible.

Fixing the value of A, the number of protons Z_0 for which $m(Z, A)$ is minimum is obtained by

$$\left| \frac{\partial m(Z, A)}{\partial Z} \right|_{A=\text{const}} = 0. \tag{5.13}$$

From Eq. (5.11), we get

$$Z_0 = \frac{A}{2} \left(\frac{m_n - m_p + a_A}{a_C A^{2/3} + a_A} \right) = \frac{A}{1.98 + 0.015 A^{2/3}}. \tag{5.14}$$

We see from Eq. (5.14) that the stability is obtained with $Z_0 < A/2$, that is, with a number of neutrons larger than that of protons. We know that this in fact happens and Figure 5.8 exhibits that the stability line obtained with Eq. (5.14) accompanies perfectly the valley of stable nuclei.

From Eq. (5.11) we can also ask if a given nucleus is or not unstable against the emission of an α-particle. For this it is necessary that

$$E_\alpha = [m(Z, A) - m(Z - 2, A - 4) - m_\alpha]c^2 > 0, \tag{5.15}$$

what happens for $A \gtrsim 150$. We can also verify the possibility of a heavy nucleus to fission, that is, of breaking in two pieces of approximately the same size. This will be possible if

$$E_f = [m(Z, A) - 2m(Z/2, A/2)] c^2 > 0, \tag{5.16}$$

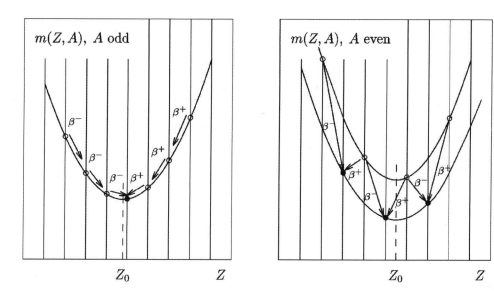

FIGURE 5.7 Mass of nuclei with a fixed A. The stable nuclei are represented by full circles.

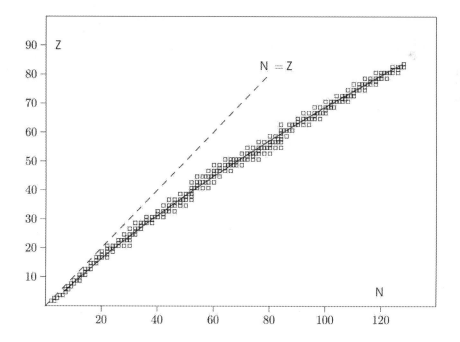

FIGURE 5.8 Location of the stable nuclei in the $N - Z$ plane. The full line is the curve of Z_0 against $N = A - Z_0$, obtained from Eq. (5.14).

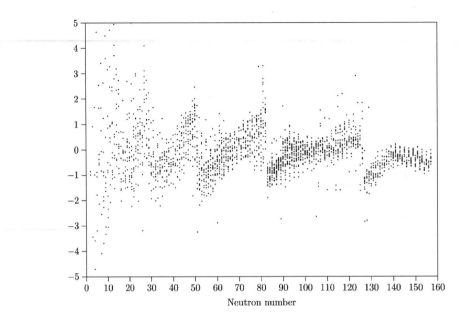

FIGURE 5.9 Difference (in MeV) between the experimental separation energy and the one calculated with the liquid drop model for about 2000 nuclei.

a relation valid for $A \gtrsim 90$.

The liquid drop model describes well the average behavior of the binding energy with the mass number but it has nothing to say on other effects as, for example, the existence of magic number of nucleons. In Figure 5.9 we show the difference between the experimental separation energy of a neutron and the calculated one with the liquid drop model for about of 2000 nuclei. It is notorious the existence of values of N where the difference is positive and then drops abruptly, becoming negative. We note that these values are the magic numbers, that appear in several other experiments of different nature. The presence of these numbers are due to a shell structure that cannot be obtained by the model.

5.3   General considerations on fusion reactions

The fusion of two nuclei to form a heavier nucleus is a very important reaction. Besides, it is a way to produce nuclei with very high spins and, accordingly, to study the rapid rotation phenomena of nuclei. At very low energies nuclear fusion is of paramount importance for stellar energy production and nuclear synthesis as originally proposed by H. Bethe [9]. Further, they are very sensitive to nuclear structure when measured below the Coulomb barrier energies.

We present below a simple picture of fusion reactions namely, the tunneling through a local one-dimensional real potential barrier formed from the addition of an attractive nuclear and a repulsive Coulomb potential. We further assume that absorption into the fusion channels (compound nucleus) ensues in the inside region to the left of the barrier after the tunnelling through the barrier (from the right) has occurred.

The one-dimensional potential for the l partial wave is

$$V_l(r) = V_N(r) + V_C(r) + \frac{\hbar^2 l(l+1)}{2\mu r^2}, \tag{5.17}$$

where $V_N(r)$ is the attractive nuclear potential taken to be, e.g., of a Woods-Saxon form

$$V_N(r) = \frac{-V_0}{1 + \exp\left[(R - R_0)/a\right]}. \tag{5.18}$$

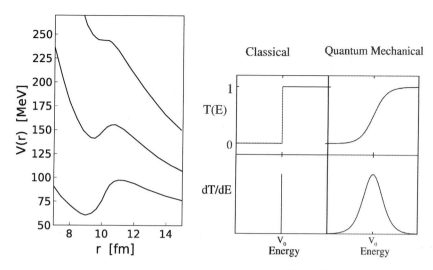

FIGURE 5.10 (a) One dimensional potential of Eq. 5.17 for the ^{64}Ni+^{64}Ni system for several l values. The lowest barrier is for $l = 0$ (the bare barrier). The middle and top barriers are for $l = 100$ and $l = 150$, respectively. (b) Classical and quantum-mechanical transmission probabilities for a one-dimensional potential barrier.

$V_C(r)$ is the Coulomb potential given by

$$V_c(r) = \begin{cases} Z_1 Z_2 \mathrm{e}^2 / r, & r \geq R_c \\ Z_1 Z_2 \mathrm{e}^2 / 2R_c \left[3 - (r/R_c)^2 \right], & r \leq R_c, \end{cases} \tag{5.19}$$

and the centrifugal potential is given as usual by $\hbar^2 l(l+1)/2\mu r^2$.

In Figure 5.10(a) we show the effective potentials for ^{64}Ni+^{64}Ni for three values of l. The Coulomb barrier (peak) is clearly seen in the lowest two curves and it becomes less conspicuous for $l = 150$ as the centrifugal repulsion starts dominating. If the c.m. energy is 160 MeV, then all partial waves up to about $l = 100$ will contribute. In the region below $l = 95$ the tunnelling probability, $T_l = 1$; for $l > 160$ it is zero. Then one expects that the tunnelling probability (*fusion transmission coefficient*) to be close to a Fermi function (being unity for small l and going gradually to zero at larger l values).

The fusion cross-section is just the sum of all the T_l weighted with the factor $2l + 1$, i.e.,

$$\sigma_F(E) = \frac{\pi}{k^2} \sum_{\ell=0}^{\infty} (2l+1) T_l(E) = \sum_{\ell=0}^{\infty} \sigma_F(l, E), \tag{5.20}$$

where $\sigma_F(l, E)$ is the partial fusion cross-section. In the following sections we show how one obtains T_l.

5.4   The one dimensional WKB approximation

We want to solve the Schrödinger equation for a particle:

$$\frac{\hbar^2}{2m} \frac{d^2}{dx^2} \psi(x) + [E - V(x)] \psi(x) = 0. \tag{5.21}$$

For such, we will define a function σ and rewrite $\psi(x)$ as

$$\psi(x) = e^{i\sigma(x)/\hbar}. \tag{5.22}$$

This leads to

$$\frac{1}{2m}\left(\frac{d}{dx}\sigma(x)\right)^2 - \frac{i\hbar}{2m}\frac{d^2}{dx^2}\sigma(x) = E - V(x). \tag{5.23}$$

In a more compact notation we have

$$\sigma'^2 - i\hbar\sigma'' = 2m(E - V). \tag{5.24}$$

The above equation reduces to the classical equation

$$\frac{1}{2m}\left(\frac{d}{dx}\sigma(x)\right)^2 = E - V(x)$$

in the limit $\hbar \longrightarrow 0$. We expect that if quantum corrections are small (i.e. $\hbar \longrightarrow 0$) then the solution of Eq. (5.21) will be very close to the classical solution. The WKB approximation explores this fact*.

Following the above argument we write σ in the form of a series of powers of $\hbar$:

$$\sigma = \sigma_0 + \frac{\hbar}{i}\sigma_1 + \left(\frac{\hbar}{i}\right)^2 \sigma_2 + \mathcal{O}(\hbar^3). \tag{5.25}$$

Replacing this in Eq. (5.24) and retaining the lowest powers of $\hbar$, we obtain

$$\sigma_0'^2 + \frac{\hbar}{i}\left(2\sigma_1'\sigma_0' + \sigma_0''\right) + \left(\frac{\hbar}{i}\right)^2 \left(\sigma_1'^2 + 2\sigma_2'\sigma_0' + \sigma_1''\right) + \mathcal{O}(\hbar^3) = 2m(E - V). \tag{5.26}$$

Let us first neglect any term in Eq. (5.26) that is proportional to any power of $\hbar$:

$$\sigma_0'^2 = 2m(E - V), \tag{5.27}$$

what leads to

$$\sigma_0' = \pm\sqrt{2m(E - V)}.$$

We identify the term under the root as being the square of the classical momentum. Thus,

$$\sigma_0'(x) = \pm p(x), \tag{5.28}$$

or

$$\sigma_0 = \pm \int p(x)dx. \tag{5.29}$$

This is the expression that we will use for σ_0.

If the term of Eq. (5.26) proportional to $\hbar$ is the dominant one, the functions σ_0 and σ_1 should satisfy to the following condition:

$$2\sigma_1'\sigma_0' + \sigma_0'' \approx 0. \tag{5.30}$$

That is,

$$\sigma_1' = -\frac{\sigma_0''}{2\sigma_0'} = -\frac{p'}{2p}, \tag{5.31}$$

where we used Eq. (5.28) and its derivative. Integrating this equation one obtains an expression for σ_1:

$$\sigma_1 = -\frac{1}{2}\ln(p) = \ln\left(p^{-1/2}\right). \tag{5.32}$$

Let us analyze, now, the term proportional to $\hbar^2$. If we are going to neglect it, it should satisfy the following condition:

$$2\sigma_0'\sigma_2' + \sigma_1'^2 + \sigma_1'' \approx 0. \tag{5.33}$$

---

*The acronym WKB is for Wentzel, Kramers and Brillouin.

Using Eq. (5.28), Eq. (5.31), as well as the derivative of the latter one, we obtain the following expression for σ_2':

$$\sigma_2' = \frac{p''}{4p^2} - \frac{2p'^2}{8p^3}. \tag{5.34}$$

Integrating the first term by parts and introducing the force acting on the particle ($F = pp'/m$), we obtain a final expression for σ_2[§]:

$$\sigma_2 = \frac{1}{4}m\frac{F}{p^3} + \frac{1}{8}m^2\int\frac{F^2}{p^5}dx. \tag{5.35}$$

5.4.1 Conditions of validity

Given the values of σ_0, σ_1 and σ_2, we can determine the expression of the wave function and which conditions should be imposed so that it assumes a simple form. From Eqs. (5.22) and (5.25), we obtain

$$\psi(x) = \exp\left(\frac{i}{\hbar}\sigma_0 + \sigma_1 + \frac{\hbar}{i}\sigma_2\right). \tag{5.36}$$

With the use of the σ_i's (Eqs. (5.29), (5.32) and (5.35)), we obtain

$$\psi\pm = \frac{1}{\sqrt{p}}\exp\left(\pm\frac{i}{\hbar}\int pdx\right)\left[\exp\left(-i\hbar\frac{mF}{4p^3}\right)\exp\left(-i\hbar\frac{m^2}{8}\int\frac{F^2}{p^5}\right)\right], \tag{5.37}$$

where the terms inside the brackets are the contributions of σ_2.

However, this expression is still very complicated. To simplify it, we will impose that the terms associated to σ_2 are negligible. This leads us to two conditions, where the first of them is

$$\hbar m^2\int\frac{F^2}{p^5}dx \ll 1. \tag{5.38}$$

So that this condition is satisfied, F should be small within distances of the order of λ, the particle's wavelength.

The second condition should be:

$$m\hbar\frac{F}{p^3} \ll 1. \tag{5.39}$$

This condition can be rewritten in an alternative way. Recalling the relationship between F and p ($F = pp'/m$), Eq. (5.39) becomes

$$\hbar\frac{p'}{p^2} \ll 1. \tag{5.40}$$

That is the same as

$$\frac{d}{dx}\left(\frac{\hbar}{p}\right) \ll 1, \tag{5.41}$$

and because $\lambda = 2\pi\hbar/p$, we obtain

$$\frac{d}{dx}\left(\frac{\lambda}{2\pi}\right) \ll 1. \tag{5.42}$$

This condition is of highest importance. A variation of λ should be small along distances of the order of one wave length.

Given these conditions (Eqs. (5.38) and (5.42)), the wave function (Eq. (5.37)) assumes the simplified expression

$$\psi\pm = \frac{1}{\sqrt{p}}\exp\left[\pm\frac{i}{\hbar}\int pdx\right]. \tag{5.43}$$

[§]The expression of F is obtained starting from the derivative of Eq. (5.28):

$$p' = \frac{d}{dx}\sqrt{2m(E-V)} = -\frac{m}{p}\frac{dV}{dx} = m\frac{F}{p} \ , \quad \text{i.e.,} \quad F = pp'/m.$$

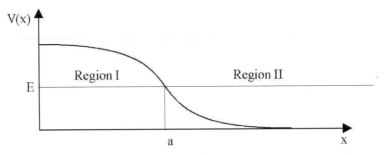

FIGURE 5.11   Classically forbidden (I) and classically allowed (II) regions.

We have, here, two solutions that should be superimposed. Let us consider a situation as the one of the Figure 5.11.

In the *classically allowed region* (region II), close to the turning point, $x = a$, the wave function should be a linear combination of ψ_+ and ψ_-:

$$\psi^{II} = \frac{C_1}{\sqrt{p}} \exp\left[\frac{i}{\hbar} \int_a^x p\,dx\right] + \frac{C_2}{\sqrt{p}} \exp\left[-\frac{i}{\hbar} \int_a^x p\,dx\right], \qquad x > a. \tag{5.44}$$

In the *classically forbidden region* (region I), to the left of $x = a$, we have a similar solution. The differences will be two. The classical momentum p will be complex ($p = i|p|$) and, in order that ψ is quadratically integrable, C_2 should be zero. With that, the wave function in the classically forbidden region is:

$$\psi^{I} = \frac{C}{\sqrt{|p|}} \exp\left[\frac{1}{\hbar} \int_a^x |p|dx\right], \qquad x < a. \tag{5.45}$$

Notice that the above integral will always give a negative number, guaranteeing that the wave function is quadratically integrable.

Using the connection formulas discussed in the next section and combining Eqs. (5.44), (5.53) and (5.55), we obtain the wave function in the classically allowed region:

$$\psi^{II} = \frac{C}{\sqrt{k}} \exp\left[i\left(\int_a^x k\,(x)\,dx - \pi/4\right)\right] + \frac{C}{\sqrt{k}} \exp\left[-i\left(\int_a^x k\,(x)\,dx - \pi/4\right)\right], \quad x > a,$$

$$= \frac{C}{\sqrt{k}} \sin\left(\int_a^x k\,(x)\,dx + \pi/4\right), \tag{5.46}$$

where $k(x) = p(x)/\hbar$. In the classically forbidden region, the wave function is:

$$\psi^{I} = \frac{C}{\sqrt{|k|}} \exp\left[\int_a^x |k\,(x)|\,dx\right], \quad x < a. \tag{5.47}$$

5.5   Connection formulas in WKB

We found above two wave functions, ψ^{I} and ψ^{II}. None of them is well defined in the proximity of the classical turning point $x = a$. In this region, the classical momentum tends to zero and the condition imposed by Eq. (5.39) is not satisfied. In order to establish a continuity condition, we need to equalize ψ_I and ψ_{II} at a point different from a, for example at a point x' belonging to the region I. This will only be possible if we can obtain the form of ψ_{II} in that region. However, on the real axis this task cannot be done, therefore that is only possible if we find a way to continuate the wave function over a path not too close to a in passing from region II to region I. That is, we have to perform an analytic continuation of the wave function at the complex plane.

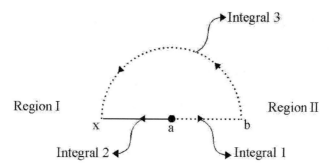

FIGURE 5.12 Integration path in the complex plane for the connection formulas.

Let us consider the path which will bypass the point $x = a$. This path will be a semi-circle that will leave a point b belonging to the region II and arrive at the point x of the region I with a radius ρ. The order of magnitude of this radius should be large enough so that the conditions of validity of the WKB approximation is valid, but small enough so that we can expand $\sqrt{V(x) - E}$ in a Taylor series so that it converges quickly. This path can be made either at the upper semi-plane, or at the lower semi-plane and, in fact, we will do both.

Given the condition on ρ discussed above, we can expand the function $V(x) - E$ in the following way:

$$V(x) - E = V(a) - E + \frac{d(V(x) - E)}{dx}(x - a) + \mathcal{O}(x^2) = V(a) - E + \frac{dV(x)}{dx}(x - a) + \mathcal{O}(x^2).$$

From the definition of a (turning point), $V(a) = E$. Identifying the derivative in the second term as the force upon the particle at the point a, we obtain

$$V(x) - E = F(a)(x - a) + \mathcal{O}(x^2) \approx F_0(x - a). \tag{5.48}$$

With this, we obtain the following expressions for ψ^I and ψ^{II} along the complex path:

$$\psi^I(x) = \frac{C}{(2m|F_0|(a - x))^{1/4}} \exp\left(\frac{1}{\hbar}\sqrt{2m|F_0|}\int_a^x \sqrt{a - x}\,dx\right), \quad x < a$$

$$\psi^{II}(x) = \frac{C_1}{(2m|F_0|(x - a))^{1/4}} \exp\left(\frac{i}{\hbar}\sqrt{2m|F_0|}\int_a^x \sqrt{x - a}\,dx\right)$$

$$+ \frac{C_2}{(2m|F_0|(x - a))^{1/4}} \exp\left(-\frac{i}{\hbar}\sqrt{2m|F_0|}\int_a^x \sqrt{x - a}\,dx\right), \quad \text{any } x.$$

Now we define γ as $2m|F_0|$, and obtain

$$\psi^I(x) = \frac{C}{(\gamma(a - x))^{1/4}} \exp\left(\frac{1}{\hbar}\sqrt{\gamma}\int_a^x \sqrt{a - x}\,dx\right), \quad x < a$$

$$\psi^{II}(x) = \frac{C_1}{(\gamma(x - a))^{1/4}} \exp\left(\frac{i}{\hbar}\sqrt{\gamma}\int_a^x \sqrt{x - a}\,dx\right)$$

$$+ \frac{C_2}{(\gamma(x - a))^{1/4}} \exp\left(-\frac{i}{\hbar}\sqrt{\gamma}\int_a^x \sqrt{x - a}\,dx\right), \quad \text{any } x.$$

As one bypasses the point $x = a$, one should determine the form that will assume the terms below the $C's$ as well as the integrands in the arguments of the exponentials in ψ^{II}. In both cases we have to find the function of $(x - a)$, that in the region II is larger than zero, but that in the region I is smaller than zero. This term always appears inside of a square root, so that it is interesting to maintain its sign explicitly.

In region I, as we by-pass the point $x = a$ on the upper semi-plane, we have:

$$(x - a) = -1(a - x) = e^{i\pi}(a - x), \tag{5.49}$$

what leads to two substitutions in the wave function ψ^{II}:

$$\frac{1}{(x-a)^{1/4}} = \frac{1}{(e^{i\pi}(a-x))^{1/4}} = \frac{e^{-i\pi/4}}{(a-x)^{1/4}},$$

$$\int_a^x \sqrt{x-a}\,dx = \int_a^x \sqrt{e^{i\pi}}\sqrt{a-x}\,dx = e^{i\pi/2}\int_a^x \sqrt{a-x}\,dx = i\int_a^x \sqrt{a-x}\,dx. \tag{5.50}$$

Replacing this in ψ^{II}, we obtain

$$\psi^{II}(x) = \frac{C_1 e^{-i\pi/4}}{(\gamma(a-x))^{1/4}} \exp\left(\frac{i^2}{\hbar}\sqrt{\gamma}\int_a^x \sqrt{a-x}\,dx\right) + \frac{C_2 e^{-i\pi/4}}{(\gamma(x-a))^{1/4}} \exp\left(-\frac{i^2}{\hbar}\sqrt{\gamma}\int_a^x \sqrt{a-x}\,dx\right).$$

This can be rewritten as:

$$\psi^{II}(x) = \frac{C_1}{(\gamma(a-x))^{1/4}} \exp\left(-\frac{1}{\hbar}\sqrt{\gamma}\int_a^x \sqrt{a-x}\,dx - i\frac{\pi}{4}\right)$$

$$+ \frac{C_2}{(\gamma(x-a))^{1/4}} \exp\left(\frac{1}{\hbar}\sqrt{\gamma}\int_a^x \sqrt{a-x}\,dx - i\frac{\pi}{4}\right).$$

As we equalize this equation with ψ^I we observe that all the denominators of the fractions are the same, and we thus obtain the following equation:

$$C_1 \exp\left(-\frac{1}{\hbar}\sqrt{\gamma}\int_a^x \sqrt{a-x}\,dx - i\frac{\pi}{4}\right) + C_2 \exp\left(\frac{1}{\hbar}\sqrt{\gamma}\int_a^x \sqrt{a-x}\,dx - i\frac{\pi}{4}\right)$$

$$= C \exp\left(\frac{1}{\hbar}\sqrt{\gamma}\int_a^x \sqrt{a-x}\,dx\right). \tag{5.51}$$

Now, as x increases, the term that multiplies C_1 can be neglected before the other ones, what leads to the following equation:

$$C_2 \exp\left(\frac{1}{\hbar}\sqrt{\gamma}\int_a^x \sqrt{a-x}\,dx - i\frac{\pi}{4}\right) = C \exp\left(\frac{1}{\hbar}\sqrt{\gamma}\int_a^x \sqrt{a-x}\,dx\right), \tag{5.52}$$

or

$$C_2 = C \exp\left(i\frac{\pi}{4}\right). \tag{5.53}$$

Now it is necessary to determine C_1. For this, we have to use an integration path along the inferior semi-plane and to repeat all the steps we did above. It is straightforward to show that

$$C_1 \exp\left(\frac{1}{\hbar}\sqrt{\gamma}\int_a^x \sqrt{a-x}\,dx + i\frac{\pi}{4}\right) = C \exp\left(\frac{1}{\hbar}\sqrt{\gamma}\int_a^x \sqrt{a-x}\,dx\right), \tag{5.54}$$

or

$$C_1 = C \exp\left(-i\frac{\pi}{4}\right). \tag{5.55}$$

With this we obtained C_2 and C_1 as a function of C. This explains Eqs. (5.46) and (5.47).

5.6   The three-dimensional WKB approximation

The WKB solution in three dimensions is simplified if we express ψ as a combination of simultaneous functions of the operators H, L^2 and L_z, as we saw in Chapter 2 (see Eq. (2.8)). Defining an effective potential, by means of

$$V_{eff} = V(r) + \frac{\hbar^2}{2\mu}\frac{l(l+1)}{r^2}. \tag{5.56}$$

the radial wave equation becomes

$$\frac{\hbar^2}{2\mu}\frac{d^2}{dr^2}u_l(k,r) + [E - V_{eff}(r)]\,u_l(k,r) = 0. \tag{5.57}$$

This equation is identical to the unidimensional equation and thus its solutions should be the equations (5.46) and (5.47) by means of the substitution of k for $k_{eff} = \sqrt{2m(E - V_{eff})}/\hbar$. However, the results obtained in the unidimensional case were associated to a variable that could assume values from $-\infty$ to $+\infty$ and that didn't possess singularities in the proximity of the classical turning point. With the increment of the contribution of the angular momentum term, we have a singularity at $r = 0$. Also, r is limited to the real positive semiaxis. In frontal collisions (of low angular momentum) the turning point a can be closely enough to $r = 0$ so that it is impossible to perform the necessary analytic continuation for obtaining the connection formulas.

To solve this problem it will be necessary a change of variables that takes the singular point $r = 0$ to $-\infty$. With this change, the connection formula can be established for a new variable x, so that $r = e^x$. Given this change, Eq. (5.57) becomes

$$\frac{\hbar^2}{2\mu}\frac{d^2}{dx^2}u_l(k,e^x) + \frac{\hbar^2}{2\mu}\frac{d}{dx}u_l(k,e^x) + e^{2x}(E - V_{eff}(e^x))u_l(k,e^x) = 0. \tag{5.58}$$

To eliminate the first derivative of u_l, we will also do a change in the wave function:

$$u_l(e^x) = g(x)e^{x/2}. \tag{5.59}$$

This change, when substituted in the radial equation, gives us

$$\frac{\hbar^2}{2\mu}\frac{d^2}{dx^2}g(x) + \left[e^{2x}\left(E - V_{eff}(r) - \frac{e^{-2x}}{4}\right)\right]g(x) = 0. \tag{5.60}$$

This equation does not possess singularities for finite x and just has one isolated turning point. We can make use in the connection approach developed for the unidimensional case and we obtain an identical solution to the one of the previous sections, with the term between brackets of the Eq. (5.60) taking the role of $p^2/2m$. That is, the solution will be

$$g_\pm(x) = \frac{1}{\sqrt{w}}\exp\left[\pm i\int w(x)dx\right]$$

$$w(x) = \int e^x\sqrt{\frac{2m}{\hbar^2}(E-V) - \left[l(l+1) + \frac{1}{4}\right]e^{-2x}}dx.$$

However, $l(l+1) + 1/4 = (l+1/2)^2$. Substituting this in the above equation and returning to the variables r and u_l, we finally obtain the radial wave function, that will be the same as the wave function of the unidimensional case, but with of the replacement of $V(x)$ by $V_{eff}(x)$, that incorporates the centrifugal barrier and that makes use of the substitution obtained above: $l(l+1) \to (l+1/2)^2$.

For convenience, we summarize all results below:

$$u_l(r) = \frac{C}{\sqrt{k_{eff}(r)}}\sin\left(\int_{r_0}^{r}k_{eff}(r')dr' + \pi/4\right), \qquad r > r_0 \tag{5.61}$$

$$u_l(r) = \frac{C}{\sqrt{|k_{eff}(r)|}}\exp\left(-\int_{r_0}^{r}|k_{eff}(r')|dr'\right), \qquad r < r_0 \tag{5.62}$$

$$k_{eff}(r) = \sqrt{\frac{2\mu}{\hbar^2}(E - V_{eff}(r))}, \qquad V_{eff} = V(r) + \frac{\hbar^2}{2\mu}\frac{(l+1/2)^2}{r^2}. \tag{5.63}$$

5.6.1   Phase-shift for short-range potentials

We know that the radial equation has an exact solution that can be written in terms of a combination of the Ricatti-Bessel functions $\hat{\jmath}$ and $\hat{n}$.:

$$u_l(k, r) = \alpha_l[\hat{\jmath}_l(kr) + \beta_l \hat{n}_l(k, r)]. \tag{5.64}$$

The influence of the potential and the consequent appearance of the phase-shift is related to the constant β_l. For a short-range potential, we know that in the limit that r tends to infinite, the wave function becomes:

$$u_l(k, r) = \alpha_l \left[\sin\left(kr - l\pi/2\right)\cos\delta_l + \sin\delta_l\cos\left(kr - l\pi/2\right)\right] = \quad \alpha_l' \sin\left(kr - l\pi/2 + \delta_l\right). \tag{5.65}$$

where $\delta_l = \arctan\beta_l$ is the phase-shift. To determine δ_l in the WKB approximation, we should compare Eq. (5.65) with the radial wave function of the WKB approximation that was deduced above, Eq. (5.61). For such, some changes need to be done. We shall rewrite the argument of Eq. (5.61) in the following form:

$$\int_{r_0}^{r} k_{eff}(r')dr' + \pi/4 = \int_{r_0}^{r} kdr' + \left(\int_{r_0}^{r} k_{eff}dr' - \int_{r_0}^{r} kdr'\right) + \pi/4$$

$$= \int_{r_0}^{r} \Delta k(r')dr' + \int_{r_0}^{r} kdr' + \pi/4$$

$$= \int_{r_0}^{r} \Delta k(r')dr' + kr - kr_0 + \pi/4,$$

where $k = \sqrt{2mE}/\hbar$ and

$$\Delta k(r) = k_{eff}(r) - k. \tag{5.66}$$

With that, Eq. (5.61) becomes

$$u_l(k, r) = \frac{C}{\sqrt{k_{eff}}}\sin\left(\int_{r_0}^{r} \Delta k(r)dr + kr - kr_0 + \pi/4\right). \tag{5.67}$$

Knowing that when r tends to infinity the function $k_{eff}(r)$ tends to a constant, we can absorb the term $1/\sqrt{k_{eff}(r)}$ in the constant C, to equal the argument of the above equation to that of Eq. (5.65):

$$\int_{r_0}^{r} \Delta k(r)dr + kr - kr_0 + \pi/4 = kr - l\pi/2 + \delta_l.$$

Solving this equation for δ_l, we find:

$$\delta_l^{WKB} = \int_{r_0}^{\infty} \Delta k(r)dr - kr_0 + \pi/4 + l\pi/2. \tag{5.68}$$

When rewriting this equation, we obtain the final expression of the WKB phase-shift associated to nuclear potentials:

$$\delta_l^{WKB} = \int_{r_0}^{\infty} \left[k_{eff}(r) - k\right]dr - kr_0 + \frac{\pi}{2}\left(l + \frac{1}{2}\right), \tag{5.69}$$

where

$$k_{eff}(r) = \sqrt{\frac{2\mu}{\hbar^2}(E - V(r)) + \frac{(l + 1/2)^2}{r^2}}.$$

5.6.2 Phase-shift for long-range potentials

Let us determine the expression of $k_{eff}(r)$ when $V(r)$ is, for example, a Coulomb potential. In this case we shall call k_{eff} by k_C, with

$$k_C(r) = k\sqrt{1 - \frac{2a_0}{r} - \frac{b^2}{r^2}}, \tag{5.70}$$

where $a_0 = a/2$ is the Coulomb parameter (half of the classical distance of closest approach in a head-on collision[‡]) and b is the WKB-modified expression for the impact parameter ($k^2 b^2 = \lambda^2 = (l + \frac{1}{2})^2$). Defining r_C as the Coulomb turning point, we can calculate the integral of $k_C(r)$, obtaining the following result:

$$\int_{r_C}^r k_C(r')dr' = rk_C(r) - \eta \ln\left(rk_C(r) + rk - \eta\right) + \frac{1}{2}\eta \ln\left(\eta^2 + \lambda^2\right)$$

$$+ \lambda \sin^{-1}\left[\frac{\eta kr + \lambda^2}{kr(\eta^2 + \lambda^2)^{1/2}}\right] - \frac{1}{2}\pi\lambda, \tag{5.71}$$

where η is the Sommerfeld parameter (see Section 2.14).

With this result we are capable of, using Eqs. (5.61) and (5.62), to determine the form of the wave function for a particle subject to a Coulomb potential in the WKB approximation. We should compare this result to the exact solution of the Schrödinger equation, which possesses an asymptotic behavior under these circumstances given by

$$u_l(r) \sim \sin\left(kr - \eta \ln 2kr - \frac{1}{2}l\pi + \sigma_l\right), \tag{5.72}$$

where σ_l is the Coulomb phase-shift. Taking the limit of r tending to infinity in the expression (5.71) we obtain the value of σ_l given by

$$\sigma_l^{WKB} = \frac{1}{2}\eta \ln(\eta^2 + \lambda^2) - \eta + \lambda \tan^{-1}\left(\frac{\eta}{\lambda}\right). \tag{5.73}$$

We can compare this result with the exact solution of the Coulomb phase-shift, that is given by

$$\sigma_l = \arg \Gamma(l + 1 + i\eta). \tag{5.74}$$

We observe that both will be the same asymptotically if $\eta \gg 1$ and $\lambda \gg \eta$.

5.6.3 Short-range plus Coulomb potential

We now combine the Coulomb interaction with a short range (nuclear) one. We need to determine the form of the WKB wave function and to compare it with the asymptotic behavior of the exact solution, which in this case is given by

$$u_l(r) \sim \sin\left(kr - \eta \ln 2kr - \frac{1}{2}l\pi + \sigma_l + \delta_{nl}\right), \tag{5.75}$$

where δ_{nl} is the nuclear phase-shift.

As we have a clear separation between the Coulomb contribution and the nuclear one, it will be of great usefulness if we write k_{eff} so that it also distinguishes these two contributions. For such, we rewrite it in the following way:

$$\int_{r_0}^r k_{eff}(r')dr' = \int_{r_0}^r k_C(r')dr' + \left[\int_{r_0}^r k_{eff}(r')dr' - \int_{r_0}^r k_C(r')dr'\right]. \tag{5.76}$$

[‡]a_0 is half the value of the parameter a, defined by Eq. (5.2).

When we replace this in Eq. (5.61), we see that the first term to the right of the above equation is responsible for the Coulomb phase-shift (σ_l) and that the term between brackets is what was added to this phase-shift when we included the short-range interaction. Thus,

$$\delta_{nl}^{WKB} = \int_{r_0}^{r} k_{eff}(r')dr' - \int_{r_0}^{r} k_C(r')dr', \tag{5.77}$$

where,

$$k_{eff}(r) = \sqrt{\frac{2\mu}{\hbar^2}(E - V_{eff}(r))}$$

$$V_{eff} = V_n(r) + V_C(r) + \frac{\hbar^2}{2\mu}\frac{(l+1/2)^2}{r^2}.$$

This is the expression for the nuclear phase-sift in the presence of a short range interaction. The Coulomb phase-shift is given by Eq. (5.73).

5.7   Heavy ion fusion reactions

The calculation of $T_l(E)$ in equation (5.20) must rely on solving the appropriate Schrödinger equation with *incoming wave boundary condition* (IWBC). Here IWBC enforces the condition that once the system is in the potential pocket, it does not return. An alternative approximate way of calculating T_l is through the use of the WKB, or semiclassical approximation as we explained in the previous sections. One gets [10]

$$T_l(E) = \{1 + \exp[2S_l(E)]\}^{-1}, \tag{5.78}$$

where the WKB integral $S_l(E)$ is given by

$$S_l(E) = \sqrt{\frac{2\mu}{\hbar^2}} \int\limits_{r_1(l)}^{r_2(l)} dr \left[V_0(r) + \frac{\hbar^2(l+1/2)^2}{2\mu r^2} - E\right]^{1/2}. \tag{5.79}$$

Here $r_1(l)$ and $r_2(l)$ are the outermost classical turning points determined by setting

$$V_0 + \frac{\hbar^2(l+1/2)^2}{2\mu r^2(l)} = E. \tag{5.80}$$

The "width" of the barrier is just $r_2(l) - r_1(l)$.

For energies that correspond to a trajectory close, but below, the top of the barrier, the barrier looks like an inverted parabola. Then one expresses $V_0(r)$ in the form $V_N + V_C$, i.e.

$$V_0(r) = V_B - \frac{1}{2}\mu\omega^2(r - R_B)^2, \tag{5.81}$$

where R_B is the position of the top of the barrier.

A more precise statement concerning Eq. (5.81) should involve the centrifugal potential. However the inclusion of this latter potential only changes V_{B_0} to $V_{B_l} \equiv V_{B_0} + \hbar^2\left(l + \frac{1}{2}\right)^2/2\mu R_B^2$. The change in the curvature of the barrier, measured by ω, is very small and is thus neglected.

With the parabolic approximation for $V_0 + \hbar^2(l+1/2)/2\mu r^2 = V_{B_l} - \frac{1}{2}\mu\omega^2\left(r - R_B^2\right)^2$, the integral $S_l(E)$ can be performed straight forwardly and when inserted in T_ℓ, gives the well-known *Hill-Wheeler formula* [10]

$$T_l = \frac{1}{1 + \exp[2\pi(V_{B_l} - E)/\hbar\omega]}, \tag{5.82}$$

with

$$V_{B_l} = V_{B_0} + \frac{\hbar^2(l+1/2)^2}{2\mu R_B^2}. \tag{5.83}$$

The above Hill-Wheeler expression for T_l, Eq. (5.82), is valid only in the proximity of the barrier (E slightly above or slightly below V_{B_l}). For energies that do not satisfy this condition the parabolic approximation is a bad one. One has to evaluate $S_l(E)$ exactly.

With the form of $T_l(E)$ determined (at least in the vicinity of the height of the Coulomb barrier), one can now evaluate the fusion cross-section, using Eq. (5.20),

$$\sigma_F(E) = \frac{\pi}{k^2} \sum_{l=0} (2l+1) \frac{1}{1 + \exp\left[2\pi\left(V_{B_0} + \hbar^2\left(l+1/2\right)^2 / 2\mu R_B^2 - E\right)/\hbar\omega\right]}. \tag{5.84}$$

Since many partial waves contribute to the sum, one may replace the sum by an integral. Call $(l+1/2)^2 = \chi$, and write

$$\sum_l (2l+1) = \sum_l (2l+1)\,\Delta l = \sum_l \Delta\chi \longrightarrow \int d\chi,$$

or

$$\sigma_F(E) = \frac{2\pi}{k^2} \int_{1/4}^{\infty} d\chi \frac{1}{1 + \exp\left\{2\pi\left[(V_{B_0} - E) + \hbar^2\chi/2\mu R_B^2\right]/\hbar\omega\right\}}. \tag{5.85}$$

The integral yields

$$\sigma_F(E) = \frac{\hbar\omega R_B^2}{2E} \ln\left[1 + \exp\left[\frac{2\pi}{\hbar\omega}(E - V_{B_0})\right]\right]. \tag{5.86}$$

The above expression for $\sigma_F(E)$ is referred to as the *Wong formula* [11]. At high energies, $E \gg V_{B_0}$, the exponential dominates over unity and one finds the geometrical formula for $\sigma_F(E)$

$$\sigma_F(E) = \pi R_B^2 \left(1 - \frac{V_{B_0}}{E}\right). \tag{5.87}$$

The above form can also be obtained in the classical limit where $T_l(E)$ is a step function and accordingly by using $\omega = 0$.

Before we concentrate our discussion on σ_F at sub-barrier energies, we indulge a bit on fusion at energies above the barrier. Here it is seen that after an initial rise, $\sigma_F(E)$ reaches a maximum followed by a drop (see the two middle figures in Figure 5.13a). The drop in $\sigma_F(E)$ is attributed to the competing deep inelastic processes (where a large amount of relative energy is converted into internal excitation of the two fragments). To account for such a structure in σ_F, one can multiply T_l by the fusion probability [12]. Thus T_l takes into account the tunnelling into the strong absorption radius, R_{sa}, which is situated slightly to the left of the barrier. Part of the flux that reaches R_{sa} would fuse while the other part is lost to deep inelastic processes.

Accordingly,

$$\sigma_F(E) = \frac{\pi}{k^2} \sum_{\ell=0}^{\infty} (2l+1)\, T_l(E)\, P_l^{(F)}(E). \tag{5.88}$$

Since $T_l(E)$ contains reference to the barrier radius R_{B_0} only, $P_l^{(F)}(E)$ must contain reference to $R_{sa} \equiv R_{critical} = R_c$. The sum over l now extends to $\ell_c(R_c)$, the *critical angular momentum* associated with R_c.

The calculation of $\sigma_F(E)$ using the integral form, Eq. (5.85), with an upper limit in the integral $\chi_c \equiv (l_c + 1/2)^2$, can be done easily and one obtains the Glas-Mosel formula [12]

$$\begin{aligned}
\sigma_F &= \frac{\hbar\omega R_B^2}{2E} \ln\left\{\left[1 + \exp\left[\frac{2\pi}{\hbar\omega}(V_{B_0} - E)\right]\right]\right. \\
&\quad \times \left.\left[\exp\left[-\frac{2\pi}{\hbar\omega}\left(\frac{\hbar^2(l_c + 1/2)^2}{2\mu R_c^2}\right)\right] + \exp\left[\frac{2\pi}{\hbar\omega}(V_{B_0} - E)\right]\right]^{-1}\right\}. \tag{5.89}
\end{aligned}$$

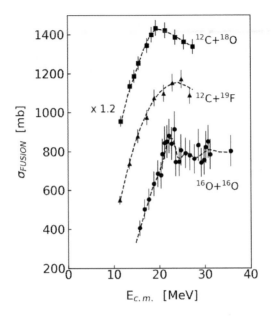

FIGURE 5.13 Fusion data for several systems exhibiting regions I and II. See text for details. The data for ^{12}C+^{18}O was multiplied by 1.2 for clarity. The theoretical calculation are within the statistical yrast line model of Ref. [13].

Calling $E - V_{B_0} = \hbar^2 \left(l_B + \frac{1}{2}\right)^2 / 2\mu R_B^2 \cong \hbar^2 l_B^2 / 2\mu R_B^2$, the above expression can be reduced further into

$$\sigma_F = \frac{\hbar \omega R_B^2}{2E} \ln \left\{ \frac{\exp\left(\alpha l_B^2\right) + 1}{\exp\left(\alpha \left(l_B^2 - l_c^2\right)\right) + 1} \right\}, \qquad \alpha \equiv \frac{2\pi}{\hbar \omega} \frac{\hbar^2}{2\mu R_B^2}. \tag{5.90}$$

Clearly, if σ_F of Eq. (5.86) is considered very close to the total reaction cross-section (since peripheral processes have much smaller cross-sections than σ_F at lower energies), Eq. (5.90), shows clearly that $\sigma_F < \sigma_R$. It has become common to call the region where $\sigma_F \sim \sigma_R$ as region I while that where $\sigma_F < \sigma_R$ as region II. In this later region, deep inelastic collisions (DIC) constitute a considerable part of σ_R. Figure 5.13 shows the data of systems exhibiting regions I and II*.

A possible physical interpretation for l_{cr}, besides the one related to the competition between fusion and DIC, may reside in the compound nucleus formed in the process. The *yrast line model* states that the angular momentum imparted to the compound nucleus (CN) can be associated with the yrast line which is the maximum angular momentum for a given excitation energy. Since the excitation energy of the CN is $E_{c.m.} + Q$ where Q is the Q-value of the fusion $A_1 + A_2 \rightarrow (A_1 + A_2)$, then the yrast line is

$$E_{c.m.} + Q = \frac{\hbar^2 J_y \left(J_y + 1\right)}{2\Im}, \tag{5.91}$$

where $\Im$ is the moment of inertia of the CN. For a rigid spherical body, $\Im = 2M R_{CN}^2 / 5$.

Then

$$J_y^2 \simeq \frac{2\Im E_{c.m.}}{\hbar^2} \left(1 + \frac{Q}{E_{c.m.}}\right). \tag{5.92}$$

---

*Region I is where σ_F goes through a maximum and region II is at larger energies.

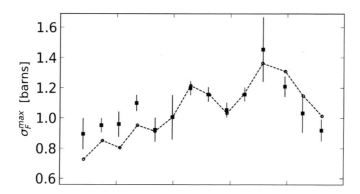

FIGURE 5.14 The maximum values of σ_F (in barns) for several systems, from ^{12}C $+^{12}$ C (leftmost point) to ^{40}Ca $+^{40}$ Ca (rightmost point). The theoretical calculation are within the statistical yrast line model of Ref. [13].

One then makes the identification $l_{cr} = J_y$. At high energies,

$$\sigma_F \simeq \frac{\pi \hbar^2}{2\mu E_{c.m.}} l_{cr}^2 = \frac{\pi \Im}{\mu} \left(1 + \frac{Q}{E_{c.m.}} \right). \tag{5.93}$$

A variance of the CN model is the statistical yrast line model of Lee at al. [13], which says that part of the excitation energy $E_{c.m.} + Q$ is statistical (thermal) in nature. Calling this energy ΔQ one has

$$\frac{\hbar^2 J_y' \left(J_y' + 1 \right)}{2\Im} + \Delta Q = E_{c.m.} + Q, \tag{5.94}$$

and thus

$$\sigma_F = \frac{\pi \Im}{\mu} \left(1 + \frac{Q - \Delta Q}{E_{c.m.}} \right). \tag{5.95}$$

The above expression can account for a wide range of experimental data if $\Delta Q \simeq 10$ MeV and $r_0 = 1.2$ fm, where r_0 is the radius parameter that enters in the calculation of the moment of inertia $\Im$. An example of how the above formula works is shown in Figure 5.14 which exhibits the maximum value of σ_F as a function of the system (Q-value), obtained with $\Delta Q = 5$ MeV.

5.8 Sub-barrier fusion

At energies close or below the $l = 0$ barrier (the Coulomb barrier), the collision time becomes longer and specific effects of the structure of the participating nuclei in the fusion process become important. In such cases the one-dimensional barrier penetration model does not work anymore and one has to resort to *coupled channels* treatments. A case which illustrates this effect is supplied by the fusion of ^{16}O with the well deformed ^{154}Sm target [14]. Here, since the rotational band is well formed in ^{154}Sm, one may freeze the rotation axis and perform a one-dimensional calculation of the type described before. An average over all orientations would then supply σ_F,

$$\sigma_F \left(E_{c.m.} \right) = \int d\Omega \, \sigma_F \left(E_{c.m.}, \Omega \right), \tag{5.96}$$

where Ω represents the solid angle that describes the direction of the rotation axis of the assumed rigid rotor ^{154}Sm.

One can easily convince oneself that the "equivalent sphere" σ_F of Eq. (5.96), is larger than σ_F calculated for a spherical ^{154}Sm. The sub-barrier enhancement of σ_F is found to occur in many systems. The calculation of σ_F according to Eq. (5.96) assumes an infinite moment of inertia of the rotor (degenerate or sudden limit). A more realistic way of calculating σ_F which takes into account the energy loss (Q-value) must rely on coupled channels theory.

Another simple model that can be evaluated analytically is that of a *two-level system*. Ignoring the angular momentum of the excited state, and its excitation energy, one has the two-coupled equations

$$\left[E - \left(-\frac{\hbar^2}{2\mu} \frac{d^2}{dr^2} + V(r) \right) \right] \Psi_1 = F_{12} \, \Psi_2$$

$$\left[E - \left(-\frac{\hbar^2}{2\mu} \frac{d^2}{dr^2} + V(r) \right) \right] \Psi_2 = F_{21} \, \Psi_1. \tag{5.97}$$

If the coupling $F_{12} = F_{21} = F$ is taken to be constant, one can solve for σ_F, by diagonalizing the above two equations. One gets

$$\sigma_F = \frac{1}{2} \left[\sigma_F (V_B + F) + \sigma_F (V_B - F) \right], \tag{5.98}$$

which is always larger than $\sigma_F(V_B)$.

The inclusion of the non-zero excitation energy can be done without great difficulty*. The result of diagonalization gives for σ_F the following

$$\sigma_F = A_+ \, \sigma_+ + A_- \, \sigma_-, \tag{5.99}$$

where

$$A_{\pm} = \frac{2F^2}{4F^2 + Q^2 \mp Q\sqrt{4F^2 + Q^2}}, \qquad \sigma_{\pm} = \sigma_F(V + \lambda_{\pm}),$$

$$\text{and} \quad \lambda_{\pm} = \frac{1}{2} \left(-Q \pm \sqrt{Q^2 + 4F^2} \right). \tag{5.100}$$

When $Q = 0$, one recovers Eq. (5.98).

The above results can be extended to several channels. One introduces the eigenchannels $|C\rangle$ that diagonalize the many-channels Schrödinger equation. The eigenvalues are denoted by λ_c. Then

$$\sigma_F = \sum_C |\langle C|0\rangle|^2 \, T_C(E, V + \lambda_c), \tag{5.101}$$

where $|0\rangle$ is the entrance channel. For more details on sub-barrier enhancement of fusion see, e.g., Ref. [15].

An interesting observation can be made concerning the extraction of the barrier or eigenbarriers directly from the data. It can be seen from the form of T_l given by Hill-Wheeler, Eq. (5.82), that $T_l(E)$ may be written as $T_0 \left[E - l(l+1)\hbar^2/2\mu \, R^2 \right]$. Therefore it is not surprising that the approximation $T_l(E) \simeq T_0[E - l(l+1)\hbar^2/2\mu R^2(E)]$, with $R(E)$ being a slowly varying function of E, may turn out to be reasonably accurate. Then, with $\lambda \equiv l+1/2 \cong \sqrt{l(l+1)}$ and $E' \equiv E - \lambda^2/2\mu R^2(E)$,

$$\sigma_F = \frac{\pi}{k^2} \sum_{l=0}^{\infty} (2l+1) T_l(E) \simeq \frac{2\pi}{k^2} \int_0^{\infty} \lambda \, d\lambda \, T(\lambda; E) = \frac{\pi R^2(E)}{E} \int_{-\infty}^{E} dE' \, T_0(E'). \tag{5.102}$$

Therefore the second derivative $d^2 \left[E \, \sigma_F(E) \right]/dE^2$ can be directly related to the first derivative $dT_0(E)/dE$, up to corrections coming from the energy dependence of $R(E)$

$$\frac{d \, T_0(E)}{dE} \sim \frac{1}{\pi R^2(E)} \frac{d^2}{dE^2} \left[E\sigma_F(E) \right] + \mathcal{O}\left(\frac{dR}{dE} \right). \tag{5.103}$$

---

*In this case, $E_1 = E$, $E_2 = E + Q$.

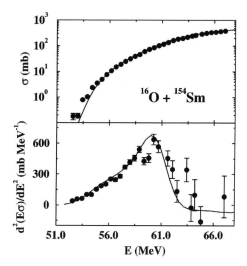

FIGURE 5.15 Fusion cross section and barrier distribution for the ^{16}O+^{154}Sm system by Leiht et al. [16]. Compare to Figure 5.10b.

In Figure 5.10 (right) we show an example of dT_0/dE for the one-dimensional and the two-channel barrier tunnelling problem. The actual data, e.g. ^{16}O+^{154}Sm, show a clear deviation from the one-dimensional case when dT/dE is examined, Figure 5.15.

Contrary to the sub-barrier fusion enhancement found in light heavy systems, the fusion of very heavy systems of the type employed in the production of *superheavy elements* (SHE), shows hindrance when compared to the simple one-dimensional barrier penetration model. The incident energy has to be much higher than the fusion barrier which accounts for fusion of lighter systems, in order for heavy element production to proceed. One needs an *"extra push"*, as proposed by Bjornholm and Swiatecki [17]. This extra energy needed for fusion to occur comes about from the fact that the fission barrier for massive systems is located inside the potential barrier in the entrance channel. Ref. [17] introduced the "extra push"concept; the energy needed to overcome the saddle point in the potential energy surface under the constraint of mass asymmetry, and the *"extra extra push"* which is the energy needed to carry the system beyond (inside) the fission saddle point.

5.9 Superheavy elements

The heaviest element found in the nature is $^{238}_{92}$U. It is radioactive, but it survived from its formation in the supernovae explosions since it has a decay half-life of the order of the age of the Earth. Elements with larger atomic number (transuranic) have shorter half-lives and have disappeared. They are created artificially through nuclear reactions using as target heavy elements or transuranic elements obtained previously. The projectiles were initially light particles: protons, deuterons, α-particles and neutrons. The use of neutron is justified because the β^--emission of the compound nucleus increases the value of Z and it was in this way [18] that the first transuranic element, the neptunium, was obtained:

$$\text{n} + \ ^{238}\text{U} \rightarrow \ ^{239}\text{U} \rightarrow \ ^{239}\text{Np} + \beta^-. \tag{5.104}$$

Reactions with light particles could produce isotopes until the mendelevium ($Z = 101$), but with this way of production it is not possible to go beyond that; the half-lives for α-emission or spontaneous fission become extremely short when this region of the periodic table is reached, turning impracticable the preparation of a target. The alternative is to place a heavy element under a flux of very intense neutrons. This can be done using special reactors of high flux or using the rest material of nuclear explosions. The element einsteinium ($Z = 99$) and fermium ($Z = 100$) were discovered in this way in 1955 but the increasing competition that the beta decay has with the alpha decay and with the spontaneous fission prevents the formation of elements with larger Z.

Starting in 1955, heavy ion accelerators began to deliver beams with high enough intensity and energy to compete in the production of transuranic isotopes. The first positive result was the production of two californium isotopes ($Z = 98$) in the fusion of carbon and uranium nuclei:

$$\,^{12}_{6}\mathrm{C} + \,^{238}_{92}\mathrm{U} \ \rightarrow \ \,^{244}_{98}\mathrm{Cf} + 6\mathrm{n}, \tag{5.105}$$

$$\,^{12}_{6}\mathrm{C} + \,^{238}_{92}\mathrm{U} \ \rightarrow \ \,^{246}_{98}\mathrm{Cf} + 4\mathrm{n}. \tag{5.106}$$

This opened the possibility of reaching directly the nucleus one wants to create from the fusion of two appropriately chosen smaller nuclei. The difficulty of such task is that the cross sections for the production of heavy isotopes are extremely low. As example, the reaction $^{50}_{22}\mathrm{Ti} + \,^{208}_{82}\mathrm{Pb} \rightarrow \,^{257}_{104}\mathrm{Rf} + \mathrm{n}$, which produces the element rutherfordium, has a cross section of 5 nb. A small increase in the charge reduces drastically that value: the cross section for the fusion reaction $^{58}_{26}\mathrm{Fe} + \,^{208}_{82}\mathrm{Pb} \rightarrow \,^{265}_{108}\mathrm{Hs} + \mathrm{n}$ is 4 pb. As comparison, the typical cross sections of DIC for heavy nuclei are in the range 1-2 b.

In spite of the experimental refinement that these low cross sections demand, one is able to produce an isotope as heavy as $^{277}112$. The understanding of the mechanisms that lead to fusion is, however, not fully understood. According to the traditionally accepted model, the fusion of two nuclei proceeds in two stages: the formation of a compound nucleus and the de-excitation of the compound nucleus by evaporation of particles, preferentially neutrons. The difficulties for the materialization of the process in very heavy nuclei reside in both stages and they will be discussed next.

When two light nuclei are in contact the attractive nuclear force overcomes the repulsive Coulomb force and the final fate of the system is fusion. In heavier systems the opposite happens and the projectile should have a certain minimum kinetic energy to penetrate the target. At this stage an appreciable amount of the kinetic energy is transferred to the internal excitation energy and this has an effect equivalent to the increase of the Coulomb barrier. An increment in the incident energy, or as is said, an "extra-push", is necessary to take the fusion ahead. Effects of nuclear structure can also be present, favoring or hindering the process.

But, an increase in the kinetic energy of the projectile is conditioned to the survival of the nucleus to the de-excitation process. In each stage of the neutron emission there is a possibility of fission. If the excitation energy is very high, let us say, about 50 MeV, it will be much larger than the fission barrier. The fission will very probably win the competition and the final stage will not be a cold residual nucleus of atomic number $Z = Z_1 + Z_2$ as one wants. One should also keep in mind the limitation, already mentioned, of the value of the angular momentum l transferred to the nucleus: it cannot surpass l_{crit}, so that the allowed values of l are located inside a small window around $l = 0$. In other words, only central collisions have some chance of producing a compound nucleus that de-excites without fission.

A guide for the choice of more appropriate projectiles and targets to produce a given nucleus should come from an examination of a graph as the one of Figure 5.16, where we show, as an example, the minimum value of the excitation energy of the compound nucleus $^{258}_{104}\mathrm{Rf}^{\,*}$ created from several projectile/target configurations. This value was calculated [19] adding the negative value of Q of the reaction to the value of the interaction barrier $B = Z_1 Z_2 e^2 / [1.45(A_1^{1/3} + A_2^{1/3})]$. It is evident that projectiles around $A = 50$ hitting targets of Pb and Bi produce the smallest excitation energy. This "cold fusion" has a larger probability of happening. The excitation function for a reaction of this type is seen in Figure 5.17, where the element $_{104}\mathrm{Rf}$ is formed by the irradiation of $^{208}_{82}\mathrm{Pb}$ with $^{50}_{22}\mathrm{Ti}$ [20]. The maximum probability of fusion with emission of one or no neutron happens for a center of mass energy smaller than the Coulomb barrier! In a model that one can envisage for an event of this type, the projectile has a central collision with the target and stops before reaching the maximum of the Coulomb barrier. The idea is that at this point the external shells are in contact and the nucleon transfer between projectile and target can decrease the Coulomb barrier and allow the nuclei to overcome it.

The favorable conditions provided by the method of cold fusion imply that, for the creation of the heaviest elements, it is preferable to do reactions of this kind than with actinide targets and lighter projectiles.

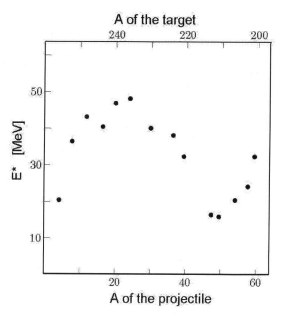

FIGURE 5.16 Minimum excitation energy of the compound nucleus $^{258}_{104}$Rf formed by several projectile/target combinations.

In fact, it was through the reactions

$$^{54}_{24}\text{Cr} + {}^{209}_{83}\text{Bi} \rightarrow {}^{262}_{107}\text{Bh} + \text{n},$$

$$^{58}_{26}\text{Fe} + {}^{208}_{82}\text{Pb} \rightarrow {}^{265}_{108}\text{Hs} + \text{n},$$

$$^{58}_{26}\text{Fe} + {}^{209}_{83}\text{Bi} \rightarrow {}^{266}_{109}\text{Mt} + \text{n},$$

$$^{62}_{28}\text{Ni} + {}^{208}_{82}\text{Pb} \rightarrow {}^{269}110 + \text{n},$$

$$^{64}_{28}\text{Ni} + {}^{209}_{83}\text{Bi} \rightarrow {}^{272}111 + \text{n},$$

$$^{70}_{30}\text{Zn} + {}^{208}_{82}\text{Pb} \rightarrow {}^{277}112 + \text{n}, \tag{5.107}$$

that for the first time isotopes of the elements $Z = 107$ to $Z = 112$ [22, 20] were synthesized, the first of the list in 1976 and the last in 1996. In spite of that, reactions with actinide targets were not abandoned. In 1996, Lazarev and collaborators [23] synthesized the isotope $^{273}110$, bombarding ^{244}Pu with ions of ^{34}S. The isotope $^{273}110$ was obtained starting from the evaporation of 5 neutrons of a nucleus of $^{278}110$ excited with approximately 50 MeV. The cross section for the process was 0.4 pb!

Figure 5.18 shows the cross-section systematic for the production of the elements from Nobelium to 116. The production cross sections for the elements 104 to 112 decrease by a factor of about 10 per two elements down to 1 picobarn for element 112 as shown in Figure 5.18. The competing processes of barrier penetration through the Coulomb –, or more precisely, the fusion-barrier and the survival of the compound nucleus can be inferred from the excitation functions. For element 104 the two-neutron (2n) and the three-neutron (3n) evaporation channels are still present; the 1n and 2n channels section show similar cross section. For element 108 the 2n channel is already rather weak. It completely disappears for element 110. Moreover, the 1n channel is shifted towards low excitation energies, and comes close to the binding energy of the last neutron. The synthesis of element 112 is the coldest fusion ever observed. The width of the excitation functions shrink towards element 110 where the half width becomes only 5 MeV. All these observations point towards a strong competition between barrier transmission and survival probability.

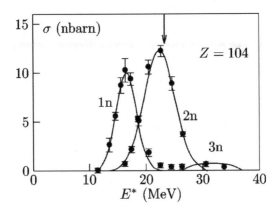

FIGURE 5.17 Excitation function for the production of $_{104}$Rf from the fusion of ^{208}Pb with ^{50}Ti. The arrow indicates the value of the interaction barrier, as calculated by Bass [21]. The abscissa is the dissipated energy, calculated starting from the energy in the CMS at half of the thickness of the target.

5.10   Occurrence of fission

The discovery of the neutron in 1932 led to intense investigations of nuclear reactions induced by this new projectile. Of particular interest was the possibility of obtaining a transuranic element in neutron capture by uranium, the heavier element that occurs in nature and with the highest atomic number known at the time. In fact, E. Fermi and his team have soon experimentally verified that the element neptunium, $Z = 93$, can be produced in this way, after the capture of a neutron by uranium and the emission of an electron.

But the new discovery revealed a much more complex situation. Besides the presence of neptunium, activities were detected that could not be clearly attributed to any element in the neighborhood of uranium. O. Hahn and F. Strassmann showed, in 1939 [24], after a careful radiochemical study, that the observed activities were due to several elements with roughly half the mass of uranium. This lead L. Meitner and O. R. Frisch [25] to finally give the correct interpretation of the phenomenon. They proposed that, in capturing the neutron, the uranium can divide into two fragments of comparable masses. Meitner and Frisch called this process *fission*, a term employed in biology to describe a cellular division. These same authors soon recognized, by an analysis of the binding energy of the participants, that the process releases a large amount of energy, close to 200 MeV per fission. Still in 1939, as the result of several works, it was established that fission does not occur preferentially into two equal fragments but with masses distributed around $A = 95$ and $A = 138$.

Nuclear fission is nowadays a very well-established phenomenon, and one knows that it can be induced by a vast combination of projectile-target-energy, and that it can also occur spontaneously in some elements. Moreover, the large amount of energy released in the process gave place to the development of large power explosive artifacts, as the so called "atomic bomb" and also to the use of nuclear reactors as an alternative source of energy.

The first factor to be examined in a possible nuclear disintegration is the energy balance. If we observe Figure 5.6 we see that the division of a heavy nucleus into two fragments of comparable masses is a highly exoergic process. Let us examine the typical case, the fission of ^{236}U* into two fragments with mass numbers 95 and 138, with the release of 3 neutrons. We can evaluate for ^{236}U* a total binding energy of about 1770 MeV and a value close to 1976 MeV for the sum of the total binding energies of the daughter nuclei. Since the final products are more bound that the initial one, the resulting binding energy is released in the form of kinetic energy of the fragments and of the neutrons, being available about 200 MeV for that.

Then, a question arises: if fission is an exoergic process, why it does not occur spontaneous and immediately? The answer is the same as for the α-disintegration of a heavy nucleus: before the process can release energy there is a barrier to overcome. In the case of fission this barrier

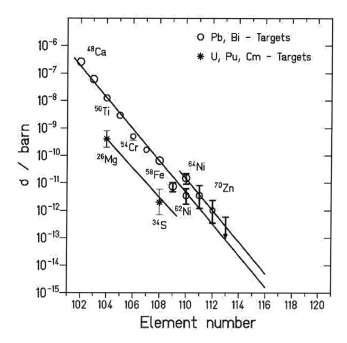

FIGURE 5.18 Cross section systematics for the production of heavy and superheavy elements with cold fusion and actinide based fusion.

is manifest through the increase in the potential energy of the nucleus in response to a small deformation of its surface. Figure 5.19 is an schematic representation of what occurs. It shows the deformation energy of a charged drop as a function of a parameter that measures the deformation (a simplistic view of the problem since it is necessary more than one parameter to better describe the form of the nucleus). The behavior of a nucleus like ^{236}U can be represented by the solid line. A small deformation starting from $\alpha = 0$ increases its energy, and the restoration force tends to bring it back to the initial situation. However, when the deformation is large enough, an increase in deformation decreases the energy, and the curve begins to fall after passing to a maximum. From that point on the process follows spontaneously and the nucleus splits into two fragments. The tail of the curve represents the electrostatic potential energy of the two charged fragments, with a distance α between its centers.

How can a fission process occur? There are two possibilities: the first one is to give the nucleus an excitation energy greater than E_a. This allows the onset of oscillations that can lead to the necessary deformation to overcome the barrier. This is the case of ^{236}U*, excited by a neutron capture in ^{235}U. The other possibility is spontaneous fission. In this case the barrier is surpassed by tunnel effect starting from the nucleus ground state.

The above processes have variable importance along the periodic table. For intermediate mass nuclei the activation energy is very large: induced fission only proceeds with high-energy projectiles, and spontaneous fission is practically nonexistent. On the other side, very heavy nuclei can fission without any barrier, as shown by the dashed curve of Figure 5.19. Elements that are close to this situation have very small spontaneous fission half-lives. This situation can be approximately established imagining that with a small deformation a spherical nucleus assumes the form of an ellipsoid with major semi-axis $a = R(1 + \epsilon)$, where R is the radius of the original spherical nucleus. If the deformation process proceeds with constant volume, the minor semi-axis has a length $b = R(1 + \epsilon)^{-1/2}$; with this choice, the volume part of the binding energy (5.10) remains constant. One shows, on the other side [26], that the surface and Coulomb energy terms change with deformation

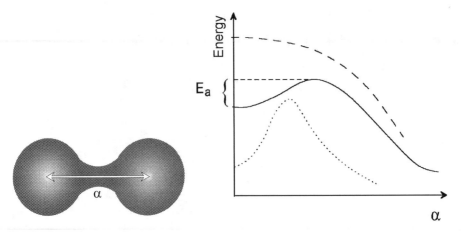

FIGURE 5.19 Deformation energy of a nucleus as a function of a parameter α that characterizes the distance between the pre-fragments. The solid line refers to a heavy nucleus, the dotted line to a lighter one and the dashed line represents a situation where a fission barrier no more exists for a very heavy nucleus. E_a is the activation energy.

according to

$$E_S = E_{S_0}\left(1 + \frac{2}{5}\epsilon^2 + \cdots\right), \qquad E_C = E_{C_0}\left(1 - \frac{1}{5}\epsilon^2 + \cdots\right), \tag{5.108}$$

where $E_{S_0} = B_2$ and $E_{C_0} = B_3$ are the respective zero deformation energies, given by (5.5) and (5.6). Thus the total energy gained with the deformation is

$$\Delta E = E_S - E_{S_0} + E_C - E_{C_0}$$
$$= -a_S A^{2/3}\left(\frac{2}{5}\epsilon^2 + \cdots\right) - a_C Z^2 A^{-1/3}\left(-\frac{1}{5}\epsilon^2 + \cdots\right). \tag{5.109}$$

The searched limit occurs when $\Delta E = 0$, i.e., when

$$\frac{Z^2}{A} \cong \frac{2a_S}{a_C} \cong 50, \tag{5.110}$$

where, in Eq. (5.109), we neglected powers of ϵ higher than the second. The quantity Z^2/A has therefore a relevant role to estimate the probability of a nuclear spontaneous fission and is called *fissionability parameter*. Figure 5.20 shows the spontaneous fission half-life as a function of this parameter. It is clear that the half-lives decrease when one tends to the limit value (5.110), but there are no available experimental data close to that value.

5.11 Mass distribution of the fragments

The way a nucleus divides itself in a fission process is not identical for all events. Thus a nucleus like ^{252}Cf can fission spontaneously through, among others, the following modes:

$$^{252}\text{Cf} \longrightarrow {}^{112}_{46}\text{Pd} + {}^{138}_{52}\text{Te} + 2\text{n}$$
$$^{252}\text{Cf} \longrightarrow {}^{101}_{42}\text{Mo} + {}^{148}_{56}\text{Ba} + 3\text{n}$$
$$^{252}\text{Cf} \longrightarrow {}^{105}_{45}\text{Rh} + {}^{146}_{53}\text{I} + \text{n, etc.} \tag{5.111}$$

with different probabilities for each mode. In this way we can, with a large number of events, build a curve like that of Figure 5.21, where the probability of occurrence of each fragment is represented.

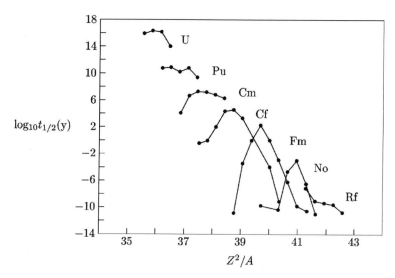

FIGURE 5.20   Spontaneous fission half-lifes of even-even nuclei.

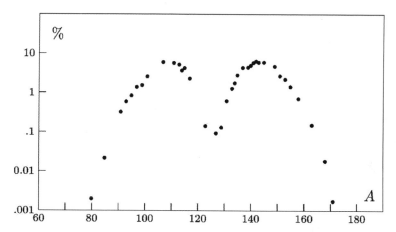

FIGURE 5.21   Percent yield of fragments in the spontaneous fission of ^{252}Cf [27].

The curve shows two peaks, one corresponding to lighter fragments centered at $A \cong 110$ and the other to heavier fragments, with a maximum at $A \cong 140$. We see that symmetrical fission corresponds to a minimum of the curve and is about 600 times less probable to occur than the most favorable asymmetric division. The reason of such a behavior is due to the influence of the shell structure of the nucleus. Hence the effect should be less important for large excitation energies, where the high-level density has to include an average over a large number of states and not just a few isolated ones. It is in fact experimentally verified that the asymmetric fission tends to the symmetric one when the projectile energy increases. This is seen in Figure 5.22 for the case of ^{238}U bombarded with protons from 10 MeV to 200 MeV. We see that from 150 MeV on, the minimum in the symmetric fission disappears, giving place to a plateau where the production of $A = 100$ to $A = 130$ fragments are equally probable.

The behavior of ^{238}U is typical of a highly fissionable nuclide, as normally are the ones of very large mass. Lighter nuclei as radium and actinium have, near the threshold energy for fission, a mass distribution whose characteristic is the presence of three peaks of approximately the same height, indicating that the symmetric fission and the very asymmetric one are the most probable fission modes. A typical example is seen in Figure 5.23, where the yield of each fragment in the

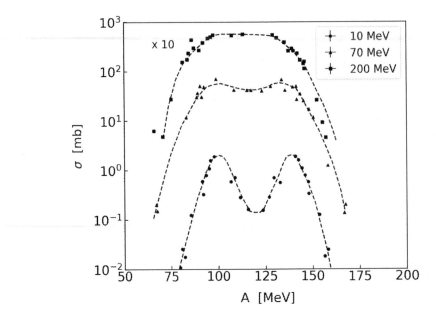

FIGURE 5.22   Cross section for the production of a mass number A fragment in the ^{238}U fission induced by protons of several energies [28]. For clarity, the data at 200 MeV were multiplied by a factor of 100.

fission of ^{226}Ra induced by 11 MeV protons is plotted. For even lighter nuclei, like lead and bismuth, symmetric fission is already dominant from threshold, the mass distribution having a single peak near the mass number $A = 100$.

The shell effect, that leads to asymmetric spontaneous fission of nuclei with $A \gtrsim 256$, can produce the opposite effect when the mass number goes beyond that value. Nuclei like ^{258}Fm, ^{259}Fm, ^{259}Md, ^{260}Md, ^{258}No, ^{262}No, and ^{262}Rf have a prominent symmetric fission peak in the mass distribution of the fragments [23]. In these cases, the proximity of the ^{264}Fm nucleus is responsible for the effect, since this nucleus can divide into two double magic fragments ^{132}Sn.

For the spontaneous fission, or the fission induced by low-energy projectiles, the division into two fragments is the only experimentally observed. The nuclear fission into three fragments of comparable masses (*ternary fission*), was announced for several nuclei [29, 30] but later work [31] put serious doubts on those results.

A different situation occurs when the third fragment is an α-particle. The α-emission during fission is a relatively common process, occurring in 0.2% – 0.4% (depending on the nucleus) of the total of fissions. This α-particle has, on the average, more energy (peak at 15 MeV) than on the ordinary emission and its occurrence is explained by the decrease of the barrier due to the formation of a neck between the two fragments.

High energy projectiles can leave the nucleus with a high-excitation energy. If this energy is greater than about 5 MeV/nucleon (approximate value of the nucleus total binding energy), a process of nuclear rupture into several fragments (*nuclear fragmentation*) can occur. However, the fragmentation involves mechanisms that have no similarity with the ones present in the deformation of a drop in low-energy fission.

5.12   Neutrons emitted in fission

Fission fragments are necessarily rich in neutrons. This is expected as the fissioning nucleus has a neutron to proton ratio larger than the necessary to balance the fragments, which have much smaller atomic number. As a result, the fission fragments are β^--radioactive. In a short time

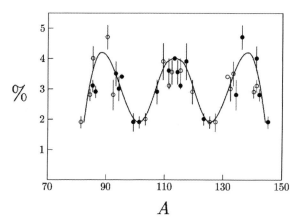

FIGURE 5.23 Mass distribution in the ^{226}Ra fission induced by 11 MeV protons [32]. The open circles are the reflection of the experimental points (full circles) relative to $A = 111$.

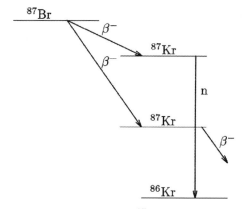

FIGURE 5.24 Emission of a delayed neutron in the ^{87}Br decay.

after the division it can also go towards the β-stability line emitting a neutron. Contrary to these *prompt neutrons*, emitted by the fragments in a time of at most 10^{-16} s after fission, are the *delayed neutrons*, whose emission proceeds after at least one β^--emission by the fragment. Since the decay by neutron emission is a very fast process, the emission time is conditioned to the half-life of the β-decay that precedes it.

Figure 5.24 shows an event of this type: ^{87}Br can decay by two ways into ^{87}Kr, having the upper one enough energy to release one neutron, decaying into the stable isotope ^{86}Kr. The delayed neutrons are only 1% of all the neutrons emitted in fission but have an important role in the control of a reactor.

The number of neutrons emitted in fission is of outmost importance in *chain reactions*, where neutrons proceeding from one fission are used to induce new fissions. This number can vary from 1 to 8, but this upper limit occurs very rarely, with an average between 2 and 4 for the great majority of events produced by spontaneous and thermal neutrons induced fission. This average increases a little with the energy of the incident particle reaching, for example, 4.75 for ^{238}U fission induced by 15 MeV neutrons.

As important as the neutron number is the neutron kinetic energy distribution. In Figure 5.25 we see that this distribution is wide, with a maximum around 0.5 MeV. The fitted curve [33]

$$N(E) = \exp(-1.036E) \sinh \sqrt{2.29E} \qquad (5.112)$$

represents well the experimental values from 0 to 10 MeV.

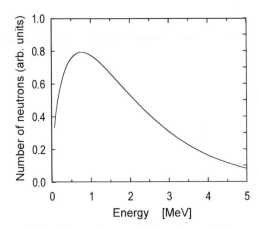

FIGURE 5.25   Kinetic energy distribution of prompt neutrons.

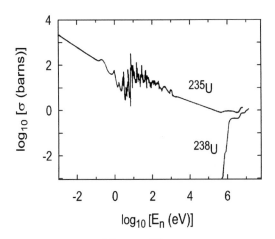

FIGURE 5.26   Fission cross sections of ^{235}U and ^{238}U bombarded by 0 to 10 MeV neutrons.

5.13   Fission cross sections

The behavior of the fission cross section will be presented for the important examples of ^{235}U and ^{238}U bombarded by neutrons from 0 to 1 MeV. In Figure 5.26 we see that neutrons with energies close to zero are already able to fission ^{235}U, which indicates that the ^{236}U activation energy is lower than the neutron binding energy. In this way the fission barrier is overcome and the compound nucleus fissions. Increasing the energy the cross section goes down until reaching, at 1 MeV, a value 600 times less than the corresponding thermal neutron value, after crossing a region of resonances, between 0.1 eV and 1 keV.

On the other hand ^{238}U is not able to fission by slow neutrons, with its fission threshold near 1 MeV. The energy gained by the binding of the incident neutron is not enough, in this case, to reach the summit of the barrier, and it is also necessary the contribution of the neutron kinetic energy to reach the activation energy.

How to explain this difference in behavior in nuclei very close to each other? The reason is the pairing energy, that contributes with the term (5.8) to the mass formula. This energy implies that an even-even nucleus is, on the average, more bound than its odd neighbors. When ^{235}U absorbs

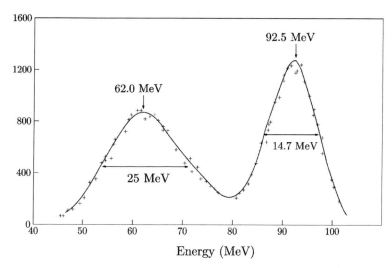

FIGURE 5.27 Kinetic energy distribution of the fragments in the fission of ^{235}U. The vertical axis indicates the number of counts of each experimental point [34].

a neutron, the binding energy per nucleon increases by δ in (5.10), being converted in excitation energy; in this way the summit of the fission barrier is more easily reached; the absorption of a neutron by ^{238}U, on the other side, decreases the binding energy per nucleon by δ, and part of the excitation energy will be used increasing the ^{239}U mass per nucleon; this difference has to be delivered by the neutron incident energy and this explains why there is no fission of ^{238}U by neutrons of energy lower than 1 MeV.

From the above argumentation it is clear that this effect is not restricted to uranium, but this element is important because ^{235}U is the only isotope found in nature that is fissionable by slow neutrons, although existing in a small proportion (0.7%) regarded to ^{238}U (99.3%) in a natural uranium sample.

5.14 Energy distribution in fission

The fission of a heavy nucleus releases about 200 MeV. The larger part of this energy ($\cong 165$ MeV) is spent as kinetic energy of the fragments, the light fragment carrying the larger part. This is due to linear momentum conservation, which implies that the ratio between the energies of the fragments

$$\frac{\frac{1}{2}m_1 v_1^2}{\frac{1}{2}m_2 v_2^2} = \frac{v_1}{v_2} = \frac{m_2}{m_1} \tag{5.113}$$

is equal to the inverse ratio of the masses. Thus the kinetic energy distribution of the fragments has the aspect of Figure 5.27, where the 92.5 MeV peak corresponds to the light fragment and the 62.0 MeV peak to the heavy one.

The rest of the available energy is shared by neutrons, gammas and also among the electrons of β^--decay. Typical numbers are: the neutrons consume about 5 MeV, result of an average energy of 2 MeV multiplied by an average number of 2.5 neutrons by fission. The β^--decay releases an approximate energy of 7 MeV, while the prompt γ-rays and the products of the fragment decay carry a total of 15 MeV. We have to also mention the neutrinos emitted together with the electrons and that are responsible for 12 MeV of the total energy. However, owing to the weak interaction of these particles, they cross all the experimental apparatus without being detected.

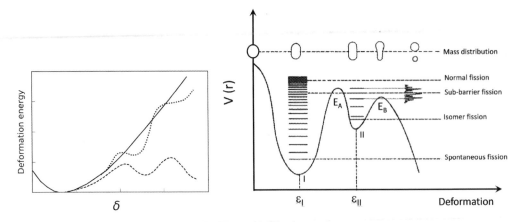

FIGURE 5.28 (1) Potential energy as a function of deformation. (a) For small deformations of an ellipsoidal nucleus. (b) With the addition of the shell correction. (c) With the existence of a neck. (2) The specific decay rate for fission is determined by the density of states at the fission barrier. The energy released in the descent from saddle point of scission is assumed to appear as kinetic energy of the fragments (adiabatic limit).

5.15   Isomeric fission

In 1962, S. M. Polikanov and colleagues published [35] results from a spontaneous fission study of very heavy and short half-lives isotopes. The isotopes were formed by the bombardment of ^{238}U with ions of ^{22}Ne and ^{16}O of several energies around 100 MeV. The motivation for the work was the appearance of a fission half-life of about 14 ms coming from a product of the reaction $^{242}_{94}$Pu + $^{22}_{10}$Ne, studied previously by the same group with the idea of creating the $Z = 104$ element. In the experimental apparatus used, the reaction products, of high linear momentum, were pulled out from the uranium target, reaching a high velocity disk that transported them rapidly to two ionization chambers able to detect fission fragments. The authors could measure fission half-lives of (0.02 ± 0.01) s in the irradiations with neon and (0.013 ± 0.008) s in the irradiations with oxygen. These values were, taking into account the experimental error, identical to the half-life of 14 ms obtained previously, what discarded the element 104 as the responsible for the fission events. In fact, a further work has shown that the detected fragments had their origin on ^{242}Am. The half-life in question was several orders of magnitude lower than the expected for the ground state of possible reaction products and the conclusion was that the fission occurred from an isomeric excited state. It is hard to understand, however, how this state resists to gamma decay for so long time and how it exhibits a so drastic reduction in the fission half-life compared with that of the ground state. A possible explanation, suggested by the authors, and that would shown to be correct latter, is that the isomery was connected to a strong nuclear deformation.

The complete understanding of the mechanism came in 1967, with a work of V. M. Strutinsky [36] about the shell effects in the determination of nuclear masses and deformation energy. There was the notion that the liquid drop model described well the nuclear properties that depend smoothly on the number of nucleons like, for instance, the general trend of the curve of binding energy versus A (Figure 5.6). The shell model, by its turn, explained the nuclear properties that depend strongly on the arrangement of the last nucleons, that is, the nucleons in the neighborhood of the Fermi surface. It was natural the attempt to build a model that incorporates both effects. Such a model starts with the nucleus total energy

$$E = E_{\text{LDM}} + E_{\text{SM}}, \tag{5.114}$$

where the first term gives the contribution of the liquid drop model and the second one accounts for the small correction due to shell effects. Strutinsky elaborated an elegant and efficient method to evaluate this correction, assuming that it should be obtained from the difference

$$E_{\text{SM}} = U - \tilde{U} \tag{5.115}$$

between the total energy given by the shell model,

$$U = \sum_i E_i, \tag{5.116}$$

where the sum must be extended to all occupied states, and the energy corresponding to an average situation

$$\tilde{U} = \int_{-\infty}^{E_F} E\tilde{g}(E)\,dE, \tag{5.117}$$

and $\tilde{g}(E)$ is a function that tries to reproduce the average behavior of the level density $g(E)$, which in the shell model is given by

$$g(E) = \sum_i \delta(E - E_i), \tag{5.118}$$

and that has as boundary condition the conservation of the number of particles N:

$$\int_{-\infty}^{E_F} \tilde{g}(E)dE = N, \tag{5.119}$$

where E_F is the Fermi energy. Summarizing, (5.114) and (5.115) propose to calculate the total energy of the nucleus using the shell model energy, but with the care of exchanging the average value of this energy by the energy coming from the liquid drop model.

The Strutinsky method is of a great value in the precise determination of the ground state masses. More than this, it established that the potential energy as a function of deformation has more than one minimum. The first minimum corresponds to the ground state and the second would be the reason for the existence of the shape isomeric states.

A qualitative understanding of the presence of the second well can be obtained in the following way: when we plot the deformation energy of a nucleus in the actinide region versus a parameter that measures the deformation, what one obtains in principle is a function described by the curve (a) of Figure 5.28. To this simple curve we can add small shell corrections. For this goal we must remember that the distribution of levels filled by the nucleons change with deformation (Nilsson model – see Appendix A). When the deformation increases, the contribution of the shell part changes since the level distribution changes. This contribution is more significant when the deformation reaches such a point that energy levels that were above the last filled level begin to cross it and go below it, creating new options for the nucleons at lower energies. This tends to decrease the energy of the system and decreases the potential energy slope. Increasing the deformation other level crossings cause new oscillations and the aspect of the function can be that of the curve (b) of Figure 5.28. To draw the real curve we have yet to consider the appearance of a neck that makes the deformation easier reducing therefore the deformation energy. In a simple liquid drop treatment the decrease in potential energy by the presence of the neck leads to a single peak, as in Figure 5.19. But, with the shell correction the final aspect is that of the curve (c) of Figure 5.28. We see the appearance of a second minimum in the curve, which is the origin of the isomeric fission.

When a nucleus in the actinide region absorbs a nucleon and receives an excitation energy above its fission barrier, the most common result is its immediate fission. With a lesser probability the process can end in a *radiative capture* (capture of a nucleon followed by a γ-ray emission), leading the nucleus to a state of the first well. But, a third possibility is the imprisonment in the region of the second minimum, that occurs in 0.0001% to 0.01% of the cases. The state of lowest energy of the second minimum is an excited isomeric state of the nucleus, that can decay to the first well or fission spontaneously. The fission barrier in this case is much narrower than the barrier that the nucleus feel in the first well and this explains the much shorter fission half-life than the half-life of the ground state.

5.16   The nuclear reactor

The facts that fission release neutrons and that neutrons are able to induce fission point to the possibility of producing *chain reactions*, where neutrons released by fission events produce new fissions, maintaining a self-sustained process. The *nuclear reactor* is a machine that works based on this principle. For understanding how it operates it is necessary to establish the conditions for supporting the chain reactions. Let us first examine the case where the fuel is natural uranium contained in a very large volume. The fission events will come mainly from thermal neutrons incident on ^{235}U, a situation where the fission cross section is very large (see Figure 5.26). Each neutron incident on ^{235}U can produce a fission that, by its turn, releases on the average 2 to 3 neutrons. The neutrons released are fast and need to be thermalized to increase the cross section. At the end of thermalization we shall have a certain number of neutrons able to produce new fission events. Let us define the *neutron reproduction factor*, k_∞, for a very large volume where one can neglect the losses through the surface. This factor will measure the ratio between the number of thermal neutrons from one generation and that of the previous one. On the average each thermal neutron produces k_∞ new thermal neutrons. To sustain the chain reaction we must have, naturally, $k_\infty > 1$.

Let us see which factors are important in the determination of k_∞. Let us suppose that we start with N thermal neutrons. These neutrons will not produce N fissions since there is a competition with the radiative capture process (n,γ), where a neutron is absorbed in ^{235}U or ^{238}U and the deexcitation proceeds by a γ-ray emission. If ν is the average number of fast neutrons produced by fission, the available number of fast neutrons per incident neutron will be

$$\eta = \nu \frac{\sigma_f}{\sigma_f + \sigma_r}, \tag{5.120}$$

where σ_f and σ_r are the fission and radiative capture cross sections by thermal neutrons in natural uranium. Their values are obtained weighting the respective cross section values, of ^{235}U, $\sigma_f = 584$ b and $\sigma_r = 97$ b, and of ^{238}U, $\sigma_f = 0$ b and $\sigma_r = 2.75$ b. The weighting is dictated by the isotopic composition of natural uranium, 0.7 % of ^{235}U and 99.3 % of ^{238}U. Using $\nu = 2.5$ results in $\eta = 1.33$. This value can be increased if we use enriched uranium (larger contents of ^{235}U) instead of natural uranium.

Another factor that has to be considered is the small gain with the fast neutron fission of ^{238}U ($\sigma_f \cong 1$ b). This *fast fission factor* ϵ is about 1.03 for natural uranium. The number of fast neutrons per incident neutron is now $\eta\epsilon$.

The fast neutrons should be moderated, i.e., they have to be transformed in thermal neutrons. This is achieved through collisions with light nuclei mixed with uranium, since the lighter the target the larger the energy loss in the collision. Cheap and abundant candidates are carbon in the form of graphite and water. In water there is the tendency of hydrogen to absorb neutrons and form the deuteron, taking off neutrons from the process. For that reason, it can only be used when the uranium is enriched and the cross section increased. However, a natural uranium reactor can be moderated by heavy water, D_2O, where D is a deuterium atom. The deuteron plays a good moderation role and has a very small cross section for neutron capture. The disadvantage is the very low abundance of D_2O and its onerous production.

In Figure 5.29 we show a pressurized water reactor. The water serves both as the moderator and the heat transfer material. It is isolated from the water used to produce the steam that drives the turbines.

On the way to thermalization there are obstacles, the ^{238}U neutron capture resonances, in the same region of ^{235}U fission resonances (Figure 5.26) but even greater. Localized between 5 eV and 300 eV, they would be insurmountable traps for neutrons, that must make about one hundred collisions to thermalize, if the moderator were completely mixed to the fuel. The artifice is to use the fuel in the form of rods inserted in the moderator, constituting a *pile*. A reactor of this type is called *heterogeneous*, opposite to an *homogeneous* reactor, where the fuel and moderator are mixed. The configuration of an heterogeneous reactor allows the neutron to do a long path in the moderator and to thermalize before entering again the fuel region. However, there is still the possibility that before the neutron is completely thermalized it can be captured. To take into account these cases one introduces a new factor p, the *resonance escape probability*. It is about 0.9.

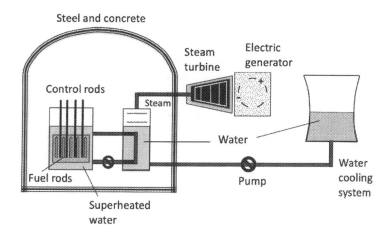

FIGURE 5.29   Schematic representation of a pressurized water reactor.

A last contribution to the neutron reproduction factor takes into account the capture of thermal neutrons by the moderator. In the carbon case the cross section is very small (3.4 mb) and even with the large quantity of moderator present this *thermal utilization factor f* is about 0.9. The neutron reproduction factor has the final expression:

$$k_\infty = \eta \epsilon p f, \tag{5.121}$$

which is known as the *four factor formula*. The last three depend on the geometry and can be optimized by an adequate project of the pile.

With the estimate done for the four factors, k_∞ is a little greater than 1, but one still needs to compute the losses through the surface, that is, what counts in practice is the reproduction factor k for a finite pile. The fraction of thermal and fast neutrons lost through the surface drops when the ratio between surface and volume decreases. There is a size which corresponds to the *critical* value $k = 1$. A pile with a dimension smaller than the critical value cannot sustain chain reactions. The mass of fissionable material that corresponds to the critical size is called *critical mass*. An estimate of the critical size of a cubic pile of natural uranium with graphite moderator is 5 m of side. This value can be smaller if we wrap the pile with a material that reflects neutrons, giving back to the pile neutrons that would escape through the surface.

For a continuous reactor operation, k should be kept exactly equal to the critical value. The criticality can be controlled inserting a material of high cross section for absorption of thermal neutrons in the pile, as cadmium. Cadmium rods can, by external control, be inserted or removed from the pile gradually. If the pile is a little subcritical for prompt neutrons, the small number of delayed neutrons can be used to reach the critical value. As the decay of these neutrons can have half-lives up to some minutes, there is enough time to reach the critical regime with an adequate manipulation of the rods.

For $k < 1$ (*subcritical* value) the chain reactions do not sustain themselves. With $k > 1$ we enter in the *supercritical* regime: the number of fissions increases enormously and without control a large amount of energy is released in a small fraction of time. This is the principle of operation of the atomic bomb, which was the first explosive artifact of nuclear origin to be constructed.

The several components that take part in the pile operation can be used to define the purpose of the reactor. Thus *power* reactors use the thermal energy generated to transform it in electrical energy. *Research* reactors, by their turn, take profit of the high neutron flux (10^{13} neutrons/cm^2/s is a typical value) for studies in nuclear and solid state physics. A third class of reactors, the *converter* reactors, have the purpose of producing new isotopes which are fissionable by thermal neutrons. As they are even-even nuclei, ^{238}U and ^{232}Th do not fission by thermal neutrons, i.e., they are not *fissile* isotopes. But they can acquire this condition by a neutron capture followed by two beta decays. The captures

$$^{238}\text{U} + \text{n} \rightarrow\ ^{239}\text{Np} + \beta^- + \nu,$$
$$^{239}\text{Np} \rightarrow\ ^{239}\text{Pu} + \beta^- + \nu,$$
$$^{232}\text{Th} + \text{n} \rightarrow\ ^{233}\text{Pa} + \beta^- + \nu,$$
$$^{233}\text{Pa} \rightarrow\ ^{233}\text{U} + \beta^- + \nu, \tag{5.122}$$

produce the fissile isotopes ^{239}Pu and ^{233}U. Due to this potentiality we say that ^{238}U and ^{232}Th are *fertile* isotopes. If $\eta > 2$ and other losses could be minimized, one neutron can keep the fission chain and the other to react with the fertile nucleus. A reactor operating in such a way produces more fissile material than it consumes and is called a *breeder*. The possibility of breeding and the relative abundance of ^{232}Th and ^{238}U in nature, makes that ^{233}U and ^{239}Pu are also largely employed as fuel elements in reactors. With these isotopes it is possible the building of reactors that operate with fast neutrons and, in this case, one can dispense the moderator. With a fuel like ^{239}Pu one can achieve a value of η for fast neutrons of 3. A converter can be constituted by a nucleus enriched by ^{233}U or ^{239}Pu, wrapped by a "blanket" of ^{238}U. One can in this way produce ^{239}Pu with a rate equal or greater than the consumption of ^{233}U or ^{239}Pu in the core. Thermal reactors can also be used as breeders, but it is necessary to use ^{233}U as fuel in place of ^{235}U, since the first one releases on the average more neutrons in fission than the last one.

5.17   Appendix 5.A – The Nilsson model

The fact of some nuclei are deformed in the ground state suggests the use of shell models where the central potential is no more spherically symmetric. Such a potential was developed by S. G. Nilsson in 1955 [37], who proposed the employment of the following single particle Hamiltonian:

$$H = H_0 + C\mathbf{l} \cdot \mathbf{s} + Dl^2, \tag{5.123}$$

where H_0 is an anisotropic harmonic oscillator Hamiltonian:

$$H_0 = -\frac{\hbar^2}{2m}\,\triangledown^2 + \frac{1}{2}m(\omega_x^2 x^2 + \omega_y^2 y^2 + \omega_z^2 z^2), \tag{5.124}$$

$\mathbf{l} \cdot \mathbf{s}$ is the usual spin-orbit coupling term of the spherical shell model and l^2 a term thought to simulate a flattening of the oscillator potential and to turn it closer to a real potential. C and D are constants to be determined by the adjustment to the experimental results.

If we are just interested in the particular case of axial symmetry we can write

$$\omega_x^2 = \omega_y^2 = \omega_0^2\left(1 + \frac{2}{3}\delta\right)$$
$$\omega_z^2 = \omega_0^2\left(1 - \frac{4}{3}\delta\right), \tag{5.125}$$

where the frequencies were placed as function of two parameters, ω_0 and δ; these can be related through the condition imposed by the conservation of the nuclear volume (assuming that it can be determined by an equipotential surface). The result of this condition takes us to the relation $\omega_x\omega_y\omega_z = const = \widetilde{\omega}_0^3$, where $\widetilde{\omega}_0$ is the frequency for a zero deformation. Thus,

$$\omega_0(\delta) = \widetilde{\omega}_0^3\left(1 - \frac{4}{3}\delta^2 - \frac{16}{27}\delta^3\right)^{-1/6}, \tag{5.126}$$

is a relationship that leaves us with a single parameter, called deformation parameter, δ. Thus, the resulting energy by the diagonalization of the Hamiltonian are functions of δ. The procedure employed for the calculation of the energy will not be presented here. The result is shown in Figure 5.30, where we plot the energy levels as a function of the deformation δ. A straight vertical line

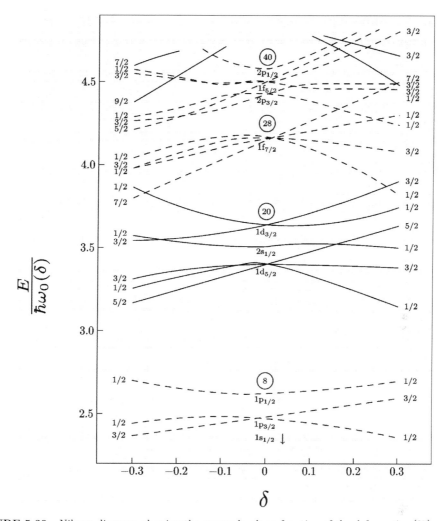

FIGURE 5.30   Nilsson diagram, showing the energy levels as function of the deformation [37].

drawn on this graphic indicates the energy levels allowed for a given value of the parameter δ. Positive values of δ correspond to a prolate nucleus and negative values to an oblate nucleus. It is also important to mention that for a deformed potential, j and l are no longer good quantum numbers, what classically corresponds to the fact that the angular momentum is no more a constant of motion for a non-spherically symmetric potential.

5.18   Exercises

1. Obtain Eqs. (5.78) and (5.79) using WKB wave functions.

2. Using the parabolic approximation deduce Eq. (5.82).

3. Show that Eqs. (5.97) can be rewritten in terms of two decoupled equations for two distinct wave functions. If one takes these wave functions as two possible states for the system, then explain why Eq. (5.98) should follow.

4. Build a reaction in which a projectile of $A = 20$ produces a compound nucleus $^{258}_{104}\mathrm{Rf}^*$. Using the approximate expression for the Coulomb barrier,

$$B = \frac{Z_1 Z_2 e^2}{1.45(A_1^{1/3} + A_2^{1/3})\ \mathrm{fm}},$$

determine the smallest possible excitation energy for the compound nucleus. Repeat the procedure for a projectile with $A = 50$ and compare the results with the one of Figure 5.16.

5. With the semi-empirical mass formula (5.10) estimate the binding energy per nucleon for $^{10}\mathrm{B}$, $^{27}\mathrm{Al}$, $^{59}\mathrm{Co}$ and $^{236}\mathrm{U}$. Compare with the experimental values found in the literature.

6. Using the semi-empirical mass formula, show that the fission of a Z, A nucleus into two equal fragments. a) Is energetically possible only for $Z^2/A \geq 18$. b) Releases more energy than in different fragments. Do not take into account the pairing energy in these calculations.

7. Show that the energy S_n required to separate a neutron from the nucleus (A,Z) is given approximately by (neglect the pairing therm in Eq. (5.10).

$$S_n \approx a_V - \frac{2}{3} a_S A^{-1/3} - a_A \left[1 - \frac{4Z^2}{A(A-1)} \right].$$

Discuss the factors which would lead to corrections to this estimate.

Some nuclei with large neutron excess (e.g., $^{31}_{11}\mathrm{Na}$) have been produced and have a relatively long lifetime (16 ms). How would such a nucleus decay? Estimate the mass number of the Na nucleus which is just unstable against neutron emission.

8. Evaluate the energy released in each one of the processes (5.111) and compare with the value for a symmetric division with the emission of 2 neutrons.

9. Use the chart of nuclides to follow the decay of each one of the fragments formed in Eqs. (5.111) and determine the stable nuclei obtained at the end of the chain.

10. Estimate the maximum relative angular momentum of the two ions when 300 MeV $^{16}\mathrm{O}$ ions bombard a target of $^{107}\mathrm{Ag}$.

11. A beam of $^{32}\mathrm{S}$ nuclei incident on a $^{24}\mathrm{Mg}$ target produces various nuclear reactions. Some are direct reactions and some involve formation of a compound nucleus (fusion reactions). Explain the terms direct reaction (see next chapter), total reaction cross section σ_R and fusion cross section σ_{fus}.

12. In a simple classical model it is assumed that the relative motion of the projectile and target is affected only by the Coulomb force between them if their distance of closest approach d is greater than a critical distance R, while they produce a reaction if $d < R$. By calculating the distance of closest approach as a function of the impact parameter for a Rutherford orbit show that the total reaction cross-section has the form:

$$\sigma_R = \sigma_0 (1 - V/E_{\mathrm{cm}}) \qquad \text{if} \quad E_{\mathrm{cm}} > V,$$

where E is the energy in the center of mass coordinate system. Find expressions for the constants σ_0 and V.

13. The table below shows some experimental values of the fusion cross-section for $^{32}\mathrm{S}$ on $^{24}\mathrm{Mg}$ as a function of the energy E_{lab} of the $^{32}\mathrm{S}$ nuclei in the laboratory coordinate system. Use these data to test the hypothesis that $\sigma_R = \sigma_{\mathrm{fus}}$ at these energies.

E_{lab} [MeV]	70	80	90	110
σ_{fus} [mb]	150	414	670	940

14. Although practically unobserved, one can admit the existence of spontaneous fission processes where the nucleus divides into more than two fragments. Using Figure 5.6, find the largest value of n for which the breaking of a $A = 240$ nucleus into n equal fragments is energetically favorable.

15. Thermal neutrons are neutrons in thermal equilibrium with the nuclei of the material that contains them. In this case the velocity of the neutrons follows a Maxwell distribution

$$n(v)dv = 4\pi n \left(\frac{m}{2\pi kT} \right)^{3/2} v^2 \exp \left(-\frac{mv^2}{2kT} \right) dv, \qquad (5.127)$$

where $n(v)dv$ is the number of neutrons with velocity between v and $v + dv$, n is the total number of neutrons, m the neutron mass, k the Boltzmann constant and T the absolute temperature. a) Show that at $20°C$ the energy corresponding to the most probable velocity of the neutrons is $E_n = 0.025$ eV. b) What is the average energy of these neutrons?

16. Use Eq. (5.112) to estimate the average energy of prompt neutrons and the energy for which the probability of neutron emission is a maximum.

17. A 1 MeV neutron incident in a natural uranium sample can fission a ^{235}U or ^{238}U nucleus. Compare the probabilities for the two cases taking into account the proportion of each isotope in the sample and the curves of Figure 5.26.

18. A thin film of 0.1 mg of uranium oxide (U_3O_8), made with natural uranium, is deposited over a 6 cm x 3 cm mica sheet, able to register fission fragments tracks coming from the film. The mica is then submitted to a flux of 10^7 thermal neutrons/cm^2/s during 1 hour. Find the number of tracks registered by the mica per cm^2, knowing that the ^{235}U fission cross section for thermal neutrons is 548 b and the proportion of it in the natural uranium is 0.7%.

19. Transforming heat into electricity with a conversion efficiency of 5 percent, evaluate the consumption rate of ^{235}U in a nuclear reactor producing electricity with a power of 1000 MW.

20. a) Using Figure 5.6, find the approximate energy released in the spontaneous fission of $^{238}_{92}$U into two fragments of equal masses. b) Which percentage is that energy relative to the $^{238}_{92}$U total binding energy?

21. Using the value 200 MeV as the energy released in each fission, evaluate the energy in Joules produced by the fission of 1g of ^{235}U. How far would 1g of ^{235}U drive a car which consumes 1 liter of gasoline (density $= 0.7$ gm/cm^3) for each 10 km? The combustion heat of octane is 5500 kJ/mole, and the combustion engine has an efficiency of 18%.

22. Estimate if fusion of deuterium into helium releases more or less energy per gram of material consumed than the fission of uranium.

23. Prove the Eqs. (5.108).

24. The operators bring a nuclear reactor to operate with a multiplication factor $k = 1.01$. How many chains of fissions are necessary to double the reactor power?

25. A nuclear explosion is triggered by a single neutron in a mass of ^{235}U with a multiplication factor 2. If in the explosion about 1 kg of uranium fissions, how many successive chains of fission are necessary.

References

1. Goldstein H, Poole C P, Safko J L 2013 *Classical Mechanics* (Pearson)
2. Wilczyński J 1973 *Nucl. Phys.* **A216** 386.
3. Kowalsky L, Jodogne J C and Miller J M 1968 *Phys. Rev.* **169** 894.
4. Natowitz J B 1970 Phys. Rev. **C1** 623.
5. Zebelman A M and Miller J M 1973 *Phys. Rev. Lett.* **30** 27
6. Cohen S, Plasil F and Swiatecki W J 1974 *Ann. Phys.* **82** 557
7. von Weizsäcker C F 1935 *Z. Phys.* **96** 431
8. Wapstra A H 1958 *Handbuch der Physik* **38** 1
9. Bethe H 1939 *Phys. Rev.* **55** 434
10. Hill D L and Wheeler J A 1953 *Phys. Rev.* **89** 1102
11. Wong C Y 1973 *Phys. Rev. Lett.* **31** 766
12. Mosel U 1981 *Comm. Nucl. Part. Phys.* **9** 213
13. Lee S M, Matsuse T and Arima A 1980 *Phys. Rev. Lett.* **45** 165
14. Stokstad R G et al. 1978 *Phys. Rev. Lett.* **41** 465
15. Balantekin A B and Takigawa N 1998 *Rev. Mod. Phys.* **70** 77
16. Leigh J R et al. 1995 *Phys. Rev.* **C52** 3151
17. Björnholm S and Swiatecki W J 1982 *Nucl. Phys.* **A391** 471
18. McMillan E and Abelson P H 1940 *Phys. Rev.* **57** 1185
19. Oganessian Yu Ts, Iljinov A S, Demin A G and Tretyakova S P 1975 *Nucl. Phys.* **A239** 353
20. Hofmann S 1996 *Nucl. Phys. News* **6** 26; Hofmann S et al. 1996 *Z. Phys.* **A354** 229
21. Bass R 1974 *Nucl. Phys.* **A231** 45
22. Armbruster P 1985 *Ann. Rev. Nucl. Sci.* **35** 135
23. Lazarev Yu A et al. 1996 *Phys. Rev.* **C54** 620
24. Hahn O and Strassmann F 1939 *Naturwiss.* **27** 11, 89
25. Meitner L and Frisch O R 1939 *Nature* **143** 239, 471
26. Bohr N and Wheeler J A 1939 *Phys. Rev.* **56** 426
27. Nervik W E 1960 *Phys. Rev.* **119** 1685
28. Hyde E K 1971 *The Nuclear Properties of the Heavy Elements, vol. 3: Fission Phenomena* (New York: Dover Publications)
29. Rosen L and Hudson A M 1950 *Phys. Rev.* **78** 533
30. Muga M L, Rice C R and Sedlacek W A 1967 *Phys. Rev. Lett.* **18** 404
31. Steinberg E P, Wilkins B D, Kaufman S B and Fluss M J 1970 *Phys. Rev.* **C1** 2046
32. Jensen R C and Fairhall A W 1958 *Phys. Rev.* **109** 942
33. Watt B E 1952 *Phys. Rev.* **87** 1037
34. Fowler J L and Rosen L 1947 *Phys. Rev.* **72** 926
35. Polikanov S M et al. 1962 *Soviet Phys. JETP* **15** 1016
36. Strutinsky V M 1967 *Nucl. Phys.* **A95** 420
37. Nilsson S G 1955 *Mat. Fys. Medd. Dan. Vid. Selsk.* **16** 29

6

Direct reactions

6.1 Introduction

We have already mentioned the existence of reactions that occur through a short duration of the projectile-target interaction. Several of these mechanisms of direct reaction is known. This reaction type becomes more probable as one increases the energy of the incident particle: the wavelength associated to the particle decreases and localized areas of the nucleus can be "probed" by the projectile. In this context, peripheral reactions, where only a few nucleons of the surface participate, become important. These direct reactions happen during a time of the order of 10^{-22} s; reactions in which there is the formation of a compound nucleus can be up to six orders of magnitude slower. We should notice that one reaction form for a given energy is not necessarily exclusive; the same final products can be obtained, part of the events in a direct way, other parts through the formation and decay of a compound nucleus.

Besides elastic scattering, which is not of interest in the present study, we can describe two characteristic types of direct reactions. In the first, the incident particle suffers inelastic scattering and the transferred energy is used to excite a collective mode of the nucleus. Rotational and vibrational bands can be studied in this way. The second type involves a modification in the nucleon composition. Examples are the transfer reactions of nucleons, as *pick-up* and *stripping* reactions. Another important reaction of the second type is the *knock-out* reaction where the incident particle

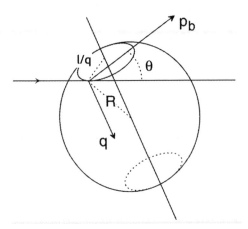

FIGURE 6.1   Representation of a direct reaction (d,n), where a deuteron with momentum $\mathbf{p}_a$ hits a nucleus, transferring a proton with momentum $\mathbf{q}$ and releasing a neutron with momentum $\mathbf{p}_b$.

(generally a nucleon) picks up a particle of the target nucleus and continues in its path, resulting in three reaction products. Reactions with nucleon exchange can also be used to excite collective states. An example is a *pick-up* reaction where a projectile captures a neutron from a deformed target and the product nucleus is in an excited state belonging to a rotational band.

Some types of direct reaction exhibits peculiar forms of angular distribution that allows us to extract information on the reaction mechanism with the employment of simple models. Typical examples are the *stripping* reactions (d,n) and (d,p), where the angular distribution of the remaining nucleon presents a forward prominent peak and smaller peaks at larger angles, with the characteristic aspect of a diffraction figure.

With the increase of the energy of the incident particle, it would be interesting initially to test the applicability of semi-classical methods in the treatment of a direct reaction. Let us take as example the reaction (d,n) (Figure 6.1): a deuteron of momentum $\mathbf{p}_a$, following a classical straight line trajectory hits the surface of the nucleus, where it suffers a *stripping*, losing a proton to the interior of the nucleus and releasing the neutron of momentum $\mathbf{p}_b$ which leaves it forming an angle θ with the incident direction. The proton is incorporated to the nucleus with momentum $\mathbf{q}$, transferring to it an angular momentum $\mathbf{l} = \mathbf{R} \times \mathbf{q}$, i.e., approximately $|\mathbf{l}|/\hbar$ units of angular momentum.

Let us concentrate on a certain energy level of the product nucleus produced by a transfer of l units of angular momentum. The energy of the emitted neutron and the vector $\mathbf{q}$ are determined for a given observation angle θ. This reduces the regions of the surface of the nucleus where the reaction takes place at circles of radius l/q, as marked in Figure 6.1. The fact that the radii of the circles cannot be larger than the radius of the nucleus imposes a minimum value for θ, a value that increases with the increase of l.

In this way we can sketch the prediction of this semi-classical approach for the behavior of the differential cross section. The starting point is to assume that in a direct reaction the particles leave the nucleus with a considerable momentum at forward angles and we should expect a decrease of the cross section at growing angles. With the mentioned restriction for the angular distribution, the valid region for observation is located above the limit angle θ_l. The value of θ_l results from the simple application of conservation of linear momentum:

$$q^2 = p_a^2 + p_b^2 - 2p_a p_b \cos\theta, \tag{6.1}$$

where p_a and p_b are obtained from the initial and final energy of the particles and $q = l/R$. For a typical situation where an incident particle of dozens of MeV excites the first low-energy states of the product nucleus, we see from (6.1) that the dependence with θ is essentially proportional to

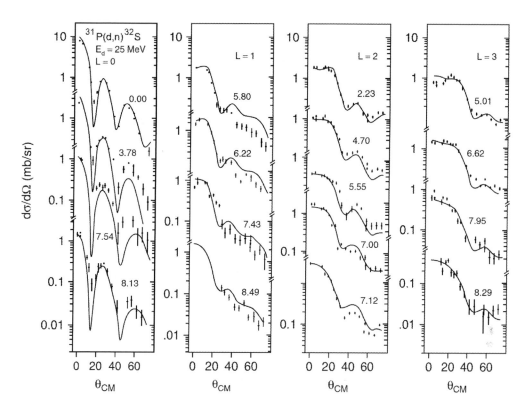

FIGURE 6.2 Angular distribution of the reaction ^{31}P(d,n)^{32}S, with the transfer of a proton to several states of ^{32}S. The curves are results of DWBA calculations for the indicated l values ([1]).

the magnitude of the transferred momentum, q, and in the limit $q = l/R$, the corresponding angle θ_l grows with l. Finally, we should expect that the angular distribution is a function with maxima and minima, resulting from the interference between events that happen in the upper and lower circles represented in Figure 6.1.

Figure 6.2 shows experimental results for the reaction ^{31}P(d,n)^{32}S [1]: angular distributions of the detected neutrons corresponding to each energy level of ^{32}S are exhibited for every corresponding angular momentum. We see that the behavior of the cross sections is in agreement with the qualitative predictions: the curves exhibit a first and more important peak at a value of θ that grows with l; starting from this other smaller peaks occur as θ increases. The increase of θ_l with l is an important characteristic that can be used, as we shall see ahead, to identify the value of the transferred momentum in a given angular distribution.

6.2 Level width and Fermi's golden rule

A quantum system, described by a wave function that is an Hamiltonian eigenfunction, is in a well defined energy state and, if it does not suffer external influences, it will remain indefinitely in that state. But this ideal situation does not prevail in excited nuclei, or in the ground state of an unstable nuclei. Interactions of several types can add a perturbation to the Hamiltonian and the pure energy eigenstates no more exist. In this situation a transition to a lower energy level of the same or of another nucleus can occur.

An unstable state normally lives a long time compared to the fastest nuclear processes, e. g., the time spent by a particle with velocity near that of the light to cross a nuclear diameter. In this way we can admit that a nuclear state is approximately stationary, and to write for its wave

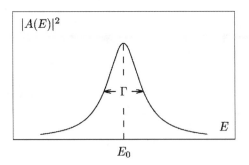

FIGURE 6.3   Form of the distribution in Eq. (6.10).

function:

$$\Psi(\mathbf{r}, t) = \psi(\mathbf{r})e^{-iWt/\hbar}. \tag{6.2}$$

$|\Psi(\mathbf{r}, t)|^2 dV$ is the probability to find the nucleus in the volume dV and, if the state described by Ψ decays with a decay constant λ, it is reasonable to write:

$$|\Psi(\mathbf{r}, t)|^2 = |\Psi(\mathbf{r}, 0)|^2 e^{-\lambda t}. \tag{6.3}$$

To obey simultaneously (6.2) and (6.3), W must be a complex quantity with imaginary part $-\lambda\hbar/2$. We can write:

$$W = E_0 - \frac{\hbar\lambda}{2}i, \tag{6.4}$$

which shows that the wave function (6.2) does not represent a well defined stationary state since the exponential contains a real part $-\lambda t/2$. However, we can write the exponent of (6.2) as a superposition of values corresponding to well defined energies E (for $t \geq 0$):

$$e^{-(iE_0/\hbar + \lambda/2)t} = \int_{-\infty}^{+\infty} A(E)e^{-iEt/\hbar}dE. \tag{6.5}$$

Functions connected by a Fourier transform relate to each other as

$$f(t) = \frac{1}{\sqrt{2\pi}} \int_{-\infty}^{+\infty} g(\omega)e^{-i\omega t}d\omega, \tag{6.6}$$

$$g(\omega) = \frac{1}{\sqrt{2\pi}} \int_{-\infty}^{+\infty} f(t)e^{i\omega t}dt. \tag{6.7}$$

This allows to establish the form of the amplitude $A(E)$:

$$A(E) = \frac{1}{2\pi\hbar} \int_0^{+\infty} e^{[i(E-E_0)/\hbar - \lambda/2]t}dt, \tag{6.8}$$

where the lower limit indicates that the stationary sate was created at the time $t = 0$. The integral (6.8) is of easy solution, giving

$$A(E) = \frac{1}{\hbar\lambda/2 - 2\pi i(E - E_0)}. \tag{6.9}$$

The probability of finding a value between E and $E + dE$ in an energy measurement is given by the product

$$A^*(E)A(E) = \frac{1}{\hbar^2\lambda^2/4 + 4\pi^2(E - E_0)^2}, \tag{6.10}$$

and this function of energy has the form of a Lorentzian, with the aspect shown in Figure 6.3. Its width at half-maximum is $\Gamma = \hbar\lambda = \hbar/\tau$. The relationship

$$\tau\Gamma = \hbar \tag{6.11}$$

between the half-life and the width of a state is directly connected to the uncertainty principle and shows that the longer time a state survives the greater is the precision with which its energy can be determined. In particular, only to stable states one can attribute a single value for the energy.

The decay constant λ was presented as the probability per unit time of occurrence of a transition between quantum states, and its values were supposed to be known experimental data. Now we will show that a formula to evaluate the decay constant can be obtained from the postulates of perturbation theory.

We can describe an unstable state by the addition of a perturbation to a stationary state. Formally we can write

$$H = H_0 + V, \tag{6.12}$$

where H is the Hamiltonian of the unstable state, composed by the non-perturbed Hamiltonian H_0 and a small perturbation V. The Hamiltonian H_0 satisfies an eigenvalue equation

$$H_0\psi_n = E_n\psi_n, \tag{6.13}$$

whose eigenfunctions form a complete basis in which the total wave function Ψ, that obeys

$$H\Psi = i\hbar\frac{\partial\Psi}{\partial t}, \tag{6.14}$$

can be expanded:

$$\Psi = \sum_n a_n(t)\psi_n e^{-iE_nt/\hbar}. \tag{6.15}$$

Using (6.12) and (6.15) in (6.14), and with the aid of (6.13), we obtain

$$i\hbar\sum_n \dot{a}_n\psi_n e^{-iE_nt/\hbar} = \sum_n Va_n\psi_n e^{-iE_nt/\hbar}, \tag{6.16}$$

with $\dot{a}_n \equiv da_n(t)/dt$. Using the orthogonalization properties of the ψ_n, let us multiply (6.16) by ψ_k^* and integrate it in the coordinate space. From this, results the coupled-channels equations

$$\dot{a}_k(t) = -\frac{i}{\hbar}\sum_n a_n(t)V_{kn}(t)e^{i\frac{E_k-E_n}{\hbar}t}, \tag{6.17}$$

where we introduced the matrix element ($d\tau$ is the volume element)

$$V_{kn} = \int \psi_k^* V\psi_n d\tau. \tag{6.18}$$

Let us make the following assumptions about the perturbation V: (a) It begins to act at time $t = 0$, when the unperturbed system is described by an eigenstate ψ_m and, (b) It stays at a very low value and after a short time interval becomes zero at $t = T$. These assumptions allow us to say that the conditions

$$a_m = 1, \qquad \text{and} \qquad a_m = 0 \quad \text{if} \ (m \neq n) \tag{6.19}$$

are rigorously verified for $t < 0$ and also works approximately for $t > 0$. Thus, (6.17) has only one term and the value of the amplitude is obtained from

$$a_k = -\frac{i}{\hbar}\int_0^T V_{km}e^{i(E_k-E_m)t/\hbar}dt, \tag{6.20}$$

whose value must be necessarily small by the assumption that followed (6.19). The above approach is also known as *first order perturbation theory*. The integral of (6.20) gives

$$a_k = \frac{V_{km}\left[1 - \exp\left(i\frac{E_k-E_m}{\hbar}T\right)\right]}{E_k - E_m}. \tag{6.21}$$

We need now to interpret the meaning of the amplitude a_k. The quantity $a_k^*a_k$ measures the probability of finding the system in the state k. This characterizes a transition occurring from the initial state m to the state k, and the value of $a_k^* a_k$ divided by the interval T should be a measure of the decay constant λ_k relative to the state k. The total decay constant is obtained by the sum over all states:

$$\lambda = \sum_{k \neq m} \lambda_k = \frac{\sum |a_k|^2}{T}. \tag{6.22}$$

Let us now suppose that there is a large number of available states k. We can in this case replace the summation in (6.22) by an integral. Defining $\rho(E)$ as the density of available states around the energy E_k, we write

$$\lambda = \frac{1}{T} \int_{-\infty}^{+\infty} |a_k|^2 \rho(E_k) dE_k = \frac{4}{T} \int_{-\infty}^{+\infty} |V_{km}|^2 \frac{\sin^2 \left[\left(\frac{E_k - E_m}{2\hbar} \right) T \right]}{(E_k - E_m)^2} \rho(E_k) dE_k. \tag{6.23}$$

The function $\sin^2 x / x^2$ only has significant amplitude near the origin. In the case of (6.23), if we suppose that V_{km} and ρ do not vary strongly in a small interval of the energy E_k around E_m, both these quantities can be taken outside of the integral, and we obtain the final expression

$$\lambda = \frac{2\pi}{\hbar} |V_{km}|^2 \rho(E_k). \tag{6.24}$$

Eq. (6.24) is known as the golden rule n$^{\underline{o}}$2 (also known as *Fermi golden rule*), and allows to determine the decay constant if we know the wave functions of the initial and final states.

6.3 Direct reactions: a simple approach

In the next section we will make a detailed account of the theory of direct reactions. But, for the moment, let us see how we can apply a simple quantum treatment to a direct reaction. For this purpose let us initially obtain an expression for the differential cross section using the concept of transition probability which was used to establish the golden rule no. 2, expression (6.24).

For the application of (6.24) the entrance channel will be understood as an initial quantum state, constituted of particles of mass m_a hitting a target A of mass m_A. The final quantum state is the exit channel, where particles of mass m_b move away from the nucleus B of mass m_B. The essential hypothesis for the application of the golden rule is that the interaction responsible for the direct reaction can be understood as a perturbation among a group of interactions that describe the system as a whole.

If we designate as λ the transition rate of the initial quantum state to the final quantum state, we can write a relationship between λ and the total cross section:

$$\lambda = v_a \sigma = \sigma \frac{k_a \hbar}{m_a}, \tag{6.25}$$

where v_a is now the velocity of the particles of the incident beam. If $d\lambda$ is the part relative to the emission in the solid angle $d\Omega$, then

$$d\sigma = \frac{m_a}{k_a \hbar} d\lambda, \tag{6.26}$$

and, using (6.24),

$$d\sigma = \frac{m_a}{k_a \hbar} \frac{2\pi}{\hbar} |V_{fi}|^2 d\rho(E_f), \tag{6.27}$$

where the indices i and f refer to the initial and final stages of the reaction.

A calculation similar to the one in Section 4.5 allows us to use the expression (4.44) to obtain the value of ρ:

$$\rho = \frac{dN}{dE} = \frac{p m_b s^3}{2\pi^2 \hbar^3}. \tag{6.28}$$

If we assume an isotropic distribution of momentum for the final states, the fraction $d\rho$ that corresponds to the solid angle $d\Omega$ is

$$d\rho = \frac{p m_b s^3}{2\pi^2 \hbar^3} \frac{d\Omega}{4\pi} = \frac{k_b m_b s^3}{8\pi^3 \hbar^2} d\Omega. \tag{6.29}$$

Using (6.29) in (6.27), we obtain an expression for the differential cross section:

$$\frac{d\sigma}{d\Omega} = \frac{m_a m_b k_b}{(2\pi\hbar^2)^2 k_a} |V_{fi}|^2, \tag{6.30}$$

involving the matrix element

$$V_{fi} = \int \Psi_b^* \Psi_B^* \chi_\beta^{(-)*}(\mathbf{r}_\beta) V \Psi_a \Psi_A \chi_\alpha^{(+)}(\mathbf{r}_\alpha) d\tau. \tag{6.31}$$

$\Psi_a, \Psi_b, \Psi_A, \Psi_B$ are the internal wave functions of the nuclei a, b, A and B. $\chi_\alpha^{(+)}, \chi_\beta^{(-)}$ are the wave functions of the relative momentum in the entrance channel α and in the exit channel β. The volume s^3 was not written in (6.30) because it can be chosen as the unity, since the wave functions that appear in (6.31) are normalized to 1 particle per unit volume. V is the perturbation potential that causes the "transition" from the entrance to the exit channel. It can be understood as an additional interaction to the average behavior of the potential and, in this sense, can be written as the difference between the total potential in the exit channel and the potential of the optical model in that same channel.

The simplest treatment than we can do for a direct reaction is to consider the incident beam as a plane wave whose only interaction with the target is through the perturbation potential that causes the reaction. The emerging beam is also treated as a plane wave. The use of plane waves for $\chi_\alpha^{(+)}$ and $\chi_\beta^{(-)}$ in (6.31) is denominated *first Born approximation*. With it we can arrive to an approximate expression for the behavior of the differential cross section. Thus, we shall start with the fact that the nuclear forces are of short range, what allows to restrict the integral (6.31) to regions where $\mathbf{r}_\alpha \cong \mathbf{r}_\beta = \mathbf{r}$. This leads to

$$V_{fi} \cong \int d\mathbf{r} \exp(i\mathbf{q} \cdot \mathbf{r}) \left\{ \int \Psi_b^* \Psi_B^* V \Psi_a \Psi_A d\tau' \right\}, \tag{6.32}$$

where $\mathbf{q} = \mathbf{k}_\alpha - \mathbf{k}_\beta$. The global variables in $d\tau$ were separated into variables $d\mathbf{r}$ and $d\tau'$.

The choice of the interaction V in the formula above depends on the direct reaction process. The physics involved appears more transparent if we discuss the pick-up process. Consider the A(p,d)B reaction for definiteness; then d = n + p and A = B + n, and the neutron is transferred. The proton picks up the neutron through their mutual interaction V_{pn}. The transition amplitude $T_{p,d}$ is given by Eq. 6.31 with $a \longrightarrow$ p, $b \longrightarrow$ d and $V = V_{pn}$.

Now, we shall use the expansion (2.36) for the plane wave of (6.32)

$$V_{fi} \cong \sum_{l=0}^{\infty} i^l (2l+1) \int j_l(qr) P_l(\cos\theta) F(\mathbf{r}) d\mathbf{r}, \tag{6.33}$$

where $F(\mathbf{r}) = \int \Psi_b^* \Psi_B^* V \Psi_a \Psi_A d\tau'$ concentrates all the internal properties and can be considered as the *form factor* of the reaction. Let us restrict the action of V to the surface of the nucleus, what is reasonable: outside the nucleus the action of V is limited by the short range of the nuclear forces and inside the nucleus there is a strong deviation to the absorption channel. The expression (6.33) then should be rewritten with $r = R$ and we obtain

$$V_{fi} \cong \sum_{l=0}^{\infty} c_l j_l(qR), \tag{6.34}$$

where the coefficients c_l include the constants that multiply the spherical Bessel function in (6.33), besides the integral over the form factor $F(\mathbf{r})$. The calculation of this factor can be laborious even for simpler cases [2], but what we want to retain is the dependence in j_l. The index l can be identified as the angular momentum transferred and, for a reaction that involves a single value of l, we can write for the differential cross section

$$\frac{d\sigma}{d\Omega} \propto |j_l(qR)|^2, \tag{6.35}$$

where the dependence in θ is contained here again in q,

$$q^2 = p_\alpha^2 + p_\beta^2 - 2p_\alpha p_\beta \cos\theta. \tag{6.36}$$

We have an oscillatory behavior for the angular distribution, the maximum separated by π from each other in the axis qR. Incidentally, this same result is obtained with the previously developed semi-classical approach (Figure 6.1), in the case of small scattering angles and small radii of the circles. It is, in fact, the result corresponding to the interference of two slits separated by $2R$.

The Born approximation with plane waves predicts for certain cases the correct place of the first peaks in the angular distribution but without reproducing correctly the intensities. A considerable progress can be done in the perturbative calculations if, instead of plane waves in (6.31), we use *distorted* waves (3.79) that contain, besides the plane wave, the part dispersed elastically by the optical potential of the target. The Born approximation with distorted waves, or DWBA (*distorted wave Born approximation*), became a largely employed tool in the analysis of experimental results of direct reactions. With it one can try to extract with a certain reliability the value of the angular momentum l transferred to the nucleus, or removed from it: stripping or pick-up. For this aim it is enough to compare the experimental angular distributions for a given exit energy with the DWBA calculations and to verify if a given value of l reproduces better the experimental data. An example of this is the already mentioned stripping reaction, $^{31}P(d,n)^{32}S$, for deuterons of 25 MeV. For the energy levels shown in Figure 6.2 the attribution of the value of l for the level is, in most cases, univocal.

6.4   Direct reactions: detailed calculations

In this section we will describe in more details the theory of direct reactions. The Figure 6.4 shows a schematic description of the reaction we will study. If we denote the momentum transfer, which is the difference between the final and initial momenta, by $\mathbf{q}_k$, for the k-particle where $k = 1, 2, 3$, we have

$$\mathbf{q}_1 = \mathbf{P}_f + \frac{m_1}{m_1 + m_2}\mathbf{P}_i \tag{6.37}$$

$$\mathbf{q}_2 = -\frac{m_2}{m_2 + m_3}\mathbf{P}_f + \frac{m_2}{m_1 + m_2}\mathbf{P}_i \tag{6.38}$$

$$\mathbf{q}_3 = -\frac{m_3}{m_2 + m_3}\mathbf{P}_f - \mathbf{P}_i. \tag{6.39}$$

Because the mass arrangements are not the same in the initial and the final states, the reduced masses μ_i and μ_f associated with the initial and the final momenta are different. μ_i and μ_f are given by

$$\frac{1}{\mu_i} = \frac{1}{m_1 + m_2} + \frac{1}{m_3}, \qquad\qquad \frac{1}{\mu_i} = \frac{1}{m_1} + \frac{1}{m_2 + m_3}. \tag{6.40}$$

The total energy E is given by

$$E = \frac{\mathbf{P}_i^2}{2\mu_i} - \epsilon_{12} = \frac{\mathbf{P}_f^2}{2\mu_f} - \epsilon_{23}, \tag{6.41}$$

where $P_i^2/2\mu_i$ and $P_f^2/2\mu_f$ are the initial and final kinetic energies and ϵ_{12} and ϵ_{23} are, respectively, the binding energies of the initially and finally bound systems, (1+2) and (2+3). The Q-value for the reaction is given by

$$Q = \frac{P_f^2}{2\mu_f} - \frac{P_i^2}{2\mu_i} = \epsilon_{23} - \epsilon_{12}. \tag{6.42}$$

The center of mass of particles (1+2) and (2+3) are given by

$$\mathbf{R}_{12} = \frac{m_1\mathbf{r}_1 + m_2\mathbf{r}_2}{m_1 + m_2}, \qquad\qquad \mathbf{R}_{23} = \frac{m_2\mathbf{r}_2 + m_3\mathbf{r}_3}{m_2 + m_3}, \tag{6.43}$$

respectively. Since the center of mass remains fixed, the independent sets of coordinates are both the relative coordinates of the particles in the bound system and the differences between the coordinates

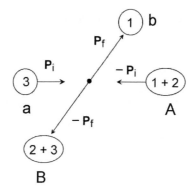

FIGURE 6.4 Schematic representation of a rearrangement direct reaction in the center of mass system of reference.

of the free particle and the center of mass of these particles. Thus,

$$
\begin{aligned}
\mathbf{r}_{12} &= \mathbf{r}_1 - \mathbf{r}_2, & \mathbf{r}_i &= \mathbf{r}_3 - \mathbf{R}_{12}, & \text{(initially)} \\
\mathbf{r}_{23} &= \mathbf{r}_2 - \mathbf{r}_3, & \mathbf{r}_f &= \mathbf{r}_1 - \mathbf{R}_{23}, & \text{(finally)}
\end{aligned}
\tag{6.44}
$$

are the set of independent coordinates initially and finally, with the condition

$$
m_1 \mathbf{r}_1 + m_2 \mathbf{r}_2 + m_3 \mathbf{r}_3 = 0,
\tag{6.45}
$$

that is, the center of mass of the whole system is at rest.

The momenta associated with these coordinates can be calculated by using the definition of canonical momentum, i.e., $\mathbf{P}_k = (1/i)\, \partial/\partial \mathbf{r}_k$, with $k = i$, f, 12, 23 (in this section we use $\hbar = 1$). We obtain

$$
\begin{aligned}
\mathbf{P}_{12} &= \mu_{12} \left(\frac{\mathbf{P}_1}{m_1} - \frac{\mathbf{P}_2}{m_2} \right), & \mathbf{P}_i &= \mu_i \left(-\frac{\mathbf{P}_1 + \mathbf{P}_2}{m_1 + m_2} + \frac{\mathbf{P}_3}{m_3} \right), \\
\mathbf{P}_{23} &= \mu_{23} \left(\frac{\mathbf{P}_2}{m_2} - \frac{\mathbf{P}_3}{m_3} \right), & \mathbf{P}_i &= \mu_f \left(\frac{\mathbf{P}_1}{m_1} - \frac{\mathbf{P}_2 + \mathbf{P}_3}{m_2 + m_3} \right),
\end{aligned}
\tag{6.46}
$$

where μ_{12} and μ_{23} are the reduced masses of $(1{+}2)$ and $(2{+}3)$, respectively.

The kinetic energy operator, T_{cm}, in the center of mass system is therefore

$$
T_{\mathrm{cm}} = \frac{\mathbf{P}_{12}^2}{2\mu_{12}} + \frac{\mathbf{P}_i^2}{2\mu_i} = \frac{\mathbf{P}_{23}^2}{2\mu_{23}} + \frac{\mathbf{P}_f^2}{2\mu_f}.
\tag{6.47}
$$

The total Hamiltonian for the system is

$$
H = T_{\mathrm{cm}} + V_{12} + V_{13} + V_{23},
\tag{6.48}
$$

and the wave functions for the states of the initial $(1{+}2)$ and final $(2{+}3)$ bound systems are

$$
\begin{aligned}
\left(\frac{\mathbf{P}_{12}^2}{2\mu_{12}} + V_{12} \right) \phi_{12}(\mathbf{r}_{12}) &= -\epsilon_{12} \phi_{12}(\mathbf{r}_{12}), \\
\left(\frac{\mathbf{P}_{23}^2}{2\mu_{23}} + V_{23} \right) \phi_{23}(\mathbf{r}_{23}) &= -\epsilon_{23} \phi_{23}(\mathbf{r}_{23}).
\end{aligned}
\tag{6.49}
$$

We denote by $\chi_i(\mathbf{r}_i)$ and $\chi_f(\mathbf{r}_f)$ the wave functions of the free particle 3 with respect to the center of mass of $(1{+}2)$ in the initial state, and of 1 with respect to the center of mass of $(2{+}3)$ in the final state, respectively. The total wave functions $\psi(\mathbf{r}_{12}, \mathbf{r}_i)$ and $\psi(\mathbf{r}_{23}, \mathbf{r}_f)$ can be expanded as follows

$$
\psi(\mathbf{r}_{12}, \mathbf{r}_i) = \sum_\alpha \phi_\alpha(\mathbf{r}_{12}) \chi_\alpha(\mathbf{r}_i), \qquad \psi(\mathbf{r}_{23}, \mathbf{r}_f) = \sum_\alpha \phi_\alpha(\mathbf{r}_{23}) \chi_\alpha(\mathbf{r}_f),
\tag{6.50}
$$

where α stands for a complete set of states of the bound system and the free particle. From Eqs. (6.50) we get

$$
\begin{aligned}
\chi_i\left(\mathbf{r}_i\right) &= \int \phi_{12}^*\left(\mathbf{r}_{12}\right) \psi\left(\mathbf{r}_{12}, \mathbf{r}_i\right) d^3 r_{12}, \\
\chi_f\left(\mathbf{r}_f\right) &= \int \phi_{23}^*\left(\mathbf{r}_{23}\right) \psi\left(\mathbf{r}_{23}, \mathbf{r}_f\right) d^3 r_{23}.
\end{aligned}
\tag{6.51}
$$

To obtain the equation satisfied by $\chi_i\left(\mathbf{r}_i\right)$, we note that $P_f^2=-\nabla_f^2$. From (6.41), (6.47) and (6.48) we have

$$
H \psi\left(\mathbf{r}_{23}, \mathbf{r}_f\right)=\left(\frac{\mathbf{p}_{23}^2}{2 \mu_{23}}+\frac{\mathbf{P}_f^2}{2 \mu_f}+V_{12}+V_{13}+V_{23}\right) \psi\left(\mathbf{r}_{23}, \mathbf{r}_f\right)=\left(\frac{\mathbf{P}_f^2}{2 \mu_f}-\epsilon_{23}\right) \psi\left(\mathbf{r}_{23}, \mathbf{r}_f\right), \tag{6.52}
$$

or

$$
\frac{1}{2 \mu_f}\left(\nabla_f^2+P_f^2\right) \psi\left(\mathbf{r}_{23}, \mathbf{r}_f\right)=\left(\frac{\mathbf{p}_{23}^2}{2 \mu_{23}}+V_{12}+V_{13}+V_{23}+\epsilon_{23}\right) \psi\left(\mathbf{r}_{23}, \mathbf{r}_f\right). \tag{6.53}
$$

Applying the operator $\left(\nabla_f^2+P_f^2\right)$ to $\chi_f\left(\mathbf{r}_f\right)$ in Eq. (6.51) and using Eqs. 6.49 and (6.53), we obtain

$$
\left(\nabla_f^2+P_f^2\right) \chi_f\left(\mathbf{r}_f\right)=2 \mu_f \int \phi_{23}^*\left(\mathbf{r}_{23}\right)\left[V_{12}+V_{13}\right] \psi\left(\mathbf{r}_{23}, \mathbf{r}_f\right) d^3 r_{23}. \tag{6.54}
$$

As we have seen in Chapter 3, the solution for $\chi_f\left(\mathbf{r}_f\right)$ can be obtained by using the outgoing Green's function

$$
-\frac{\exp\left[i P_f\left|\mathbf{r}_f-\mathbf{r}\right|\right]}{4 \pi\left|\mathbf{r}_f-\mathbf{r}\right|}.
$$

Thus

$$
\chi_f\left(\mathbf{r}_f\right)=-\frac{\mu_f}{2 \pi} \int \frac{\exp\left[i P_f\left|\mathbf{r}_f-\mathbf{r}\right|\right]}{4 \pi\left|\mathbf{r}_f-\mathbf{r}\right|} \phi_{23}^*\left(\mathbf{r}_{23}\right)\left[V_{12}+V_{13}\right] \psi\left(\mathbf{r}_{23}, \mathbf{r}\right) d^3 r_{23} d^3 r, \tag{6.55}
$$

and the reaction scattering amplitude is

$$
f(\theta, \phi)=-\frac{\mu_f}{2 \pi} \int \phi_{23}^*\left(\mathbf{r}_{23}\right) \exp\left(-i \mathbf{P}_f \cdot \mathbf{r}_f\right)\left[V_{12}+V_{13}\right] \psi\left(\mathbf{r}_{23}, \mathbf{r}_f\right) d^3 r_{23} d^3 r_f. \tag{6.56}
$$

In the Born approximation (also called *Butler theory* [3]for direct reactions) one replaces ψ by

$$
\phi_{12}\left(\mathbf{r}_{12}\right) \exp\left[-i \mathbf{P}_i \cdot \mathbf{R}_{12}+i \mathbf{P}_i \cdot \mathbf{r}_3\right]=\phi_{12}\left(\mathbf{r}_{12}\right) \exp\left[i \mathbf{P}_i \cdot \mathbf{r}_i\right], \tag{6.57}
$$

and the reaction scattering amplitude becomes

$$
\begin{aligned}
f(\theta, \phi) &\simeq-\frac{\mu_f}{2 \pi} \int \phi_{23}^*\left(\mathbf{r}_{23}\right) \exp\left(-i \mathbf{P}_f \cdot \mathbf{r}_f\right)\left[V_{12}+V_{13}\right] \exp\left[i \mathbf{P}_i \cdot \mathbf{r}_i\right] \phi_{12}\left(\mathbf{r}_{12}\right) d^3 r_{23} d^3 r_f \\
&=-\frac{\mu_f}{2 \pi} \int \phi_{23}^*\left(\mathbf{r}_{23}\right) \exp\left(i \mathbf{q}_3 \cdot \mathbf{r}_{23}-i \mathbf{q}_1 \cdot \mathbf{r}_{12}\right)\left[V_{12}+V_{13}\right] \phi_{12}\left(\mathbf{r}_{12}\right) d^3 r_{12} d^3 r_{23}, \tag{6.58}
\end{aligned}
$$

where we made use of the equalities

$$
-\mathbf{P}_f \cdot \mathbf{r}_f+\mathbf{P}_i \cdot \mathbf{r}_i=\mathbf{q}_3 \cdot \mathbf{r}_{23}-\mathbf{q}_1 \cdot \mathbf{r}_{12}, \quad \text{and} \quad d^3 r_{23}\ d^3 r_f=d^3 r_{12}\ d^3 r_{23}.
$$

The rearrangement reaction cross section is thus given by

$$
\sigma(\theta, \phi)=\frac{\mu_i \mu_f}{(2 \pi)^2} \frac{P_f}{P_i}\left|\int \phi_{23}^*\left(\mathbf{r}_{23}\right) \exp\left(i \mathbf{q}_3 \cdot \mathbf{r}_{23}-i \mathbf{q}_1 \cdot \mathbf{r}_{12}\right)\left[V_{12}+V_{13}\right] \phi_{12}\left(\mathbf{r}_{12}\right) d^3 r_{12} d^3 r_{23}\right|^2. \tag{6.59}
$$

This justifies our discussions related to Eqs. (6.31) and (6.32), with the appropriate change of notations.

Eqs. (6.58) and (6.59) can be further simplified by neglecting the interaction V_{13} between particles 1 and 3. This is done because in the reaction $(1 + 2) + 3 \rightarrow 1 + (2 + 3)$ particles 1 and 3 never appear in a bound state. In the stripping reaction X^A (d,p) X^{A+1}, $V_{13} = V_{p,A}$ and in the pick-up reaction X^A (p,d) X^{A-1}, $V_{13} = V_{p,A-1}$. The interaction V_{12} equals V_{np} for both stripping reaction (d,p) and the pick-up reaction (p,d) if the latter is considered as a process inverse to that of the former.

In Eq. (6.59) the V_{12} term contributes a non-vanishing result provided the final state contains components of the core, the target nucleus, left in its ground state. On the other hand, the contribution arising from the interaction $V_{13} = V_{pA}$ will be non-vanishing if the final state corresponds to excitation of the core. Generally, in a stripping reaction the contribution of the V_{13} interaction is much less than that of V_{12} unless the final state involves almost purely the excitation of the core. Hence, we can use

$$\sigma\left(\theta,\phi\right) \simeq \frac{\mu_i \mu_f}{(2\pi)^2} \frac{P_f}{P_i} \left| \int \exp\left(i\mathbf{q}_3 \cdot \mathbf{r}_{23}\right) \phi_{23}^*\left(\mathbf{r}_{23}\right) d^3 r_{23} \int \exp\left(-i\mathbf{q}_1 \cdot \mathbf{r}_{12}\right) V_{12}\left(\mathbf{r}_{12}\right) \phi_{12}\left(\mathbf{r}_{12}\right) d^3 r_{12} \right|^2 . \tag{6.60}$$

Now, using the identity

$$\left(-\frac{\boldsymbol{\nabla}^2}{2\mu_{12}} - \frac{q_1^2}{2\mu_{12}} \right) \exp\left(-i\mathbf{q}_1 \cdot \mathbf{r}_{12}\right) = 0, \tag{6.61}$$

and the first of Eq. (6.49) it is straightforward to show that

$$\int \exp\left(-i\mathbf{q}_1 \cdot \mathbf{r}_{12}\right) V_{12}\left(\mathbf{r}_{12}\right) \phi_{12}\left(\mathbf{r}_{12}\right) d^3 r_{12} = -\left(\epsilon_{12} + \frac{q_1^2}{2\mu_{12}} \right) \int \exp\left(-i\mathbf{q}_1 \cdot \mathbf{r}_{12}\right) \phi_{12}\left(\mathbf{r}_{12}\right) d^3 r_{12}. \tag{6.62}$$

Thus,

$$f\left(\theta,\phi\right) \simeq -\frac{\mu_f}{2\pi} \left(\epsilon_{12} + \frac{q_1^2}{2\mu_{12}} \right) \mathcal{G}_1\left(\mathbf{q}_1\right) \mathcal{G}_3\left(\mathbf{q}_3\right)$$

$$\sigma\left(\theta,\phi\right) \simeq \frac{\mu_i \mu_f}{(2\pi)^2} \frac{P_f}{P_i} \left(\epsilon_{12} + \frac{q_1^2}{2\mu_{12}} \right)^2 \left| \mathcal{G}_1\left(\mathbf{q}_1\right) \mathcal{G}_3\left(\mathbf{q}_3\right) \right|^2, \tag{6.63}$$

where

$$\mathcal{G}_1\left(\mathbf{q}_1\right) = \int \exp\left(-i\mathbf{q}_1 \cdot \mathbf{r}_{12}\right) \phi_{12}\left(\mathbf{r}_{12}\right) d^3 r_{12}$$

$$\mathcal{G}_3\left(\mathbf{q}_3\right) = \int \exp\left(i\mathbf{q}_3 \cdot \mathbf{r}_{23}\right) \phi_{23}^*\left(\mathbf{r}_{23}\right) d^3 r_{23}. \tag{6.64}$$

If the bound nuclei (1+2) and (2+3) are in definite orbital-angular momentum states (l_{12}, m_{12}) and (l_{23}, m_{23}) respectively, we can write

$$\phi_{12}\left(\mathbf{r}_{12}\right) = \frac{u_{12}\left(r_{12}\right)}{r_{12}} Y_{l_{12}m_{12}}\left(\widehat{\mathbf{r}}_{12}\right), \qquad \phi_{23}\left(\mathbf{r}_{23}\right) = \frac{u_{23}\left(r_{23}\right)}{r_{23}} Y_{l_{23}m_{23}}\left(\widehat{\mathbf{r}}_{23}\right).$$

The angular part of the integrals in Eq. (6.64) can be done by using the plane wave expansion. Averaging over the initial magnetic quantum number m_{12} and summing over the final magnetic quantum number m_{23}, we obtain

$$f\left(\theta,\phi\right) \simeq 8\pi i^{l_{23}-l_{12}} \mu_f \left(\epsilon_{12} + \frac{q_1^2}{2\mu_{12}} \right) Y_{l_{12}m_{12}}\left(\widehat{\mathbf{q}}_{12}\right) Y_{l_{23}m_{23}}\left(\widehat{\mathbf{q}}_{23}\right) \mathcal{F}_1\left(q_1\right) \mathcal{F}_3\left(q_3\right), \tag{6.65}$$

and

$$\sigma\left(\theta,\phi\right) \simeq 4\mu_i \mu_f \frac{P_f}{P_i} \left(\epsilon_{12} + \frac{q_1^2}{2\mu_{12}} \right)^2 \left(2l_{23}+1\right) \mathcal{F}_1^2\left(q_1\right) \mathcal{F}_3^2\left(q_3\right), \tag{6.66}$$

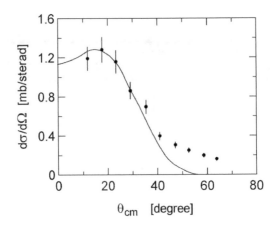

FIGURE 6.5 Angular distributions of deuterons from $^{16}O(p,d)^{15}O$ ground state. The solid line is a plane wave Butler curve with $l = 1$ and $r_0 = 5.2$ fm. The proton energy was 20 MeV. Data are from Ref. [4].

where

$$\mathcal{F}_1(q_1) = \int_0^\infty r_{12} u_{12}(r_{12}) j_{l_{12}}(q_1 r_{12}) dr_{12},$$

$$\mathcal{F}_3(q_3) = \int_0^\infty r_{23} u_{23}(r_{23}) j_{l_{23}}(q_3 r_{23}) dr_{23}. \tag{6.67}$$

The angular dependence of the differential cross section (6.66) occurs in the momentum transfers q_1 and q_3, which according to Eqs. (6.37) and (6.39) depend on the angle θ between the vectors $\mathbf{P}_i$ and $\mathbf{P}_f$.

In the (d,p) reaction, the notation of Figure 6.4 has the interpretation $1 = p$, $2 = n$, $3 = X^A$, and the experimentally observed quantities are the momentum $-\mathbf{P}_i$ of the lighter incident particle $d = (2 + 3)$ and the momentum $\mathbf{P}_f$ of the lighter outgoing particle. When the proton is emitted in the forward direction, that is, in the direction of the incident deuteron, the values of the two momentum transfers $\mathbf{q}_1$ and $\mathbf{q}_3$ are at a minimum. The maximum values occur when the proton is emitted in the backward direction.

The integrals $\mathcal{F}_1(q_1)$ and $\mathcal{F}_3(q_3)$ are the Fourier transforms and $\mathcal{F}_1^2(q_1)$ indicates the probability that the internal momentum of particle 1 will have the value q_1 when it is in a bound state with particle 2.

We note that if the process $1 + (2 + 3) \to (1 + 2) + 3$ describes a stripping reaction then the inverse reaction $(1 + 2) + 3 \longrightarrow 1 + (2 + 3)$ describes a pick-up reaction. It is necessary to consider only one of these. Using the principle of detailed balance, as explained in Chapter 4, we have

$$\sigma_{1+(2+3)\longrightarrow(1+2)+3} = \frac{P_f^2}{P_i^2} \sigma_{(1+2)+3\longrightarrow 1+(2+3)}. \tag{6.68}$$

The reverse or pick-up reactions such as (p,d) and (p,t) are similar to the stripping processes such as (d,p) and (t,p) and their mathematical treatments are alike. One expects, and in fact obtains, similar angular distributions of disintegration products.

In the theory developed by Butler [3] the wave functions u_{12} and u_{23} which are needed to solve the integrals in Eq. (6.67) are obtained by assuming that the nuclear interaction vanishes for distances $r \geq r_0$. Integrating Eqs. (6.67) by parts and using the recursion relations of the spherical Bessel functions, one can show [3] that all one needs are the wave functions u_{12} and u_{23} at $r = r_0$, which can be estimated from a knowledge of the potential. This additional simplification is however not needed since a good solution of the Schrödinger equations (6.49) can be obtained in most cases.

Figure 6.5 shows the angular distributions of deuterons from $^{16}O(p,d)^{15}O$ ground state [4]. The solid line is a plane wave Butler calculation with $l = 1$ and $r_0 = 5.2$ fm. The agreement with the data is quite good in view of the simplicity of the theory. For such reasons the direct reactions are a powerful tool to obtain information on the shell model of the nucleus, e.g. the spin assignment, energy, and spectroscopic factors associated with a given nucleon. We will show this in a series of examples. But before that we will describe in more details the applications of the shell model to the nucleus.

6.5 Applications of the shell model

6.5.1 The extreme shell model

Each level of Figure 4.21, characterized by the quantum numbers n, l, j, contains $2j+1$ nucleons of a same type and is also referred to as a *sub-shell*. We establish what one denominates by *single particle model* or *extreme shell model*. In this version, the model admits that an odd nucleus is composed of an inert even-even core plus an unpaired nucleon, and that this last nucleon determines the properties of the nucleus. This idea was already exposed in the appendix A of Chapter 4; what we can do now is to determine, starting from the Figure 4.21, in which state we find the unpaired nucleon.

Let us take ^{17}O as example. This nucleus has a shell closed with 8 protons but has a remaining neutron above the closed core of 8 neutrons. A fast examination of Figure 4.21 indicates that this neutron finds itself in the level $1d_{5/2}$. We can say that the neutron *configuration* of this nucleus is

$$(1s_{1/2})^2(1p_{3/2})^4(1p_{1/2})^2(1d_{5/2})^1,$$

being evident that in this definition we list the filled levels for the neutrons, with the upper indexes, equal to $2j+1$, indicating the number of particles in each one of them. It is also common to restrict the configuration to the levels partially filled, the sub-shells being completely ignored. Thus, the configuration of neutrons in ^{17}O would be $(1d_{5/2})^1$. The prediction of the model is, in this case, that the spin of the ground state of ^{17}O is $\frac{5}{2}$ and the parity is positive ($l = 2$). This prediction is in agreement with experiments. For similar reasons, the model predicts that the ground state of ^{17}F is a $\frac{5}{2}^+$ state and this is indeed the measured value.

The extreme shell model works well when we have a nucleon above a closed shell, as in the examples above. It also works well for a hole (absence of a nucleon) in a closed shell. Examples of this case are the nuclei ^{15}O and ^{15}N for which the model predicts correctly a $\frac{1}{2}^-$ ground state. There are situations, however, in which the model needs a certain adaptation. Such is the case, for example, of the stable nuclei ^{203}Tl and ^{205}Tl. They have 81 protons, existing, therefore, a hole in $h_{11/2}$. Its ground state is, however, $\frac{1}{2}^+$ instead of $\frac{11}{2}^-$. In order to understand what happens it is necessary to recall that the model, in the simplest form we are using, totally neglects the individualized interaction of the nucleons; a correction of the model would be to take into account certain nucleon-nucleon interactions that we know are present and that are part of the residual interaction, Eq. (4.225).

A class of interactions of special interest is the one that involves a proton (or a neutron) pair of equal orbits n, l and j and with symmetric values of m_j. A collision of these particles can take the pair to other orbits of same quantum numbers n, l, j but with new projections m'_j and $-m'_j$. These collisions conserve energy (the $2j+1$ states of a same sub-shell are degenerate), angular momentum and parity and we expect that there is a permanent alternation between the several possible values of the pair $m_j, -m_j$. The interaction between two nucleons in those circumstances is commonly called the *pairing force*. It leads to an increase of the binding energy of the nucleus, since, belonging both the nucleons to a same orbit, their wave functions have the same space distribution and the average proximity between them are maximum. As the nuclear force is attractive, this leads to an increase in the binding energy. The pairing force is responsible for the pairing term (5.9) of the mass formula (5.10). It is the same type of force that, acting between the conduction electrons of a metal, in special circumstances and at low temperatures, yields the superconductivity phenomenon [5].

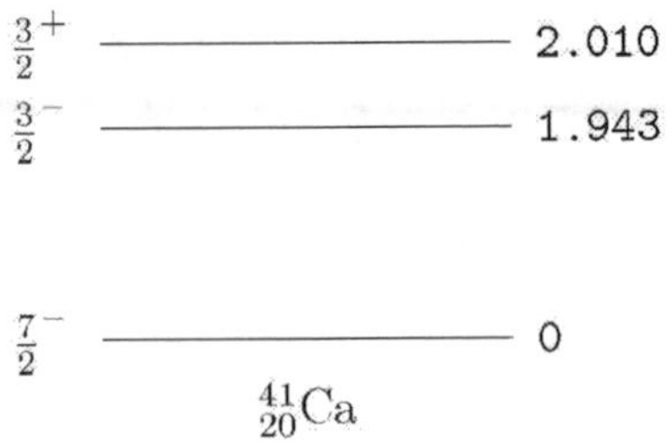

FIGURE 6.6 Level scheme of ^{41}Ca. The values to the right are the energies in MeV.

Nucleus	Z	State (p)	N	State (n)	$\mathcal{N}$	j_{pred}	j_{exp}
^{14}N	7	$p_{1/2}$	7	$p_{1/2}$	-1	0 or 1	1^+
^{42}K	19	$d_{3/2}$	23	$f_{7/2}$	0	2	2^-
^{80}Br	35	$p_{3/2}$	45	$p_{1/2}$	0	1	1^+
^{208}Tl	81	$s_{1/2}$	127	$g_{9/2}$	1	4 or 5	5^+

TABLE 6.1 Determination of the spin of odd-odd nuclei. The 7th column lists the possible values of j predicted by the Nordheim rules and the last column lists the experimental values with the respective parities.

The pairing force increases with the value of j since for larger angular momentum, the larger the location of the wave function of the nucleon around a classical orbit will be and the argument of the previous paragraph will be stronger. This implies that it is sometimes energetically more advantageous that the isolated nucleon is not in the last level, but below it, leaving the last level for a group of paired nucleons. This happens in our example of Tl; the hole is not located in $h_{11/2}$ but in $s_{1/2}$, leaving the orbital $h_{11/2}$ (of high j) occupied by a pair of protons. Another example is ^{207}Pb, for which the hole in the closed shell of 126 neutrons is not in $i_{13/2}$ but in $p_{1/2}$, resulting in the value $\frac{1}{2}^-$ for the spin of its ground state.

An example of another kind is ^{23}Na. This nucleus has the last 3 protons in the orbital $1d_{5/2}$. The value of its spin is, however, $\frac{3}{2}$. This is an example of a flaw in the predictions of the extreme shell model. Here, it is the coupling between the three nucleons that determines the value of the spin and not separately the value of j of each one of them. This type of behavior will be analyzed in the following subsection.

Having established the outline of operation of the shell model, it is easy to apply it to the determination of the excited states of nuclei. Let us see the case of the ^{41}Ca (Figure 6.6): the ground state is a $\frac{7}{2}^-$, since the extra neutron occupies the orbital $f_{7/2}$. The first excited level corresponds to a jump of that neutron to $p_{3/2}$, generating the state $\frac{3}{2}^-$. The second excited state, $\frac{3}{2}^+$, is obtained by the passage of a neutron from $1d_{3/2}$ to $1f_{7/2}$, leaving a hole in $1d_{3/2}$. It should be noticed, however, that in very few cases the single particle model gets to a reasonable prediction of spins and energy of excited states. For such a purpose it is necessary to have a more sophisticated version of the shell model, that it will be described in the following subsection.

As a last topic of this section we will examine the situation of one odd-odd nucleus. In this case, two nucleons, a proton and a neutron, are unpaired. If $\mathbf{j}_p$ and $\mathbf{j}_n$ are the respective angular momenta of the nucleons, the angular momentum of the nucleus j can have values from $|j_p - j_n|$ to $j_p + j_n$. L. W. Nordheim proposed, in 1950 [6], rules to determine the most probable value of the spin of an odd-odd nucleus. Defining the *Nordheim number*

$$\mathcal{N} = j_p - l_p + j_n - l_n, \tag{6.69}$$

those rules establish that:

a) If $\mathcal{N} = 0$, $j = |j_p - j_n|$; this is called the *strong rule*.

b) If $\mathcal{N} = \pm 1$, $j = j_p + j_n$ or $j = |j_p - j_n|$, constituting the *weak rule*.

The Nordheim rules reveal a tendency (not widespread, because there are exceptions) to an alignment of the intrinsic spins, as in the state $j = 1$ of the deuteron. Table 6.1 exhibits some practical examples.

6.5.2 Extension of the shell model: contribution of more than one particle

The recipe of the single particle model for the determination of the spin of an odd nucleus assumes that the value of j is given by the level (or sub-shell) that has the odd number of nucleons. Although this idea works well when it has only one nucleon (or only one hole) in the level, it is mistaken several times, as in the case of the ^{23}Na of previous section, when the level is partially full. Flaws as that are foreseen given the extreme simplicity of the single particle model. A refinement of the model can be made trying somehow to include the residual interactions, that is, the part of the real interactions acting on a nucleon that are not represented in the average potential (see Eq. (4.192)).

When we have an incomplete sub-shell, the states that it can form with k nucleons is degenerate. The presence of residual forces among those nucleons separates the states in energy, that is, the degeneracy is removed. The angular momentum of each state is one of the possible values that result of adding k angular momenta **j**. The ground state will be the lowest energy of the group and its angular momentum will not necessarily be equal j *.

Before we examine the possible values of the angular momentum it is convenient to discuss about the composition of the valence level. When we filled out the levels of the shell model we placed $2j + 1$ protons in a level n, l, j and also $2j + 1$ neutrons, since the Pauli principle does not restrict the presence of different nucleons in the same quantum state. We can, in this way, think of a "proton well" and of a "neutron well", filled in an independent way. When we have a heavy nucleus, the Coulomb force implies that the proton well is more shallow than the neutron well. We have a situation similar to the one shown in Figure 4.7 for the case of the Fermi gas. Thus, the shell where we find the last protons and the shell where we find the last neutrons, both close to the Fermi level, can be different, corresponding to quite different wave functions and of small space overlap. It is expected that a residual proton-neutron interaction is not important in this case. For light nuclei, on the other hand, the Coulomb effect is small and the inclusion of particle-particle interactions should treat protons and neutrons within a single context.

We will analyze the first case initially and imagine that k last nucleons, say, protons, reside in a sub-shell defined by n, l, j. The configuration for the proton well would have the form

$$(n_1 l_1 j_1)^{2j_1+1} (n_2 l_2 j_2)^{2j_2+2} \cdots (nlj)^k.$$

If we do not take into consideration the interaction forces, the $2j + 1$ states that compose the last level, where we find k valence protons, are degenerate. The presence of the interaction removes the degeneracy; thus, for example, if we have two protons ($k = 2$) in a level of $j = \frac{5}{2}$, we can follow the picture shown in Table 6.2. The first two columns list the possible values of m_j for the two particles allowed by the Pauli principle and the last column lists the value $m_j = m_j(1) + m_j(2)$. We know that the possible values of the resulting angular momentum are located in the range $|j_1 - j_2| < j < j_1 + j_2$, that is, j could, in this example, take the values 0, 1, 2, 3, 4 or 5. If, however, we accept the rule that each value of m_j of Table 6.2 is the projection of a single value of j, we will see that only the momenta $j = 0$, 2 and 4 can exist.

The absence of odd values in the sum of the angular momenta of two identical nucleons of same j can be demonstrated from angular momentum coupling, i.e.,

$$\chi_J^M = \sum_m \langle jjmM - m|JM \rangle \chi_j^m(1)\chi_j^{M-m}(2). \tag{6.70}$$

---

*It can be shown that the forces between nucleons inside of the closed shells and the ones that act between the valence nucleons and the closed shells do not modify the ordering of the energy levels.

m_1	m_2	m	m_1	m_2	m
5/2	3/2	4	3/2	-5/2	-1
5/2	1/2	3	1/2	-1/2	0
5/2	-1/2	2	1/2	-3/2	-1
5/2	-3/2	1	-1/2	-5/2	-2
5/2	-5/2	0	-1/2	-3/2	-2
3/2	1/2	2	-1/2	-5/2	-3
3/2	-1/2	1	-3/2	-5/2	-4
3/2	-3/2	0			

TABLE 6.2 Possible values of the projection of the total angular momentum when two identical nucleons occupy the level $j = \frac{5}{2}$.

j	k	
$\frac{1}{2}$	1	$\frac{1}{2}$
$\frac{3}{2}$	1	$\frac{3}{2}$
	2	$0,2$
$\frac{5}{2}$	1	$\frac{5}{2}$
	2	$0,2,4$
	3	$\frac{3}{2},\frac{5}{2},\frac{9}{2}$
$\frac{7}{2}$	1	$\frac{7}{2}$
	2	$0,2,4,6$
	3	$\frac{1}{2},\frac{3}{2},\frac{5}{2},\frac{7}{2},\frac{9}{2},\frac{11}{2},\frac{15}{2}$
	4	$0,2(2),4(2),5,6,8$
$\frac{9}{2}$	1	$\frac{9}{2}$
	2	$0,2,4,6,8$
	3	$\frac{3}{2},\frac{5}{2},\frac{7}{2},\frac{9}{2}(2),\frac{11}{2},\frac{15}{2},\frac{17}{2},\frac{21}{2}$
	4	$0(2),2(2),3,4(3),5,6(3),7,8(2),9,10,12$
	5	$\frac{1}{2},\frac{3}{2},\frac{5}{2}(2),\frac{7}{2}(2),\frac{9}{2}(3),\frac{11}{2}(2),\frac{13}{2}(2),\frac{15}{2}(2),\frac{17}{2}(2),\frac{19}{2},\frac{21}{2},\frac{25}{2}$

TABLE 6.3 Possible values for the total angular momentum of k identical nucleons placed in a sub-shell of angular moment j.The numbers between parenthesis indicate the number of times that the value repeats.

If we now exchange the particles 1 and 2 we get

$$\chi_J^M = \sum_m \langle jjmM - m|JM\rangle \chi_j^m(2)\chi_j^{M-m}(1), \tag{6.71}$$

or, with the transformation $m' = M - m$,

$$\chi_J^M = \sum_{m'} \langle jjM - m'm'|JM\rangle \chi_j^{m'}(1)\chi_j^{M-m'}(2). \tag{6.72}$$

Using the property of the Clebsch-Gordan coefficients,

$$\langle j_1 j_2 m_1 m_2|jm\rangle = (-1)^{j_1+j_2-j} \langle j_1 j_2 m_2 m_1|jm\rangle, \tag{6.73}$$

Eq. (6.72) can be written as

$$\chi_J^M = \sum_{m'} (-1)^{2j-J} \langle jjm'M - m'|JM\rangle \chi_j^{m'}(1)\chi_j^{M-m'}(2), \tag{6.74}$$

and, as $2j$ is odd, Eqs. (6.70) and (6.74) only differ by a factor $(-1)^{J+1}$. Therefore, only even values of J are allowed since they turn χ_J^M antisymmetric for the exchange of the two nucleons.

When we have more than two particles in a level the situation can become much more complex. The practical method of finding the possible values of the resulting angular momentum, shown in Table 6.2, can always be used, but it is a quite difficult way when we have many nucleons. Table 6.3 shows the possibilities for several configurations, where we see new restrictions imposed by the antisymmetrization, or, what is equivalent, by the Pauli principle, for given values of the total angular momentum J.

6.5.3 Isobaric analog states

We will now examine the case in which k nucleons of the last sub-shell is a mixture of protons and neutrons. In the same way that spins are added to form the total spin of a group of nucleons, the isospin of a nucleus is obtained by coupling the isospins of each nucleon. The algebra for the two procedures is the same. Thus, the total isospin of a nucleus is defined by

$$\mathbf{T} = \sum_{i=1}^{A} \mathbf{t}^{(i)}, \tag{6.75}$$

being their z-component

$$T_z = \sum_{i=1}^{A} t_z^{(i)}, \tag{6.76}$$

where $\mathbf{t}^{(i)}$ and $t_z^{(i)}$ are the isospin of the nucleon i and its z-component, respectively. The isospin is part of the Hamiltonian of the system through the Coulomb potential

$$\sum_{ij} V_{ij}^c = \sum_{ij} \frac{e^2}{r_{ij}} \left(\frac{1}{2} - t_z^{(i)} \right) \left(\frac{1}{2} - t_z^{(j)} \right), \tag{6.77}$$

and it is easy to see that T_z commutes with $\sum_{ij} V_{ij}^c$ and, consequently with H. Its eigenvalue equation,

$$T_z \Psi = \left(\frac{1}{2} + \frac{1}{2} + \cdots - \frac{1}{2} - \frac{1}{2} + \cdots \right) \Psi = \frac{1}{2}(N - Z)\Psi, \tag{6.78}$$

shows that T_z is a measure of the neutron excess in the nucleus. We know that the operator $\mathbf{T}^2$ does not commute with H, because it contains components t_x and t_y that do not commute with t_z, and the total isospin of a nucleus is not conserved. On the other hand, the Coulomb energy varies with the square of the charge and it is, therefore, much less important in light nuclei. For these nuclei we can, in a reasonable approximation, ignore the Coulomb term of the Hamiltonian. With these considerations H commutes with $\mathbf{T}^2$ and one can define a quantum number of isospin T, obtained from the eigenvalue equation

$$\mathbf{T}^2 \Psi = T(T + 1)\Psi, \tag{6.79}$$

that becomes an additional label for the quantum state of the nucleus.

If we have k nucleons in an orbital around a light core of $Z = N$ we can form $k + 1$ nuclei varying the neutron and proton numbers in the level. The value of T for each nucleus is located inside the limits

$$|T_z| \leq T \leq \frac{k}{2}. \tag{6.80}$$

Let us take as example the magic core $Z = N = 4$ above which two nucleons are placed. We can form the nuclei $^{10}_{4}\text{Be}$, $^{10}_{5}\text{B}$ and $^{10}_{6}\text{C}$. As $k = 2$, T can assume the values 0 and 1. The component T_z for each nucleus has the values -1, 0 and $+1$, respectively. Figure 6.7 exhibits the level diagrams of each one of these nuclei. We can see that states of T equal to 1 form a triplet with a representative in each one of the nuclei and states of $T = 0$ exist only in the ^{10}B, of component $T_z = 0$. The states that compose each triplet have the same approximate energy in the three nuclei and they are denominated *isobaric analog states*. As the effect of the Coulomb energy was subtracted, this result is another indication that the nuclear forces are independent of charge. In other words, if the p-p, n-n and p-n nuclear forces are identical, systems of A nucleons should have the same properties and, in particular, if we discount the Coulomb force, the same energy levels. The existence of states of $T = 0$ for ^{10}B that do not have partners in ^{10}Be and ^{10}C can be understood by the Pauli principle: ^{10}B has five n-p pairs while ^{10}Be and ^{10}C have four n-p pairs plus a pair of identical nucleons. Thus, the last pair of ^{10}B can produce states not allowed for the last pair of ^{10}Be and ^{10}C.

It is now useful to observe that for the two-nucleon system, the systems p-p, n-n and the excited state of the deuteron form a triplet of states similar to $T = 1$ above of the ground state of the deuteron with $T = 0$. From this point of view it is directly justified that the p-p and n-n systems do not exist, since the excited state of the deuteron is not bound.

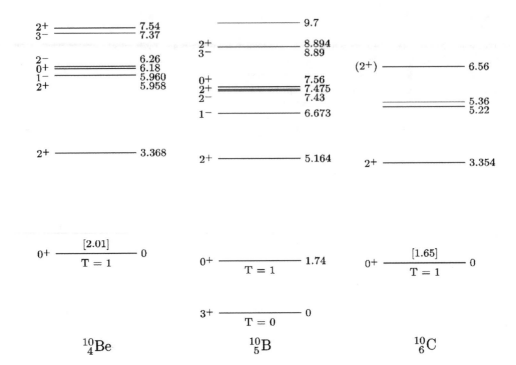

FIGURE 6.7 States in $\mathbf{T} = 1$ of $^{10}_{4}\text{Be}$, $^{10}_{5}\text{B}$ and $^{10}_{6}\text{C}$, above the ground state of $^{10}_{5}\text{B}$ with $\mathbf{T} = 0$. The effect of the Coulomb energy (Eq. 4.1c) is cancelled by adding its value to the binding energy of each isobar. The energy values are given in MeV [7].

A nucleus characterized by Z and N is also characterized by the isospin component $T_z = (N - Z)/2$. The isospin value T of a given state of this nucleus cannot be smaller than $|T_z|$, since T_z is the projection of $\mathbf{T}$. Observing the example of Figure 6.7 we see that the ground states have all the smallest possible value of T for each case, that is, $T = |T_z|$. One verifies that the rule works well for all light nuclei.

6.5.4 Energy levels with residual interaction

Let us take as practical example $^{18}_{8}\text{O}$ for which we use a (residual) interaction potential $V_{1,2}$ of valence nucleons as in Eq. (4.225). For the shell model this nucleus is constituted of a doubly closed core of 8 protons and 8 neutrons plus two valence neutrons in the level $1d_{5/2}$. In agreement with the Table 6.3 the neutron interaction will split the level $j = \frac{5}{2}$ into three states, of $j = 0, 2$ and 4, all with isotopic spin $T = 1$ (two neutrons) and positive parity ($l = 2$). The corresponding levels appear in the experimental spectrum of the $^{18}_{8}\text{O}$ (Figure 6.8).

In Ref. [8] the matrix elements for this nucleus were calculated using a modified surface delta interaction (MSDI),

$$V(1, 2) = A\delta(\mathbf{r}_1 - \mathbf{r}_2)\delta(|\mathbf{r}1| - R_0) + B[\mathbf{t}^{(1)}.\mathbf{t}^{(2)}] + C, \tag{6.81}$$

which is obtained from SDI by adding a constant part and another with the Heisenberg operator (4.8), what substantially increases the agreement with the experience. The parameters A, B and C are given by the use of Eq. (6.81) to determine a great number of binding and excitation energy in a certain region of masses, assuming that these parameters are constant in that region. The values of the matrix elements determined for the three levels

$$\left\langle (d_{5/2})^2 |V_{12}|(d_{5/2})^2 \right\rangle_{J=0,T=1} = -2.78 \text{ MeV}$$

$$\left\langle (d_{5/2})^2 |V_{12}|(d_{5/2})^2 \right\rangle_{J=2,T=1} = 0.43 \text{ MeV}$$

$$\left\langle (d_{5/2})^2 |V_{12}|(d_{5/2})^2 \right\rangle_{J=4,T=1} = 0.99 \text{ MeV} \qquad (6.82)$$

establish the values of the excitation energy of the states 2^+, $E_{2^+} = 0.43 - (-2.78) = 3.21$ MeV and 4^+, $E_{4^+} = 0.99 - (-2.78) = 3.77$ MeV, that appear in part (b) of Figure 6.8. The value of the binding energy of ^{18}O can be obtained through the experimental value of the binding energy of ^{16}O:

$$B(^{18}\text{O}) = B(^{16}\text{O}) + 2S_n(d_{5/2}) - \left\langle (d_{5/2})^2 |V_{12}|(d_{5/2})^2 \right\rangle_{J=0}, \qquad (6.83)$$

where S_n is the separation energy of a neutron of the level $d_{5/2}$ that can, by its turn, be determined by

$$S_n = B(^{17}\text{O}) - B(^{16}\text{O}), \qquad (6.84)$$

from the experimental values $B(^{17}\text{O}) = 131.77$ MeV and $B(^{16}\text{O}) = 127.62$ MeV. With these values the result of (6.83) is $B(^{18}\text{O}) = 138.7$ MeV, in good agreement with the experimental value $B(^{18}\text{O})_{\text{exp}} = 139.8$ MeV.

The above results were obtained supposing the two neutrons permanently in $1d_{5/2}$, that is, the *configuration space* was only limited to that configuration. If we compare figures 6.8(a) and (b) we will see that the excitation energy are poorly determined, not leading to a group of parameters that reproduce the relative distance of the levels 2^+ and 4^+. This is a direct consequence of the limitation of our configuration space and, in fact, there is no reason why we do not admit that the two neutrons can spend part of their time in other configurations close in energy. When we take into account a group of possible configurations for the valence nucleons we are making a *configuration mixing* and, with that resource, the results are substantially better. The part (c) of the Figure 6.8 exhibits the same excited states of ^{18}O but this time using a mixture of the configurations $1d_{5/2}$, $2s_{1/2}$ and $1d_{3/2}$. The agreement with experiment is clearly better.

That approach has, however, a price. As we increase the number of configurations, especially if there are more than two valence nucleons, we have to solve the problem of the diagonalization of matrices of dimensions that can reach the millions. This requires sufficiently fast computers and the progress of the shell model calculations has, in fact, been conditioned to the progress in computer science.

6.6   Direct reactions as probe of the shell model

The angular momentum l transferred in a direct reaction generally modifies the value of the total angular momentum of the nucleus. If J_i is the spin of the target nucleus, the spin J_f of the product nucleus is limited to the values

$$\left| |J_i - l| - \frac{1}{2} \right| \leq J_f \leq J_i + l + \frac{1}{2}, \qquad (6.85)$$

and the initial and final parities obey the relationship

$$\pi_i \pi_f = (-1)^l. \qquad (6.86)$$

The relationship (6.86) allows, with the knowledge of the target nucleus and of the transferred angular momentum, the determination of the parity of the product state formed and Eq. (6.85) is a guide for the determination of its spin.

The knowledge of the transferred angular momentum value in a direct reaction opens the possibility to test the predictions of the shell model for the structure of ground states of nuclei. In a direct reaction one assumes that the nucleon is located in an orbit of the nucleus with the same angular momentum as the transferred momentum in the reaction. Table 6.4 exhibits a series of nuclei in the region $28 < A < 43$, where the deformations are small and the simple shell model works well. The nuclei are formed starting from *pick-up* or *stripping* reactions and the respective value of

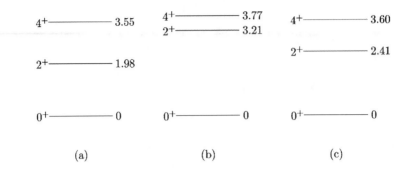

FIGURE 6.8 Ground state and first excited states of $^{18}_{8}$O: (a) experimental values of the energy (MeV); (b) calculated with the configuration $1d_{5/2}$; (c) calculated with a mixture of the configurations $1d_{5/2}$, $2s_{1/2}$ and $1d_{3/2}$.

^{28}Si	+	p	2	^{40}Ca	−	n	3
^{29}P	+	p	0	^{41}Sc	+	p	3
^{31}S	−	n	0	^{42}K	+	n	3
^{35}P	−	p	0	^{42}Ca	+	p	2
^{37}K	+	p	2	^{43}K	−	p	2
^{40}Ar	−	p	2	^{43}Sc	+	p	3

TABLE 6.4 Products formed by stripping (+) or pick-up (−) reactions involving the transfer of a proton (p) or neutron (n). The last column indicates the experimental value of the angular orbital momentum carried by the transferred nucleon.

l is identified by identical calculations to those leading to the curves of Figure 6.2. A comparison with Figure 4.21 exhibits that in all cases the value corresponds exactly to the predicted by the shell model. To illustrate, we shall examine the two examples of the first line. In the first, ^{28}Si is formed from a *stripping* reaction with the transfer of a proton and of two units of angular momentum. If we observe Figure 4.21 we shall see that the 14th proton will locate in the orbit $1d_{\frac{5}{2}}$, which corresponds to $l = 2$. In the second case, ^{40}Ca is formed by a *pick-up* reaction that picks up the 21st neutron, and Figure 4.21 exhibits that it was located in $1f_{\frac{7}{2}}$, justifying the value $l = 3$ found experimentally. The success of a comparison as this not only evidences once again the correctness of the basic assumptions of the shell model as well as gives a guarantee to the simple ideas of the model that describes the direct reactions.

If the shell model were the final theory about the distribution of the energy levels in the nucleus, the single-particle states above Z (or N) should appear as states of energy of the product of a *stripping* reaction that adds a proton (or neutron) to target of mass A. In fact, the spins of the final states could be obtained from the quantum numbers j and l that identify the single particle states. If, for example, a proton is deposited above the closed shell of $Z = 50$ of an even-even nucleus and we identified the values of l associated to final states as being 0, 4 and 5, we could say that the single-particle states $3s\frac{2}{2}$, $1g\frac{7}{2}$ and $1h\frac{11}{2}$ (see Figure 4.21) would appear as $\frac{1}{2}^{+}$, $\frac{7}{2}^{+}$ and $\frac{11}{2}^{-}$ states in the residual nucleus. For $l = 2$ an uncertainty would exist between $2d\frac{5}{2}$ and $2d\frac{3}{2}$, which can give place to states $\frac{5}{2}^{+}$ and $\frac{3}{2}^{+}$. In these cases, the spins of the states should be determined from other processes.

We know, however, that the real situation is more complicated, due to the presence of the residual interactions that give place to configuration mixing. The consequence is that a single particle state $1g\frac{7}{2}$, for example, will not necessarily generate a single final state $\frac{7}{2}^{+}$. It can mix with other configurations of same angular momentum and parity (in this process even states of

E(keV)	l	S'	S	J^π
0	3	7.328	0.916	$\frac{7}{2}^-$
646	1	2.796	0.699	$\frac{3}{2}^-$
1398	2	0.212	0.053	$\frac{3}{2}^+$
1992	1	0.3	0.075	$\frac{3}{2}^-$
2023	(3)	0.08	(0.01; 0.013)	$(\frac{7}{2}^-, \frac{5}{2}^-)$
2515	(3)	0.272	(0.034; 0.045)	$(\frac{7}{2}^-, \frac{5}{2}^-)$
2638	1	1.662	0.831	$\frac{1}{2}^-$
3262	1	0.568	0.142	$\frac{3}{2}^-$
3355	1	0.120	(0.03; 0.06)	$(\frac{3}{2}^-, \frac{1}{2}^-)$
3442	3	0.598	(0.075; 0.1)	$(\frac{7}{2}^-, \frac{5}{2}^-)$
3493	1	0.340	(0.085; 0.17)	$(\frac{3}{2}^-, \frac{1}{2}^-)$

TABLE 6.5   Energy (E), transferred angular momentum l and spectroscopic strength S' for the first levels of ^{37}S populated by the reaction ^{36}S(d,p) ^{37}S. The corresponding spectroscopic factor S and the spin and parities $J(\pi)$ are shown in the last columns [9, 10]. Parenthesis indicate doubtful values.

different nature can be involved, as collective states) to generate a series of states $\frac{7}{2}^+$ of the final nucleus, each one of them spending just part of their time in the configuration $1g\frac{7}{2}$. As result, the cross section for the formation of a state i of the product nucleus is related to the calculated one with DWBA for the formation from a single-particle state by

$$\left(\frac{d\sigma}{d\Omega}\right)_{exp} = \frac{2J_f + 1}{2J_i + 1}\mathcal{S}_{ij}\left(\frac{d\sigma}{d\Omega}\right)_{DWBA}, \tag{6.87}$$

where the *spectroscopic factor* $\mathcal{S}_{ij}$ measures the weight of the configuration j used in the DWBA calculation, in the final state i measured experimentally, with the sum of the contributions limited to

$$\sum_i \mathcal{S}_{ij} = n_j. \tag{6.88}$$

The sum (6.88) embraces all the states i of the product nucleus that contains a given configuration j, with total number of nucleons equal to n_j. The statistical weight $(2J_f+1)/(2J_i+1)$ that appears in the DWBA calculation*, involving the angular momentum of the target nucleus J_i and final nucleus J_f, is explicitly given in (6.87), because the spectroscopists prefer to work with the product

$$\mathcal{S}' = \frac{2J_f + 1}{2J_i + 1}\mathcal{S}, \tag{6.89}$$

denominated *spectroscopic strength*. The advantage of the definition (6.89) is that $\mathcal{S}'$ can be determined by (6.87) even when the final angular momentum J_f is not known. Usually, the values of $\mathcal{S}'$ enter as parameters for a better adjust between the calculated curves and the experimental ones.

To illustrate we shall show the results of the reaction ^{36}S(d,p)^{37}S, where a neutron is located above the closed shell $N = 20$ of $^{36}_{16}$S. Table 6.5 exhibits the energy values of the first levels of ^{37}S formed by the reaction, as well as the value of the transferred angular momentum and of the spectroscopic strength calculated from (6.87). The spectroscopic factor and the spin and parities deduced are also shown.

Let us examine Table 6.5 using the information of Figure 4.21. In agreement with the figure, the neutron occupies the orbits above the closed shell of 20 neutrons and values of the transferred angular momentum $l = 1$ and $l = 3$ are expected. Column 2 confirms, with one exception, this expectation. Let us notice that now it is not possible to determine univocally the spins of the final states only with the values of l, since 2 values of j exist for each l. Thus, we can resort to

---

*In a *pick-up* reaction the statistical weight is equal to 1.

the spectroscopic strength $\mathcal{S}'$. Let us take the ground state, that is a $\frac{7}{2}^-$ or a $\frac{5}{2}^-$, formed by the transfer of a neutron with $l = 3$ to the orbit $1f\frac{7}{2}$ or $1f\frac{5}{2}$. The value of $\mathcal{S}'$ is 7.328 and to this value corresponds the values 0.916 and 1.221 for $\mathcal{S}$. The last one overruns the sum (6.88) by an amount above the experimental uncertainty, what allows to affirm that the ground state of ^{37}S is $\frac{7}{2}^-$. It is an almost pure $1f\frac{7}{2}$ state, because $\mathcal{S}$ almost exhausts the sum (6.88). In the same way, the state with energy 646 keV can be deduced as being a $\frac{3}{2}^-$. These attributions were confirmed by experiments where the degree of polarization of the emerging proton is measured and that are capable to distinguish contributions of different values of j for the same l. Similarly one determined the spins of the states with 1992 keV, 2638 keV and 3262 keV, which could not be obtained only from the value of the spectroscopic factor.

6.7 Nuclear vibrations

We shall now study collective excitation modes, that is, what affects the nucleus as a whole and not just a few nucleons. The model of this section assumes that the nuclear surface can accomplish oscillations around an equilibrium form in the same way as it happens with a liquid drop. The starting point is to imagine that a point on the surface of the nucleus is now defined by its radial coordinate $R(\theta, \phi, t)$, a function of the polar and azimuthal angles and of the time. Let us consider, for simplicity, oscillations around a spherical form that do not alter the volume and the nuclear density; to describe the form of the nucleus at each instant it will be very convenient to use the property that a function of two variables can be expanded in an infinite series of spherical harmonics and write:

$$R(\theta, \phi, t) = R_0 \left[1 + \sum_{\lambda=0}^{\infty} \sum_{\mu=-\lambda\mu}^{\mu=+\lambda} \alpha_{\lambda\mu}(t) Y_{\lambda\mu}(\theta, \phi) \right], \tag{6.90}$$

where the dependence with time is transferred to the coefficients of the expansion.

The application of (6.90) to our problem imposes, immediately, conditions on the possible values of λ. Thus, a first examination of this equation shows that the term $\lambda = 0$ just corresponds to a change in the radius of a spherical form and that is against our hypothesis of volume conservation. This term must, therefore, be removed from the sum. In the same way, the terms with $\lambda = 1$ are not relevant to the description of our model because they just correspond to the center of mass motion of the nucleus. In fact, let X, Y, Z be the coordinates of the center of mass, that we will assume to be fixed at the origin. Thus $(dv = d^3r)$,

$$X = \int x dv = 0 \qquad Y = \int y dv = 0 \qquad Z = \int z dv = 0. \tag{6.91}$$

But

$$X = \int x dv = \int r \sin\theta \cos\phi \, r^2 dr d\Omega =$$
$$= \int \sin\theta \cos\phi \left(\int_0^R r^3 dr \right) d\Omega = \frac{1}{4} \int R^4 \sin\theta \cos\phi d\Omega. \tag{6.92}$$

In a similar way,

$$Y = \frac{1}{4} \int R^4 \sin\theta \sin\phi d\Omega \quad \text{and} \quad Z = \frac{1}{4} \int R^4 \cos\theta d\Omega. \tag{6.93}$$

If we admit now that the oscillations are small compared with the radius, we can write:

$$R = R_0[1 + \epsilon(\theta, \phi)], \qquad R^4 = R_0^4(1 + \epsilon)^4 \cong R_0^4(1 + 4\epsilon). \tag{6.94}$$

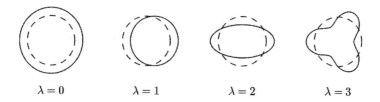

$\lambda = 0 \qquad \lambda = 1 \qquad \lambda = 2 \qquad \lambda = 3$

FIGURE 6.9 The first vibration modes of the nuclear surface, showing the form of the nucleus for each mode (full line) in comparison to the original spherical nucleus.

Thus,

$$X = 0 \quad \rightarrow \int (1 + 4\epsilon)(Y_{11} + Y_{11}^*) \, d\Omega = 0 \, ,$$

$$Y = 0 \quad \rightarrow \int (1 + 4\epsilon)(Y_{11} - Y_{11}^*) \, d\Omega = 0 \, ,$$

$$Z = 0 \quad \rightarrow \int (1 + 4\epsilon)Y_{10} d\Omega = 0. \tag{6.95}$$

Expanding the first equation:

$$\int (1 + 4\epsilon)(Y_{11} + Y_{11}^*) d\Omega = \int \left(1 + 4\sum_{\lambda\mu} \alpha_{\lambda\mu} Y_{\lambda\mu}\right)(Y_{11} + Y_{11}^*) d\Omega$$

$$= \int (Y_{11} + Y_{11}^*) d\Omega + 4\sum_{\lambda\mu} \alpha_{\lambda\mu} \left[\int Y_{\lambda\mu}(-Y_{1,-1}^*) d\Omega + \int Y_{\lambda\mu} Y_{11}^* d\Omega\right]$$

$$= 4(\alpha_{11} - \alpha_{1-1}) = 0. \tag{6.96}$$

Thus, $\alpha_{11} = \alpha_{1-1}$. The condition $Y = 0$ implies $\alpha_{11} = -\alpha_{1-1}$; thus, $\alpha_{11} = \alpha_{1-1} = 0$. The condition $Z = 0$ yields $\alpha_{10} = 0$. Therefore, by fixing the center of mass at the origin one gets $\alpha_{11} = \alpha_{10} = \alpha_{1-1} = 0$, what eliminates the terms of $\lambda = 1$ from our sum.

Therefore, the vibration modes that are of our interest begin with $\lambda = 2$; they are the quadrupole oscillations, where the deformations take the nucleus to a form similar to an ellipsoid. For $\lambda = 3$ we have the octupole oscillations, for $\lambda = 4$ the hexadecapole oscillations, etc. Figure 6.9 shows the main aspects of the vibration modes.

The energy problem of an oscillating liquid drop was solved by Lord Rayleigh in 1877 [11]. He considered an incompressible and irrotational fluid, and arrived to the expression

$$T = \frac{1}{2} \sum_{\lambda\mu} B_\lambda \mid \dot{\alpha}_{\lambda\mu}(t) \mid^2 \tag{6.97}$$

for the kinetic energy, where

$$B_\lambda = \frac{\rho R_0^5}{\lambda}, \tag{6.98}$$

with ρ being the density and R_0 the radius of the droplet. For the potential energy it resulted a value

$$V = \frac{1}{2} \sum_{\lambda\mu} C_\lambda \mid \alpha_{\lambda\mu} \mid^2, \tag{6.99}$$

with

$$C_\lambda = SR_0^2(\lambda - 1)(\lambda + 2) - \frac{3}{2\pi} \frac{Z^2 e^2}{R_0} \frac{\lambda - 1}{2\lambda + 1}, \tag{6.100}$$

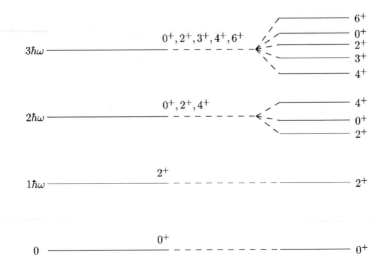

FIGURE 6.10   Vibrational states induced by up to 3 quadrupole phonons.

where one subtracted the Coulomb energy of the protons reduced by the distortion from the first part of Rayleigh's calculations. The factor S is the surface tension and it can be calculated by the mass formula:

$$4\pi R_0^2 S = a_S A^{2/3} \text{ MeV}. \tag{6.101}$$

To adapt the above calculation to our problem it is necessary that we write the Hamiltonian starting from the classical expressions (6.97) and (6.99). If we define the coefficients $\alpha_{\lambda\mu}$ as generalized coordinates, the canonical conjugated momentum of $\alpha_{\lambda\mu}$ is

$$P_{\lambda\mu} = \frac{\partial \mathcal{L}}{\partial \dot{\alpha}_{\lambda\mu}} = B_\lambda \dot{\alpha}_{\lambda\mu}, \tag{6.102}$$

where $\mathcal{L} = T - V$ is the Lagrangian. Thus, the Hamiltonian is given by

$$H = T + V = \sum_{\lambda\mu} \left(\frac{1}{2B_\lambda} P_{\lambda\mu}^2 + \frac{1}{2} C_\lambda \alpha_{\lambda\mu}^2 \right), \tag{6.103}$$

which is a sum of harmonic oscillator Hamiltonians

$$H_{\text{osc}} = \frac{p^2}{2m} + \frac{1}{2} k q^2, \tag{6.104}$$

whose frequency ω is given by $\sqrt{k/m}$. Thus, the vibrations that we are considering can be understood as a sum of harmonic oscillators of frequency

$$\omega_\lambda = \sqrt{\frac{C_\lambda}{B_\lambda}}, \tag{6.105}$$

and our quantization of energy scheme is composed of levels built from these oscillators. When the nucleus passes from a level to a higher or a lower one it is said that it absorbed or emitted a *phonon* of vibration. Thus, quadrupole, octupole, etc., vibration phonons contribute to build the vibration spectrum of the nuclei starting from the ground state. This spectrum is denominated *vibrational band*.

$\mu_{1,2}$															
+2	xx	x	x	x	x										
+1		x				xx	x	x	x						
0			x				x			xx	x	x			
−1				x				x			x		xx	x	
−2					x				x			x		x	xx
μ	+4	+3	+2	+1	0	+2	+1	0	−1	0	−1	−2	−2	−3	−4

TABLE 6.6 The several possible configurations for the states of 2 phonons, with $\lambda = 2$.

It can be shown that a phonon characterized by $\lambda\mu$ has an associated angular momentum of module $\sqrt{\lambda(\lambda+1)}$, with z-component μ and parity $(-1)^{\lambda}$. The vibrational band associated to a given λ in an even-even nucleus must therefore be built up of a series of equidistant levels of energy, with each one of them being possibly associated with more than one state. This situation is illustrated in the diagram of Table 6.6. To the left of the diagram the idealized situation is shown for the excitation by 1, 2 or 3 quadrupole phonons. To the right, the situation in a real nucleus is shown, where the degeneracy is removed by residual interactions. Notice that the spins of each level, resulting from the coupling of the angular momenta of the phonons of that level, do not contain all the integer values from zero to the maximum possible value.

There is a practical way of establishing which spins result from the coupling of a given phonon number. To see, for example, why in the coupling of 2 phonons (with $\lambda = 2$) the spins 1 and 3 are absent, consider the scheme of Table 6.6. In principle, the total spins could be $I = 0, 1, 2, 3$ and 4. But, in this scheme all the possible combinations of the quantum numbers μ_1 and μ_2 of the two phonons are shown, whose sum is μ. Take the spin $I = \lambda = 4$; its possible projections are $\mu = +4, +3, \cdots, -4$. For the spin $I = 3$, the maximum projection would be $+3$. But there is no other value $\mu = 3$ besides the one attributed to the spin 4; thus it is not possible the existence of the spin 3. That is, we only obtain 3 groups of values of μ, which correspond to spins 0, 2 and 4. In the same way $I = 1$ is prohibited. A similar method can be used for the combination of 3 phonons to justify the absence of $I = 1$ and $I = 5$. It should be kept in mind in this picture that the phonons, having integer spins, do not obey the Pauli principle and, therefore, there is no restriction to two phonons having the same quantum numbers.

A more formal justification for the results of the previous paragraph can be obtained directly from the resulting wave function for the coupling of the states that describe the two phonons. If we use the appropriate Clebsch-Gordan coefficients we will see, for example, that the coupling of two phonons of $\lambda = 2$ results in symmetrical functions for the spins 0, 2 and 4 and antisymmetric functions for the spins 1 and 3. The last ones are inadmissible for the description of a composed system of bosons.

The vibrational model predicts, in the ideal case, a ratio equal to 2 between the energy of the second and first excited states. Figure 6.11 shows the experimental values of this ratio for even-even nuclei, as a function of the neutron number N. There is an oscillation around the value 2 for $40 < N < 80$, but in other regions, around 100 and 140, they show a ratio close to 3.3. Those last regions are characteristic of large deformations of the nucleus and the low-energy levels are rotational states.

To conclude, it is worth to mention that odd nuclei also admit vibrational bands. If the nucleus is composed of a spherical core plus an extra nucleon, states can be formed by the coupling of the individual orbit j with the vibrational states of the core. Thus, for example, it can be created in ^{63}Cu, whose ground state is $\frac{3}{2}^-$, a quadruplet of states $\frac{1}{2}^-$, $\frac{3}{2}^-$, $\frac{5}{2}^-$ and $\frac{7}{2}^-$, resulting from the coupling of $j = \frac{3}{2}^-$ with $\lambda = 2$.

6.8   Photonuclear reactions – giant resonances

A photonuclear reaction is a reaction resulting from the interaction of the electromagnetic radiation with a nucleus or, more specifically, with the protons of the nucleus.

The essential difficulty in the study of this reaction type is to obtain a beam of monoenergetic photons with high intensity. Thus, the first experimental works used primarily *bremsstrahlung*

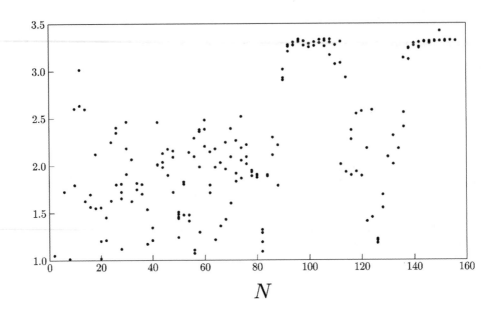

N

FIGURE 6.11   Ratio between the energy of the second and first excited states of even-even nuclei as a function of the neutron number N. The sampling is composed of all the even-even nuclei plus the ones of long lifetime for the isotopes with $Z > 82$.

photons (radiation from decelerated charged particles). This radiation has a continuous distribution of energy, from zero to the energy of the incident particles that produce it. The study of the behavior of a reaction for a given energy E of the photons is obtained by the subtraction of two irradiations: with energy E and $E + \Delta E$.

When the energy of the photons is located a little above the separation energy of a nucleon, the cross section of photo-absorption reveals the presence of characteristic sharp resonances. But, when the incident energy reaches the range of 15-25 MeV, a new behavior appears in the cross section, with the presence of a wide and large peak, called by *giant electric dipole resonance*.

The first indication of the presence of a giant resonance appeared in W. Bothe and W. Gentner's work [12], who used photons of 17.6 MeV from the reaction $^7\mathrm{Li}(\mathrm{p},\gamma)$ in several targets. Baldwin and Kleiber [13] confirmed these observations with photons from a betatron. Figure 6.12 exhibits the excitation function of photoabsorption of $^{120}\mathrm{Sn}$ around the electric dipole giant resonance at 15 MeV.

The giant resonance happens in nuclei along the whole periodic table, with the resonance energy decreasing with A without large oscillations (see Figure 6.13) starting at $A = 20$. This shows that the giant resonance is a property of the nuclear matter and not a characteristic phenomenon of nuclei of a certain type. The widths of the resonances are almost all in the range between 3.5 MeV and 5 MeV. It can reach 7 MeV in a few cases.

The first proposal of explanation of the resonance came from Goldhaber and Teller in 1948 [14]. The photon, through the action of its electric field on the protons, takes the nucleus to an excited state where a group of protons oscillates in opposite phase against a group of neutrons. In such an oscillation, those groups interpenetrate, keeping constant the incompressibility of each group separately. A classic calculation using this hypothesis leads to a vibration frequency that varies with the inverse of the squared root of the nuclear radius, i.e., the resonance energy varies with $A^{-1/6}$.

Later on, H. Steinwedel and J. H. Jensen [15] developed a classic study of the oscillation in another way, already suggested by Goldhaber and Teller, in which the incompressibility is abandoned. The nucleons move inside of a fixed spherical cavity with the proton and neutron densities being a function of the position and time. The nucleons at the surface have fixed position with

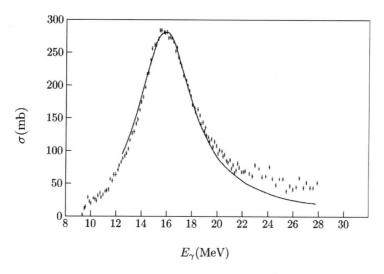

FIGURE 6.12   Giant resonance in the absorption of photons by ^{120}Sn [16].

respect to each other and the density is written in such a way that, at a given instant, the excess of protons on one side of the nucleus coincides with the lack of neutrons on that same side, and vice-versa. Such a model leads to a variation of the resonance energy with $A^{-1/3}$.

The behavior of the energy of the giant electric dipole resonance E_{GDR} does not agree exactly with none of the two models. But, a calculation as developed in reference [18], which assumes a simultaneous contribution of the two models, obtains an expression for E_{GDR} as function of the mass number A,

$$E_{GDR}(\text{MeV}) = 112 \left[A^{2/3} + (A_0 A)^{1/3} \right]^{-1/2}, \tag{6.106}$$

where $A_0 \cong 274$. This expression, with the exception of some very light nuclei, reproduces the behavior of the experimental values very well, as we can see in Figure 6.13. An examination of Eq. (6.106) shows that the Gamow-Teller mode prevails broadly in light nuclei, where the contribution of the Steinwedel-Jensen mode is negligible. The latter one increases with A but it only becomes predominant at the end of the periodic table, at $A = A_0$.

The giant electric dipole resonance arises from an excitation that transmits 1 unit of angular momentum to the nucleus ($\Delta l = 1$). If the nucleus is even-even it is taken to a 1^- state. What one verifies is that the transition also changes the isospin of 1 unit ($\Delta T = 1$) and, due to that, it is also named an *isovector resonance*. For many years this was the only known giant resonance. In the decade of 70, giant isoscalar resonances ($\Delta T = 0$) of electric quadrupole ($\Delta l = 2$) [19] and electric monopole ($\Delta l = 0$) [20] were observed in reactions with charged particles. The first is similar to the vibrational quadrupole state created by the absorption of a phonon of $\lambda = 2$, since both are, in even-even nuclei, states of 2^+ vibration. But the giant quadrupole resonance has a much larger energy. This resonance energy, in the same way that the one for the dipole, decreases smoothly with A, obeying the approximate formula

$$E_{GQR}(\text{MeV}) \cong 62 A^{-1/3}. \tag{6.107}$$

In the state of giant electric quadrupole resonance the nucleus oscillates between the spherical (supposing that this is the form of the ground state) and ellipsoidal form. If protons and neutrons act in phase for that aim, we have an isoscalar resonance ($\Delta T = 0$) and if they are oscillate in opposite phase the resonance is isovector ($\Delta T = 1$). Figure 6.14 illustrates these two possible vibration modes. The existence of isoscalar electric quadrupole resonances are firmly established with a large number of measured cases. The electric quadrupole isovector resonance has been identified in numerous experimental works.

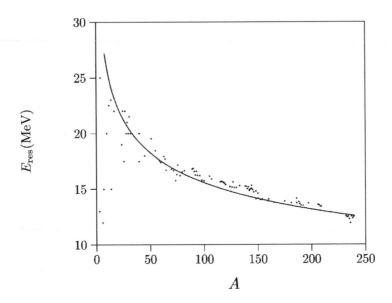

FIGURE 6.13   Location of the energy of the giant electric dipole resonance given by Eq. (6.106) (continuous curve), compared with experimental points [17].

The giant monopole resonance is a very special way of nuclear excitation where the nucleus contracts and expands radially, maintaining its original form but changing its volume. It is also called *breathing mode*. It can also happen in the isoscalar and isovector forms. It is an important way to study the compressibility of nuclear matter. Again here, the isoscalar form has a reasonable number of measured cases, the location of the resonance energy being given by the approximate expression

$$E_{GMR}(\text{MeV}) \cong 80A^{-1/3}. \tag{6.108}$$

In 1996 evidence for the existence of a giant isoscalar electric dipole resonance was presented, using the scattering of α-particles of 200 MeV on ^{208}Pb [21]. Results for the observation of octupole giant resonances were also published [22].

Besides the electric giant resonances, associated to a variation in the form of the nucleus, magnetic giant resonances exist, involving what one calls by *spin vibrations*. In these, nucleons with spin upward move out of phase with nucleons with spin downward. The number of nucleons involved in the process cannot be very large because it is limited by the Pauli principle.

The magnetic resonances can also separate in isoscalar resonances, where protons and neutrons of same spin vibrate against protons and neutrons of opposite spin, and isovector, where protons with spin up and neutrons with spin down vibrate against their corresponding ones with opposite spins. These last cases, originate from reactions of charge exchange. They are also called by *giant Gamow-Teller resonances*. Figure 6.15 exhibits a schematic diagram of the two types of magnetic resonance for dipole vibrations. Magnetic resonances of monopole and quadrupole type were also already observed.

Another important aspect of the study of the giant resonances is the possibility that they can be induced in already excited nuclei. This possibility was analyzed theoretically by D. M. Brink and P. Axel [23] for giant resonances excited "on top" of nuclei rotating with high angular momentum, resulting the suggestion that the frequency and other properties of the giant resonances are not affected by the excitation. A series of experiences in the decade of 80 (see Reference [24]) gave support to this hypothesis.

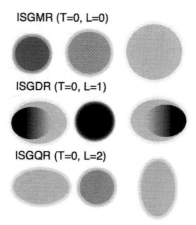

ISGMR (T=0, L=0)

ISGDR (T=0, L=1)

ISGQR (T=0, L=2)

FIGURE 6.14 Macroscopic view of vibration cycles of giant monopole and quadrupole resonances. In an isoscalar resonance, protons and neutrons vibrate in phase. In an isovector resonance, in opposite phase, i.e., when the protons are at the stage (b), the neutrons will be at the stage (d), and vice-versa.

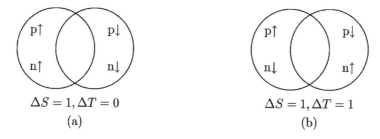

FIGURE 6.15 Magnetic dipole resonances: (a) isoscalar, (b) isovector.

6.9 Coulomb excitation at low and intermediate energies

Coulomb excitation is a process of inelastic scattering in which a charged particle transmits energy to the nucleus through the electromagnetic field. This process can happen at a much lower energy than the necessary for the particle to overcome the Coulomb barrier; the nuclear force is, in this way, excluded in the process.

The simplest treatment that one can give to the problem is a semi-classical calculation, where the incident particle describes a well defined trajectory, which is a classic hyperbolic trajectory of Rutherford scattering. According to Eq. (6.20) the probability of exciting the nucleus to a state f above the ground state i is

$$a_{if} = -\frac{i}{\hbar} \int V_{if} e^{i\omega t} dt, \qquad (6.109)$$

with $\omega = (E_f - E_i)/\hbar$, is the probability amplitude that there will be a transition $i \to f$. The matrix element

$$V_{if} = \int \Psi_f^* V \Psi_i d\tau \qquad (6.110)$$

contains a potential V of interaction of the incident particle with the nucleus. The square of a_{if} measures the transition probability from i to f and this probability should be integrated along the trajectory.

A simple calculation can be done in the case of the excitation of the ground state $J = 0$ of a deformed nucleus to an excited state with $J = 2$ as a result of a frontal collision with scattering angle of $\theta = 180°$. The perturbation V comes, in this case, from the interaction of the charge $Z_p e$ of the projectile with the quadrupole moment of the target nucleus. This quadrupole moment should work as an operator that acts between the initial and final states. The way of adapting (6.239) of Appendix A is writing

$$V = \frac{1}{2} \frac{Z_p e^2 Q_{if}}{r^3},$$

(6.111)

with

$$Q_{if} = \sum_i \int \Psi_f^* (3z_i^2 - r_i^2) \Psi_i d\tau,$$

(6.112)

where the sum extends to all protons. The excitation amplitude is then written as

$$a_{if} = \frac{Z_p e^2 Q_{if}}{2i\hbar} \int \frac{e^{i\omega t}}{r^3} dt.$$

(6.113)

At an scattering of $\theta = 180°$ a relationship exists between the separation r, the velocity v, the initial velocity v_0 and the distance of closest approach a:

$$v = \frac{dr}{dt} = \pm v_0 \left(1 - \frac{a}{r}\right),$$

(6.114)

which is obtained easily from the conservation of energy. Besides, if the excitation energy is small, we can assume that the factor $e^{i\omega t}$ in (6.113) does not vary much during the time that the projectile is close to the nucleus. Thus, this factor can be placed outside the integral and it does not contribute to the cross section. One gets

$$a_{if} = \frac{Z_p e^2 Q_{if}}{2i\hbar v_0} \times 2 \int \frac{dr}{r^3 (1 - a/r)^{1/2}}.$$

(6.115)

The integral is solved easily by the substitution $u = 1 - a/r$, in what results

$$a_{if} = \frac{4 Z_p e^2 Q_{if}}{3i\hbar v_0 a^2} = \frac{4 Q_{if} E^2}{3 Z_p e^2 \hbar v_0 Z_T^2},$$

(6.116)

where the conservation of energy, $E = \frac{1}{2} m_0 v_0^2 = Z_p Z_T e^2 / a$ was used, with m_0 being the reduced mass of the projectile+target system and Z_T the atomic number of the target. The differential cross section is given by the product of the Rutherford differential cross section at $180°$ and the excitation probability along the trajectory, measured by the square of a_{if}:

$$\left. \frac{d\sigma}{d\Omega} \right|_{\theta=180°} = \left. \frac{d\sigma_R}{d\Omega} \right|_{\theta=180°} \times |a_{if}|^2.$$

(6.117)

The Rutherford differential cross section is the classic expression

$$\frac{d\sigma_R}{d\Omega} = \left(\frac{Z_p Z_T e^2}{4E}\right)^2 \sin^{-4}\left(\frac{\theta}{2}\right)$$

(6.118)

and, at $\theta = 180°$, we obtain

$$\left. \frac{d\sigma}{d\Omega} \right|_{\theta=180°} = \frac{m_0 E |Q_{if}|^2}{18 \hbar^2 Z_T^2},$$

(6.119)

an expression that is independent of the charge of the projectile. It is, on the other hand, proportional to the mass of the projectile, indicating that heavy ions are more effective for Coulomb excitation.

The quadrupole moment operator Q_{if} uses, as we saw, the wave functions Ψ_i and Ψ_f of the initial and final states. If those two wave functions are similar, as is the case of an excitation to the first level of a rotational band, the operator Q_{if} can be replaced by the intrinsic quadrupole moment Q. The expression (6.119) translates, in this way, the possibility to evaluate the quadrupole moment from a measurement of a value of the cross section.

A more accurate description of Coulomb excitation for all scattering angles requires the correct treatment of magnetic interactions and the Coulomb recoil of the classical trajectories. Following the derivation of the electromagnetic interaction presented in the Appendix B the excitation a target nucleus from an initial rate $|i\rangle$ to a final state $|f\rangle$ is, to first order, given by

$$a_{fi} = \frac{1}{i\hbar} \int_{-\infty}^{\infty} dt e^{i(E_f - E_i)t/\hbar} \langle f \mid \mathcal{H}_{int} \mid i\rangle, \tag{6.120}$$

where

$$\langle f \mid \mathcal{H}_{int} \mid i\rangle = \int \left[\rho_{fi}\phi(\mathbf{r}, t) - \frac{1}{c}\mathbf{j}_{fi}(\mathbf{r}) \cdot \mathbf{A}(\mathbf{r}, t)\right] d^3r, \tag{6.121}$$

with

$$\rho_{fi}(\mathbf{r}) = \varphi_i(\mathbf{r})\varphi_f^*(\mathbf{r}) \tag{6.122}$$

$$\mathbf{j}_{fi}(\mathbf{r}) = \frac{\hbar}{2im}\left[\varphi_i^*(\mathbf{r})\boldsymbol{\nabla}\varphi_f(\mathbf{r}) - \varphi_f(\mathbf{r})\boldsymbol{\nabla}\varphi_i^*(\mathbf{r})\right]. \tag{6.123}$$

Thus,

$$a_{fi} = \frac{1}{i\hbar} \int_{-\infty}^{\infty} dt e^{i\omega t}\left[\rho_{fi}(\mathbf{r}) - \frac{\mathbf{v}(t)}{c^2} \cdot \mathbf{j}_{fi}(\mathbf{r})\right] \phi(\mathbf{r}, t) d^3r. \tag{6.124}$$

where $\mathbf{v}$ is the velocity of the projectile, and we used $\mathbf{A}(\mathbf{r}, t) = \dfrac{\mathbf{v}(t)}{c}\phi(\mathbf{r}, t)$, valid for a spinless projectile following a classical trajectory.

The Green's function, appropriate for Coulomb potential is $\dfrac{e^{i\kappa|\mathbf{r} - \mathbf{r}'(t)|}}{|\mathbf{r} - \mathbf{r}'(t)|}$ where $\kappa = \omega/c$ [25, 26]. Thus,

$$\phi(\omega, \mathbf{r}) = Z_p e \int_{-\infty}^{\infty} e^{i\omega t}\frac{e^{i\kappa|\mathbf{r} - \mathbf{r}'(t)|}}{|\mathbf{r} - \mathbf{r}'(t)|}dt \tag{6.125}$$

$$\mathbf{A}(\omega, \mathbf{r}) = \frac{Z_p e}{c} \int_{-\infty}^{\infty} \mathbf{v}'(t)e^{i\omega t}\frac{e^{i\kappa|\mathbf{r} - \mathbf{r}'(t)|}}{|\mathbf{r} - \mathbf{r}'(t)|}dt \tag{6.126}$$

are the energy spectrum of the scalar and vector potentials generated by a projectile with charge Z_p following a Coulomb trajectory, described by its time-dependent position $\mathbf{r}(t)$.

We now use the expansion

$$\frac{e^{i\kappa|\mathbf{r} - \mathbf{r}'|}}{|\mathbf{r} - \mathbf{r}'|} = 4\pi i\kappa \sum_{\lambda\mu} j_\lambda(\kappa r_<)Y_{\lambda\mu}^*(\hat{\mathbf{r}}_<)h_\lambda(\kappa r_>)Y_{\lambda\mu}(\hat{\mathbf{r}}_>), \tag{6.127}$$

where j_λ (h_λ) denotes the spherical Bessel (Hankel) functions (of first kind), $\mathbf{r}_>$ ($\mathbf{r}_<$) refers to whichever of $\mathbf{r}$ and $\mathbf{r}'$ has the larger (smaller) magnitude. Assuming that the projectile does not penetrate the target, we use $\mathbf{r}_>$ ($\mathbf{r}_<$) for the projectile (target) coordinates.

The time dependence for a particle moving along the Rutherford trajectory can be directly obtained by solving the equation of angular momentum conservation (see [27]) for a given scattering angle ϑ in the center of mass system (see Figure 1.10). Introducing the parametrization

$$r(\chi) = a_0\left[\epsilon\cosh\chi + 1\right], \tag{6.128}$$

where

$$a_0 = \frac{Z_p Z_T e^2}{m_0 v^2} = \frac{a}{2}, \qquad \text{and} \qquad \epsilon = 1/\sin(\vartheta/2), \tag{6.129}$$

one obtains [27]

$$t = \frac{a_0}{v}\left[\chi + \epsilon\sinh\chi\right]. \tag{6.130}$$

Using the scattering plane perpendicular to the Z-axis, one finds that the corresponding components of $\mathbf{r}$ may be written as

$$x = a_0\left[\cosh\chi + \epsilon\right], \tag{6.131}$$

$$y = a_0\sqrt{\epsilon^2 - 1}\sinh\chi, \tag{6.132}$$

$$z = 0. \tag{6.133}$$

The impact parameter b in Figure 1.10 of Chapter 1 is related to the scattering angle ϑ by

$$b = a_0 \cot(\vartheta/2).$$

In the limit of straight-line motion $\epsilon \simeq b/a_0 \gg 1$, and the equations above reduce to the simple straight-line parametrization,

$$y = vt, \qquad x = b, \quad \text{and} \quad z = 0. \tag{6.134}$$

Using the continuity equation, $\nabla \cdot \mathbf{j}_{fi} = -i\omega\rho_{fi}$, for the nuclear transition current, we can show that the expansion (6.127) can be expressed in terms of spherical tensors (see, e.g., Ref. [28], Vol. II) and Eq. (6.124) becomes

$$a_{fi} = \frac{Z_p e}{i\hbar} \sum_{\lambda\mu} \frac{4\pi}{2\lambda+1} (-1)^\mu \Big[S(E\lambda,\mu)\mathcal{M}_{fi}(E\lambda,-\mu) + S(M\lambda,\mu)\mathcal{M}_{fi}(M\lambda,-\mu) \Big], \tag{6.135}$$

where $\mathcal{M}(\pi\lambda,\mu)$ are the matrix elements for electromagnetic transitions in the target nucleus, defined as

$$\mathcal{M}_{fi}(E\lambda,\mu) = \frac{(2\lambda+1)!!}{\kappa^{\lambda+1}c(\lambda+1)} \int \mathbf{j}_{fi}(\mathbf{r}) \cdot \nabla \times \mathbf{L}\left[j_\lambda(\kappa r) Y_{\lambda\mu}(\hat{\mathbf{r}}) \right] d^3r, \tag{6.136}$$

$$\mathcal{M}_{fi}(M\lambda,\mu) = -i\frac{(2\lambda+1)!!}{\kappa^\lambda c(\lambda+1)} \int \mathbf{j}_{fi}(\mathbf{r}) \cdot \mathbf{L}\left[j_\lambda(\kappa r) Y_{\lambda\mu}(\hat{\mathbf{r}}) \right] d^3r. \tag{6.137}$$

The *orbital integrals* $S(\pi\lambda,\mu)$ are given by

$$S(E\lambda,\mu) = -\frac{i\kappa^{\lambda+1}}{\lambda(2\lambda-1)!!} \int_{-\infty}^{\infty} \frac{\partial}{\partial r'}\left\{ r'(t) h_\lambda\left[\kappa r'(t)\right] \right\} Y_{\lambda\mu}\left[\theta'(t), \phi'(t)\right] e^{i\omega t} dt$$
$$- \frac{\kappa^{\lambda+2}}{c\lambda(2\lambda-1)!!} \int_{-\infty}^{\infty} \mathbf{v}'(t) \cdot \mathbf{r}'(t) h_\lambda[\kappa r'(t)] Y_{\lambda\mu}[\theta'(t), \phi'(t)] e^{i\omega t} dt, \tag{6.138}$$

and

$$S(M\lambda,\mu) = -\frac{i}{m_0 c} \frac{\kappa^{\lambda+1}}{\lambda(2\lambda-1)!!} \mathbf{L}_0 \cdot \int_{-\infty}^{\infty} \nabla'\left\{ h_\lambda[\kappa r'(t)] Y_{\lambda\mu}[\theta'(t), \phi'(t)] \right\} e^{i\omega t} dt, \tag{6.139}$$

where $\mathbf{L}_0$ is the angular momentum of relative motion, which is constant:

$$L_0 = a_0 m_0 v \cot\frac{\vartheta}{2}. \tag{6.140}$$

In the *long-wavelength approximation*

$$\kappa r' = \frac{\omega r'}{c} < \frac{E_x R}{\hbar c} \ll 1, \tag{6.141}$$

where E_x is the excitation energy and R is the nuclear radius. Then one uses the limiting form of h_λ for small values of its argument [32] to show that

$$S(E\lambda,\mu) \simeq \int_{-\infty}^{\infty} r'^{-\lambda-1}(t) Y_{\lambda\mu}\left\{ \theta'(t), \phi'(t) \right\} e^{i\omega t} dt, \tag{6.142}$$

and

$$S(M\lambda,\mu) \simeq -\frac{1}{\lambda m_0 c} \mathbf{L}_0 \cdot \int_{-\infty}^{\infty} \nabla'\left\{ r'^{-\lambda-1}(t) Y_{\lambda\mu}\left[\theta'(t), \phi'(t)\right] \right\} e^{i\omega t} dt, \tag{6.143}$$

which are the usual orbital integrals in the non-relativistic Coulomb excitation theory with hyperbolic trajectories.

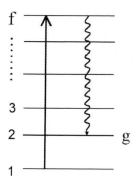

FIGURE 6.16 Schematic of a first-order Coulomb excitation of a nucleus from an initial state $|i\rangle$ to a final state $|f\rangle$ and its subsequent decay to a state $|g\rangle$.

Performing a translation of the integrand by $\chi \to \chi + i\,(\pi/2)$, the orbital integrals $S(\pi\lambda, \mu)$ become

$$S(E\lambda, \mu) = \frac{\mathcal{C}_{\lambda\mu}}{v a_0^\lambda} I(E\lambda, \mu),$$

$$S(M\lambda, \mu) = -\frac{\mathcal{C}_{\lambda+1,\mu}}{\lambda c a_0^\lambda}[(2\lambda + 1)/(2\lambda + 3)]^{1/2}[(\lambda + 1)^2 - \mu^2]^{1/2} \cot(\vartheta/2)\, I(M\lambda, \mu), \qquad (6.144)$$

with

$$\mathcal{C}_{\lambda\mu} = \begin{cases} \sqrt{\dfrac{2\lambda + 1}{4\pi}}\, \dfrac{\sqrt{(\lambda - \mu)!(\lambda + \mu)!}}{(\lambda - \mu)!!(\lambda + \mu)!!}(-1)^{(\lambda+\mu)/2} & , \quad \text{for} \quad \lambda + \mu = \text{even} \\ 0 & , \quad \text{for} \quad \lambda + \mu = \text{odd} \end{cases}, \qquad (6.145)$$

where, in the long-wavelength approximation, Eq. (6.141) [26],

$$I(E\lambda, \mu) = e^{-\pi\eta/2} \int_{-\infty}^{\infty} d\chi \exp\left[-\eta\epsilon \cosh\chi + i\eta\chi\right] \frac{(\epsilon + i \sinh\chi - \sqrt{\epsilon^2 - 1}\cosh\chi)^\mu}{(i\epsilon \sinh\chi + 1)^{\lambda+\mu}},$$

$$I(M\lambda, \mu) = I(E\lambda + 1, \mu), \qquad (6.146)$$

where

$$\eta = \omega a_0/v.$$

The square modulus of Eq. (6.135) gives the probability of exciting the target from the initial state $|\,I_i M_i\rangle$ to the final state $|\,I_f M_f\rangle$ in a collision with the center of mass scattering angle ϑ. If the orientation of the initial state is not specified, the cross section for exciting the nuclear state of spin I_f is

$$d\sigma_{i \to f} = \frac{a_0^2 \epsilon^4}{4} \frac{1}{2I_i + 1} \sum_{M_i, M_f} |\,a_{fi}\,|^2\, d\Omega. \qquad (6.147)$$

where $a_0^2 \epsilon^4 d\Omega/4$ is the elastic (Rutherford) cross section. Using the Wigner-Eckart theorem

$$\mathcal{M}(\pi\lambda, -\mu) = (-1)^{I_f - M_f} \begin{pmatrix} I_f & \lambda & I_i \\ -M_f & \mu & M_i \end{pmatrix} \langle I_f \,\|\mathcal{M}(\pi\lambda)\|\, I_i\rangle, \qquad (6.148)$$

and the orthogonality properties of the Clebsch-Gordan coefficients, one gets (for more details, see [29])

$$\frac{d\sigma_{i \to f}}{d\Omega} = \frac{4\pi^2 Z_P^2 e^2}{\hbar^2} a_0^2 \epsilon^4 \sum_{\pi\lambda\mu} \frac{B(\pi\lambda, I_i \to I_f)}{(2\lambda + 1)^3}\,|\, S(\pi\lambda, \mu)\,|^2, \qquad (6.149)$$

where $\pi = E$ or M stands for the electric or magnetic multipolarity, and

$$B(\pi\lambda, I_i \longrightarrow I_f) = \frac{1}{2I_i + 1} \sum_{M_i, M_f} \mid \mathcal{M}(\pi\lambda, \mu) \mid^2 = \frac{1}{2I_i + 1} |\langle I_f \|\mathcal{M}(\pi\lambda)\| I_i\rangle|^2 \qquad (6.150)$$

is the *reduced transition probability*.

Integration of (6.149) over all energy transfers $\varepsilon = \hbar\omega$, and summation over all possible final states of the projectile nucleus leads to

$$\frac{d\sigma_C}{d\Omega} = \sum_f \int \frac{d\sigma_{i\to f}}{d\Omega} \rho_f(\varepsilon) d\epsilon, \qquad (6.151)$$

where $\rho_f(\varepsilon)$ is the density of final states of the target with energy $E_f = E_i + \varepsilon$. Inserting Eq. (6.149) into Eq. (6.151) one finds

$$\frac{d\sigma_C}{d\Omega} = \sum_{\pi\lambda} \frac{d\sigma_{\pi\lambda}}{d\Omega} = \sum_{\pi\lambda} \int \frac{d\varepsilon}{\varepsilon} \frac{dn_{\pi\lambda}}{d\Omega}(\varepsilon)\sigma_\gamma^{\pi\lambda}(\varepsilon), \qquad (6.152)$$

where $\sigma_\gamma^{\pi\lambda}$ are the *photonuclear cross sections* for a given multipolarity $\pi\lambda$, given by

$$\sigma_\gamma^{\pi\lambda}(\varepsilon) = \frac{(2\pi)^3(\lambda + 1)}{\lambda \left[(2\lambda + 1)!!\right]^2} \sum_f \rho_f(\varepsilon)\kappa^{2\lambda-1} B(\pi\lambda, I_i \to I_f). \qquad (6.153)$$

The *virtual photon numbers*, $n_{\pi\lambda}(\varepsilon)$, are given by [30]

$$\frac{dn_{\pi\lambda}}{d\Omega} = \frac{Z_p^2\alpha}{2\pi} \frac{\lambda \left[(2\lambda + 1)!!\right]^2}{(\lambda + 1)(2\lambda + 1)^3} \frac{c^2 a_0^2 \epsilon^4}{\kappa^{2(\lambda-1)}} \sum_\mu \mid S(\pi\lambda, \mu) \mid^2 . \qquad (6.154)$$

where $\alpha = 1/137$.

In terms of the orbital integrals $I(E\lambda, \mu)$, given by Eq. (6.146), and using the Eq. (6.154), we find for the electric multipolarities

$$\frac{dn_{E\lambda}}{d\Omega} = \frac{Z_p^2\alpha}{8\pi^2}(\frac{c}{v})^{2\lambda} \frac{\lambda \left[(2\lambda + 1)!!\right]^2}{(\lambda + 1)(2\lambda + 1)^2} \epsilon^4 \eta^{-2\lambda+2} \sum_{\substack{\mu \\ \lambda+\mu=even}} \frac{(\lambda - \mu)!(\lambda + \mu)!}{[(\lambda - \mu)!!(\lambda + \mu)!!]^2} \mid I(E\lambda, \mu) \mid^2 . \qquad (6.155)$$

In the case of magnetic excitations one obtains

$$\frac{dn_{M\lambda}}{d\Omega} = \frac{Z_p^2\alpha}{8\pi^2}(\frac{c}{v})^{2(\lambda-1)} \frac{\left[(2\lambda + 1)!!\right]^2}{\lambda(\lambda + 1)(2\lambda + 1)^2} \eta^{-2\lambda+2} \epsilon^4(\epsilon^2 - 1)$$

$$\times \sum_{\substack{\mu \\ \lambda+\mu=odd}} \frac{\left[(\lambda + 1)^2 - \mu^2\right](\lambda + 1 - \mu)!(\lambda + 1 + \mu)!}{[(\lambda + 1 - \mu)!!(\lambda + 1 + \mu)!!]^2} \mid I(M\lambda, \mu) \mid^2 . \qquad (6.156)$$

Only for the $E1$ multipolarity the orbital integrals, Eqs. (6.146), can be performed analytically and one gets the closed expression

$$\frac{dn_{E1}}{d\Omega} = \frac{Z_p^2\alpha}{4\pi^2}\left(\frac{c}{v}\right)^2 \epsilon^4 \eta^2 e^{-\pi\eta} \left\{\frac{\epsilon^2 - 1}{\epsilon^2} [K_{i\eta}(\epsilon\eta)]^2 + \left[K'_{i\eta}(\epsilon\eta)\right]^2\right\}, \qquad (6.157)$$

where $K_{i\eta}$ is the modified Bessel function with imaginary index, $K'_{i\eta}$ is the derivative with respect to its argument.

Since the impact parameter is related to the scattering angle by $b = a_0 \cot\vartheta/2$, we can also write

$$n_{\pi\lambda}(\varepsilon, b) \equiv \frac{dn_{\pi\lambda}}{2\pi bdb} = \frac{4}{a_0^2\epsilon^4} \frac{dn_{\pi\lambda}}{d\Omega}, \qquad (6.158)$$

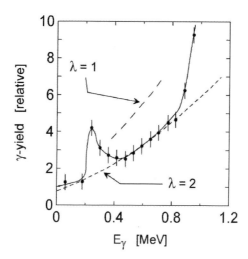

FIGURE 6.17 Coulomb excitation of sodium by protons. The yield of the 446 keV γ-rays is shown [36]. The dashed curves correspond to the cross sections expected for $\lambda = 1$ and 2 on the basis of the observed cross section for the excitation with α-particles.

which are interpreted as the number of equivalent photons of energy $\varepsilon = \hbar\omega$, incident on the target per unit area, in a collision with impact parameter b [31].

The total cross section for Coulomb excitation is obtained by integrating Eq. (6.152) over b from a minimum impact parameter b_{min}, or equivalently, integrating it over ϑ up to a maximum scattering angle ϑ_{max}. This condition is necessary for collisions at high energies to avoid the situation in which the nuclear interaction with the target becomes important. At very low energies, below the Coulomb barrier, $b_{min} = 0$ and $\vartheta_{max} = 180^0$.

We stress the usefulness of the concept of virtual photon numbers, specially in high-energy collisions. In such collisions the momentum and the energy transfer due to the Coulomb interaction are related by $\Delta p = \Delta E/v \simeq \Delta E/c$. This means that the virtual photons are almost real. One usually explores this fact to extract information about real photon processes from the reactions induced by relativistic charges, and vice-versa. This is the basis of the *Weizsäcker-Williams method* [33, 34], used to calculate cross sections for Coulomb excitation, particle production, Bremsstrahlung, etc (see, e.g., Ref. [35, 31]). In the case of Coulomb excitation, even at low energies, although the equivalent photon numbers should not be interpreted as (almost) real ones, the cross sections can still be written as a product of them and the cross sections induced by real photons, as we have shown above. The reason for this is the assumption that Coulomb excitation is a process which involves only collisions for which the nuclear matter distributions do not overlap at any point of the classical trajectory. The excitation of the target nucleus thus occurs in a region where the divergence of the magnetic field is zero, i.e. $\nabla \cdot \mathbf{B}_p(t) = 0$, where $\mathbf{B}_p(t)$ is the magnetic field generated by the projectile at the target's position. This condition implies that the electromagnetic fields involved in Coulomb excitation are exactly the same as those involved in the absorption of a real photon [28].

The results presented above are only valid if the excitation is of first-order. For higher-order excitations one has to use the coupled-channels equations (6.17) with the excitation amplitudes in each time interval given by Eq. (6.135).

The semiclassical Coulomb excitation method described in this section is valid for sub-barrier collisions. As one increases the bombarding energy one can still use the semiclassical approach, but only for scattering angles where no influence of the nuclear interaction occurs. The correct quantum calculation for Coulomb excitation in first-order perturbation theory uses the DWBA Eq. (3.79) with U_{int} being the Coulomb interaction potential of Eqs. (6.125) and (6.126) and the center of mass scattering waves $\chi^{(\pm)}$ calculated with the sum of the Coulomb potential between two charges

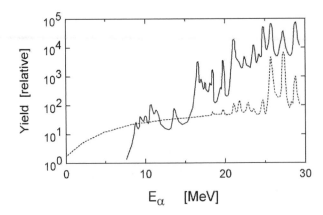

FIGURE 6.18 γ-rays from Coulomb excitation and compound nucleus formation in ^{19}F bombarded with α-particles. The dashed curve shows the yields of the 114 keV γ-ray from the first excited state in ^{19}F and the solid curve shows the 1.28 MeV γ-ray from the first excited state of ^{22}Ne formed by an (α, p') process on ^{19}F [37].

and a nuclear optical potential. The scattering wave functions $\chi^{(\pm)}$ enter in the calculation of $\phi(\mathbf{r})$ and $\mathbf{A}(\mathbf{r})$.

Coulomb excitation is a useful method to obtain static quadrupole moments as well as the reduced probabilities for several nuclear transitions. In order to identify the multipolarity of the excitation it is often necessary to study the de-excitation of the excited state by measuring a γ-ray from its decay (see Figure 6.18). The angular distribution of the gamma rays emitted into solid angle Ω_γ, as a function of the scattering angle of the projectile ϑ, is given by

$$W(\theta_\gamma) = 1 + \sum_{k=2,4,\cdots} b_k^{\pi\lambda}(\vartheta) P_k(\cos\theta_\gamma), \tag{6.159}$$

where $P_k(\cos\theta_\gamma)$ are the Legendre polynomials. The coefficients $b_k^{\pi\lambda}(\vartheta)$ are related to the Coulomb excitation amplitudes (6.135) and to the B-values (6.150) for the transition from the excited state f to a final state g.

Figure 6.17 shows the Coulomb excitation of sodium by protons. The yield of the 446 keV γ-rays is shown [36]. Between the resonances due to compound nucleus formation one observes a smoothly rising background yield which may be ascribed to Coulomb excitation. It is possible to determine the multipole order of the Coulomb excitation by comparing with the yield observed in the Coulomb excitation with α-particles [36]. The dashed curves correspond to the cross sections expected for $\lambda = 1$ and 2 on the basis of the observed cross section for the excitation with α-particles. The close agreement of the measured cross section with the theoretical curve for $E2$ excitation also confirms that the yield away from resonances is primarily due to Coulomb excitation.

Figure 6.18 shows the γ-ray yield from Coulomb excitation and compound nucleus formation in ^{19}F bombarded with α-particles. The dashed curve shows the yields of the 114 keV γ-ray from the first excited state in ^{19}F and the solid curve shows the 1.28 MeV γ-ray from the first excited state of ^{22}Ne formed by an (α, p) process on ^{19}F [37]. For bombarding energies below 1.2 MeV, the penetration of the α-particle through the Coulomb barrier is very small and the cross section for compound nucleus formation is small compared to that for Coulomb excitation. With increasing energy, σ_{CN} increases rapidly and soon becomes larger than σ_C. However, even for $E_\alpha \sim 2$ MeV, at which energy the average value of σ_{CN} is an order of magnitude larger than σ_C, the yield of the 114 keV γ-ray is only very little affected by the compound nucleus formation, since the probability that the compound nucleus decays by inelastic α-emission is small. Finally, for $E_\alpha \geq 2$ MeV, the Coulomb excitation yield of the 114 keV γ-ray is overshadowed by the resonance yield from compound nucleus formation.

As the bombarding energy increases Coulomb excitation predominantly favors the excitation of high lying states, e.g. giant resonances. This feature will be discussed in more details in Chapter 8 when we discuss high-energy collisions.

6.10 Electromagnetic transition probabilities for nuclear vibrations

6.10.1 Electromagnetic transition probabilities

In Section 6.7 we saw that the Hamiltonian for small amplitude oscillations of the nuclear surface is equal to

$$H = \frac{1}{2} \sum_{\lambda\mu} \left[B_\lambda |\dot{\alpha}_{\lambda\mu}|^2 + C_\lambda |\alpha_{\lambda\mu}|^2 \right]. \tag{6.160}$$

The classical solution for this Hamiltonian is given by

$$\omega_\lambda = \left(\frac{C_\lambda}{B_\lambda} \right)^{1/2}, \qquad \alpha_{\lambda\mu} = \epsilon_{\lambda\mu} \cos(\omega_\lambda t), \qquad E = \sum_{\lambda\mu} \frac{1}{2} |\epsilon_{\lambda\mu}|^2 \omega_\lambda^2 B_\lambda. \tag{6.161}$$

We now quantize the Hamiltonian by using the quantization rules

$$\pi_{\lambda\mu} = \frac{\partial H}{\partial \dot{\alpha}_{\lambda\mu}} = B_\lambda \dot{\alpha}_{\lambda\mu}^*, \qquad [\alpha_{\lambda\mu}, \pi_{\lambda\mu}] = i\hbar, \tag{6.162}$$

which implies that

$$\pi_{\lambda\mu} = -i\hbar \frac{\partial}{\partial \alpha_{\lambda\mu}}. \tag{6.163}$$

If we now introduce the boson operators

$$O_{\lambda\mu}^+ = \left(\frac{\omega_\lambda B_\lambda}{2\hbar} \right)^{1/2} \left[\alpha_{\lambda\mu} - \frac{i}{\omega_\lambda B_\lambda} (-1)^\mu \pi_{\lambda,-\mu} \right]$$

$$O_{\lambda\mu} = \left(\frac{\omega_\lambda B_\lambda}{2\hbar} \right)^{1/2} \left[(-1)^\mu \alpha_{\lambda,-\mu} + \frac{i}{\omega_\lambda B_\lambda} \pi_{\lambda\mu} \right] \tag{6.164}$$

we get the commutation relation

$$\left[O_{\lambda\mu}, O_{\lambda\mu}^+ \right] = 1. \tag{6.165}$$

The Hamiltonian (6.160) can be rewritten as

$$H = \sum_{\lambda\mu} \hbar\omega_\lambda \left(O_{\lambda\mu}^+ O_{\lambda\mu} + \frac{1}{2} \right). \tag{6.166}$$

The operator $O_{\lambda\mu}^+$ obeys the equations of motion

$$\left[H, O_{\lambda\mu}^+ \right] = \hbar\omega_\lambda O_{\lambda\mu}^+. \tag{6.167}$$

The ground-state is defined as

$$O_{\lambda\mu}\phi_0(\alpha) = 0; \qquad \text{all } \lambda\mu, \tag{6.168}$$

and the number operator as $N_{\lambda\mu} = O_{\lambda\mu}^+ O_{\lambda\mu}$. Using these definitions and Eq. 6.165, the excited states can be found by means of standard techniques for the harmonic oscillator model [38].

In the long-wavelength approximation the electric excitation matrix elements, Eq. (6.136) become [26]

$$\mathcal{M}(E\lambda\mu) = \int r^\lambda Y_{\lambda\mu}^*(\theta) \delta\rho_\alpha^c(\mathbf{r}) d^3r, \tag{6.169}$$

where $\delta\rho_\alpha^C = \varphi_i^C(\mathbf{r})\varphi_f^{C*}(\mathbf{r})$ is the transition density for the nuclear charge. For a collective surface oscillation, as explained in Section 6.7, $\delta\rho_\alpha^C$ will be peaked at the surface. The equidensity surfaces are given by Eq. (6.90), i.e.,

$$r_\theta = r\left[1 + \sum_{\lambda\mu}\alpha_{\lambda\mu}^* Y_{\lambda\mu}(\theta)\right] \tag{6.170}$$

for constant r. In other words,

$$\rho_\alpha^C(r_\theta, \theta) = \rho_0^C(r), \tag{6.171}$$

where $\rho_0^C(r)$ is the non-deformed density, or ground state density.

Conversely,

$$\rho_\alpha^C(r, \theta) = \rho_0^C\left(\frac{r}{1 + \sum_{\lambda\mu}\alpha_{\lambda\mu}^* Y_{\lambda\mu}(\theta)}\right) \cong \rho_0^C(r) - r\frac{d\rho_0^C}{dr}(r)\sum_{\lambda\mu}\alpha_{\lambda\mu}^* Y_{\lambda\mu}(\theta) + \mathcal{O}(\alpha^2). \tag{6.172}$$

Thus, for small oscillations, the transition density in Eq. (6.169) is given by

$$\delta\rho_\alpha^c(\mathbf{r}) = \rho_\alpha^C(r, \theta) - \rho_0^C(r) = -r\frac{d\rho_0^C}{dr}(r)\sum_{\lambda\mu}\alpha_{\lambda\mu}^* Y_{\lambda\mu}(\theta). \tag{6.173}$$

The charge density maybe related to the matter density by means of

$$\rho_0^C(r) = \frac{Ze}{A}\rho_0(r). \tag{6.174}$$

Inserting (6.173) and (6.174) into (6.169) one obtains

$$\mathcal{M}(E\lambda\mu) = -\frac{Ze}{A}\int r^\lambda Y_{\lambda\mu}^*(\theta)\left\{r\frac{d\rho_0(r)}{dr}\sum_{\ell m}\alpha_{\ell m}^* Y_{\ell m}(\theta)\right\}d^3r = -\frac{Ze}{A}\alpha_{\lambda\mu}^*\int r^{\lambda+3}\frac{d\rho_0(r)}{dr}dr$$

$$= Ze\alpha_{\lambda\mu}^*\frac{(\lambda+3)}{4\pi}\left\{\frac{4\pi}{A}\int r^{\lambda+2}\rho_0(r)dr\right\}. \tag{6.175}$$

That is,

$$\mathcal{M}(E\lambda, \mu) = \frac{\lambda+3}{4\pi}Ze\langle r^\lambda\rangle_0\alpha_{\lambda\mu}^*. \tag{6.176}$$

Using the definitions of the creation and annihilation operators given in Eq. (6.164) we can rewrite Eq. (6.176) as

$$\mathcal{M}(E\lambda\mu) = \frac{\lambda+3}{4\pi}Ze\langle r^\lambda\rangle_0\left(\frac{\hbar}{2\omega_\lambda B_\lambda}\right)^{1/2}\left[O_{\lambda\mu} + (-1)^\mu O_{\lambda,-\mu}^+\right]. \tag{6.177}$$

In going from (6.176) to (6.177) we have quantized the matrix element $\mathcal{M}(E\lambda\mu) = r^\lambda Y_{\lambda\mu}^*(\theta)$ in form of an operator. It now connects states through collective vibrations of $\lambda\mu$ multipolarity.

The electric multipole operator (6.177) links states which differ by a unit of phonons. That is,

$$\Delta n_{\lambda\mu} = \pm 1. \tag{6.178}$$

For transitions from the ground state, $|0\rangle \to |\lambda\mu\rangle$, we obtain using (6.177),

$$B(E\lambda, 0 \to \lambda) = (2\lambda+1)\left|\langle 0||\mathcal{M}(E\lambda)||\lambda\rangle\right|^2 = (2\lambda+1)\left[\frac{\lambda+3}{4\pi}Ze\langle r^\lambda\rangle_0\right]^2\frac{\hbar}{2\omega_\lambda B_\lambda}. \tag{6.179}$$

For a constant matter distribution with radius R_0

$$\langle r^\lambda\rangle_0 = \frac{3R_0^\lambda}{\lambda+3}, \tag{6.180}$$

and

$$B(E\lambda, 0 \to \lambda) = (2\lambda+1)\left[\frac{3}{4\pi}ZeR_0^\lambda\right]^2\frac{\hbar}{2\omega_\lambda B_\lambda}. \tag{6.181}$$

We now define the *deformation length* as

$$\delta_\lambda = \left[(2\lambda+1)\frac{\hbar}{2\omega_\lambda B_\lambda}\right]^{1/2} R_0. \tag{6.182}$$

It measures the amplitude of the surface oscillations for a given mode λ. In terms of δ_λ

$$B(E\lambda, 0 \to \lambda) = \left[\frac{3}{4\pi}ZeR_0^{\lambda-1}\right]^2 \delta_\lambda^2. \tag{6.183}$$

In this macroscopic model, the magnetic transitions are predicted to be zero since the deformation of a uniformly charged fluid induces no magnetic moment. In fact, $M1$ transitions are inhibited by factors of order $1/100$ for good vibrational nuclei.

In lowest order in $\alpha_{\lambda\mu}$ the transition density conserves particle number. This can be seen from Eq. (6.172). An integration over volume implies that

$$\int \rho_\alpha^c(r,\theta)d^3r = \int \rho_0^C(r)d^3r. \tag{6.184}$$

This relationship fails if $\lambda = 0$. The $\lambda = 0$ modes are interesting and correspond to monopole excitations often called by "breathing" modes. To conserve particle number we have to add a correction to (6.172) for the $\lambda = 0$ case. It is easy to check that a transition density given by

$$\delta\rho_{\alpha(\lambda=0)}^C(\mathbf{r}) = -\alpha_0^* \left[3\rho_0^C(r) + r\frac{d\rho_0^C}{dr}\right]Y_{00}(\hat{\mathbf{r}}) \tag{6.185}$$

conserves particle number, since

$$\int \delta\rho_{\alpha(\lambda=0)}^C(r)d^3r = 0.$$

For electromagnetic monopole transitions,

$$\mathcal{M}(E0) = \frac{Ze}{A}\int r^2 \delta\rho_{\alpha(\lambda=0)}(\mathbf{r})Y_{00}(\hat{\mathbf{r}})d^3r \tag{6.186}$$

is the matrix element of interest.

Repeating the same procedure as before, we get in terms of α_0,

$$\mathcal{M}(E0) = \frac{\alpha_0^*}{2\pi}Ze\langle r^2\rangle,$$

and

$$B(E0) = \left[\frac{Ze\langle r^2\rangle}{2\pi}\right]^2 \frac{\hbar}{2\omega_0 B_0},$$

or, else,

$$B(E0) = \left[\frac{3ZeR_0^2}{10\pi}\right]^2 \alpha_0^2. \tag{6.187}$$

6.10.2 Sum rules

Sum rules are very useful to calculate the cross sections when the reduced matrix elements $B(E\lambda)$ are not known. For a system governed by a Hamiltonian H, we can define the sum

$$S(F) = \sum_a (E_a - E_0)\left|\langle a|F|0\rangle\right|^2 \tag{6.188}$$

for the excitation $|0\rangle \to |a\rangle$ induced by the operator F. If F is Hermitian

$$S(F) = \sum_a \langle a|F|0\rangle (E_a - E_0)\langle 0|F|a\rangle = \frac{1}{2}\sum_a \langle 0|F|a\rangle \{\langle a|[H,F]|0\rangle - \langle 0|[H,F]|a\rangle\}\langle a|F|0\rangle$$

$$= \frac{1}{2}\langle 0|[F,[H,F]]|0\rangle. \tag{6.189}$$

For a microscopic operator (interaction) acting on each nucleon k, we can write

$$F = \sum_k F(\mathbf{r}_k), \tag{6.190}$$

and

$$H = \sum_k \left[-\frac{\hbar^2 \nabla_k^2}{2m_k} + V(\mathbf{r}_k) \right]. \tag{6.191}$$

Then, Eq. (6.189) becomes

$$S(F) = \frac{1}{2} \left\langle 0 \left| \sum_k \frac{\hbar^2}{2m_k} \left[F(\mathbf{r}_k), \left[\nabla_k^2, F(\mathbf{r}_k) \right] \right] \right| 0 \right\rangle. \tag{6.192}$$

Since the expectation value of HF^2 and F^2H are the same, $\langle 0|F^2\nabla^2|0\rangle = \langle 0|(\nabla^2 F)F|0\rangle$, and

$$\left\langle 0 \left| \left[F, \left[\nabla^2, F \right] \right] \right| 0 \right\rangle = \left\langle 0 \left| F\nabla^2 F - 2(\nabla^2 F)F + F\nabla^2 F \right| 0 \right\rangle = \left\langle 0 \left| 2F\nabla^2 F - 2(\nabla^2 F)F \right| 0 \right\rangle.$$

But

$$\boldsymbol{\nabla} \cdot (F\boldsymbol{\nabla} F) = (\nabla F)^2 + F\nabla^2 F,$$

and

$$\left\langle 0 \left| \left[F, \left[\nabla^2, F \right] \right] \right| 0 \right\rangle = 2 \left\langle 0 \left| (\boldsymbol{\nabla} F)^2 \right| 0 \right\rangle. \tag{6.193}$$

Thus,

$$S(F) = \left\langle 0 \left| \sum_k \frac{\hbar^2}{2m_k} \left[\boldsymbol{\nabla}_k F(\mathbf{r}_k) \right]^2 \right| 0 \right\rangle. \tag{6.194}$$

For nuclear and Coulomb excitation one often encounters operators of the form

$$F \equiv F_{\lambda\mu} = f(r) Y_{\lambda\mu}(\hat{\mathbf{r}}). \tag{6.195}$$

Then, one can show that (see Ref. [39], p. 400-401)

$$\sum_\mu \boldsymbol{\nabla} \left[f(r) Y_{\lambda\mu}^*(\hat{\mathbf{r}}) \right] \cdot \boldsymbol{\nabla} \left[f(r) Y_{\lambda\mu}(\hat{\mathbf{r}}) \right] = \frac{2\lambda+1}{4\pi} \left[\left(\frac{df}{dr} \right)^2 + \lambda(\lambda+1) \left(\frac{f}{r} \right)^2 \right].$$

Inserting these results in Eq. (6.188), we find

$$S(F_\lambda) \equiv \sum_{\alpha\mu} (E_\alpha - E_0) |\langle \alpha | F_{\lambda\mu} | 0 \rangle|^2 = \left\langle 0 \left| \frac{\hbar^2}{2m} \sum_{\mu,k} |\boldsymbol{\nabla}_k F_{\lambda\mu}(\mathbf{r}_k)|^2 \right| 0 \right\rangle$$

$$= \frac{2\lambda+1}{4\pi} \frac{\hbar^2}{2m} A \left\langle 0 \left| \left(\frac{df}{dr} \right)^2 + \lambda(\lambda+1) \left(\frac{f}{r} \right)^2 \right| 0 \right\rangle. \tag{6.196}$$

The electric multipole matrix elements are given by

$$\mathcal{M}(E\lambda;\mu) = e \sum_{k=\text{protons}} \left[r^\lambda Y_{\lambda\mu} \right]_k \equiv e \sum_k \left[\left(\frac{1}{2} - \tau_z \right) r^\lambda Y_{\lambda\mu} \right]_k,$$

where

$$\tau_z \psi_{\text{proton}} = -\frac{1}{2} \psi_{\text{proton}}, \qquad \tau_z \psi_{\text{neutron}} = \frac{1}{2} \psi_{\text{neutron}}. \tag{6.197}$$

It is important to include recoil corrections for $E1$-transitions. The matrix elements in (6.197) include not only internal displacement of the protons but also a spurious displacement of the center of mass. For $\lambda = 1$, Eq. (6.197) can be written as

$$\mathcal{M}(E1,\mu) = \left(\sum_{i=1}^A e_i \mathbf{r}_i \right)_\mu$$

($e_i = e$ for protons, 0 for neutron). Subtracting the coordinates $\mathbf{r}_i$ from the center of mass $\mathbf{R} = \frac{1}{A}\sum_{i=1}^{A}\mathbf{r}_i$, one gets (we drop the index μ for the moment)

$$
\begin{aligned}
\mathcal{M}(E1) &= \sum_{i=1}^{A} e_i\,(\mathbf{r}_i - \mathbf{R}) = e\sum_{i=1}^{Z}\mathbf{r}_i - \frac{eZ}{A}\sum_{i=1}^{A}\mathbf{r}_I \\
&= e\sum_{i=1}^{Z}\left(1 - \frac{Z}{A}\right)\mathbf{r}_i - \frac{eZ}{A}\sum_{i=Z+1}^{A}\mathbf{r}_i = \frac{eN}{A}\sum_{i=1}^{Z}\mathbf{r}_i - \frac{eZ}{A}\sum_{i=Z+1}^{A}\mathbf{r}_i.
\end{aligned} \tag{6.198}
$$

This expression shows that the inclusion of the center of mass recoil correction can be accomplished from the beginning by assuming that neutron have an *effective charge* eN/A and protons an effective charge $(-eZ/A)$. It can be shown that in the general case (any λ) (see Ref. [28])

$$
e_p^{\text{eff}} = e\frac{1}{A^\lambda}\left[(A-1)^\lambda + (-1)^\lambda(Z-1)\right], \tag{6.199}
$$

and

$$
e_n^{\text{eff}} = eZ\left(-\frac{1}{A}\right)^\lambda. \tag{6.200}
$$

For multipoles higher than $E1$ the effective charges involve corrections of order of λ/A or smaller. Therefore, they are small for heavy nuclei.

Including recoil corrections, the $E1$-matrix element is given by

$$
\mathcal{M}(E1,\mu) = e\sum_k\left[\left(\frac{N-Z}{2A} - \tau_z\right)rY_{\lambda\mu}\right]_k. \tag{6.201}
$$

Using Eq. (6.196), one gets

$$
S(E1) = \frac{3}{4\pi}\frac{\hbar^2}{2m}\cdot 3e^2\left\{\left(\frac{N-Z}{2A} - \frac{1}{2}\right)^2 N + \left(\frac{N-Z}{A} + \frac{1}{2}\right)^2 Z\right\}
$$

$$
= \frac{9}{4\pi}\frac{\hbar^2}{2m}\frac{NZ}{A}e^2 = 14.8\frac{NZ}{A}e^2\ \text{fm}^2\text{MeV}. \tag{6.202}
$$

For $\lambda \geq 2$ and isoscalar excitations, using again Eq. (6.196),

$$
S(E\lambda) = \frac{\lambda(2\lambda+1)^2}{4\pi}\frac{\hbar^2}{2m}Ze^2\langle r^{2\lambda-2}\rangle_{\text{prot.}}, \tag{6.203}
$$

where $\langle r^{2\lambda-2}\rangle_{\text{prot}}$ is defined in terms of the one-particle $|\psi_p|^2$ as

$$
\langle r^{2\lambda-2}\rangle_{\text{prot}} = \sum_p\int|\psi_p|^2 r^{2\lambda-2}d^3r = \frac{Z}{A}\int\rho(r)r^{2\lambda-2}d^3r = \frac{Z}{A}\langle r^{2\lambda-2}\rangle, \tag{6.204}
$$

where $\rho(r)$ is the total particle density. Thus,

$$
S(E\lambda) = \frac{\lambda(2\lambda+1)^2}{4\pi}\frac{\hbar^2}{2m}\frac{Z^2e^2}{A}\langle r^{2\lambda-2}\rangle. \tag{6.205}
$$

In the case of a uniform charge distribution we use (6.180):

$$
\langle r^{2\lambda-2}\rangle = \frac{3}{(2\lambda+1)}R^{2\lambda-2} \tag{6.206}
$$

and

$$
S(E\lambda) = \lambda(2\lambda+1)\frac{\hbar^2}{2m}\frac{3Z^2}{4\pi A}R^{2\lambda-2}e^2. \tag{6.207}
$$

For ($E0$) monopole transitions

$$\mathcal{M}(E0) = e \sum_k \left[\left(\frac{1}{2} - \tau_z \right) r^2 \right]_k Y_{00}, \tag{6.208}$$

and

$$S(E0) = \frac{\hbar^2}{2\pi m} Ze^2 \langle r^2 \rangle_{\text{prot}} = \frac{\hbar^2}{2\pi m} \frac{Z^2}{A} e^2 \langle r^2 \rangle. \tag{6.209}$$

For a uniform charge distribution

$$S(E0) = \frac{3\hbar^2}{10\pi m} \frac{Z^2}{A} R^2 e^2. \tag{6.210}$$

Using the *sum rules*, Eqs. (6.205), (6.207) and (6.210), and the relation between the transition matrix elements and the deformation parameters, Eqs. (6.183) and (6.187), we get

$$\delta_{\lambda \geq 2}^2 = \frac{4\pi}{3} \lambda(2\lambda + 1) \frac{\hbar^2}{2m} \frac{1}{AE_x}, \tag{6.211}$$

and

$$\alpha_0^2 = \frac{10\pi}{3} \frac{\hbar^2}{m} \frac{1}{AR^2 E_x}, \tag{6.212}$$

where E_x is the energy of a state assuming to exhaust the full sum rule, i.e.,

$$S(E\lambda) \equiv \sum_i E_i B(E\lambda, 0 \to \lambda) \cong E_x B(E\lambda, 0 \to \lambda). \tag{6.213}$$

This approximation is good for giant resonance states, which exhaust most part of the sum rule. For the $E1$ case, recoil corrections amount in the replacement of Z by $2NZ/A$ in Eq. (6.183). That is,

$$B(E1, 0 \to 1) = \left[\frac{3}{2\pi} \frac{NZ}{A} e \right]^2 \delta_1^2. \tag{6.214}$$

Using Eq. (6.202), one gets

$$\delta_1^2 = \pi \frac{\hbar^2}{2m} \frac{A}{NZ} \frac{1}{E_x}. \tag{6.215}$$

6.11   Nuclear excitation in the deformed potential model

The asymptotic form of the scattered wave for an unbound particle is

$$\phi_{\mathbf{k},\alpha}^{(\pm)} = \left[e^{i\mathbf{k} \cdot \mathbf{r}} + f_{\mathbf{k}}^{(\pm)}(\theta) \frac{e^{\pm ikr}}{r} \right] \chi_s, \tag{6.216}$$

where χ_s is a spin-isospin wave function.

Assuming that a residual interaction U_{int} between the projectile and target exists and is weak we can use the DWBA formula (3.79) for the inelastic amplitude. For a nuclear excitation $|0\rangle \to |\lambda_\mu\rangle$, where $\lambda\mu$ is the final angular momentum (and projection), it is convenient to define

$$\langle \mathbf{k}_\lambda | T_{\lambda\mu} | \mathbf{k}_0 \rangle \equiv \langle \psi_{\lambda\mu} \phi_{\mathbf{k}_\lambda}^{(-)} | U_{\text{int}} | \psi_0 \phi_{\mathbf{k}_0}^{(+)} \rangle, \tag{6.217}$$

where k_λ is defined as

$$E_{k_\lambda} = \frac{\hbar^2 k_\lambda^2}{2M} = \frac{\hbar^2 k_0^2}{2M} - \hbar\omega_\lambda, \tag{6.218}$$

and $\hbar\omega_\lambda$ is the excitation energy. $\mathbf{k}_\lambda$ is the vector $\mathbf{k}_\lambda = k_\lambda \mathbf{r}/r$.

For small excitation energies $\hbar\omega_\lambda \ll E_{k_\lambda}$, one obtains the useful relation

$$k_\lambda = \left(k_0^2 - 2M\omega_\lambda/\hbar \right)^{1/2} \cong k_0 \left(1 - \frac{M\omega_\lambda}{\hbar k_0^2} \right), \tag{6.219}$$

or

$$\Delta k = k_\lambda - k_0 = \frac{\omega_\lambda}{v}, \tag{6.220}$$

where v is the projectile velocity.

From Eq. (3.52) (see footnote in that page) the scattering amplitude is given by

$$f_{\lambda\mu}(\theta) = -\frac{M}{2\pi\hbar^2}\langle \mathbf{k}_\lambda | T_{\lambda\mu} | \mathbf{k}_0\rangle, \tag{6.221}$$

and the differential cross section for inelastic scattering is given by

$$\frac{d\sigma_{\lambda\mu}(\theta)}{d\Omega} = \frac{k_\lambda}{k_0}|f_{\lambda\mu}(\theta)|^2 = \left(\frac{M}{2\pi\hbar^2}\right)^2 \frac{k_\lambda}{k_0}|\langle \mathbf{k}_\lambda | T_{\lambda\mu} | \mathbf{k}_0\rangle\rangle|^2. \tag{6.222}$$

For collective excitations the projectile induces small deformations of the target surface. As we saw in Section 6.10 the matter density of the target will be slightly deformed by an additional term (proportional to the derivative of the ground state density. This term in peaked at the target surface. Since microscopically the interaction potential U_{int} can be regarded as a folding of the nucleon-nucleon interaction and the matter densities, one expects that U_{int} is also peaked at the target surface. This contains the spirit of the *Bohr-Mottelson model* for collective vibrations. The approximation is valid for light projectiles, mainly proton, He, C, O, etc.

In this model, the optical potential is not spherically symmetric, but is slightly deformed. The approach is similar to the one developed in Section 6.10. The equipotential surfaces of this field $U_\alpha(r)$ are given by Eq. (6.170) for constant r. In other words,

$$U_\alpha(r_\theta, \theta) = U_0(r), \tag{6.223}$$

where $U_0(r)$ is the non-deformed field. Thus,

$$U_\alpha(r, \theta) = U_0\left(\frac{r}{1 + \sum_{\lambda\mu}\alpha^*_{\lambda\mu}Y_{\lambda\mu}(\hat{\mathbf{r}})}\right) = U_0(r) - r\frac{dU_0(r)}{dr}\sum_{\lambda\mu}\alpha^*_{\lambda\mu}Y_{\lambda\mu}(\hat{\mathbf{r}}) + \mathcal{O}(\alpha^2), \tag{6.224}$$

and the residual interaction is given by

$$U_{int} = -r\frac{dU_0(r)}{dr}\sum_{\lambda\mu}\alpha^*_{\lambda\mu}Y_{\lambda\mu}(\hat{\mathbf{r}}) \cong -R_0\frac{dU_0(r)}{dr}\sum_{\lambda\mu}\alpha^*_{\lambda\mu}Y_{\lambda\mu}(\hat{\mathbf{r}}), \tag{6.225a}$$

where R_0 is the peak position of $dU_0(r)/dr$.

Thus, for isoscalar excitations we can write ($\lambda \geq 2$)

$$f^{IS}_{\lambda\mu}(\theta) = \frac{M}{2\pi\hbar^2}\langle \Psi_{\lambda\mu} | \alpha^*_{\lambda\mu} | \Psi_0\rangle R_0 \left\langle \phi^{(-)}_{\mathbf{k}} \left| \frac{dU_0(r)}{dr}Y_{\lambda\mu}(\hat{\mathbf{r}}) \right| \phi^{(+)}_{\mathbf{k}_0} \right\rangle. \tag{6.226}$$

Using the results of Section 6.10, we can rewrite it as

$$f^{IS}_{\lambda\mu}(\theta) = \frac{M}{2\pi\hbar^2}\frac{1}{\sqrt{2\lambda+1}}\delta_\lambda \left\langle \phi^{(-)}_{\mathbf{k}} \left| \frac{dU_0(r)}{dr}Y_{\lambda\mu}(\hat{\mathbf{r}}) \right| \phi^{(+)}_{\mathbf{k}_0} \right\rangle, \tag{6.227}$$

where δ_λ is the deformation parameter for the nuclear excitation. When a single state exhausts the sum rule we can use Eq. (6.211) of Section 6.10.

Following the same reasoning as in Section 6.10, the monopole $|0\rangle \to |\lambda = 0\rangle$ transition amplitude is given by

$$f_0(\theta) = \frac{M}{2\pi\hbar^2}\alpha_0 \left\langle \phi^{(-)}_{\mathbf{k}} \left| \left(3U_0(r) + r\frac{dU_0(r)}{dr}\right)Y_{00} \right| \phi_{\mathbf{k}_0} \right\rangle. \tag{6.228}$$

The sum-rule for α_0 is given in Section 6.10.

Assuming charge independence of the nuclear interaction, the isovector excitations by the projectile nuclear field arise when the target has a number of protons which is different from that of the neutrons.

From Eq. (6.225a) the surface potential which induces isovector excitations is given by

$$-R_n \frac{dU_0^{(n)}}{dr} \sum_{\lambda\mu} \alpha_{\lambda\mu}^{(n)*} Y_{\lambda\mu}(\hat{\mathbf{r}}) - R_p \frac{dU_0^{(p)}}{dr} \sum_{\lambda\mu} \alpha_{\lambda\mu}^{(p)*} Y_{\lambda\mu}(\hat{\mathbf{r}}). \qquad (6.229)$$

Center of mass corrections imply that

$$d_{\lambda\mu}^{(n)} \equiv R_n \alpha_{\lambda\mu}^{(n)*} Y_{\lambda\mu}(\hat{\mathbf{r}}) = Z \left(-\frac{1}{A}\right)^{\lambda} R \alpha_{\lambda\mu}^* Y_{\lambda\mu}(\hat{\mathbf{r}})$$

$$d_{\lambda\mu}^{(p)} \equiv R_p \alpha_{\lambda\mu}^{(p)} Y_{\lambda\mu}(\hat{\mathbf{r}}) = \left[\left(1-\frac{1}{A}\right)^{\lambda} + (-1)^{\lambda}\frac{(Z-1)}{A^{\lambda}}\right] R \alpha_{\lambda\mu}^* Y_{\lambda\mu}(\hat{\mathbf{r}}). \qquad (6.230)$$

where $d^{(n)} \left(d^{(p)}\right)$ is the vibrational amplitude of the neutron (proton) fluid. R is the mean radius of the total (proton + neutron) density.

The isovector potential becomes

$$-R \sum_{\lambda\mu} \alpha_{\lambda\mu}^* Y_{\lambda\mu}(\hat{\mathbf{r}}) \left\{ \overbrace{Z \left(-\frac{1}{A}\right)^{\lambda}}^{Q_{\lambda}^{(n)}} \frac{dU_0^{(n)}}{dr} + \overbrace{\left[\left(1-\frac{1}{A}\right)^{\lambda} + (-)^{\lambda}\frac{Z-1}{A^{\lambda}}\right]}^{Q_{\lambda}^{(p)}} \frac{dU_0^{(p)}}{dr} \right\}. \qquad (6.231)$$

Note that if $U_0^{(n)} = U_0^{(p)}$ and $Q_{\lambda}^{(p)} = -Q_{\lambda}^{(n)}$ there will be no isovector excitations.

If the radii of the neutron and proton potentials are slightly different

$$U_0^{(n)} = U_0 \left(r + R - R_n\right) \cong U_0(r) + (R - R_n)\frac{dU_0(r)}{dr},$$

$$U_0^{(p)} = U_0 \left(r + R - R_p\right) \cong U_0(r) + (R - R_p)\frac{dU_0(r)}{dr},$$

where $U_0(r)$ is a mean potential with mean radius R.

Inserting (6.231) into (6.229) the isovector potential becomes

$$\Delta U = -R \sum_{\lambda\mu} \alpha_{\lambda\mu}^* Y_{\lambda\mu}(\hat{\mathbf{r}}) \left\{ \left(Q_{\lambda}^{(n)} + Q_{\lambda}^{(p)}\right) \left[\frac{dU_0}{dr} + R\frac{d^2U_0}{dr^2}\right] \left(Q_{\lambda}^{(n)} R_n + Q_{\lambda}^{(p)} R_p\right) \frac{d^2U_0(r)}{dr^2} \right\}$$

$$\cong -R \sum_{\lambda\mu} \alpha_{\lambda\mu}^* Y_{\lambda\mu}(\hat{r}) \left(Q_{\lambda}^{(n)} + Q_{\lambda}^{(p)}\right) \frac{dU_0}{dr}. \qquad (6.232)$$

Thus, for isovector excitations,

$$f_{\lambda\mu}^{IV}(\theta) = \frac{M}{2\pi\hbar^2} \frac{1}{\sqrt{2\lambda+1}} \delta_{\lambda} \left(Q_{\lambda}^{(n)} + Q_{\lambda}^{(p)}\right) \left\langle \phi_k^{(-)} \left| \frac{dU_0}{dr} Y_{\lambda\mu}(\hat{\mathbf{r}}) \right| \phi_{\mathbf{k}0}^{(+)} \right\rangle. \qquad (6.233)$$

A more detailed calculation [40] leads to small changes in the above equation. For example, in the case of isovector dipole excitations one should keep the neutron and proton radius dependence and the formula above is modified according to the replacement

$$\delta_1 \left(Q_1^{(n)} + Q_1^{(p)}\right) \longrightarrow \delta_1 \frac{3\Delta R}{2R},$$

where $\Delta R = R_n - R_p$ is the difference between the neutron and the proton matter radii of the target (and $R_0 = (R_p + R_n)/2$). Thus the strength of the nuclear induced isovector excitation increases with the difference of the neutron and the proton matter radii, or *neutron skin*.

6.12 Appendix 6.A – Multipole moments and the electromagnetic interaction

6.12.1 Multipole moments

A given distribution of charges $\rho(\mathbf{r}')$ confined to a certain region, produces at each point $\mathbf{r}$ of space an electrostatic potential

$$V = \int \frac{\rho(\mathbf{r}')}{|\mathbf{r} - \mathbf{r}'|} d\mathbf{r}'. \tag{6.234}$$

The factor $1/|\mathbf{r} - \mathbf{r}'| = [(x - x')^2 + (y - y')^2 + (z - z')^2]^{-1/2}$ in Eq. (6.234) can be expanded in a *Mac-Laurin series* for 3 variables,

$$f(t, u, v) = \sum_{n=0}^{\infty} \frac{1}{n!} \left[t\frac{\partial}{\partial t} + u\frac{\partial}{\partial u} + v\frac{\partial}{\partial v} \right]^n f(t, u, v), \tag{6.235}$$

the derivatives being calculated at the point $t = 0$, $u = 0$, $v = 0$. The convention for the power is such that $(t\partial/\partial t)^n = t^n \partial^n/\partial t^n$, etc. Using (6.235) for the source coordinates ($t = x'$, etc.), (6.234) becomes

$$V = \frac{\int \rho(\mathbf{r}')d\mathbf{r}'}{r} + \frac{x_i \int x_i'\rho(\mathbf{r}')d\mathbf{r}'}{r^3} + \frac{1}{2}\frac{x_ix_j \int (3x_i'x_j' - r'^2\delta_{ij})\rho(\mathbf{r}')d\mathbf{r}'}{r^5} + \cdots, \tag{6.236}$$

where $(x_1, x_2, x_3) \equiv (x, y, z)$. In Eq. (6.236) the sum convention was used, where the repetition of an index in the same term indicates a sum over the index, that is, $x_ix_i' \equiv \sum_{i=1}^{3} x_ix_i'$, etc.

The first term of Eq. (6.236) is identical to the potential of a charge $q = \int \rho(\mathbf{r}')d\mathbf{r}'$ (monopole) placed at the origin. The second has the form of the potential of a dipole (two charges of the same magnitude and opposite signs placed one near the other), the integrals representing each component i of the electric dipole moment vector

$$\mathbf{p} = \int \mathbf{r}'\rho(\mathbf{r}')d\mathbf{r}'. \tag{6.237}$$

The third term of Eq. (6.236) represents the contribution of a quadrupole, the six integrals

$$Q_{ij} = \int (3x_i'x_j' - r'^2\delta_{ij})\rho(\mathbf{r}')d\mathbf{r}' \tag{6.238}$$

being the components of the electric quadrupole moment tensor. Using Eq. (6.237) and Eq. (6.238) we can rewrite (6.236) more compactly as

$$V = \frac{q}{r} + \frac{\mathbf{p} \cdot \mathbf{x}}{r^3} + \frac{1}{2}\frac{Q_{ij}x_ix_j}{r^5} + \cdots. \tag{6.239}$$

The increasing powers of the denominator makes the contribution of higher order multipoles (octupole, hexadecapole, etc.) less and less important.

The expansion into multipoles can be employed for the Coulomb potential created by the protons in nuclei but it is necessary that the above treatment be adapted to a quantum system. For this purpose, the charge density $\rho(\mathbf{r}')$ must be understood as Ze times the probability density $|\psi|^2 = \psi^*(\mathbf{r}')\psi(\mathbf{r}')$ to find a proton at the point $\mathbf{r}'$. One consequence for the nuclear case is that the electric dipole moment Eq. (6.237) vanishes. In fact, the wave function $\psi(\mathbf{r})$ represents nuclear states of definite parity and $|\psi|^2$ must necessarily be an even function, that turns the integral Eq. (6.237) taken over the entire space identically zero. The first important information about the charge distribution in nucleus must come therefore from the quadrupole term, the third term of the expansion Eq. (6.236).

In the same way that a charge distribution gives place to an expansion of the type Eq. (6.234), a localized current distribution $\mathbf{J}(\mathbf{r}')$ produces a vector field, with the vector potential $\mathbf{A}$ being expanded in a sum of multipole terms. In this case, however, the monopole term does not exist and the other even terms (even powers of $\mathbf{r}$) cancel each other in the nuclear case by similar considerations to that presented in the former case. Thus, our expansion reduces to a single important term [35],

$$\mathbf{A}(\mathbf{r}) = \frac{\mu \times \mathbf{r}}{r^3} + \cdots \qquad (6.240)$$

with μ, the magnetic dipole moment, being given by

$$\mu = \frac{1}{2c} \int \mathbf{r}' \times \mathbf{J}(\mathbf{r}') d^3 \mathbf{r}'. \qquad (6.241)$$

In any way, the quantum operator corresponding to this quantity will not be obtained from Eq. (6.241) but from the relationship between μ and the corresponding angular momentum.

6.12.2   The electromagnetic interaction

The equation of motion of a charge in an electromagnetic field is given by

$$\frac{d}{dt} \frac{m\mathbf{v}}{\sqrt{1 - \mathrm{v}^2/c^2}} = q(\mathbf{E} + \frac{\mathbf{v}}{c} \times \mathbf{B}). \qquad (6.242)$$

Since one can write

$$\mathbf{B} = \boldsymbol{\nabla} \times \mathbf{A} ; \qquad \mathbf{E} = -\boldsymbol{\nabla}\phi - \frac{1}{c} \frac{\partial \mathbf{A}}{\partial t}, \qquad (6.243)$$

we get

$$\frac{d}{dt} \frac{m\mathbf{v}}{\sqrt{1 - \mathrm{v}^2/c^2}} = q\left[-\boldsymbol{\nabla}\phi - \frac{1}{c}\frac{\partial \mathbf{A}}{\partial t} + \frac{\mathbf{v}}{c} \cdot (\boldsymbol{\nabla} \times \mathbf{A})\right] = \boldsymbol{\nabla}\left(-q\phi + q\frac{\mathbf{v}}{c} \cdot \mathbf{A}\right) - \frac{q}{c}\left(\frac{\partial}{\partial t} + \mathbf{v}.\boldsymbol{\nabla}\right)\mathbf{A}$$

$$= \boldsymbol{\nabla}\left(-q\phi + q\frac{\mathbf{v}}{c} \cdot \mathbf{A}\right) - \frac{q}{c}\frac{d\mathbf{A}}{dt}, \qquad (6.244)$$

where we have used

$$\frac{d\mathbf{A}[\mathbf{r}(t), t]}{dt} = \left(\frac{\partial}{\partial t} + \mathbf{v}.\boldsymbol{\nabla}\right)\mathbf{A}[\mathbf{r}(t), t]. \qquad (6.245)$$

We can rewrite Eq. (6.244) as

$$\frac{d}{dt}\left(\frac{m\mathbf{v}}{\sqrt{1 - \mathrm{v}^2/c^2}} + \frac{q}{c}\mathbf{A}\right) + \boldsymbol{\nabla}\left(q\phi - \frac{q}{c}\mathbf{v} \cdot \mathbf{A}\right) = 0. \qquad (6.246)$$

In terms of the Lagrangian equation,

$$\frac{d}{dt}(\boldsymbol{\nabla}_{\mathbf{r}}\mathcal{L}) - \boldsymbol{\nabla}_{\mathbf{r}}\mathcal{L} = 0,$$

the suitable Lagrangian for (6.246) is

$$\mathcal{L}(\mathbf{r}, \mathbf{v}) = -mc^2\sqrt{1 - \mathrm{v}^2/c^2} + \frac{q}{c}\mathbf{v} \cdot \mathbf{A}(r, t) - q\phi(r, t). \qquad (6.247)$$

The canonical momentum is

$$\mathbf{p} = \boldsymbol{\nabla}_{\mathbf{v}}\mathcal{L} = \frac{m\mathbf{v}}{\sqrt{1 - \mathrm{v}^2/c^2}} + \frac{q}{c}\mathbf{A}(\mathbf{r}, t) = \mathbf{P} + \frac{q}{c}\mathbf{A}(\mathbf{r}, t), \qquad (6.248)$$

where

$$\mathbf{P} = m\mathbf{v}/\sqrt{1 - \mathrm{v}^2/c^2} \qquad (6.249)$$

is the kinetic momentum and $q\mathbf{A}/c$ is the momentum carried by the electromagnetic field.
The Hamiltonian is

$$\mathcal{H} = \mathbf{p}.\mathbf{v} - \mathcal{L} = \frac{mc^2}{\sqrt{1 - \mathrm{v}^2/c^2}} + q\phi. \qquad (6.250)$$

We rewrite Eq. (6.250) as

$$\mathcal{H}(\mathbf{r}, \mathbf{p}) = c \left\{ \left[\mathbf{p} - \frac{q}{c}\mathbf{A}(\mathbf{r}, t) \right]^2 + (mc)^2 \right\}^{1/2} + q\phi(\mathbf{r}, t). \tag{6.251}$$

For non-relativistic particles,

$$| \mathbf{P} | = | \mathbf{p} - \frac{q}{c}\mathbf{A} | << mc, \tag{6.252}$$

and

$$H(\mathbf{r}, \mathbf{p}) = mc^2 + \frac{(\mathbf{p} - q\mathbf{A}/c)^2}{2m} + q\phi. \tag{6.253}$$

The second term has as part of its contribution the quantity $(q\mathbf{A})^2/2mc^2$, which is relevant only in processes where two photons are involved and may be ignored. The remaining terms yield the *electromagnetic interaction* Hamiltonian

$$\mathcal{H}_{int} = q\phi - \frac{q}{c}\mathbf{v}.\mathbf{A}, \tag{6.254}$$

where the rest + kinetic energy of the particle was subtracted.

For systems involving a charge density $\rho(\mathbf{r}, t)$ and current density $\mathbf{j}(\mathbf{r}, t)$, H can be generalized to

$$\mathcal{H}_{int} = \int \left[\rho\phi - \frac{1}{c}\mathbf{j}.\mathbf{A} \right] d^3r. \tag{6.255}$$

Elastic electron scattering

At the electron energies $E_e \gg m_e c^2$, the electron can be treated as a point-like particle, its interaction with the nucleus is extremely well described by *quantum electrodynamics* (QED), and its momentum transfer can be used as a variable quantity for a given energy transfer, they are ideal probes of nuclear charge distributions, transition densities, and nuclear response functions.

Under the conditions of validity of the plane-wave Born approximation (PWBA), the elastic electron scattering cross section in the laboratory is

$$\left(\frac{d\sigma}{d\Omega} \right)_{\text{PWBA}} = \frac{\sigma_M}{1 + (2E/M_A)\sin^2(\theta/2)} |F_{ch}(q)|^2, \tag{6.256}$$

where

$$\sigma_M = \frac{e^4}{4E^2}\cos^2\left(\frac{\theta}{2}\right)\sin^{-4}\left(\frac{\theta}{2}\right) \tag{6.257}$$

is known as the *Mott cross section*. The denominator accounts for a recoil correction, E is the electron total energy, M_A is the nucleus mass and θ is the electron scattering angle. $q = 2k\sin(\theta/2)$ is the momentum transfer, $\hbar k$ is the electron momentum, and $E = \sqrt{\hbar^2 k^2 c^2 + m_e^2 c^4}$. The Mott cross section has a dependence on the momentum transfer as

$$\sigma_M \sim \frac{E}{q^4}. \tag{6.258}$$

The charge form factor $F_{ch}(q)$ is, for a spherical charge distribution,

$$F_{ch}(q) = \int_0^\infty dr\, r^2 j_0(qr)\, \rho_{ch}(r). \tag{6.259}$$

Elastic electron scattering essentially measures the Fourier transform of the charge distribution through the form factor F_{ch}. The physics displayed in the cross section is better understood if we take for simplicity the one-dimensional case. For light nuclei, well described by a Gaussian distribution, we have from Eq. (4.29)

$$F_{ch}(q) \sim \int e^{iqx}\rho(x)dx \sim \frac{\pi}{a}e^{-qa}, \tag{6.260}$$

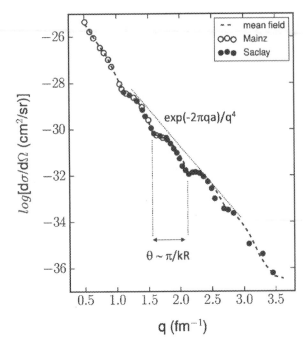

FIGURE 6.19   Data on elastic electron scattering off Pb from the Saclay and Mainz experimental groups. The dashed lines are form factors calculated with Skyrme-Hartree-Fock theory [44]. The dotted lines show that the exponential decay of the cross sections by many orders of magnitude are due to the nuclear diffuseness while the dips are reflect the nuclear radius.

While, for a heavy nucleus, the density ρ is better described by a Fermi function or Woods-Saxon charge distribution, yielding from Eq. (4.31),

$$F_{ch}(q) \sim (4\pi)\sin(qR)e^{-\pi qa}, \tag{6.261}$$

valid for $R \gg a$, and $qa \gg 1$. Upgrading the plane waves to eikonal wave function (see text ahead), one has

$$F_{ch}(q) \sim \int dbb J_0(qb)[1 - e^{i\chi(b)}] \sim \int dbb \frac{J_0(qb)}{1 + \exp[(\frac{b-R}{a})]} \sim \frac{R}{q} J_1(qR)\exp(-\pi qa), \tag{6.262}$$

where J_n are Bessel functions of order n. Therefore, in the differential cross sections for elastic electron scattering the distance between minima is a direct measure of the nuclear size, whereas their exponential decay reflects the surface diffuseness of the charge distribution. This is shown in Figure 6.19 for the electron scattering off lead. The dashed lines are form factors calculated with Skyrme-Hartree-Fock theory [44]. The dotted lines show that the exponential decay of the cross sections by many orders of magnitude are due to the nuclear diffuseness while the dips are reflect the nuclear radius.

One can also deduce that, at low-momentum transfers, Eq. (6.259) yields to leading order

$$F_{ch}(q)/Z = 1 - \frac{q^2}{3!}\left\langle r_{ch}^2 \right\rangle + \frac{q^4}{5!}\left\langle r_{ch}^4 \right\rangle + \mathcal{O}(q^6), \tag{6.263}$$

where $\langle r^n \rangle$ is the n-th moment of the charge density distributions,

$$\langle r^n \rangle = \int r^n \rho(r) d^3 r. \tag{6.264}$$

At the lowest order, elastic electron scattering at low-momentum transfers obtains the root mean squared radius of the charge distribution, $\langle r_{ch}^2 \rangle^{1/2}$. With increasing data at higher momentum transfers, more details of the charge distribution are probed.

6.13 Exercises

1. Derive Eq. (6.58) from Eq. (6.56).

2. Work out the details involved in the derivation of Eq. (6.55) from Eqs. (6.49) and (6.53).

3. It was first pointed out by Bethe and Butler [42] that one can verify and determine the shell-model single-particle level spectrum through the study of stripping reactions. Using the appropriate selection rules, illustrate this in the case of $^{10}B(d,p)^{11}B$ and compare with the actual experimental data [43].

4. A typical direct reaction is $^{44}Ca(d,p)^{45}Ca$. Assuming the Q-value of the reaction to be 3.30 MeV and the energy of the incident deuteron to be 7.0 MeV, calculate the momentum transfer and cross section using expressions given in this chapter.

5. The table below shows nuclei with their respective experimental values of spin and parity of the ground state. Compare with the predictions of the extreme shell model for those nuclei and try to justify the discrepancies.

^{7}Be	^{17}F	^{61}Cu	^{91}Zr	^{93}Nb	^{123}Sb	^{159}Tb	^{183}Ta	^{199}Tl	^{209}Pb
$\frac{3}{2}^-$	$\frac{5}{2}^+$	$\frac{3}{2}^-$	$\frac{5}{2}^+$	$\frac{9}{2}^+$	$\frac{7}{2}^+$	$\frac{3}{2}^+$	$\frac{7}{2}^+$	$\frac{1}{2}^+$	$\frac{1}{2}^-$

6. The table below exhibits the orbits attributed to the extra proton and neutron for a series of odd-odd nuclei. (a) Try to justify these properties using Figure 4.21. (b) Determine the spins and parities of these nuclei with help of the Nordheim rule and compare with the experimental values, also shown in the table.

Nucleus	p	n	Spin^P	Nucleus	p	n	Spin^P
^{16}N	$p_{1/2}$	$d_{5/2}$	2^-	^{70}Ga	$p_{3/2}$	$p_{1/2}$	1^+
^{34}Cl	$d_{3/2}$	$d_{3/2}$	0^+	^{90}Y	$p_{1/2}$	$d_{5/2}$	2^-
^{38}Cl	$d_{3/2}$	$f_{7/2}$	2^-	^{92}Nb	$g_{9/2}$	$d_{5/2}$	7^+
^{41}Sc	$f_{7/2}$	$f_{7/2}$	0^+	^{206}Tl	$s_{1/2}$	$p_{1/2}$	0^-
^{62}Cu	$p_{3/2}$	$f_{5/2}$	1^+	^{202}Bi	$h_{9/2}$	$f_{5/2}$	5^+

7. The spin and parity of ^{9}Be and ^{9}B are $\frac{3}{2}^-$ for both nuclei. Admitting that those values are given by the last nucleon, justify the observed value, 3^+, of the ^{10}B. What other combinations of spin-parity can appear? Verify in a nuclear chart the presence of excited states of ^{10}B that could correspond to those combinations.

8. ^{13}C and ^{13}N have, both, a ground state $\frac{1}{2}^-$ and three excited states below 4 MeV, of spin-parity $\frac{1}{2}^+$, $\frac{3}{2}^-$ and $\frac{5}{2}^+$. The other states are located above 6 MeV. Interpret these four states using the shell model.

9. Using an identical construction to the one of the Table 6.3, show that the possible values of j resulting from the coupling of 3 nucleons in a $j = \frac{5}{2}$ level are $\frac{3}{2}$, $\frac{5}{2}$ and $\frac{9}{2}$.

10. Find the volume of the nucleus whose surface is described by (6.90), with $\lambda = 2$ and $\mu = 0$.

11. Building a table similar to the one of Table 6.6, show that the allowed final states, resulting from the coupling of three quadrupole phonons, is 0^+, 2^+, 3^+, 4^+ and 6^+.

12. Find which are the possible states (spin and parity) resulting from the coupling of a quadrupole phonon ($\lambda = 2$) with an octupole phonon ($\lambda = 3$). Justify.

13. The second state 2^+ of a vibrational band has an energy of 3.92 keV and the first state 4^+ has an energy of 3.55 keV. Estimate the energy of the first 2^+ state and of the first 6^+ state.

14. For an axially symmetric nucleus with density given by

$$\rho(r) = \begin{cases} \rho_0, & \text{for} \quad r \leq R_0 \left(1 + \beta Y_{20}(\theta, \phi)\right) \\ 0, & \text{otherwise}, \end{cases}$$

show that the intrinsic (charge) quadrupole moment up to second order in the deformation parameter β is given by

$$Q_0 = \frac{3}{\sqrt{5\pi}} Z R_0^2 \beta \left(1 + 0.36\beta\right),$$

and the moment of inertia about the z-axis is given by

$$I = \frac{2}{5} M R_0^2 \left(1 + 0.31\beta\right),$$

to first order in β. Here M is the mass of the nucleus and R_0 is the radius of a sphere having the same volume, and ρ_0 may be found from normalization

15. For a harmonic oscillator, the expectation value of the kinetic energy of a given state is equal to the expectation value of the potential energy. Thus, the sum of the energies of the occupied states in a nucleus of mass A is

$$E = m_N \omega^2 A \left\langle r^2 \right\rangle,$$

where m_N is the nucleon mass. One can estimate $\left\langle r^2 \right\rangle \sim 3R^2/5$, with $R \sim 1.2A^{1/3}$ fm. Assume $N = Z$ and use the harmonic oscillator model for the nucleus, developed in the Appendix A of Chapter 4, to show that the frequency of the harmonic oscillator for a nucleus is related to its mass by the equation

$$\hbar\omega \simeq 41 A^{-1/3} \text{ MeV},$$

which has the correct mass dependence for the energy location of the giant resonances.

16. For collisions above the Coulomb barrier the conditions $\epsilon \sim b/a_0 \gg 1$ and $\eta \ll 1$ are often met. Obtain an expression for $dn_{E1}/2\pi bdb$ under these circumstances using Eq. (6.157). Show that the approximation

$$x^2 \left[K_1^2(x) + K_0^2(x)\right] \simeq \begin{cases} 1, & \text{for} \quad x \leq 1 \\ 0, & \text{for} \quad x > 1 \end{cases}$$

is a reasonable one. Discuss the consequences of the use of this approximation to explain the energy spectrum of the virtual photon numbers $dn_{E1}/2\pi bdb$.

17. Use the mass dependence energy of the giant resonances given in Section 6.8 and plot the mass dependence of the deformation parameters δ_1 and δ_2 as given by Eqs. (6.215) and (6.211), respectively.

18. Assume that in Eq. (6.227)

$$\phi_{\mathbf{k}}^{(-)*} \phi_{\mathbf{k}_0}^{(+)} \simeq \begin{cases} \exp\left(-i\mathbf{q} \cdot \mathbf{R}\right) & \text{for} \quad R \geq R_0 \\ 0, & \text{for} \quad R < R_0 \end{cases}$$

and that

$$\frac{dU}{dr} \simeq \begin{cases} U_0 \exp\left(-R/\alpha\right), & \text{for} \quad R \geq R_0 \\ 0, & \text{for} \quad R < R_0 \end{cases}$$

and obtain an expression for $f_{\lambda\mu}^{IS}(\theta)$, where $\mathbf{q} = \mathbf{k} - \mathbf{k}_0$. Make a plot (in arbitrary units) of $f_{\lambda\mu}^{IS}(\theta)$ as a function of θ for a reasonable value of α (α is a measure of the diffuseness of the optical potential at the surface).

References

1. Miura K et al. 1987 *Nucl Phys.* **A467** 79
2. Bhatia A B, Huang KR., Huby R and Newns H C 1952 *Phil. Mag.* **43** 485
3. Butler S T 1950 *Phys. Rev.* **80** 1095; *Nature* **166** 709; 1951 *Proc. Roy. Soc.* **208A** 559
4. Legg J C 1963 *Phys. Rev.* **129** 272
5. Bardeen J, Cooper I N and Schrieffer J R 1957 *Phys. Rev.* **108** 1175
6. Nordheim L W 1950 *Phys. Rev.* **78** 294
7. Wang L et al. 1993 *Phys. Rev.* **C47** 2123
8. Brussaard P J and M. Glaudemans P W 1977 *Shell-Model Applications in Nuclear Spectroscopy,* (Amsterdam: North-Holland)
9. Piskou and Schäferlingová W 1990 *Nucl. Phys.* **A510** 301
10. Thorn C E, Olness J W, Warburton E K and Raman S 1984 *Phys. Rev.* **C30** 1442
11. Lord Rayleigh 1877 *Theory of Sound* vol. II 364 (London: Macmillan) London; 1879 *Proc. Roy. Soc.* **A29** 71
12. Bothe W and Gentner W 1937 *Z. Phys.* **106** 236
13. Baldwin G C and Klaiber G S 1947 *Phys. Rev.* **71** 3; 1948 *Phys. Rev.* **73** 1156
14. Goldhaber M and Teller E 1948 *Phys. Rev.* **74** 1046
15. Steinwedel H and Jensen J H D 1950 *Z. Naturforsch.* **52** 413
16. Lepretre A et al. 1974 *Nucl. Phys.* **A219** 39
17. Dietrich S S and Berman B L 1988 *At. Data and Nucl. Data Tables* **38** 199
18. Myers W D, Swiatecki W J, Kodama T, El-Jaick L J and Hilf E R 1977 *Phys. Rev.* **C15** 2032
19. Pitthan R and Walcher Th 1971 *Phys. Lett.* **36B** 563
20. Marty N, Morlet M, Willis A, Comparat V and Frascaria R 1975 *Nucl. Phys.* **A238** 93.
21. Davis B F et al. 1996 *Nucl. Phys.* **A599** 277c
22. Van der Woude A 1987 *Prog. Part. Nucl. Phys.* **18** 217
23. Axel P 1962 *Phys. Rev.* **126** 671
24. Bertsch G F and Broglia R A 1986 (August) *Physics Today* p. 44
25. Alder K and Winther A 1966 *Coulomb Excitation* (New York: Academic Press)
26. Alder K and Winther A 1975 *Electromagnetic Excitation* (Amsterdam: North-Holland)
27. Goldstein H, Poole C P, Poole Jr. C P and Safko J L 2002 *Classical Mechanics* (Prentice Hall, 3rd edition)
28. Eisenberg J M and Greiner W 1987 *Excitation Mechanisms of the Nuclei* (Amsterdam: North-Holland) third edition, p. 227
29. Bertulani C A and Ponomarev V Yu 1999 *Phys. Rep.* **321** 139
30. A.N.F. Aleixo and C.A. Bertulani, Nucl. Phys. **A505** 448 (1989)
31. Bertulani C A and Baur G 1988 *Phys. Rep.* **163** 299
32. Olver F W J, Lozier D W, Boisvert R F and Clark C W 2010 *NIST Handbook of Mathematical Functions Paperback* (Cambrige University Press)
33. Fermi E 1924 *Z. Physik* **29** 315
34. Weizsacker K F 1934 *Z. Physik* **612**; Williams E J 1934 *Phys. Rev.* **45** 729
35. Jackson J D 1999 *Classical Electrodynamics* (Wiley)
36. Temmer G M and Heydenburg N P 1955 *Phys. Rev.* **98** 1198
37. Sherr R, Li C W and Christy R F 1954 *Phys. Rev.* **96** 1258
38. Messiah A 2014 *Quantum Mechanics* (Dover Publications)
39. Bohr A and Mottelson B R 1998 *Nuclear Structure* (World Scientific)
40. Satchler G R 1987 *Nucl. Phys.* **A472** 215
41. Beene J et al. 1991 *Phys. Rev.* **C44** 128
42. Bethe H A and Butler S T 1952 *Phys. Rev.* **85** 1045
43. Ajzenberg-Selove F 2013 *Nuclear Spectroscopy* (Academic Press)
44. Cavedon J M, , et al 1987 *Phys. Rev. Lett.* **58** 195

7

Nuclear reactions in the cosmos

7.1   Cosmic rays

The terrestrial surface is bombarded by a constant flux of charged particles coming from the cosmos. A measurement of the intensity of this radiation shows that it decreases until a certain altitude, but it grows quickly again for larger altitudes. This discards the hypothesis that the radiation is coming from the radioactive decay of elements in the terrestrial ground and it is indeed proven that this activity originates from particles accelerated by astrophysical sources (*primary radiation*) and of collisions of these particles with the interstellar gas and with atoms of the upper part of the atmosphere (*secondary radiation*). These particles, known as *cosmic rays*, were studied in several places of the Earth – from the deepness of mines to the heights of satellites - and in the regions of the solar system explored by the man.

Cosmic rays can be broadly defined as the massive particles, photons (γ rays, X-rays, ultraviolet and infrared radiation, $\cdots$), neutrinos, and exotics (WIMPS, axions, $\cdots$) striking the earth. Cosmic rays can be of either galactic (including solar) or extragalactic origin. The intensity of the primary radiation is approximately $2 - 4$ particles per second and per square centimeter. This radiation consists essentially of nuclei (98% in number) and of electrons (2%). Among the nuclei, the predominance is of protons (87% of the mass) and of α-particles (12% of the mass). The remaining 1% includes all nuclei, in smaller percent the larger the mass of the nucleus is.

7.1.1 Composition

The chemical composition and kinetic energy distribution of (primary) cosmic rays is shown in Figure 7.1. This distribution is approximately independent of energy, at least over the dominant energy range of 10 MeV/nucleon through several GeV/nucleon. One recognizes that the main component of cosmic rays are protons, with additionally around 10% of helium and an even smaller admixture of heavier elements.

The figure also shows the chemical distribution of the elements in our solar system, which differs from that of the cosmic rays in some remarkable ways. The most dramatic of these is an enormous enrichment in the cosmic rays of the elements Li/Be/B. Note also that there is enrichment is even Z elements relative to odd Z. Finally the cosmic rays are relatively enriched in heavy elements relative to H and He.

Although not shown in the figure, many elements heavier than the iron group have been measured with typical abundances of 10^{-5} of iron. Much of this information was gained from satellite and spacecraft measurements over the last decade. Some of the conclusions:

1) Abundances of even Z elements with $30 \lesssim Z \lesssim 60$ are in reasonable agreement with solar system abundances.

2) In the region $62 \lesssim Z \lesssim 80$, which includes the platinum-lead region, abundances are enhanced relative to solar by about a factor of two. This suggests an enhancement in r-process elements, which dominate this mass region.

There are obvious connections between other astrophysics we have discussed (e.g., if the r-process site is core-collapse supernovae, then one would expect enrichment in r-process nuclei as supernovae are also believed to be the primary acceleration mechanism for lower energy cosmic rays) and possible deviations from solar abundances in the cosmic rays.

For some nuclei, like Li, Be, and B, the percent of cosmic rays, albeit small, is several orders of magnitude larger than the average of these elements in the Universe. That is due to the fact that these elements are not nucleosynthesis products in stars, and they should be part of the secondary radiation. The proportion of the several isotopes of an element in the cosmic rays can also be different from the universal average. The *isotopic ratio* ^{3}He/^{4}He, for example, is 200 times larger in the cosmic rays than in the same average isotopic ratio in all the cosmos.

In the electron component the positron percentage is about 10%. On the other hand, the antiproton/proton ratio is of the order of 10^{-4}. These antiparticles belong, probably, to the secondary radiation. We can calculate the minimal energy of a proton able to produce anti-protons scattering on another proton at rest. Because of baryon number conservation, anti-proton production is first possible in $pp \rightarrow ppp\bar{p}$ with $s = 2m_p^2 + 2E_p m_p \geq 16m_p^2$, or $E_p \geq 7m_p$. Furthermore, the cross section of this reaction is small close to the threshold. Hence the anti-proton flux at small energies should be strongly suppressed, if anti-protons are only produced as secondaries in cosmic ray interactions.

7.1.2 Energy distribution

The energy spectrum of the primary cosmic rays has an average of 10 GeV/nucleon. For the nuclei the significant contribution begins at 10 MeV/nucleon and it grows until reaching a maximum at 300 MeV/nucleon. At this maximum, the proton flux is 2 protons/(m^2·s·MeV· steradian). The intensity of cosmic rays of energy 1 Gev/ nucleon or greater is about 1/cm^{2}sec. The energy density corresponding to this is thus about 1 ev/cm^3. This can be compared to the energy density of stellar

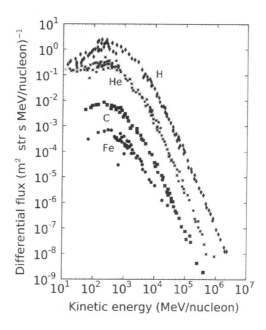

FIGURE 7.1 Main elements in the composition of cosmic rays (data collected from Ref. [1]).

light of 0.3 eV/cm^3. If we confine ourselves to the particle constituents (protons, nuclei, leptons), their motion in the Galaxy has been roughly randomized by the *galactic magnetic field*. Thus they provide very little information about the direction of the source.

The primary cosmic rays are approximately isotropic and are also constant in a very long time scale ($\sim 10^9$ y).

The energy distribution of cosmic rays from about 10^{10} eV to about 10^{15} eV has a power-law distribution

$$N(E) \propto E^{-(1.6-1.7)}.$$

However there is a break or "*knee*" in the curve at about 10^{15} eV. The slope sharpens above this knee (see the Figure 7.2), falling as

$$N(E) \propto E^{-(2.0-2.2)},$$

eventually steeping to an exponent of above -2.7. The knee and *ankle* are more visible in Figure 7.3 where the spectrum is divided by $E^{-2.5}$. The knee is generally attributed to the fact that supernovae acceleration of cosmic rays is limited to about this energy. This would argue that the cosmic rays above this energy either have a different origin, or where further accelerated after production. However the sharpness of the knee has troubled many of the experts: it is very difficult to find natural models producing such a defined break.

7.1.3 Propagation and origin

The solar wind impedes and slows the incoming cosmic rays, reducing their energy and preventing the lowest energy ones from reaching the Earth. This effect is known as *solar modulation*. The Sun has an 11-year activity cycle which is reflected in the ability of the solar wind to modulate cosmic rays. As a result, the cosmic ray intensity at Earth is anti-correlated with the level of solar activity, i.e., when solar activity is high and there are lots of sunspots, the cosmic ray intensity at Earth is low, and vice versa.

The properties listed in the previous sections do not allow us to obtain a definitive answer on the origin of the cosmic rays. The most accepted hypothesis is the that supernovas and neutron

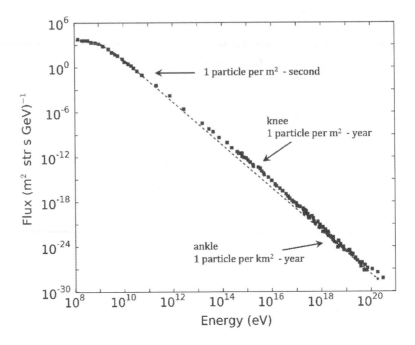

FIGURE 7.2   The energy distribution of cosmic rays.

stars in our Galaxy are the responsible agents for this radiation and it can have a relation with the fact that in our Galaxy a supernova explodes every 30 years on the average. Each one of these explosions liberates about $10^{44} - 10^{45.5}$ J of energy. To explain the energy spectrum of the cosmic rays it is also necessary to suppose that it is influenced by the re-acceleration of the particles by galactic magnetic fields.

The most common assumption is that the cosmic rays are confined within the galactic disk, where the mass density is high, but with some gradual leaking out of the disk. This model does a good job in explaining the energy dependence of the life of cosmic rays. Other models are also popular, including closed models where cosmic rays are fully confined, then explaining lifetimes through devices such as a combination of a few nearby and many distant cosmic ray sources. The relatively large presence of light and heavy nuclei would be a consequence of fragmentation processes, that is, of the interaction of protons and of α-particles with remnant nuclei in gases in the sidereal space. In modeling the origin of cosmic rays, the first conclusion, given their richness in metals, is that they must come from highly evolved stars such as those that undergo supernovae. The abundance of r-process nuclei, which could be taken as an evidence of supernova dominance, for those who accept that supernovae are the r-process site. But studies of the isotopic composition as the knee is approached shows that the composition changes: the spectrum of protons becomes noticeably steeper in energy, while the iron group elements do not show such a dramatic change. Above the knee - at energies above 10^{16} eV the galactic magnetic field is too weak to appreciably trap particles. Thus it is probable that at these high energies the character of the cosmic rays changes from primarily galactic to primarily extragalactic. This also suggests a natural explanation for the relative enrichment is heavy nuclei in the vicinity of the knee: the upper energy of confined particles should vary as Z, since the cyclotron frequency for particles of the same velocity varies as Z versus B. Above 10^{18} eV/n the cosmic rays free stream through Galaxies.

Assuming a density of particles in the space as $1/cm^3$, the mean free path of a cosmic ray is about 3×10^{22} m, or a mean time path of 3×10^6 y. This last figure is taken as the lifetime of a particle present in the cosmic rays in our Galaxy. During this time, several other particles at different points in the Universe are produced, with a wide energy spectrum, and emitted in several directions. Therefore, the large lifetime would explain the isotropy and the temporal constancy

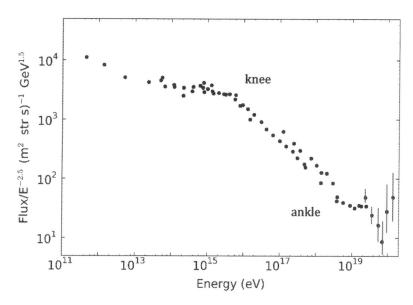

FIGURE 7.3 The energy spectrum of cosmic rays (divided by $E_9^{-2.5}$) showing a "*kink*", or "*knee*" and an "*ankle*". The energy E_9 is given in GeV.

of the cosmic rays, since fluctuations in the place of emission, in the production time, and in the intensity of the source, are smoothed out.

The conventional explanation for the most dramatic isotopic anomaly in the cosmic ray, the enrichment in Li/Be/B by about six orders of magnitude, is that they are produced in the interstellar medium when accelerated protons collide with C, N, and O. The enrichment of odd-A nuclei (these also tend to be relatively rare in their solar distribution since stellar processes tend to favor production of more stable even A nuclei) is usually also attributed to spallation reactions off more abundant even-A nuclei. These associations immediately lead to some interesting physics conclusions because, from the known density of cosmic rays (at least in the earth's vicinity) and from known spallation cross sections, one can estimate the amount of material through which a typical cosmic ray propagates. Although the estimates are model dependent, typical values are $4 - 6$ g/cm^2 for the effective thickness. Now, the mass density within intragalactic space is about 1 protron/cm^3, or about 1.7×10^{-24} g/cm^3. Thus, taking a velocity of c, we can crudely estimate the cosmic ray lifetime,

$$1.7 \times 10^{-24} \text{g/cm}^3 \times (3 \times 10^{10} \text{cm/sec}) \times t = (4 - 6) \text{g/cm}^2.$$

So this gives $t \sim 3 \times 10^6$ y, in accordance with the estimate mentioned above.

This calculation assumes an average galactic mass density that is not known by direct measurement. Thus it is nice that a more direct estimate of the galactic cosmic ray lifetime is provided by cosmic ray radioactive isotopes. The right chronometer is one that has a lifetime in the ballpark of the estimate above. ^{10}Be, with a lifetime of 1.51×10^6 y, is thus quite suitable. It is a cosmic ray spallation product: this guarantees that it is born as a cosmic ray. Its abundance can be normalized to those of the other, stable Li/Be/B isotopes: the spallation cross sections are known. Thus the absence of ^{10}Be in the cosmic ray spectrum would indicate that the typical cosmic ray lifetime is much larger than 1.51×10^6 y. The survival probability should also depend on the ^{10}Be energy, due to time dilation effects. One observes a reduction in ^{10}Be to about $(0.2 - 0.3)$ of its expected instantaneous production, relative to other Li/Be/B isotopes. From this one concludes $t \sim (2 - 3) \times 10^7$ years. This suggests that the mass density estimate used above (in our first calculation) may have been too high by a factor of $5 - 10$.

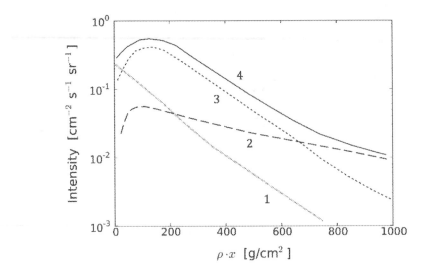

FIGURE 7.4 Intensity of the several components of the secondary cosmic rays as a function of the distance crossed in the atmosphere: 1 - nuclear, 2 - electron-positron, 3 - muons, 4 - total.

During certain short periods, a small part of the cosmic rays that reach the Earth is coming from the Sun. The emission of α-particles and of energetic protons by the Sun happens in the period of maximum solar activity, during which up to 10 eruptions in the chromosphere are observed in a period of one year. The energy of the emitted particles is of the order of 400 MeV, or smaller. However, this radiation of short period can reach 10^8 particles/cm^2/s. The acceleration of these particles is possibly due to the magnetic fields originated by the large currents in the solar plasma.

The magnetic field of the Earth has a large effect on the cosmic rays during their entrance in the atmosphere. In first place, this field avoids that particles with momentum smaller than a certain value p_{min} penetrate the atmosphere. p_{min} is related to the geomagnetic latitude at the arrival point of the cosmic ray. At the equator, this value is of the order of 15 GeV/c, while at the poles it is of order of zero.

In second place, due to the electromagnetic Lorentz force, the magnetic field of the Earth impedes the entrance of cosmic rays in the atmosphere inside a given solid angle of incidence with respect to the horizon. For positively charged particles coming from the east, there is a totally forbidden angular area above the horizon. Thus, this leads to an east-west asymmetry in the direction of incidence of the cosmic rays, what can be easily observed with a simple detectors.

Finally, the magnetic field of the Earth creates a trap around the Earth for particles with certain values of energy and momentum. In these areas, the accumulation of particles is the responsible for the effect of belts of particles around the Earth (*van Allen belts*). Another consequence of this effect is the *aurora borealis*, or *australis*, a light that appears in the sky close to the Earth poles due to the scattering of the sunlight on these particles. Figure 7.4 shows the intensity of the several components of the secondary cosmic rays as a function of the distance crossed in the atmosphere: 1 - nuclear, 2 - electron-positron, 3 - muons, 4 - total.

The cosmic rays that penetrate in the terrestrial atmosphere collide with the atoms of oxygen and nitrogen, originating an effect of cascade collisions. The first collisions yield secondary protons, as well as neutrons and pions. In small amounts kaons, α-particles and secondary nuclei are also produced. Due to the beta-decay of unstable nuclei, muons, neutrinos and gamma-rays are created. The muons are particles of large penetrability in matter and they constitute the so-called "hard component" of the cosmic rays. The photons give origin to electrons and free positrons ("soft" component). In Figure 7.4 the relative intensity of the three components are presented as a function of the mass per unit of area of the atmosphere penetrated by them (at the sea altitude this corresponds to 1000 g/cm^2). Neither the primary cosmic rays, nor the secondary nuclear component, can reach the terrestrial surface with an appreciable probability. Here only electrons, photons and muons are registered. Most of the energy of the primary cosmic rays is transformed in ionization energy of the molecules of gas in the atmosphere.

7.1.4 Scattering by cosmic background photons

There are three main energy loss processes for protons propagating over cosmological distances: Adiabatic energy losses due to the expansion of the Universe, $-(dE/dt)/E = H_0$, e^+e^- pair-, and pion-production on photons of the cosmic microwave background (CMB). The relative energy loss per time of a particle (due to interactions with the CMB) can be estimated as

$$\frac{1}{E}\frac{dE}{dt} = \langle y\sigma n_\gamma\rangle,\tag{7.1}$$

where $y = (E - E')/E$ is the energy fraction lost per interaction, $n_\gamma \approx 410/\text{cm}^3$ is the density of CMB photons with temperature $T \approx 2.7\,\text{K}$ and the brackets $\langle\ldots\rangle$ remind us that we should perform an average of the differential cross section with the momentum distribution of photons. Their typical energy is about 10^{-3} eV.

Since the produced e^+e^- pair in the process $p + \gamma \to p + e^+ + e^-$ is light, this energy loss process has a low threshold energy but leads in turn only to a small energy loss per interaction, $y = 2m_e/m_p \approx 10^{-3}$. The threshold energy on CMB photons follows from $(k_p+k_\gamma)^2 \geq (m_p+2m_e)^2$ as $E_p \geq m_e m_p/E_\gamma \sim 2 \times 10^{18}$ eV. The cross section of the reaction is a factor $\sim 2 \times 10^{-4}$ smaller than Compton scattering.

Let $p_\mu = (\epsilon, \mathbf{p})$ be the photon four-momentum. Then $|\mathbf{p}| = \epsilon$. Let $P_\mu = (\omega, \mathbf{P})$ be the proton four-momentum. The center-of-mass energy is

$$(P + p)^\mu(P + p)_\mu = (\omega + \epsilon)^2 = \epsilon^2_{CM}.$$

If ϵ_{CM} exceeds $m_\pi + M_N$ then the reaction $\gamma + N \to \pi + N$ is possible, degrading the nucleon energy.

The center-of-mass energy is a Lorentz invariant quantity, so it can be evaluated in the laboratory frame

$$\epsilon^2_{CM} = M^2_N + \omega\epsilon - \mathbf{P}\cdot\mathbf{p}.$$

As the cosmic background photons are moving in all directions, we are free to maximize the expression by taking $\cos\theta \sim -1$. As the incident nucleon is highly relativistic $(\epsilon^{max}_{CM})^2 \sim M^2_N + 2\omega\epsilon$. Thus the requirement for photoproduction is $\epsilon^{max}_{CM} \gtrsim m_\pi + M_N$, or

$$\omega \gtrsim \frac{m^2_\pi + 2M_N m_\pi}{2\epsilon} \sim 1.4 \times 10^{20}\text{eV}.\tag{7.2}$$

This results in a mean free path for protons of about 10^8 light years for protons at $\sim 10^{20}$ eV or higher energies, a distance substantially smaller than the horizon. Thus if the origin of such cosmic rays is all of extragalactic space, there should be a very sharp cutoff in the cosmic ray flux at about this energy. This is called the *Greisen-Zatsepin-Kuzmin cutoff* (GZK cutoff for short). Yet evidence for such a cutoff has not found experimentally.

For a more detailed estimate of the above result, we consider the cross section for pion production using a Breit-Wigner formula with the lowest lying nucleon resonance Δ^+ as intermediate state,

$$\sigma_{\text{BW}}(E) = \frac{(2J+1)}{(2s_1+1)(2s_2+1)}\frac{\pi}{p^2_{\text{cms}}}\frac{b_{\text{in}}b_{\text{out}}\Gamma^2}{(E - M_R)^2 + \Gamma^2/4},\tag{7.3}$$

where $p^2_{\text{cms}} = (s - m^2_N)^2/(4s)$, $M_R = m_\Delta = 1.230\,\text{GeV}$, $J = 3/2$, $\Gamma_{\text{tot}} = 0.118\,\text{GeV}$, $b_{\text{in}} \simeq b_{\text{out}} = 0.55\%$. At resonance, we obtain $\sigma_{\text{BW}} \sim 0.4$ mbarn in good agreement with experimental data. We estimate the energy loss length well above threshold $E_{\text{th}} \sim 4 \times 10^{19}$ eV with $\sigma \sim 0.1$ mb and $y = 0.5$ as

$$l^{-1}_{\text{GZK}} = \frac{1}{E}\frac{dE}{dt} \approx 0.5 \times 400/\text{cm}^3 \times 10^{-28}\,\text{cm}^2 \approx 2 \times 10^{-26}\,\text{cm}^{-1},\tag{7.4}$$

or $l_{\text{GZK}} \sim 17\,\text{Mpc}$. Thus the energy loss length of a proton with $E \gtrsim 10^{20}$ eV is comparable to the distance of the closest Galaxy clusters and we should see only local sources at these energies.

The cutoff for nuclei is more severe. At a center-of-mass energy considerably lower nuclei can absorb (in their rest frame) a photon of energy ~ 10 MeV, resulting in photodistintegration. This leads to a GZK cutoff for iron nuclei of $\sim 10^{19}$eV. In fact, The dominant loss process for nuclei of energy $E \gtrsim 10^{19}$ eV is photodisintegration $A + \gamma \rightarrow (A-1) + N$ in the CMB and the infrared background due to the giant dipole resonance. The threshold for this reaction follows from the binding energy per nucleon, ~ 10 MeV. Photodisintegration leads to a suppression of the flux of nuclei above an energy that varies between 3×10^{19} eV for He and 8×10^{19} eV for Fe.

7.2 Thermonuclear cross sections and reaction rates

In this section we will consider nuclear fusion reactions occurring in plasmas, such as in stellar interiors. For more detail on some passages, we recommend the reader to consult Refs. [2, 3, 4, 6, 7, 8, 9, 10, 11, 12, 13, 14, 15, 16, 17, 18, 19, 20, 21, 22, 23, 24]. The nuclear cross section for a reaction between target j and projectile k is defined by

$$\sigma = \frac{\text{number of reactions target}^{-1}\text{sec}^{-1}}{\text{flux of incoming projectiles}} = \frac{r/n_j}{n_k v}, \qquad (7.5)$$

where the target number density is given by n_j, the projectile number density is given by n_k , and v is the relative velocity between target and projectile nuclei. Then r, the number of reactions per cm^3 and sec, can be expressed as $r = \sigma v n_j n_k$, or, more generally,

$$r_{j,k} = \int \sigma |v_j - v_k| d^3 n_j d^3 n_k. \qquad (7.6)$$

The evaluation of this integral depends on the type of particles and distributions which are involved. For nuclei j and k in an astrophysical plasma, obeying a Maxwell-Boltzmann distribution (MB),

$$d^3 n_j = n_j \left(\frac{m_j}{2\pi kT} \right)^{3/2} \exp \left(-\frac{m_j^2 v_j}{2kT} \right) d_j^3 v_j, \qquad (7.7)$$

Eq. (7.6) simplifies to $r_{j,k} = \langle \sigma v \rangle n_j n_k$, where $\langle \sigma v \rangle$ is the average of σv over the temperature distribution in (7.7). More specifically,

$$r_{j,k} = \langle \sigma \text{v} \rangle_{j,k} \, n_j n_k \qquad (7.8)$$

$$\langle j, k \rangle \equiv \langle \sigma \text{v} \rangle_{j,k} = \left(\frac{8}{\mu\pi} \right)^{1/2} (kT)^{-3/2} \int_0^\infty E\sigma(E)\exp\left(-\frac{E}{kT}\right) dE. \qquad (7.9)$$

Here μ denotes the reduced mass of the target-projectile system. In astrophysical plasmas with high densities and/or low temperatures, effects of electron screening become highly important. This means that the reacting nuclei, due to the background of electrons and nuclei, feel a different Coulomb repulsion than in the case of bare nuclei. Under most conditions (with non-vanishing temperatures) the generalized reaction rate integral can be separated into the traditional expression without screening, Eq. (7.8), and a screening factor

$$\langle j, k \rangle^* = f_{scr}(Z_j, Z_k, \rho, T, Y_i) \langle j, k \rangle . \qquad (7.10)$$

This screening factor is dependent on the charge of the involved particles, the density, temperature, and the composition of the plasma. Here Y_i denotes the abundance of nucleus i defined by $Y_i = n_i/(\rho N_A)$, where n_i is the number density of nuclei per unit volume and N_A Avogadro's number. At high densities and low temperatures screening factors can enhance reactions by many orders of magnitude and lead to pycnonuclear ignition.

When in Eq. (7.6) particle k is a photon, the relative velocity is always c and quantities in the integral are not dependent on $d^3 n_j$. Thus it simplifies to $r_j = \lambda_{j,\gamma} n_j$ and $\lambda_{j,\gamma}$ results from an integration of the photodisintegration cross section over a Planck distribution for photons of temperature T,

$$d^3 n_\gamma = \frac{1}{\pi^2 (c\hbar)^3} \frac{E_\gamma^2}{\exp(E_\gamma/kT) - 1} dE_\gamma \tag{7.11}$$

$$r_j = \lambda_{j,\gamma}(T) n_j = \frac{1}{\pi^2 (c\hbar)^3} \int d^3 n_j \int_0^\infty \frac{c\sigma(E_\gamma) E_\gamma^2}{\exp(E_\gamma/kT) - 1} dE_\gamma. \tag{7.12}$$

There is, however, no direct need to evaluate photodisintegration cross sections, because, due to detailed balance, they can be expressed by the capture cross sections for the inverse reaction $l + m \to j + \gamma$ [18],

$$\lambda_{j,\gamma}(T) = \left(\frac{G_l G_m}{G_j} \right) \left(\frac{m_l m_m}{m_j} \right)^{3/2} \left(\frac{m_u kT}{2\pi\hbar^2} \right)^{3/2} \langle l, m \rangle \exp\left(-\frac{Q_{lm}}{kT} \right). \tag{7.13}$$

This expression depends on the reaction Q-value Q_{lm}, the temperature T, the inverse reaction rate $\langle l, m \rangle$, the partition functions $G(T) = \sum_i (2J_i + 1) \exp(-E_i/kT)$ and the mass numbers A of the participating nuclei in a thermal bath of temperature T.

A procedure similar to Eq. (7.12) is used for electron captures by nuclei. Because the electron is about 2000 times less massive than a nucleon, the velocity of the nucleus j is negligible in the center of mass system in comparison to the electron velocity ($|v_j - v_e| \approx |v_e|$). The electron capture cross section has to be integrated over a Boltzmann, partially degenerate, or Fermi distribution of electrons, dependent on the astrophysical conditions. The electron capture rates are a function of T and $n_e = Y_e \rho N_A$, the electron number density. In a neutral, completely ionized plasma, the electron abundance is equal to the total proton abundance in nuclei $Y_e = \sum_i Z_i Y_i$ and

$$r_j = \lambda_{j,e}(T, \rho Y_e) n_j. \tag{7.14}$$

This treatment can be generalized for the capture of positrons, which are in a thermal equilibrium with photons, electrons, and nuclei. At high densities ($\rho > 10^{12}$ g.cm^{-3}) the size of the neutrino scattering cross section on nuclei and electrons ensures that enough scattering events occur to thermalize a neutrino distribution. Then also the inverse process to electron capture (neutrino capture) can occur and the neutrino capture rate can be expressed similarly to Eqs. (7.12) or (7.14), integrating over the neutrino distribution. Also inelastic neutrino scattering on nuclei can be expressed in this form. Finally, for normal decays, like beta or alpha decays with half-life $\tau_{1/2}$, we obtain an equation similar to Eqs. (7.12) or (7.14) with a decay constant $\lambda_j = \ln 2/\tau_{1/2}$ and

$$r_j = \lambda_j n_j. \tag{7.15}$$

The nuclear cross section for charged particles is strongly suppressed at low energies due to the Coulomb barrier. For particles having energies less than the height of the Coulomb barrier, the product of the penetration factor and the MB distribution function at a given temperature results in the so-called Gamow peak, in which most of the reactions will take place. Location and width of the Gamow peak depend on the charges of projectile and target, and on the temperature of the interacting plasma (see Figure 7.5).

Experimentally, it is more convenient to work with the astrophysical S factor

$$S(E) = \sigma(E) E \exp(2\pi\eta), \tag{7.16}$$

with η being the Sommerfeld parameter, describing the s-wave barrier penetration $\eta = Z_1 Z_2 e^2 / \hbar v$. In this case, the steep increase of the cross section is transformed in a rather flat energy dependent function (see Figure 7.6). One can easily see the two contributions of the velocity distribution and the penetrability in the integral

$$\langle \sigma v \rangle = \left(\frac{8}{\pi\mu} \right)^{1/2} \frac{1}{(kT)^{3/2}} \int_0^\infty S(E) \exp\left[-\frac{E}{kT} - \frac{b}{E^{1/2}} \right], \tag{7.17}$$

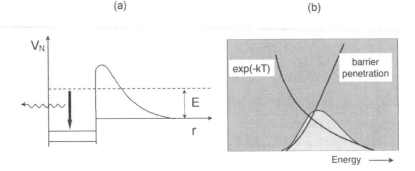

FIGURE 7.5   (a) Schematic representation of the nuclear+Coulomb potential for fusion of charged particles. (b) The integrand of Eq. (7.9) is the product of an exponentially falling distribution with a fastly growing cross section in energy.

where the quantity $b = 2\pi\eta E^{1/2} = (2\mu)^{1/2}\pi e^2 Z_j Z_k/\hbar$ arises from the barrier penetrability. Experimentally it is very difficult to take direct measurements of fusion reactions involving charged particles at very small energies. The experimental data can be guided by a theoretical model for the cross section, which can then be extrapolated to the Gamow energy, as displayed in Figure 7.6(b). The dots symbolize the experimental data points. The solid curve is a theoretical prediction, which supposedly describes the data at high energies. Its extrapolation to lower energies yields the desired value of the S-factor (and of σ) at the energy E_0. The extrapolation can be inadequate due to the presence of resonances and of subthreshold resonances, as shown schematically in Figure 7.6.

Taking the first derivative of the integrand in Eq. (7.17) yields the location E_0 of the Gamow peak, and the effective width Δ of the energy window can be derived accordingly

$$E_0 = \left(\frac{bkT}{2}\right)^{2/3} = 1.22(Z_j^2 Z_k^2 A T_6^2)^{1/3} \text{ keV},$$

$$\Delta = \frac{16 E_0 kT}{3}^{1/2} = 0.749(Z_j^2 Z_k^2 A T_6^5)^{1/6} \text{ keV}, \tag{7.18}$$

as shown in [20], carrying the dependence on the charges Z_j, Z_k, the reduced mass A of the involved nuclei in units of m_u, and the temperature T_6 given in 10^6 K.

For neutron induced reactions there is no Coulomb barrier and the transmission probability of a neutron through a nuclear potential surface is proportional to its velocity v incase one assumes a sharp potential surface. Hence, it is more appropriate to write neutron induced cross sections as

$$\sigma(E) = \frac{R(E)}{v}, \tag{7.19}$$

where $R(E)$ has also a smoother dependence on E. Since S-factors are rather smooth at the low astrophysical energies, it is common to express them in terms of an expansion about $E = 0$, i.e.

$$S = S(0)\left(1 + \frac{5kT}{36E_0}\right) + E_0 \frac{dS(0)}{dE}\left(1 + \frac{15kT}{36E_0}\right) + \cdots, \tag{7.20}$$

where $E_0/kT = (\pi Z_1 Z_2 \alpha/\sqrt{2})^{2/3}(\mu/kT)^{1/3}$ is the Gamow energy and μ the reduced mass of ions with charges Z_1 and Z_2.

Non-resonant reactions induced by the weak interaction, such $(p, e^+\nu)$, neutrino scattering or electron capture reactions, are the smallest of all cross sections. Radiative capture reactions, such as (p, γ) or (α, γ) reactions, have also small cross sections because they involve the electromagnetic interaction. The largest cross sections are for reactions induced by the strong interaction such as (p, α), (d,p), or (α, n) reactions. For example the ^{6}Li$(p,\alpha)^7$Be reaction is 4 orders of magnitude larger than the ^{6}Li$(p,\gamma)^3$He reaction.

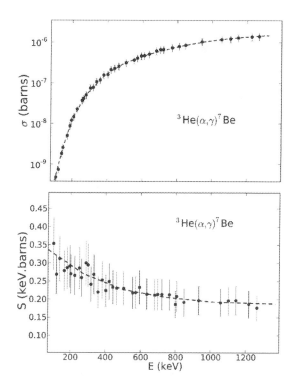

FIGURE 7.6 Upper figure: Cross section data for the reaction $^3\text{He}(\alpha, \gamma)^7\text{Be}$ as a function of the relative energy. Bottom figure: Same data expressed in terms of the astrophysical S-factor defined in Eq. (7.16).

7.3 Reaction networks

The time derivative of the number densities of each of the species in an astrophysical plasma (at constant density) is governed by the different expressions for r, the number of reactions per cm^3 and sec, as discussed above for the different reaction mechanisms which can change nuclear abundances

$$\left(\frac{\partial n_i}{\partial t}\right)_{\rho=const} = \sum_j N_j^i r_j + \sum_{j,k} N_{j,k}^i r_{j,k} + \sum_{j,k,l} N_{j,k,l}^i r_{j,k,l}. \tag{7.21}$$

The reactions listed on the right hand side of the equation belong to the three categories of reactions: (1) decays, photodisintegrations, electron and positron captures and neutrino induced reactions ($r_j = \lambda_j n_j$), (2) two-particle reactions ($r_{j,k} = \langle j, k \rangle n_j n_k$), and (3) three-particle reactions ($r_{j,k,l} = \langle j, k, l \rangle n_j n_k n_l$) like the triple-alpha process ($\alpha + \alpha + \alpha \longrightarrow {}^{12}\text{C} + \gamma$), which can be interpreted as successive captures with an intermediate unstable target ($\alpha + {}^8\text{Be}^* \longrightarrow {}^{12}\text{C} + \gamma$).

The individual N^i's are given by

$$N_j^i = N_i, \quad N_{j,k}^i = N_I \left[\prod_{m=1}^{n_m} |N_{j_m}|!\right]^{-1}, \quad \text{and} \quad N_{j,k,l}^i = N_i \left[\prod_{m=1}^{n_m} |N_{j_m}|!\right]^{-1}. \tag{7.22}$$

The $N_i's$ can be positive or negative numbers and specify how many particles of species i are created or destroyed in a reaction. The denominators, including factorials, run over the n_m different species destroyed in the reaction and avoid double counting of the number of reactions when identical particles react with each other (for example in the $^{12}\text{C} + ^{12}\text{C}$ or the triple-alpha reaction) [18]. In order to exclude changes in the number densities $\dot{n}_i$, which are only due to expansion or contraction of the gas, the nuclear abundances $Y_i = n_i/(\rho N_A)$ were introduced. For a nucleus with atomic weight A_i, $A_i Y_i$ represents the mass fraction of this nucleus, therefore $\sum A_i Y_i = 1$. In terms of nuclear abundances Y_i, a reaction network is described by the following set of differential equations

$$\frac{dY_i}{dt} = \sum_j N_j^i \lambda_j Y_j + \sum_{j,k} N_{j,k}^i \rho N_A \langle j,k \rangle Y_j Y_k + \sum_{j,k,l} N_{j,k,l}^i \rho^2 N_A^2 \langle j,k,l \rangle Y_j Y_k Y_l. \qquad (7.23)$$

Eq. (7.23) derives directly from Eq. (7.21) when the definition for the, $Y_i's$ is introduced. This set of differential equations is solved numerically. They can be rewritten as difference equations of the form $\Delta Y_i/\Delta t = f_i(Y_j(t+\Delta t))$, where $Y_i(t+\Delta t) = Y_i(t) + \Delta Y_i$. In this treatment, all quantities on the right hand side are evaluated at time $t + \Delta t$. This results in a set of non-linear equations for the new abundances $Y_i(t+\Delta t)$, which can be solved using a multi-dimensional Newton-Raphson iteration procedure [5]. The total energy generation per gram, due to nuclear reactions in a time step Δt which changed the abundances by ΔY_i, is expressed in terms of the mass excess $M_{ex,i}c^2$ of the participating nuclei

$$\Delta \epsilon = -\sum_i \Delta Y_i N_A M_{ex,i} c^2, \qquad \frac{d\epsilon}{dt} = -\sum_i \frac{dY_i}{dt} N_A M_{ex,i} c^2. \qquad (7.24)$$

The relative abundances of elements are obtained theoretically by means of these equations using stellar models for the initial conditions, as the neutron density and the temperature. The important ingredients for the calculation of nucleosynthesis, and energy generation by nuclear reactions are (a) nuclear decay half-lives, (b) electron and positron capture rates, (c) photodisintegration rates, (d) neutrino induced reaction rates, and (e) strong interaction cross sections. Additional information such as the initial electron, photon, and nuclide number densities are necessary to obtain the relative abundances of elements by means of the above equations.

7.4 Big Bang Nucleosynthesis

The observed ^{4}He abundance is one of the major predictions of the *Big Bang Nucleosynthesis* (BBN) model. ^{4}He is produced in nuclear fusion reactions in stars, but most of its presence in the Sun and in the Universe is due to BBN, and that is why it is often called by *primordial helium*. In order to explain the primordial helium abundance, the neutron to proton ratio at the moment when nucleosynthesis started should have been very close to n/p = 1/7. This is because hydrogen and helium comprise almost 100% of all elements in the Universe, and thus all neutrons existing in the primordial epoch must have in ended up inside helium. If n/p = 1/7 then for every 14 protons there were 2 neutrons and they combined with 2 protons to form helium, with 12 protons remaining. In this simple algebra, for every ^{4}He nucleus, with 4 nucleons, there are a total of 16 nucleons in the Universe and the mass ratio of helium to all mass in the Universe is 25%, whereas 75% (12/16 = 3/4) is the mass ratio for hydrogen. This is indeed what is observed in the visible Universe, with a very small trace of heavy elements. This is also shown schematically in Figure 7.7.

7.4.1 Physics of the Early Universe

The physics of the early Universe can be formulated in terms of the laws of thermodynamics and statistics and with the assumption that particles are relativistic (see, e.g., the textbooks on cosmology in Refs. [6, 7, 8, 9, 10, 11]). It also relies on the "*cosmological principle*" of a homogeneous and isotropic Universe. The number density, n, energy density, ρ, and entropy density, s, of particles in the early Universe are found to be related to the temperature as

$$n = \frac{\xi(3)}{\pi^2} g'(T) T^3, \qquad \rho = 3p = \frac{\pi^2}{30} g(T) T^4, \qquad s = \frac{p+\rho}{T} = \frac{2\pi^2}{45} g(T) T^3, \qquad (7.25)$$

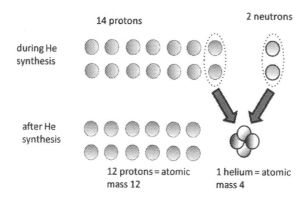

14 protons   **2 neutrons**

during He
synthesis

after He
synthesis

12 protons = atomic
mass 12

1 helium = atomic
mass 4

FIGURE 7.7   At $t \lesssim 1$ min the Universe was very hot, $T \gtrsim 10^9$ K, and protons were more abundant that neutrons in the proportion 7:1. When H and He nuclei were formed, almost all neutrons ended up inside He nuclei yielding 1 helium for every 12 hydrogen nuclei. This explains why 75% of the Universe's visible mass is in hydrogen nuclei and 25% is in helium nuclei.

where $\xi(s) = \sum_{n=1}^{\infty} 1/n^s$ is the Riemann Zeta function[§], p is the pressure and T the temperature in units of the Boltzmann constant k_B[*]. The temperature dependence in these relations is easily understood. At a temperature T, the particle density n increases with T^3 as their momentum states (k_x, k_y, k_z) will roughly fill up to energy (T, T, T) and $n \propto k^3$. For the energy density ρ, an extra power of T appears because each particle has energy T. Therefore, the energy density in the early Universe is governed by the T^4 law and is due to relativistic particles, namely, photons, electrons, positrons, and the three neutrinos.

In Eq. (7.25), the factors g and g' account for the number of degrees of freedom for bosons, b, and fermions, f, with the appropriate spin statistics weights,

$$g'(T) = g_b(T) + \frac{3}{4}g_f(T), \quad \text{and} \quad g(T) = g_b(T) + \frac{7}{8}g_f(T). \tag{7.26}$$

The factors 3/4 and 7/8 arise due to the different fermionic and bosonic energy distributions in the integrals used to obtain n, ρ and s [7]. At the early stages, the Universe is also assumed to be free from dissipative processes such as phase-transitions which can change the overall entropy. Then the following equations for the "*comoving entropy*" holds [11],

$$\frac{d(sa^3)}{dt} = 0, \quad \text{so that} \quad s \propto \frac{1}{a^3}, \quad \text{and} \quad T \propto \frac{1}{a}, \tag{7.27}$$

where a is a scale factor such that the distances in the Universe are measured as $r = a\zeta$, where r is the distance to a point and ζ is a dimensionless measure of distance. Thus, one predicts that in the early Universe the temperature decreased with the inverse of its "size".

The dynamics of the early Universe is governed by the Friedmann equation, which can be easily obtained from Newton's law of an expanding gas interacting gravitationally [7, 8, 9, 10, 11]. It reads

$$\left(\frac{\dot{a}}{a}\right)^2 = \frac{8\pi G\rho}{3}. \tag{7.28}$$

Thus the ratio of expansion of the early Universe (known as the *Hubble constant*), $H = \dot{a}/a$, is proportional to the square root of its energy density, $H \propto \sqrt{\rho}$. A more general form of the

---

[§]$\xi(3) = 1.202\ldots$ is the Apéry's constant.
[*]Here we use units $\hbar = c = k_B = 1$.

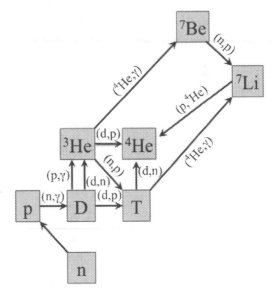

FIGURE 7.8   The 12 most important nuclear reactions (arrows) for Big Bang nucleosynthesis and the nuclei involved.

Friedmann equation is given by $\Omega_m + \Omega_k + \Omega_\Lambda = 1$, where $\Omega_m = \rho/\rho_{crit}$ is the *matter density parameter* of the Universe, $\rho_{crit} = 3H^2/8\pi G$ is the *critical density*, Ω_k is the *curvature density parameter* (for a flat Universe, $\Omega_k = 0$), and $\Omega_\Lambda = \Lambda/3H^2$ is the *vacuum energy parameter*, or *dark energy*, density. Here, Λ is Einstein's cosmological constant. Table 7.1 lists the ratio of the scale factor at the different epochs with that at the present time (denoted by index "0")*.

Equation (7.28) can be readily integrated, using Eq. (7.25), yielding the time-temperature relationship

$$t(s) \simeq \frac{2.4}{T_{10}^2}, \tag{7.29}$$

where T_X is the temperature in units of 10^X K. But in order to get the time evolution right one needs to account for temperatures when particle energies changed enough so that phase transitions occurred. For example, at about 150 to 400 MeV quarks and gluons combined to form nucleons, thus changing the value of the available degrees of freedom g of relativistic particles. This will change the time-temperature relationship, Eq. (7.29), accordingly. The observation of a perfect Planck spectrum for the *Cosmic Microwave Background* (CMB) is a testimony of our hot and dense past and directly demonstrates that the expansion of the Universe was adiabatic (with negligible energy release) back to at least $t \sim 1$ d. With the proper inclusion of nucleosynthesis processes, we can even go back to obtain the proper temperature when $t \sim 1$ s [11].

The several evolution steps of the Universe since the Big Bang to the present epoch are summarized in the following list (for more details, see Refs. [7, 8, 9, 10, 11]):

(1) *Planck era – from zero to* 10^{-43} *s (Planck time)*: During this period, all four known fundamental forces were unified, or equally strong. The distance light travels within the Planck time is 1.6×10^{-35} m and this value should be the size of the Universe. The temperature was of the order of 10^{32} K, corresponding to a particle energy of 10^{19} GeV.

(2) *Grand Unified Theory (GUT) era – up to* 10^{-38} *s*: During this time the force of gravity "freezes out" and becomes distinct from the other forces. The temperatures ran around 10^{27} K, or 10^{14} GeV.

---

*To convert Kelvins to eV: 1 eV = 11,600 K, or 1 K = 8.6×10^{-5} eV.

(3) *Electroweak era* – 10^{-38} *to* 10^{-10} *s*: The strong force "freezes out" and becomes distinct from the other forces. In this era, Higgs bosons collide to create W and Z bosons that carry the electroweak force and quarks. The temperatures are of the order of 10^{15} K, or 100 GeV.

(4) *Particle era* – 10^{-10} *to* 10^{-3} *s*: During this era, the Universe expanded and cooled down so that at its end protons and neutrons begin to form out of free quarks. Temperatures were of the order of 10^{13} K, or 1 GeV. At the end of this era, W and Z bosons were no more created and they started decaying. The electroweak force then separated from the electromagnetic force and became the short-range weak force as we know at present.

(5) *Nucleosynthesis era* – 10^{-3} *s to 3 m*: Even at 10 s, the temperature was high enough so that the energy was passed from electrons and positrons to photons. But as the temperature cooled down in this era, protons and neutrons started fusing into the deuteron, and nuclear fusion started creating helium, and tiny amounts of heavier elements. This is the era of the Big Bang that we are interested in this review.

(6) *Era of nuclei* – *3 minutes to 400,000 yrs*: During this era the temperature was still too hot ($\gtrsim 3,000$ K) so that electrons were not bound to nuclei.

(7) *Era of atoms* – *after 400,000 yrs* – Electrons and nuclei combined into neutral atoms. The first stars were then born. We are also interested in this epoch of the Universe and what comes later, as they are related to stellar nucleosynthesis and stellar evolution.

(8) *Era of galaxies* – *after 200,000 million yrs* – During this time, galaxies begin to form, but at a much higher rate than at present.

When electrons recombined with protons and helium nuclei to form neutral atoms, the photons and baryons decoupled and the Universe became transparent to radiation at $\sim 400,000$ yrs. Oscillations in the photon and baryon densities at this time would lead to tiny variations of 1 part in 10^5 in the temperature observed in the microwave background. Satellite missions have observed these anisotropies and decomposed them in terms of spherical harmonics, with the expansion coefficients describing the magnitude of the anisotropy at a certain angle and closely correlated with the cosmological parameters. These studies lead to the extraction of the values $h \simeq 0.70$ for the dimensionless Hubble constant, $h = H/(100 \text{ km.s}^{-1})$, and $\Omega_m h^2 \simeq 0.022$. One also gets from these observations, $t_0 \simeq 13.8$ Gyr for the Universe age, $\rho_{crit} \simeq 9 \times 10^{-30}$ g.cm^{-3} for the critical density, and $\Omega_b \simeq 0.044$ for the baryon density parameter.

7.4.2   Proton to neutron ratio and helium abundance

We will assume that the laws of physics were the same now and during the Big Bang so that we can predict its behavior. During the early stages ($t \sim 0.01$ s) only γ, $e^{\pm}$ and three neutrino families were present in the Universe. The only existing baryons were neutrons, n, and protons, p. To agree with present observations, calculations based on known physics need to assume a value for the *baryon-to-photon ratio* was $\eta \equiv n_B/n_\gamma$. We do not know why, but as the Universe cooled, a net baryon number remained. Right before nucleosynthesis there were protons and neutrons, not many antineutrons and antiprotons. The reasons why the Universe is made predominantly of matter, with extremely small traces of anti-matter, constitutes some of the most fascinating problems in particle cosmology.

The baryonic density of the Universe has been inferred from the observed anisotropy of the CMB radiation which reflects the number of baryons per photon, η. The number density of photons remains constant after the epoch of electron-positron annihilation, which happened during $4 - 200$ s. The η ratio remains constant during the expansion of the Universe because baryon number is also conserved. Big Bang models are consistent with astronomical observations yielding the value $\eta \simeq 6.1 \times 10^{-10}$, $T \simeq 2.72$ K , and $n_\gamma \simeq 410$ cm^{-3}.

At $t \sim 0.01$ s and $T \sim 10^{11}$ K the thermal energies were $kT \sim 10$ MeV $\gg 2m_e c^2$. Therefore, electrons, positrons and neutrinos were in chemical equilibrium by means of charged- and neutral-current interactions, that is,

$$p + e^- \longleftrightarrow n + \nu_e, \qquad n + e^+ \longleftrightarrow p + \bar{\nu}_e$$
$$p \longleftrightarrow n + e^+ + \nu_e, \qquad n \longleftrightarrow p + e^- + \bar{\nu}_e. \tag{7.30}$$

Energy	T(K)	a/a_0	$t(s)$
~ 10 MeV	10^{11}	1.9×10^{-11}	.01
~ 1 MeV	10^{10}	1.9×10^{-10}	1.1
~ 100 keV	10^9	2.6×10^{-9}	180
~ 10 keV	10^8	2.7×10^{-8}	19,000

TABLE 7.1   Properties of the early Universe according to Friedmann cosmology: particle energy, temperature, radius scale and elapsed time. The subindex 0 means present time [7, 8, 9, 10, 11].

According to the Boltzmann-Gibbs statistics, the neutron-to-proton ratio at this time should be

$$\frac{n}{p} = \frac{e^{-m_n/kT}}{e^{-m_p/kT}} = e^{-\Delta m/kT}, \tag{7.31}$$

where $\Delta m = 1.294$ MeV. At $T = 10^{11}$ K, $kT = 8.62$ MeV and n/p $= 0.86$, i.e., there were more protons than neutrons. Fermi's theory of the weak interaction predicts that the weak rates in Eqs. (7.30) drop rapidly with T^5. At some temperature the weak rates were so slow that they could not stay in pace with other changes in the Universe. A *decoupling temperature* existed when the neutron lifetime and the Hubble expansion rate, $H = \dot{a}/a$, were similar. The neutrinos start to behave as free particles, and for $T < 10^{10}$ K they are ineffective, i.e., matter became transparent to them. This temperature is equivalent to thermal energies of ~ 1 MeV, comparable to the energy for creation of electron-positron pairs. Electrons and positrons start to annihilate, and for some unclear reason a small excess of electrons remained. Neutrons and protons could be captured at this temperature, but the thermal energies were still too high and would dissociate the nuclei.

Because neutrinos were ineffective (i.e., decoupled), $e^{\pm}$ annihilation has heated up the photon background relative to them. A numerical calculation using the available degrees of freedom finds that the ratio of the two temperatures is [11] $T_\nu/T_\gamma = (4/11)^{1/3}$. The weak decay reaction n $\leftrightarrow$ p breaks out of equilibrium, and the n/p ratio becomes $n/p = \exp[-\Delta m/kT] = 0.25$, which means a large drop of the n/p ratio. This is still not 1/7, the value required to explain the observed helium abundance (see Figure 7.7). But the ν-induced reactions still favored the proton-rich side. We are now at 1 second (1 MeV) after the Big Bang and in the following 10 s the n/p ratio will decrease to about $0.17 \sim 1/6$ [11]. In a timescale of about 10 minutes the neutron β-decay will drive the n/p ratio to its magic 1/7 value. This sequence of physics processes is summarized in Figure 7.9, left, based on numerical calculations [12].

We also conclude that the first nuclei were formed within less than 10 min, as β decay cannot continue and keep the n/p ratio equal to the 1/7 steady value. The nucleon gas in the early Universe was dilute and nuclei are therefore created via two-body reactions. The lightest two-nucleon system, the deuteron, has only one bound-state, with a small binding energy of 2.24 MeV. Neutrons and protons form the deuteron by means of the radiative capture reaction

$$n + p \leftrightarrow d + \gamma. \tag{7.32}$$

This reaction is very fast because it is electromagnetic and it is therefore in equilibrium during the early Universe. Hence, the ratio of (n+p) pairs to the number of deuterons is obtained using the Boltzmann-Gibbs statistics,

$$\frac{N_{n+p\to d+\gamma}}{N_{d+\gamma}} = \frac{n_d n_\gamma \exp[-\Delta m/kT]}{n_p n_n}. \tag{7.33}$$

We can assume that at the time the deuteron was formed $n_p/n_\gamma \sim \eta$ and $n_n \sim n_d$, i.e., half of the neutrons are inside deuterium nuclei. Thus, $\exp[-2.24 \text{ MeV}/kT] \sim \eta \sim 10^{-9}$, and we get the temperature when deuterium is formed as $T_{\text{deuterium}} \simeq 1.25 \times 10^9$ K or $t_{\text{deuterium}} \sim 100$ s. Therefore, nucleosynthesis begins at about 100 seconds after the Big Bang. A numerical calculation of the ratio n/p from the time of the weak freeze-out, or decoupling (~ 1 s), to 100 s yields

$$\frac{n}{p} \sim 0.15 \sim \frac{1}{7}. \tag{7.34}$$

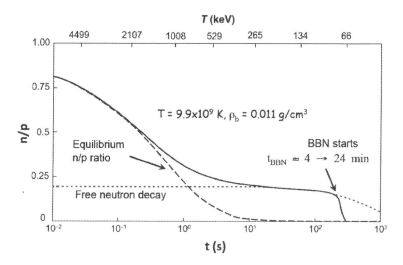

FIGURE 7.9 The neutron-to-proton ratio is shown as a function of time (lower scale) and temperature (upper scale). The n/p ratio at equilibrium, given by $\exp(-\Delta m/kT)$, is displayed by the dashed curve. The time-decay of free neutrons, $\exp(-t/\tau_n)$, is given by the dotted curve. The solid curve is the total n/p time dependence. The sudden fall-off at a few hundred seconds is due to the onset of BBN.

Hence, the n/p ratio is a "fingerprint" of the thermal history of the primordial Universe.

At the start of nucleosynthesis, $T \sim 100$ keV, whereas we would have expected it to be about 2 MeV, the deuterium binding energy. This is due to the very small value of η. Based on cross section estimates, the reaction rate for deuteron formation at $T \sim 100$ keV is

$$\sigma v(\text{p} + \text{n} \rightarrow \text{d} + \gamma) \simeq 5 \times 10^{-20} \text{ cm}^3/\text{s}. \qquad (7.35)$$

This implies a density $\rho \sim 1/\sigma v \sim 10^{17}$ cm^{-3} for $t \sim 100$ s. The density in baryons today is obtained from the observations of the density of visible matter, $\rho_0 \sim 10^{-7}$ cm^{-3}. As the baryon density n is proportional to $R^{-3} \sim T^3$, the temperature today should be $T_0 = (\rho_0/\rho)^{1/3}T_{\text{BBN}} \sim 10$ K, which is in fact close to the observed 2.72 K temperature of the CMB. Numerical calculations show that 3.5 minutes after the Big Bang the temperature decreases to 3×10^8 K and the density is 10^{-4} kg m^{-3}. At this time the Universe consists of 70% photons, 30% neutrinos, and only 10^{-7}% of other particles. From those about 70 to 80% is hydrogen and 20 to 30% is helium, and the associated number of electrons.

The $\text{p} + \text{n} \rightarrow \text{d} + \gamma$ cross section in Eq. (7.35) at the BBN energies, $0.02 < E < 0.2$ MeV, is not well known experimentally, but it can be deduced by detailed balance from its inverse reaction $\gamma + \text{d} \rightarrow \text{p} + \text{n}$, as realized in the 1940's. The deuteron photodisintegration from the threshold to 20 MeV photon energy is largely correlated with the deuterium binding energy and with the triplet effective-range of the neutron-proton interaction and the dependence of the cross section on the γ-energy is not influenced appreciably by the details of the interaction.

7.4.3   BBN of elements up to beryllium

Once deuterium is formed, reactions proceed quickly to produce helium by means of (see Figure 7.8)

$$\text{d}(\text{n},\gamma)^3\text{H}, \quad \text{d}(\text{d},\text{p})^3\text{H}, \quad \text{d}(\text{p},\gamma)^3\text{He}, \quad \text{d}(\text{d},\text{n})^3\text{He}, \quad {}^3\text{He}(\text{n},\text{p})^3\text{H}. \qquad (7.36)$$

At a longer time, ^3H will β-decay to ^3He.

When temperatures cool down to ~ 100 keV, the reactions

$$^3\text{H}(\text{p},\gamma)^4\text{He}, \quad {}^3\text{H}(\text{d},\text{n})^4\text{He}, \quad {}^3\text{He}(\text{n},\gamma)^4\text{He}, \quad {}^3\text{He}(\text{d},\text{p})^4\text{He}, \quad {}^3\text{He}({}^3\text{He},2\text{p})^4\text{He}, \qquad (7.37)$$

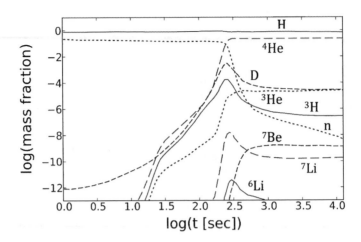

FIGURE 7.10 Mass and number fractions of nuclei during the BBN as a function of temperature and time. ^{4}He is presented as a *mass fraction* and the other elements are given as *number fractions*.

will follow, despite being strongly suppressed by Coulomb repulsion. The repulsive Coulomb barrier is given by $e^2 Z_1 Z_2/r$, which for $Z_2 = Z_2 = 2$ and $r \sim 2$ fm, yields about 3 MeV. Hence, kinetic energies of $\sim$100 keV lead to very small tunneling probabilities through the Coulomb barriers.

Once ^{3}He and ^{4}He are formed, only very tiny amounts of ^{7}Li and ^{7}Be will be produced through

$$^4\text{He}(^3\text{H},\gamma)^7\text{Li}, \qquad ^4\text{He}(^3\text{He},\gamma)^7\text{Be}. \tag{7.38}$$

When ^{7}Be becomes an atom, i.e., after electrons are captured, it will be able to capture a bound electron and to decay into ^{7}Li. That is when BBN should stop as the Universe expands, cools down, and it becomes increasingly more difficult for the nuclear fusion reactions to proceed. Only extremely small amounts of elements heavier than Li isotopes are formed compared to those produced in stars during the later evolution of the Universe.

If one increases the value of the baryon/photon ratio, η, then one gets a higher BBN temperature T_{BBN} and also a larger n/p ratio [12]. Thus, larger η yields larger ^{4}He abundance. Deuterium and ^{3}He act as "catalysts" because as soon as they are produced, they are immediately consumed, reaching an equilibrium value that depends on the rate of their production and consumption. Production starts with deuteron formation, which has no Coulomb barrier. Destruction proceeds via numerous reactions, e.g., d+d, d+p, etc, which are suppressed by Coulomb repulsion. A larger T_{BBN} enhances more destruction than production and yields less d, and ^{3}He. Numerical calculations show that ^{3}He is produced at the ^{3}He/H $\sim 10^{-5}$ level in low mass stars and slowly will become larger over the history of the Universe [12]. For small η (low T_{BBN}) ^{7}Li is created and destroyed by

$$^4\text{He}(^3\text{H},\gamma)^7\text{Li} \qquad \text{and} \qquad ^7\text{Li}(p,\alpha)^4\text{He}, \tag{7.39}$$

respectively. The second reaction has a smaller Coulomb barrier and thus a lower T_{BBN} will increase more the creation of ^{7}Li than its destruction yielding more ^{7}Li. On the other hand, for large η, ^{7}Li production changes because more ^{3}He is produced, and ^{4}He(^{3}He,γ)^{7}Be becomes important, benefited from high temperatures because of the Coulomb barrier. Also, there are very few neutrons around to deplete ^{7}Be by (n,α) reactions. Therefore, ^{7}Li produced via the decay of ^{7}Be is also occurring at high T and large η. The final reaction in the list of the most important ones for BBN is the ^{7}Be(n, p)^{7}Li reaction.

In Figure 7.10 numerical calculations are presented for the mass and number fractions of nuclei during the BBN as a function of temperature and time. ^{4}He is presented as a *mass fraction* and the other elements are given as *number fractions* [13]. The calculations are done assuming the present knowledge of the cross sections and S-factors for the reactions of relevance for BBN. The agreement between the calculations and present astronomical observations of primordial elements are very good, except for a few cases, such as the lithium isotopes ^{6}Li and ^{7}Li. The BBN predictions and observations are of broad interest for its significance in constraining cosmological models of the cosmic expansion of the early Universe. Cosmological observations of type Ia supernovae at intermediate redshift, together with additional observational restrictions at low and intermediate redshift, as well as the CMB spectrum of temperature anisotropies and polarization all suggest that the Universe is accelerating due to the influence of a dominant dark energy $\Omega_\Lambda \approx 0.7$ with a negative pressure P. The simplest interpretation for this dark energy is the existence of a cosmological constant in Einstein's field equations. The resulting the equation of state from this assumption is that the relation of pressure and density ρ is of the form $P = -\rho$.

7.5 Models for astrophysical nuclear cross sections

Nuclear burning in astrophysical environments produces unstable nuclei, which again can be targets for subsequent reactions. In addition, it involves a very large number of stable nuclei, which are not fully explored by experiments. Thus, it is necessary to be able to predict reaction cross sections and thermonuclear rates with the aid of theoretical models. Especially during the hydrostatic burning stages of stars, charged-particle induced reactions proceed at such low energies that a direct cross-section measurement is often not possible with existing techniques. Hence extrapolations down to the stellar energies of the cross sections measured at the lowest possible energies in the laboratory are the usual procedures to apply. To be trustworthy, such extrapolations should have as strong a theoretical foundation as possible. Theory is even more mandatory when excited nuclei are involved in the entrance channel, or when unstable very neutron-rich or neutron-deficient nuclides (many of them being even impossible to produce with present-day experimental techniques) have to be considered. Such situations are often encountered in the modelling of explosive astrophysical scenarios.

Various models have been developed in order to complement the experimental information.

7.5.1 Microscopic models

In these model, the nucleons are grouped into clusters, as was explained in section 3.12. Keeping the internal cluster degrees of freedom fixed, the totally antisymmetrized relative wave functions between the various clusters are determined by solving the Schrödinger equation for a many-body Hamiltonian with an effective nucleon-nucleon interaction. When compared with most others, this approach has the major advantage of providing a consistent, unified and successful description of the bound, resonant, and scattering states of a nuclear system. Various improvements of the model have been made.

The microscopic model has been applied to many important reactions involving light systems, and in particular to the various p-p chain reactions [21]. The available experimental data can generally be well reproduced. The microscopic cluster model or its variant (the microscopic potential model) has also made an important contribution to the understanding of the key ^{12}C$(\alpha, \gamma)^{16}$O reaction rate.

7.5.2 Potential models

The potential model has been known for a long time to be a useful tool in the description of radiative capture reactions. It assumes that the physically important degrees of freedom are the relative motion between the (structureless) nuclei in the entrance and exit channels, and by the introduction of spectroscopic factors and strength factors in the optical potential. The associated drawbacks are that the nucleus-nucleus potentials adopted for calculating the initial and final wave functions from the Schrödinger equation cannot be unambiguously defined, and that the spectroscopic factors cannot be derived from first principles. They have instead to be obtained from more or less rough "educated guesses."

The *potential model* can be applied successfully [25] to several radiative capture reactions, such as the p-p chain ^{3}He $(\alpha, \gamma)^7$ Be reaction, to the ^{7}Be $(p, \gamma)^8$ B, and to the ^{8}Li $(n, \gamma)^9$ Li reactions. In this model the bound state wave functions of $c = a + b$ are specified by

$$\Psi_{JM}(\mathbf{r}) = \frac{u_{lj}^J(r)}{r} \mathcal{Y}_{JM}^l, \tag{7.40}$$

where $\mathbf{r}$ is the relative coordinate of a and b, $u_{lj}^J(r)$ is the radial wave function and $\mathcal{Y}_{JM}^l$ is the spin-angle wave function

$$\mathcal{Y}_{JM}^l = \sum_{m, M_a} \langle jmI_aM_a|JM\rangle \, |jm\rangle \, |I_aM_a\rangle, \qquad \text{with} \qquad |jm\rangle = \sum_{m_l, M_b} Y_{lm_l}(\hat{\mathbf{r}}) \chi_{M_b} \tag{7.41}$$

where χ_{M_b} is the spinor wave function of particle b and $\langle jmI_aM_a|JM\rangle$ is a Clebsch-Gordan coefficient.

The ground-state wave function is normalized so that $\int d^3r \, |\Psi_{JM}(\mathbf{r})|^2 = 1$, i.e., $\int_0^\infty dr |u_{lj}^J(r)|^2 = 1$. The wave functions are usually calculated with a central + spin-orbit + Coulomb potential of the form

$$V(\mathbf{r}) = V_0(r) + V_S(r)(\mathbf{l.s}) + V_C(r), \tag{7.42}$$

which follows the same kind of parametrization of the real part of the optical potential, as described in Section 4.7. The potential parameters are adjusted so that the ground state energy E_B (or the energy of excited states) is reproduced.

The bound-state wave functions are calculated by solving the radial Schrödinger equation

$$-\frac{\hbar^2}{2m_{ab}} \left[\frac{d^2}{dr^2} - \frac{l(l+1)}{r^2} \right] u_{lj}^J(r) + [V_0(r) + V_C(r) + \langle \mathbf{s.l}\rangle V_{S0}(r)] u_{lj}^J(r) = E_i u_{lj}^J(r), \tag{7.43}$$

where $\langle \mathbf{s.l}\rangle = [j(j+1) - l(l+1) - s(s+1)]/2$. This equation must satisfy the boundary conditions $u_{lj}^J(r=0) = u_{lj}^J(r=\infty) = 0$ which is only possible for discrete energies E corresponding to the bound states of the nuclear + Coulomb potential.

The continuum wave functions are calculated with the potential model as described above. The parameters are often not the same as the ones used for the bound states. The continuum states are now identified by the notation $u_{Elj}^J(r)$, where the (continuous) energy E is related to the relative momentum k of the system $a + b$ by $E = \hbar^2 k^2/2m_{ab}$. They are normalized so as to satisfy the relation

$$\left\langle u_{Elj}^J \middle| u_{E'l'j'}^{J'} \right\rangle = \delta\left(E - E'\right) \delta_{JJ'} \delta_{jj'} \delta_{ll'}, \tag{7.44}$$

what means, in practice, that the continuum wave functions $u_{Elj}(r)$ are normalized to $-\sqrt{2m_{ab}/\pi\hbar^2 k} \, e^{i\delta_{lJ}} \sin(kr + \delta_{lJ})$ at large r.

The radial equation for the continuum wave functions can also be obtained with Eq. (7.43), but with the boundary conditions at infinity replaced by (see Section 2.14)

$$u_{Elj}^J(r \longrightarrow \infty) = i\sqrt{\frac{m_{ab}}{2\pi k\hbar^2}} \left[H_l^{(-)}(r) - S_{lJ} H_l^{(+)}(r) \right] e^{i\sigma_l(E)}, \tag{7.45}$$

where $S_{lJ} = \exp[2i\delta_{lJ}(E)]$, with $\delta_{lJ}(E)$ being the nuclear phase-shift and $\sigma_l(E)$ the Coulomb one, and $H_l^{(\pm)}(r) = G_l(r) \pm iF_l(r)$. F_l and G_l are the regular and irregular Coulomb wave functions. If the particle b is not charged (e.g., a neutron) the Coulomb functions reduce to the usual spherical Bessel functions, $j_l(r)$ and $n_l(r)$.

At a conveniently chosen large distance $r = R$, outside the range of the nuclear potential, one can define the logarithmic derivative

$$\mathcal{L}_{lJ} = \left(\frac{du_{Elj}^J/dr}{u_{Elj}^J} \right)_{r=R}. \tag{7.46}$$

The phase shifts $\delta_{lJ}(E)$ are obtained by matching the logarithmic derivative with the asymptotic value obtained with the Coulomb wave functions. This procedure yields

$$S_{lJ} = \frac{G_l' - iF_l' - \mathcal{L}_{lJ}(G_l - iF_l)}{G_l' + iF_l' - \mathcal{L}_{lJ}(G_l + iF_l)}, \tag{7.47}$$

where the primes mean derivation with respect to the radial coordinate at the position R.

As described in Chapter 6, the operators for electric transitions of multipolarity $\lambda\pi$ are given by

$$\mathcal{M}_{E\lambda\mu} = e_\lambda r^\lambda Y_{\lambda\mu}(\widehat{\mathbf{r}}), \tag{7.48}$$

where the effective charge, which takes into account the displacement of the center-of-mass, is

$$e_\lambda = Z_b e \left(-\frac{m_a}{m_c}\right)^\lambda + Z_a e \left(\frac{m_b}{m_c}\right)^\lambda. \tag{7.49}$$

For magnetic dipole transitions

$$\mathcal{M}_{M1\mu} = \sqrt{\frac{3}{4\pi}}\mu_N \left[e_M l_\mu + \sum_{i=a,b} g_i (s_i)_\mu\right], \qquad e_M = \left(\frac{m_a^2 Z_a}{m_c^2} + \frac{m_b^2 Z_b}{m_c^2}\right), \tag{7.50}$$

where l_μ and s_μ are the spherical components of order μ ($\mu = -1, 0, 1$) of the orbital and spin angular momentum ($\mathbf{l} = -i\mathbf{r} \times \nabla$, and $\mathbf{s} = \sigma/2$) and g_i are the gyromagnetic factors of particles a and b. The nuclear magneton is given by $\mu_N = e\hbar/2m_N c$.

The matrix element for the transition $J_0 M_0 \longrightarrow JM$ is given by

$$\langle JM |\mathcal{M}_{E\lambda\mu}| J_0 M_0\rangle = \langle J_0 M_0 \lambda\mu|JM\rangle \frac{\langle J \|\mathcal{M}_{E\lambda}\| J_0\rangle}{\sqrt{2J+1}}. \tag{7.51}$$

From the single-particle wave functions one can calculate the reduced matrix elements $\langle lj \|\mathcal{M}_{E\lambda}\| l_0 j_0\rangle_J$. The subscript J is a reminder that the matrix element depends on the channel spin J, because one can use different potentials in the different channels. The reduced matrix element $\langle J \|\mathcal{M}_{E\lambda}\| J_0\rangle$ can be obtained from a standard formula of angular momentum algebra. One gets,

$$\langle J \|\mathcal{M}_{E\lambda}\| J_0\rangle = (-1)^{j+I_a+J_0+\lambda} \left[(2J+1)(2J_0+1)\right]^{1/2} \begin{Bmatrix} j & J & I_a \\ J_0 & j_0 & \lambda \end{Bmatrix} \langle lj \|\mathcal{M}_{E\lambda}\| l_0 j_0\rangle_J. \tag{7.52}$$

To obtain $\langle lj \|\mathcal{M}_{E\lambda}\| l_0 j_0\rangle_J$ one needs the matrix element $\langle lj \|r^\lambda Y_\lambda\| l_0 j_0\rangle_J$ for the spherical harmonics. For $l_0 + l + \lambda =$ even. The result is

$$\langle lj \|\mathcal{M}_{E\lambda}\| l_0 j_0\rangle_J = \frac{e_\lambda}{\sqrt{4\pi}} (-1)^{l_0+l+j_0-j} \frac{\widehat{\lambda}\widehat{j_0}}{\widehat{j}} \langle j_0 \tfrac{1}{2}\lambda 0|j\tfrac{1}{2}\rangle \int_0^\infty dr r^\lambda u_{lj}^J (r)\, u_{l_0 j_0}^{J_0} (r), \tag{7.53}$$

where we use here the notation $\widehat{k} = \sqrt{2k+1}$, and $\widetilde{k} = \sqrt{k(k+1)}$. For $l_0 + l + \lambda =$ odd, the reduced matrix element is null.

The multipole strength, or response functions, for a particular partial wave, summed over final channel spins, is defined by

$$\frac{dB(\pi\lambda;\, l_0 j_0 \longrightarrow klj)}{dk} = \sum_J \frac{|\langle kJ \|\mathcal{M}_{\pi\lambda}\| J_0\rangle|^2}{2J_0+1}$$

$$= \sum_J (2J+1) \begin{Bmatrix} j & J & I_a \\ J_0 & j_0 & \lambda \end{Bmatrix}^2 |\langle klj \|\mathcal{M}_{\pi\lambda}\| l_0 j_0\rangle_J|^2, \tag{7.54}$$

where $\pi = E$, or M.

If the matrix elements are independent of the channel spin, this sum reduces to the usual single-particle strength $|\langle klj \|\mathcal{M}_{\pi\lambda}\| l_0 j_0\rangle|^2 / (2j_0+1)$. For transitions between the bound states the same formula as above can be used to obtain the reduced transition probability by replacing the continuum wave functions $u_{klj}^J (r)$ by the bound state wave function $u_{lj}^J (r)$. That is,

$$B(\pi\lambda; l_0 j_0 J_0 \longrightarrow ljJ) = (2J+1) \begin{Bmatrix} j & J & I_a \\ J_0 & j_0 & \lambda \end{Bmatrix}^2 |\langle lj \|\mathcal{M}_{\pi\lambda}\| l_0 j_0\rangle|^2. \tag{7.55}$$

For bound state to continuum transitions the total multipole strength is obtained by summing over all partial waves,

$$\frac{dB\,(\pi\lambda)}{dE} = \sum_{lj} \frac{dB\,(\pi\lambda; l_0 j_0 \longrightarrow klj)}{dE}. \tag{7.56}$$

The differential form of the response function in terms of the momentum E is a result of the normalization of the continuum waves according to Eq. (7.45).

Generalizing Eq. (6.153), the photo-absorption cross section for the reaction $\gamma + c \longrightarrow a + b$ is given in terms of the response function by

$$\sigma_\gamma^{(\lambda)}\,(E_\gamma) = \frac{(2\pi)^3\,(\lambda+1)}{\lambda\,[(2\lambda+1)!!]^2}\left(\frac{E_\gamma}{\hbar c}\right)^{2\lambda-1}\frac{dB\,(\pi\lambda)}{dE}, \tag{7.57}$$

where $E_\gamma = E + |E_B|$, with $|E_B|$ being the binding energy of the $a + b$ system. For transitions between bound states, one replaces $dB(\pi\lambda)/dE$ by $B\,(\pi\lambda; l_0 j_0 J_0 \longrightarrow ljJ)\,\delta\,(E_f - E_i - E_\gamma)$ where E_i (E_f) is the energy of the initial (final) state.

The cross section for the radiative capture process $a + b \longrightarrow c + \gamma$ can be obtained by detailed balance (see Eq. (4.179)), and one gets

$$\sigma_{(\pi\lambda)}^{(\mathrm{rc})}\,(E) = \left(\frac{E_\gamma}{\hbar c}\right)^{2\lambda-1}\frac{2\,(2I_c+1)}{(2I_a+1)(2I_b+1)}\sigma_\gamma^{(\lambda)}\,(E_\gamma). \tag{7.58}$$

The total capture cross section σ_{nr} is determined by the capture to all bound states with the single particle spectroscopic factors S_i in the final nucleus

$$\sigma_{\mathrm{nr}}\,(E) = \sum_{i,\pi,\lambda} S_i \sigma_{(\pi\lambda),i}^{(\mathrm{rc})}\,(E). \tag{7.59}$$

Experimental information or detailed shell model calculations have to be performed to obtain the spectroscopic factors S_i.

The threshold behavior of radiative capture cross sections is fundamental in nuclear astrophysics because of the small projectile energies in the thermonuclear region. For example, for neutron capture near the threshold the cross section can be written (see Chapter 2) as

$$\sigma_{if} = \frac{\pi}{k^2}\frac{-4kR\,\mathrm{Im}\mathcal{L}_0}{|\mathcal{L}_0|^2}, \tag{7.60}$$

where $\mathcal{L}_0$ is the logarithmic derivative for the s wave. Since $\mathcal{L}_0$ is only weakly dependent on the projectile energy, one obtains for low energies the well–known $1/v$–behavior.

In most astrophysical neutron–induced reactions, neutron s–waves will dominate, resulting in a cross section showing a $1/v$–behavior (i.e., $\sigma(E) \propto 1/\sqrt{E}$). In this case, the reaction rate will become independent of temperature, $R = $ const. Therefore it will suffice to measure the cross section at one temperature in order to calculate the rates for a wider range of temperatures. The rate can then be computed very easily by using

$$R = \langle\sigma v\rangle = \langle\sigma\rangle_T\, v_T = \mathrm{const}, \tag{7.61}$$

with

$$v_T = \left(\frac{2kT}{m}\right)^{1/2}. \tag{7.62}$$

The mean lifetime τ_{n} of a nucleus against neutron capture, i.e., the mean time between subsequent neutron captures is inversely proportional to the available number of neutrons n_{n} and the reaction rate $R_{\mathrm{n}\gamma}$:

$$\tau_{\mathrm{n}} = \frac{1}{n_{\mathrm{n}} R_{\mathrm{n}\gamma}}. \tag{7.63}$$

If this time is shorter than the beta–decay half–life of the nucleus, it will be likely to capture a neutron before decaying (*r-process*). In this manner, more and more neutrons can be captured to build up nuclei along an isotopic chain until the beta–decay half–life of an isotope finally becomes shorter than τ_n. With the very high neutron densities encountered in several astrophysical scenarios, isotopes very far–off stability can be synthesized.

For low $|E_B|$-values, e.g. for *halo-nuclei*, the simple $1/v$-law does not apply anymore. A significant deviation can be observed if the neutron energy is of the order of the $|E_B|$-value. In this case the response function in Eq. (7.56) can be calculated analytically under simplifying assumptions. For direct capture to weakly bound final states, the bound–state wave function $u_{lj}(r)$ decreases very slowly in the nuclear exterior, so that the contributions come predominantly from far outside the nuclear region, i.e., from the *nuclear halo*. For this asymptotic region the scattering and bound wave functions in Eq. (7.40) can be approximated by their asymptotic expressions neglecting the nuclear potential

$$u_l(kr) \propto j_l(kr), \qquad u_{l_0}(r) \propto h_{l_0}^{(+)}(i\xi r),$$

where j_l and $h_{l_0}^{(+)}$ are the spherical Bessel, and the Hankel function of the first kind, respectively. The separation energy $|E_B|$ in the exit channel is related to the parameter ξ by $|E_B| = \hbar^2 \xi^2/(2m_{ab})$.

Performing the calculations of the radial integrals in Eq. (7.53), one readily obtains the energy dependence of the radiative capture cross section for halo nuclei. For example, for a transition s⟶p it becomes

$$\sigma_{(E1)}^{(rc)}(\text{s} \to \text{p}) \propto \frac{1}{\sqrt{E}} \frac{(E + 3|E_B|)^2}{E + |E_B|}, \tag{7.64}$$

while a transition p→s has the energy dependence

$$\sigma_{(E1)}^{(rc)}(\text{p} \to \text{s}) \propto \frac{\sqrt{E}}{E + |E_B|}. \tag{7.65}$$

If $E \ll |E_B|$ the conventional energy dependence is recovered. From the above equations one obtains that the reaction rate is not constant (for s-wave capture) or proportional to T (for p-wave capture) in the case of small $|E_B|$-values. These general analytical results can be used as a guide for interpreting the numerical calculations involving neutron halo nuclei.

For the case of resonances, where E_r is the resonance energy, we can approximate $\sigma(E)$ by a Breit-Wigner resonance formula, i.e.,

$$\sigma_r(E) = \frac{\pi \hbar^2}{2\mu E} \frac{(2J_R + 1)}{(2J_a + 1)(2J_b + 1)} \frac{\Gamma_p \Gamma_\gamma}{(E_r - E)^2 + (\Gamma_{\text{tot}}/2)^2}, \tag{7.66}$$

where J_R, J_a, and J_b are the spins of the resonance and the nuclei a and b, respectively, and the total width Γ_{tot} is the sum of the particle decay partial width Γ_p and the γ-ray partial width Γ_γ. The particle partial width, or entrance channel width, Γ_p can be expressed in terms of the single-particle spectroscopic factor S_i and the single-particle width $\Gamma_{\text{s.p.}}$ of the resonance state

$$\Gamma_p = S_i \Gamma_{\text{s.p.}}. \tag{7.67}$$

The single-particle width $\Gamma_{\text{s.p.}}$ can be calculated from the scattering phase-shifts of a scattering potential with the potential parameters being determined by matching the resonance energy.

The gamma partial widths Γ_γ are calculated from the electromagnetic reduced transition probabilities $B(J_i \to J_f; L)$ which carry the nuclear structure information of the resonance states and the final bound states. The reduced transition rates are usually computed within the framework of the shell model.

Most of the typical transitions are M1 or E2 transitions. For these the relations are

$$\Gamma_{E2}[\text{eV}] = 8.13 \times 10^{-7} E_\gamma^5 \text{ [MeV] } B(E2) \text{ [e}^2\text{fm}^4], \tag{7.68}$$

and

$$\Gamma_{M1}[\text{eV}] = 1.16 \times 10^{-2} E_\gamma^3 \text{ [MeV] } B(M1) \text{ [}\mu_N^2]. \tag{7.69}$$

For the case of narrow resonances, with width $\Gamma \ll E_r$, the Maxwellian exponent $\exp(-E/k_B T)$ can be taken out of the integral, and one finds

$$\langle \sigma v \rangle = \left(\frac{2\pi}{m_{ab} kT} \right)^{3/2} \hbar^2 (\omega\gamma)_R \exp\left(-\frac{E_r}{kT} \right), \tag{7.70}$$

where the *resonance strength* is defined by

$$(\omega\gamma)_R = \frac{2J_R + 1}{(2J_a + 1)(2J_b + 1)}(1 + \delta_{ab})\frac{\Gamma_p\Gamma_\gamma}{\Gamma_{\text{tot}}}. \tag{7.71}$$

For broad resonances Eq. (7.17) is usually calculated numerically. An interference term has to be added. The total capture cross section is then given by

$$\sigma(E) = \sigma_{\text{nr}}(E) + \sigma_r(E) + 2[\sigma_{\text{nr}}(E)\sigma_r(E)]^{1/2}\cos[\delta_R(E)]. \tag{7.72}$$

In this equation $\delta_R(E)$ is the resonance phase shift. Only the contributions with the same angular momentum of the incoming wave interfere in Eq. (7.72).

7.5.3 Field theory models

Field theories adopt a completely independent approach for nuclear physics calculations in which the concept of nuclear potentials is not used. The basic method of field theories is to start with a Lagrangian for the fields. From this Lagrangian one can "read" the Feynman diagrams and make practical calculations, not without bypassing well-known complications such as regularization and renormalization. Quantum chromodynamics (QCD) is the proper quantum field theory for nuclear physics. But it is a very hard task to bridge the physics from QCD to the one in low-energy nuclear processes. *Effective field theory* (EFT) tries to help in this construction by making use of the concept of the separation of scales [26]. One can form small expansion parameters from the ratios of short and long distance scales, defined by

$$\epsilon = \frac{\text{short distance scales}}{\text{long distance scales}} \tag{7.73}$$

and try to explain physical observables in terms of powers of ϵ.

In low-energy nuclear processes, the characteristic momenta are much smaller than the mass of the pion, which is the lightest hadron that mediates the strong interaction. In this regime, one often uses the pionless effective field theory, in which pions are treated as heavy particles and are integrated out of the theory. In this theory, the dynamical degrees of freedom are nucleons. The pion and the delta resonance degrees of freedom are hidden in the contact interactions between nucleons. The scales of the problem are the nucleon-nucleon scattering length, , the binding energy, B, and the typical nucleon momentum k in the center-of-mass frame. Then, the nucleon-nucleon interactions are calculated perturbatively with the small expansion parameter

$$p = \frac{(1/a, \text{ or } B, \text{ or } k)}{\Lambda} \tag{7.74}$$

which is the ratio of the light to the heavy scale. The heavy scale Λ is set by the pion mass ($m_\pi \sim 140$ MeV).

The pionless effective Lagrangian will only involve the nucleon field $\Psi^T = (p, n)$ and its derivatives. It must obey the symmetries observed in strong interactions at low energies, such as parity, time-reversal, and Galilean invariance. The Lagrangian can then be written as a series of local operators with increasing dimensions. In the limit where the energy goes to zero, the interactions of lowest dimension dominate. To leading order (LO), the relevant Lagrangian ($\hbar = c = 1$) is given by [26]

$$\mathcal{L} = \Psi^\dagger\left(i\partial_t + \frac{\nabla^2}{2m}\right)\Psi - C_0(\Psi^T\mathcal{P}\Psi)(\Psi^T\mathcal{P}\Psi)^\dagger, \tag{7.75}$$

where m is the nucleon mass. The projection operators $\mathcal{P}$ enforce the correct spin and isospin quantum numbers in the channels under investigation. For spin-singlet interactions $\mathcal{P}_i = \sigma_2\tau_2\tau_i/\sqrt{8}$, while for spin-triplet interactions $\mathcal{P}_i = \sigma_2\sigma_i\tau_2/\sqrt{8}$.

The Feynman-diagram rules can be directly "read" from the Lagrangian at hand. In the case that the scattering length a is large, i.e., $a \gg 1/\Lambda$, as it is in the nucleon-nucleon system, the full

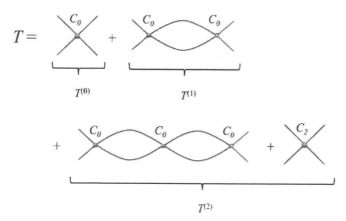

FIGURE 7.11   Feynman diagram series for NN-scattering in effective field theory.

scattering amplitude T is obtained from an infinite sum of such Feynman diagrams (see Figure 7.11), leading to a geometric series that can be written analytically as

$$T(p) = \frac{C_0}{1 - C_0 J(p)}, \qquad J(p) = \int \frac{d^3q}{(2\pi)^3} \frac{1}{E - \mathbf{q}^2/m + i\epsilon}, \qquad (7.76)$$

where $E = p^2/m$ is the total center-of-mass energy. The integral is linearly divergent but is finite using *dimensional regularization*. One gets

$$J = -(\mu + ip)\frac{m}{4\pi},$$

where μ is the regularization parameter. The scattering amplitude $T(p)$ has then the same structure as the s-wave partial-wave amplitude,

$$T = -\frac{4\pi}{p \cot\delta - ip},$$

and one obtains the effective range expansion for the phase-shift δ,

$$p \cot\delta = -\frac{1}{a} + r_0 \frac{p^2}{2} + \cdots$$

in the zero-momentum limit when the coupling constant takes the renormalized value

$$C_0(\mu) = \frac{4\pi}{m} \frac{1}{(1/a - \mu)}. \qquad (7.77)$$

In leading order, we see that the effective range vanishes, or $r_0 = 0$. The small inverse $1/a$ scattering length is given by the difference between two large quantities. For example, in proton-neutron scattering we have $a_{pn} = -23.7$ fm in the pn spin-singlet channel. Choosing the value $\mu = m_\pi$ for the regularization parameter, one obtains $C_0 = 3.54$ fm^2. Physical results should be independent of the exact value of the renormalization mass μ as long as $1/a < \mu \ll m_\pi$.

The theory is now ready for practical applications. For example, this procedure has been applied to obtain the electromagnetic form factor of the deuteron, the electromagnetic polarizability and the Compton scattering cross section for the deuteron, the radiative neutron capture on protons, and the continuum structure of halo nuclei. Based on the same effective field theory, the three-nucleon system and neutron-deuteron scattering have been investigated. Better agreement with data can be obtained in higher orders (next-to-leading order [NLO], next-to-next-to leading order [N^2LO], etc.). For nuclear processes involving momenta p comparable to m_π, the starting effective, pionfull, Lagrangian is more complicated. But the basic field theoretic method remains the same. The EFT unifies single-particle approaches in a model-independent framework, with the added power counting that allows for an a priori estimate of errors. Concepts of quantum field theory, such as regularization and renormalization, are key ingredients of the theory. In nuclear astrophysics, this theory has been applied to almost every reaction that allows the use of field theoretical concepts.

7.5.4 Parameter fits

Reaction rates dominated by the contributions from a few resonant or bound states are often extrapolated in terms of *R-* or *K-matrix* fits, which rely on quite similar strategies. A sketch of the *R*-matrix theory was presented in section 4.9. The appeal of these methods rests on the fact that analytical expressions which allow for a rather simple parametrization of the data can be derived from underlying formal reaction theories. However, the link between the parameters of the *R*-matrix model and the experimental data (resonance energies and widths) is only quite indirect. The *K*-matrix formalism solves this problem, but suffers from other drawbacks.

The *R*- and *K*-matrix models have been applied to a variety of reactions, and in particular to the analysis of the $^{12}\mathrm{C}(\alpha,\gamma)^{16}\mathrm{O}$ reaction rate.

7.5.5 The statistical models

Many astrophysical scenarios involve a wealth of reactions on intermediate-mass or heavy nuclei. This concerns the non-explosive or explosive burning of C, Ne, O and Si, as well as the s-, r- and p-process nucleosynthesis. Fortunately, a large fraction of the reactions of interest proceed through compound systems that exhibit high enough level densities for statistical methods to provide a reliable description of the reaction mechanism. In this respect, the *Hauser-Feshbach (HF) model* has been widely used with considerable success. Explosive burning in supernovae involves in general intermediate mass and heavy nuclei. Due to a large nucleon number they have intrinsically a high density of excited states. A high-level density in the compound nucleus at the appropriate excitation energy allows to make use of the statistical model approach for compound nuclear reactions which averages over resonances.

A high-level density in the compound nucleus permits to use averaged transmission coefficients T, which do not reflect a resonance behavior, but rather describe absorption via an imaginary part in the (optical) nucleon-nucleus potential. This leads to the expression similar to the one derived in Section 4.14, i.e.,

$$\sigma_i^{\mu\nu}(j,o;E_{ij}) = \frac{\pi\hbar^2/(2\mu_{ij}E_{ij})}{(2J_i^\mu+1)(2J_j+1)} \tag{7.78}$$
$$\times \sum_{J,\pi}(2J+1)\frac{T_j^\mu(E,J,\pi,E_i^\mu,J_i^\mu,\pi_i^\mu)T_o^\nu(E,J,\pi,E_m^\nu,J_m^\nu,\pi_m^\nu)}{T_{tot}(E,J,\pi)}$$

for the reaction $i^\mu(j,o)m^\nu$ from the target state i^μ to the excited state m^ν of the final nucleus, with a center of mass energy E_{ij} and reduced mass μ_{ij}. J denotes the spin, E the corresponding excitation energy in the compound nucleus, and π the parity of excited states. When these properties are used without subscripts they describe the compound nucleus, subscripts refer to states of the participating nuclei in the reaction $i^\mu(j,o)m^\nu$ and superscripts indicate the specific excited states. Experiments measure $\sum_\nu \sigma_i^{0\nu}(j,o;E_{ij})$, summed over all excited states of the final nucleus, with the target in the ground state. Target states μ in an astrophysical plasma are thermally populated and the astrophysical cross section $\sigma_i^*(j,o)$ is given by

$$\sigma_i^*(j,o;E_{ij}) = \frac{\sum_\mu (2J_i^\mu+1)\exp(-E_i^\mu/kT)\sum_\nu \sigma_i^{\mu\nu}(j,o;E_{ij})}{\sum_\mu(2J_i^\mu+1)\exp(-E_i^\mu/kT)}. \tag{7.79}$$

The summation over ν replaces $T_o^\nu(E,J,\pi)$ in Eq. (7.78) by the total transmission coefficient

$$T_o(E,J,\pi) = \sum_{\nu=0}^{\nu_m} T_o^\nu(E,J,\pi,E_m^\nu,J_m^\nu,\pi_m^\nu)$$
$$+ \int_{E_m^{\nu_m}}^{E-S_{m,o}} \sum_{J_m,\pi_m} T_o(E,J,\pi,E_m,J_m,\pi_m)\rho(E_m,J_m,\pi_m)dE_m. \tag{7.80}$$

Here $S_{m,o}$ is the channel separation energy, and the summation over excited states above the highest experimentally known state ν_m is changed to an integration over the level density ρ. The summation over target states μ in Eq. (7.79) has to be generalized accordingly.

The important ingredients of statistical model calculations as indicated in the above equations are the particle and gamma-transmission coefficients T and the level density of excited states ρ. Therefore, the reliability of such calculations is determined by the accuracy with which these components can be evaluated (often for unstable nuclei).

The gamma-transmission coefficients have to include the dominant gamma-transitions (E1 and M1) in the calculation of the total photon width. The smaller, and therefore less important, M1 transitions have usually been treated with the simple single particle approach $T \propto E^3$. The E1 transitions are usually calculated on the basis of the Lorentzian representation of the giant dipole resonance (see section 6.8). Within this model, the E1 transmission coefficient for the transition emitting a photon of energy E_γ in a nucleus ${}^A_N Z$ is given by

$$T_{E1}(E_\gamma) = \frac{8}{3} \frac{NZ}{A} \frac{e^2}{\hbar c} \frac{1+\chi}{mc^2} \sum_{i=1}^{2} \frac{i}{3} \frac{\Gamma_{G,i} E_\gamma^4}{(E_\gamma^2 - E_{G,i}^2)^2 + \Gamma_{G,i}^2 E_\gamma^2}. \tag{7.81}$$

Here $\chi (\simeq 0.2)$ accounts for the neutron-proton exchange contribution, and the summation over i includes two terms which correspond to the split of the GDR in statically deformed nuclei, with oscillations along $(i = 1)$ and perpendicular $(i = 2)$ to the axis of rotational symmetry.

7.6 Stellar evolution: hydrogen and CNO cycles

The energy production in the stars is a well known process. The initial energy which ignites the process arises from the gravitational contraction of a mass of gas. The contraction increases the pressure, temperature, and density, at the center of the star until values able to start the thermonuclear reactions, initiating the star lifetime. The energy liberated in these reactions yield a pressure in the plasma, which opposes compression due to gravitation. Thus, an equilibrium is reached for the energy which is produced, the energy which is liberated by radiation, the temperature, and the pressure.

A star is formed when the primordial clustering material reaches a mass of about $0.08 M_\odot$. Protostars with masses below this value are known as *brown dwarfs*, with insufficient temperatures to ignite hydrogen. But some brown dwarfs, heavier than about 13 times Jupiter's mass can fuse deuterium during a short period. Heavier stars have higher core temperatures making it easier for particles, starting with hydrogen, to tunnel their mutual Coulomb barrier and form heavier elements. Astrophysicists call stars with the initial stage of hydrogen burning by *main sequence* stars. Stellar masses and nuclear reactions are related by:

(a) $0.1 - 0.5 M_\odot$. In such stars there will be hydrogen but no helium burning.

(b) $0.5 - 8 M_\odot$. For these stars both hydrogen and and helium burning occur.

(c) $< 1.4 M_\odot$. Such stars end up as a white dwarf (WD). A WD is formed from the core of stars with masses $\lesssim 8 M_\odot$ after their envelope is ejected.

(d) $8 - 11 M_\odot$. They can house hydrogen, helium and carbon burning.

(e) $> 11 M_\odot$. Nuclear burning in these stars encompass all stages of thermonuclear fusion.

The basic equations of stellar structure, in the absence of convection, assuming spherical symmetry, are mass conservation and hydrostatic support,

$$\frac{dm(r)}{dr} = 4\pi r^2 \rho(r), \qquad \frac{dP(r)}{dr} = -\frac{Gm(r)}{r^2} \rho(r), \tag{7.82}$$

energy generation and radiation transport,

$$\frac{dL(r)}{dr} = 4\pi^2 \rho(r)[\epsilon(r) - \epsilon_\nu(r)], \qquad \frac{dT}{dr} = -\frac{3}{4ac} \frac{\bar{\kappa}\rho}{T^3} \frac{L(r)}{4\pi r^2}. \tag{7.83}$$

In these equations, $m(r)$ is the mass within a sphere of radius r from the center of the star, $\rho(r)$ is the local mass density at r, P the pressure, L the luminosity radiated away from distance r, T the local temperature, $\epsilon(\rho, T)$ the local energy density produced by nuclear reactions and $\epsilon_\nu(\rho, T)$ the local energy density carried away by neutrino emission. The mean opacity $\bar{\kappa}(\rho, T)$ which takes care of the radiation energy absorbed by atomic processes, such as ionization, Compton scattering, etc. a is the Stefan-Boltzmann constant. These equations are complemented by the *Equation of State* (EoS), i.e., $P \equiv P(\rho, \epsilon, T)$ which is a sum radiation pressure, ion pressure, and electron pressure. For the Sun the EoS is well described by the ideal gas law of a mixture of hydrogen and helium ions. The energy densities ϵ and ϵ_ν are obtained by solving the reaction networks as in Eq. (7.23), which is also responsible for the changes in the local elemental composition. If convection and entropy changes occur, the equations above need to be modified by including local changes due to convection mixing and the time-dependence of local entropy [17]. In a convective region we must solve the equations above with additional equations arising from convection, such as

$$\frac{P}{T}\frac{dT}{dP} = \frac{\gamma - 1}{\gamma}, \tag{7.84}$$

where $\gamma = c_p/c_v$ the ratio of the specific heat under constant pressure and the specific heat under constant volume [17].

Once these coupled equations have been solved, one can calculate L_{rad} and the luminosity due to convective transport $L_{conv} = L - L_{rad}$. The equations appropriate for a convective region must be switched on when the temperature gradient becomes equal to the adiabatic value, and switched off when all energy is transported by radiation. This method of solution may break down close to the surface of a star.

7.6.1   The pp-chain

The Sun is powered by the *pp-chain*, i.e., a chain of reactions leading to the formation of helium nuclei, as shown in Figure 7.12. There are in fact three such chains, the ppI chain,

$$p(e^-p, \nu_e)^2H, \quad {}^2H(p, \gamma)^3He, \quad \text{and,} \quad {}^3He({}^3He, 2p)^4He, \tag{7.85}$$

the ppII chain

$$p(e^-p, \nu_e)^2H, \quad {}^2H(p, \gamma)^3He, \quad {}^4He({}^3He, \gamma)^7Be, \quad {}^7Be(e^-, \nu_e)^7Be \quad \text{and} \quad {}^7Li(p, {}^4He)^4He, \tag{7.86}$$

and the ppIII chain

$$p(e^-p, \nu_e)^2H, \quad {}^2H(p, \gamma)^3He, \quad {}^4He({}^3He, \gamma)^7Be, \quad {}^7Be(p, \gamma)^8Be \quad \text{and} \quad {}^8B(, e^+\nu_e)^8Be, \tag{7.87}$$

with ${}^8Be$ promptly decaying into two α particles. The net product of these chain reactions is the consumption of 4 protons and the gain of 28 MeV by the formation of a helium nucleus. Energy is released away in the form of photons and neutrinos. The energy production in the Sun at $T_6 = 15$ is approximately given by $\epsilon \sim \epsilon_{pp}X^2T_6^4$, where X is the hydrogen mass fraction and $\epsilon_{pp} = 10^{-12}$ Jm^3 kg^{-2} s^{-1}.

The Sun is a star in its initial phase of evolution. The temperature in its surface is 6000° C, while in its interior the temperature reaches 1.5×10^7 K, with a pressure given by 6×10^{11} atm and density 150 g/cm^3. The present mass of the Sun is $M_\odot = 2 \times 10^{33}$ g and its main composition is hydrogen (70%), helium (29%) and less than 1% of more heavy elements, like carbon, oxygen, etc.

What are the nuclear processes which originate the huge thermonuclear energy of the Sun, and that has last 4.6×10^9 years (the assumed age of the Sun)? It cannot be the simple fusion of two protons, or of α-particles, or even the fusion of protons with α-particles, since neither ${}^2_2He$, ${}^8_4Be$, or ${}^5_3Li$, are stable. The only possibility is the proton-proton fusion in the form

$$p + p \longrightarrow d + e^+ + \nu_e, \tag{7.88}$$

which occurs via the β-decay, i.e., due to the weak-interaction. This reaction is non-resonant and occurs in two steps. First the protons tunnel the Coulomb barrier with height $E_c = 0.55$ MeV and in a second step one of the protons β-decays by positron and neutrino emission. The Coulomb barrier height is relatively small and the Sun would burn its fuel quite quickly if this was the only impediment for the pp reaction. But, because it is a β-decay process, the final stage occurs with a very low probability to yield a proton and a neutron system captured within a deuteron. Using Fermi's theory of β-decay, based on point interactions, one can obtain the second part. The cross section for this reaction for protons with energy below 1 MeV is very small, about 10^{-23} b. The effective energy for this reaction in the Sun is in fact much smaller than 1 MeV, i.e., about 20 keV. The mean lifetime of protons in the Sun due to the p(p,d) reaction is about 10^{10} y. Thus, the energy radiated by the Sun is nearly constant in time, and not an explosive process. Using *Fermi's Golden rule* for this reaction, we get $d\sigma = 2\pi\rho(E)|\langle f|H_\beta|i\rangle|^2/\hbar v_i$, where $\rho(E)$ is the density of final states in the interval dE and v_i is the initial relative velocity. Ψ_i is the initial wave function of the two protons and the final state wave function Ψ_f is a product of the deuteron, the positron and the neutrino wave functions, $\Psi_f = \Psi_d\Psi_e\Psi_\nu$. A plane wave can be used for the electron if its energy is large compared to ZR_∞, where the *Rydberg constant* is $R_\infty = 2\pi^2 me^4/ch^3$. The energy release in the reaction is 0.42 MeV and therefore the kinetic energy of the electron is $K_e \leq 0.42$ MeV. The mean energy of the neutrinos is $\langle E_\nu \rangle = 0.26$ MeV, which is too low and both plane wave exponentials for the electron and the neutrino wave functions can be approximated by the unity. The weak decay matrix element becomes

$$\langle f|H_\beta|i\rangle = g\sum_{m_f}\sum_j |\langle \Psi_d^*|t_\pm\sigma_j|\Psi_i\rangle|^2, \tag{7.89}$$

where $\sum_j$ is a sum over the Pauli matrices, σ_x, σ_y and σ_z, $\sum_{m_f}$ is a sum over final spins and g is the weak coupling constant. For the deuteron, $J_f^\pi = 1^+$, dominated by $l_f = 0$ in a triplet state $S_f = 1$. The maximum (super-allowed) transition probability has $\Delta l = 0$ and the initial $p+p$ wave function must have $l_i = 0$. In order for their wave function be antisymmetric in space and spin, one must have $S_i = 0$. Thus, the transition is $(S_i = 0, l_i = 0) \to (S_f = 1, l_f = 0)$, i.e., a pure Gamow-Teller transition. The full calculation yields

$$\sigma = \frac{m^5 c^4}{2\pi^3\hbar^7 v_i}f(E)g^2\frac{M_{space}^2 M_{spin}^2}{2}, \tag{7.90}$$

where $M_{spin}^2 = (2J+1)/(2J_1+1)(2J_2+1) = 3$, $M_{space} = \int_0^\infty \chi_f(r)\chi_i(r)r^2 dr$, and the dimensionless Fermi integral $f(E)$ accounts for Coulomb distortion of the electron wave function. At large energies, $f(E) \propto E^5$. The deuteron radial wave function $\chi_f(r)$ and the initial two-proton wave function at the low stellar energies involve only the s-wave parts. A numerical integration of the radial integral yields a cross section of about $\sigma = 10^{-47}$ cm^2 at $E_p = 1$ MeV, that cannot be measured experimentally. In the core of the Sun, $T_6 = 15$, and the reaction rate can be estimated based on theoretical models as $\langle \sigma v \rangle_{pp} \simeq 1.2 \times 10^{-43}$ cm^3 s^{-1} [3]. The density in the Sun's core is about $\rho = 100$ gm cm^{-3} and if one assumes an equal mixture of hydrogen and helium, then $X_H = X_{He} = 0.5$, and the average life of a hydrogen nucleus to be converted to deuterium is $\tau_H(H) = \langle \sigma v \rangle_{pp}/N_H \sim 10^{10}$ y, comparable to the age of old stars.

Once deuterium is produced in the Sun, it is promptly consumed by a nonresonant direct capture reaction to the ground state of ^{3}He, $d+p \to {}^3\text{He} + \gamma$. The deuteron induced reactions

$$\text{d(d,p)t}, \quad \text{d(d,n)}^3\text{He}, \quad \text{d}(^3\text{He,p})^4\text{He}, \quad \text{and} \quad \text{d}(^3\text{He},\gamma)^5\text{Li} \tag{7.91}$$

have a larger cross section than that for $d(p,\gamma)^3$He. But because of the much larger number of protons in stars, the process $d(p,\gamma)^3$He dominates. The rate of energy production in the pp-chain is still determined by $p + p \longrightarrow d + e^+ + \nu_e$ which is much slower than $d(p,\gamma)^3$He.

The equation for the evolution of deuterium (D) abundance is given by its production through the pp-reaction and its destruction though the pd-reaction, i.e.,

$$\frac{dY_D}{dt} = \frac{Y_H^2}{2}\rho N_A\langle \sigma v \rangle_{pp} - Y_H Y_D \rho N_A\langle \sigma v \rangle_{pd}. \tag{7.92}$$

In thermodynamical equilibrium, this yields

$$\frac{Y_D}{Y_H} = \frac{\langle \sigma v \rangle_{pp}}{2\langle \sigma v \rangle_{pd}}. \tag{7.93}$$

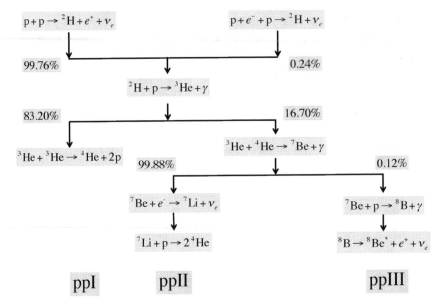

FIGURE 7.12   The p-p chain reaction (p-p cycle). The percentage for the several branches are calculated in the center of the Sun [14].

The observed (Y_D/Y_H) ratio in the Cosmos is $\sim 10^{-5}$, which is much larger due to BBN, before stars were formed. In stellar interiors deuterium is destroyed by means of the reaction $d(p,\gamma)^3He$. The equilibrium condition in Eq. (7.93) is reached in about $\tau_d = 2$ s, which is the lifetime for the consumption of deuterons. That is, deuterons are burned almost instantaneously in stellar environments.

After ^3He is formed, it can be consumed via the non-resonant capture reaction $^3He + {}^3He \rightarrow p + p + {}^4He,$. Another non-resonant capture reaction consuming ^3He is $^3He + D \rightarrow {}^4He + p$. Both reactions have comparable $S(0)$ factors, but because the deuterium concentration is very small, the first reaction dominates. ^4He synthesized in the BBN can react with ^3He and lead to the ppII and ppIII parts of the chain in Figure 7.12. The reactions leading to the formation of ^7Be and ^8B, via $^3He(\alpha,\gamma)^7Be$ and $^7Be(p,\gamma)^8B$, are responsible for the production of high energy neutrinos in the Sun [14]. The first reaction occurs 14% compared to the one in Eq. (7.6.1). In this route, ^7Be can be consumed in two alternate ways to complete the fusion process of four protons transforming into a helium nucleus, $4H \rightarrow {}^4He$, in the ppII and ppIII chains.

After ^7Be is formed, the first step of the ppII chain is the capture of an atomic electron by means of $^7Be + e^- \rightarrow {}^7Li + \nu_e$. The initial and final wave functions of the nuclei vanish rapidly outside the nuclear radius and the electron wave function within the nucleus can be approximated by its value at the origin. The neutrino wave function can taken as a plane wave normalized to the nuclear volume V, so that $H_{if} = \Psi_e(0)g \int \Psi^*_{^7Li}\Psi_{^7Be}d^3r$. The capture rate is then given by

$$\lambda_{EC} = \frac{1}{\tau_{EC}} = \left(\frac{g^2 M^2_{if}}{\pi c^3 \hbar^4}\right) E^2_\nu |\Psi_e(0)|^2, \qquad (7.94)$$

where $M_{if} = \int \Psi^*_{^7Li}\Psi_{^7Be}d^3r$ is the nuclear matrix element for the transition.

The electrons from the atomic K-shell give the dominant contribution to the capture. In the core of the Sun, $T_6 = 15$, and the ^7Be are mostly ionized. But they are immersed in a sea of free electrons and electron capture from continuum states can occur. Since the factors in the calculations with continuum wave functions are approximately the same as for atomic electron capture, except for the corresponding electron densities, the ^7Be lifetime in the Sun, τ_s, and the terrestrial lifetime, τ_t, are related by $\tau_s/\tau_t \sim 2|\Psi_t(0)|^2/|\Psi_s(0)|^2$, where $|\Psi_s(0)|^2$ can be identified as the density of free electrons in the Sun, $n_e = \rho/m_H$. The factor of 2 accounts for the two spin states in calculation of λ_t, while λ_s is calculated by averaging over these two orientations. The electron wave functions are distorted due to Coulomb interaction with hydrogen (of mass fraction X_H) and heavier nuclei in the plasma, and one has instead

$$\tau_s = \frac{2|\Psi_t(0)|^2 \tau_t}{(\rho/M_H)[(1+X_H)/2]2\pi Z\alpha(m_e c^2/3kT)^{1/2}}, \tag{7.95}$$

where we used $|\Psi_e(0)|^2 \sim (Z/a_0)^3/\pi$. This equation leads to the lifetime ^{7}Be in the Sun [14],

$$\tau_s(^7\mathrm{Be}) = 4.72 \times 10^8 \frac{T_6^{1/2}}{\rho(1+X_H)} \text{ s.} \tag{7.96}$$

The thermally averaged Coulomb corrections on the electron wave function yields the temperature dependence in this formula. One gets the continuum capture rate of $\tau_s(^7\mathrm{Be}) = 140$ d whereas on Earth $\tau_t = 77$ d [14]. Considering also partially ionized ^{7}Be atoms under solar conditions, one gets another 21% increase in the decay rate, leading to a lifetime of ^{7}Be in the Sun as $\tau_\odot(^7\mathrm{Be}) = 120$ d.

^{7}Be can also be consumed 0.12% of the time by means of $^7\mathrm{Be}(p,\alpha)^8\mathrm{B}$, completing the ppIII part of the pp-chain in the Sun. This reaction occurs at energies much smaller than the 640 keV resonance in ^{8}B. Within a lifetime of $\tau \simeq 1$ s, $^8\mathrm{B}(J^\pi = 2^+)$ decays by means of $^8\mathrm{B} \to \, ^8\mathrm{Be} + e^+ + \nu_e$, mainly to the broad ($\Gamma = 1.6$ MeV) resonance in ^{8}Be at $E_r \simeq 2.9$ MeV ($J^\pi = 2^+$), which promptly decays into two α-particles. The neutrinos acquire an average energy in the decay of $\bar{E}_\nu(^8\mathrm{B}) = 7.3$ MeV.

7.7 Tests of the solar models

7.7.1 Neutrinos as solar thermometers

The Sun serves as a very important test case for a variety of problems related to stellar structure and evolution, as well as to fundamental physics. The central temperature $T_\odot$ of the Sun is a nice example of a physical quantity which can be determined by means of solar neutrino detection, provided that the relevant nuclear physics is known (and neutrino properties are also known). Surprisingly enough for a star that has all reasons to be considered as one of the dullest astrophysical objects, the Sun has been for years at the center of various controversies. One of them is the solar neutrino problem, referring to the fact that the pioneering ^{37}Cl neutrino-capture experiments carried out over the years in the Homestake gold mine observe a neutrino flux that is substantially smaller than the one predicted by the solar models. That puzzle has led to a flurry of theoretical activities, and to the development of new detectors. These activities have transformed the original solar neutrino problem into problems. The relative levels of 'responsibility' of particle physics, nuclear physics or astrophysics in these discrepancies have been debated ever since. In Ref. [14], the discussion is conducted in particular in the light of the several experiment supporting the ideas of 'oscillations' between different neutrinos types. In the neutrino oscillation picture, the electron-neutrino, detected by the chlorine experiment, can transform into a muon-neutrino on its way to the Earth from the center of the Sun. This would explain the smaller number of electron-neutrinos observed at the Earth.

SSM calculations [14] predict $T_\odot$ with an accuracy of 1% or even better. In order to appreciate such a result, let us remind that the central temperature of Earth is known with an accuracy of about 20%. However, let us remind that this is a theoretical prediction which, as any result in physics, demands observational evidence.

The fluxes of ^{8}B and ^{7}Be neutrinos (see Figure 7.12) are given by:

$$\Phi(B) = c_B S_{17} \frac{S_{34}}{\sqrt{S_{33}}} T_\odot^{20}, \qquad \Phi(Be) = c_{Be} \frac{S_{34}}{\sqrt{S_{33}}} T_\odot^{10}, \tag{7.97}$$

where S_{ij} are the low-energy astrophysical factors for nuclear reactions between nuclei with atomic mass numbers i and j, c_B and c_{Be} are well determined constants.

The high powers of $T_\odot$ in the above equations imply that the measured neutrino fluxes are strongly sensitive to $T_\odot$, i.e. ^{7}Be and ^{8}B neutrinos in principle are good thermometers for the

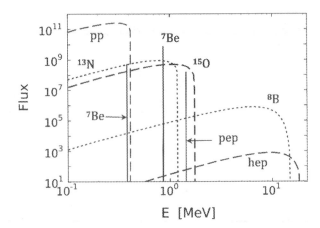

FIGURE 7.13   The solar neutrino spectrum, as calculated with the Standard Solar Model (SSM) [14].

innermost part of the Sun. On the other hand, the relevant nuclear physics has to be known, which justifies the present theoretical and experimental efforts for better determinations of the S_{ij}.

Much experimental and theoretical work has been devoted to the reactions of the p-p chains that are the main energy and neutrino producers in the Sun. In spite of that, problems remain concerning the astrophysical rates of some of the involved reactions. This is especially the case for $^7\text{Be}(\text{p},\gamma)^8\text{B}$ (which determines the value of S_{17} in Eq. (7.97)), which provides the main neutrino flux detectable by the chlorine detector, and is considered by some as one of the most important nuclear reactions for astrophysics. Improved low-energy data are also required for other reactions, like $^3\text{He}(^3\text{He},2\text{p})^4\text{He}$. Since such low-energy measurements are predominantly hampered by the cosmic-ray background, improved data could be obtained by underground measurements.

7.7.2   Neutrino-nucleus cross sections

In order to analyze the experiments with neutrino detection a knowledge of the cross sections for neutrino capture is necessary. Let us briefly discuss the main physics behind these processes. Consider the charged current reaction

$$\nu_e + (A, Z) \to e^- + (A, Z + 1).\tag{7.98}$$

The cross section for the neutrino-nucleus interaction, involves an average over initial and sum over final nuclear spins of the square of the transition amplitude. One can show that it will be proportional to the squared nuclear matrix element

$$\frac{1}{2J_i + 1}\left[\left|\langle f||\sum_{i=1}^{A}\tau_{\pm}(i)||i\rangle\right|^2 + g_A^2\left|\langle f||\sum_{i=1}^{A}\sigma(i)\tau_{\pm}(i)||i\rangle\right|^2\right].\tag{7.99}$$

where $g_A \sim 1.26$, and σ and $\tau_{\pm}$ are the spin and isospin operator, respectively. The fist matrix element inside parenthesis involves the Fermi operator which is proportional to the isospin raising/lowering operator: in the limit of good isospin, which typically is broken at the $\lesssim 5\%$ level for transitions between well-bound nuclear states, the Fermi operator only connects states in the same isospin multiplet, that is, states with a common spin-spatial structure. If the initial state has isospin (T_i, M_{Ti}), this final state has $(T_i, M_{Ti} \pm 1)$ for (ν, e^-) and $(\bar\nu, e^+)$ reactions, respectively, and is called the isospin analog state (IAS). In the limit of good isospin the sum rule for this operator in then particularly simple

$$\frac{1}{2J_i + 1}\sum_f\left|\langle f||\sum_{i=1}^{A}\tau_+(i)||i\rangle\right|^2 = \frac{1}{2J_i + 1}\left|\langle IAS||\sum_{i=1}^{A}\tau_+(i)||i\rangle\right|^2 = |N - Z|.\tag{7.100}$$

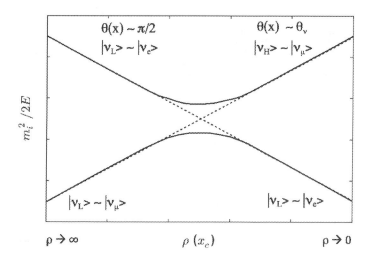

FIGURE 7.14 Schematic illustration of the MSW crossing. The dashed lines correspond to the electron-electron and muon-muon diagonal elements of the M_ν^2 matrix in the flavor basis. Their intersection defines the level-crossing density ρ_c. The solid lines are the trajectories of the light and heavy local mass eigenstates. If the electron neutrino is produced at high density and propagates adiabatically, it will follow the heavy-mass trajectory, emerging from the Sun as a ν_μ.

The excitation energy of the IAS relative to the parent ground state can be estimated accurately from the Coulomb energy difference

$$E_{IAS} \sim \left(\frac{1.728Z}{1.12A^{1/3} + 0.78} - 1.293 \right) \text{MeV}. \tag{7.101}$$

The angular distribution of the outgoing electron for a pure Fermi $(N, Z) + \nu \to (N-1, Z+1) + e^-$ transition is $1 + \beta \cos\theta_e$, and thus forward peaked. Here β is the electron velocity.

The second matrix element inside the parenthesis of Eq. (7.99) is the Gamow-Teller (GT) response (see Section 6.8) and it is more complicated, as the operator connects the ground state to many states in the final nucleus. In general we do not have a precise probe of the nuclear GT response apart from weak interactions themselves. However a good approximate probe is provided by forward-angle (p,n) scattering off nuclei. The (p,n) studies demonstrate that the GT strength tends to concentrate in a broad resonance centered at a position $\delta = E_{GT} - E_{IAS}$ relative to the IAS given by

$$\delta \sim \left(7.0 - 28.9 \frac{N - Z}{A} \right) \text{MeV}. \tag{7.102}$$

Thus while the peak of the GT resonance is substantially above the IAS for $N \sim Z$ nuclei, it drops with increasing neutron excess, with $\delta \sim 0$ for Pb. A typical value for the full width at half maximum Γ is ~ 5 MeV. The angular distribution of GT $(N, Z) + \nu_e \to (N-1, Z+1) + e^-$ reactions is $3 - \beta \cos\theta_e$, corresponding to a gentle peaking in the backward direction.

Previously we described the allowed Gamow-Teller (spin-flip) and Fermi weak interaction operators. These are the appropriate operators when one probes the nucleus at a wavelength – that is, at a size scale – where the nucleus responds like an elementary particle. We can then characterize its response by its macroscopic quantum numbers, the spin and charge. On the other hand, the nucleus is a composite object and, therefore, if it is probed at shorter length scales, all kinds of interesting radial excitations will result, analogous to the vibrations of a drumhead. For a reaction like neutrino scattering off a nucleus, the full operator involves the additional factor

$$e^{i\mathbf{k} \cdot \mathbf{r}} \sim 1 + i\mathbf{k} \cdot \mathbf{r}, \tag{7.103}$$

where the expression on the right is valid if the magnitude of $\mathbf{k}$ is not too large. Thus the full charge operator includes a "first forbidden" term

$$\sum_{i=1}^{A} \mathbf{r}_i \tau_3(i), \tag{7.104}$$

and similarly for the spin operator

$$\sum_{i=1}^{A} [\mathbf{r}_i \otimes \boldsymbol{\sigma}(i)]_{J=0,1,2} \, \tau_3(i). \tag{7.105}$$

These operators generate collective radial excitations, leading to the giant resonance excitations in nuclei (see section 6.8). The giant resonances are typically at an excitation energy of 20–25 MeV in light nuclei. One important property is that these operators satisfy the *Thomas-Reiche-Kuhn* sum rule (see Eq. 6.202), so that

$$\sum_{f} \left| \langle f | \sum_{i=1}^{A} r(i)\tau_3(i) | i \rangle \right|^2 \sim \frac{NZ}{A} \sim \frac{A}{4}, \tag{7.106}$$

where the sum extends over a complete set of final nuclear states. These first-forbidden operators tend to dominate the cross sections for scattering the high-energy supernova neutrinos (ν_μs and ν_τs), with $E_\nu \sim 25$ MeV, off light nuclei. From the sum rule above, it follows that nuclear cross sections per target *nucleon* are roughly constant.

7.7.3   Neutrino oscillations

One odd feature of particle physics is that neutrinos, which are not required by any symmetry to be massless, nevertheless must be much lighter than any of the other known fermions. For instance, the current limit on the $\bar{\nu}_e$ mass is $\lesssim 2.2$ eV. The standard model of particle physics requires neutrinos to be massless, but the reasons are not fundamental. Dirac mass terms m_D, analogous to the mass terms for other fermions, cannot be constructed because the model contains no right-handed neutrino fields. If a neutrino has a mass m, we mean that as it propagates through free space, its energy and momentum are related in the usual way for this mass. Thus if we have two neutrinos, we can label those neutrinos according to the eigenstates of the free Hamiltonian, that is, as mass eigenstates.

But neutrinos are produced by the weak interaction. In this case, we have another set of eigenstates, the *flavor eigenstates*. We can define a ν_e as the neutrino that accompanies the positron in β decay. Likewise we label by ν_μ the neutrino produced in muon decay.

The question is: are the eigenstates of the free Hamiltonian and of the weak interaction Hamiltonian identical? Most likely the answer is no: we know this is the case with the quarks, since the different families (the analog of the mass eigenstates) do interact through the weak interaction. That is, the up quark decays not only to the down quark, but also occasionally to the strange quark. Thus we suspect that the weak interaction and mass eigenstates, while spanning the same two-neutrino space, are not coincident: the mass eigenstates $|\nu_1\rangle$ and $|\nu_2\rangle$ (with masses m_1 and m_2) are related to the weak interaction eigenstates by [24]

$$\begin{aligned} |\nu_e\rangle &= \cos\theta_v |\nu_1\rangle + \sin\theta_v |\nu_2\rangle \\ |\nu_\mu\rangle &= -\sin\theta_v |\nu_1\rangle + \cos\theta_v |\nu_2\rangle, \end{aligned} \tag{7.107}$$

where θ_v is the (vacuum) mixing angle.

An immediate consequence is that a state produced as a $|\nu_e\rangle$ or a $|\nu_\mu\rangle$ at some time t – for example, a neutrino produced in β decay – does not remain a pure flavor eigenstate as it propagates away from the source. The different mass eigenstates comprising the neutrino will accumulate different phases as they propagate downstream, a phenomenon known as *vacuum oscillations* (vacuum because the experiment is done in free space). To see the effect, suppose the neutrino produced in a β decay is a momentum eigenstate. At time $t=0$

$$|\nu(t = 0)\rangle = |\nu_e\rangle = \cos\theta_v|\nu_1\rangle + \sin\theta_v|\nu_2\rangle. \tag{7.108}$$

Each eigenstate subsequently propagates with a phase

$$\exp\left[i(\mathbf{k}\cdot\mathbf{x} - \omega t)\right] = \exp\left[i(\mathbf{k}\cdot\mathbf{x} - \sqrt{m_i^2 + k^2}t)\right]. \tag{7.109}$$

But if the neutrino mass is small compared to the neutrino momentum/energy, one can write

$$\sqrt{m_i^2 + k^2} \sim k\left(1 + \frac{m_i^2}{2k^2}\right). \tag{7.110}$$

Thus, we conclude that

$$|\nu(t)\rangle = \exp\left[i(\mathbf{k}\cdot\mathbf{x} - kt - (m_1^2 + m_2^2)t/4k)\right]\left[\cos\theta_v|\nu_1\rangle e^{i\delta m^2 t/4k} + \sin\theta_v|\nu_2\rangle e^{-i\delta m^2 t/4k}\right]. \tag{7.111}$$

We see there is a common average phase (which has no physical consequence) as well as a beat phase that depends on

$$\delta m^2 = m_2^2 - m_1^2. \tag{7.112}$$

Now it is a simple matter to calculate the probability that our neutrino state remains a $|\nu_e\rangle$ at time t

$$P_{\nu_e}(t) = |\langle\nu_e|\nu(t)\rangle|^2 = 1 - \sin^2 2\theta_v \sin^2\left(\frac{\delta m^2 t}{4k}\right) \longrightarrow 1 - \frac{1}{2}\sin^2 2\theta_v, \tag{7.113}$$

where the limit on the right is appropriate for large t. (When one properly describes the neutrino state as a wave packet, the large-distance behavior follows from the eventual separation of the mass eigenstates.) Now $E \sim k$, where E is the neutrino energy, by our assumption that the neutrino masses are small compared to k. We can reinsert the implicit constants to write the probability in terms of the distance x of the neutrino from its source,

$$P_\nu(x) = 1 - \sin^2 2\theta_v \sin^2\left(\frac{\delta m^2 c^4 x}{4\hbar c E}\right). \tag{7.114}$$

If the oscillation length

$$L_o = \frac{4\pi\hbar c E}{\delta m^2 c^4} \tag{7.115}$$

is comparable to or shorter than one astronomical unit, a reduction in the solar ν_e flux would be expected in terrestrial neutrino oscillations.

The suggestion that the solar neutrino problem could be explained by neutrino oscillations was first made by Pontecorvo in 1958, who pointed out the analogy with $K_0 \longleftrightarrow \bar{K}_0$ oscillations [27]. From the point of view of particle physics, the Sun is a marvellous neutrino source. The neutrinos travel a long distance and have low energies (~ 1 MeV), implying a sensitivity down to

$$\delta m^2 > 10^{-12} eV^2. \tag{7.116}$$

In the "seesaw mechanism", $\delta m^2 \sim m_2^2$, so neutrino masses as low as $m_2 \sim 10^{-6}$ eV could be probed.

From the expressions above one expects vacuum oscillations to affect all neutrino species equally, if the oscillation length is small compared to an astronomical unit. This is somewhat in conflict with the solar neutrino data, as we have argued that the ^{7}Be neutrino flux is quite suppressed. Furthermore, there is a weak theoretical prejudice that θ_v should be small, like the Cabibbo angle of the weak interaction. The first objection, however, can be circumvented in the case of "just so" oscillations where the oscillation length is comparable to one astronomical unit. In this case the oscillation probability becomes sharply energy dependent, and one can choose δm^2 to preferentially suppress one component (e.g., the monochromatic ^{7}Be neutrinos), though the requirement for large mixing angles remains.

Below we will find that oscillations in matter can lead to nearly total flavor conversion even if the mixing angle is small. In preparation for this we first present the results above in a slightly more general way. The analog of Eq. (7.111) for an initial muon neutrino ($|\nu(t=0)\rangle = |\nu_\mu\rangle$) is

$$|\nu(t)\rangle = e^{i(\mathbf{k}\cdot\mathbf{x}-kt-(m_1^2+m_2^2)t/4k)}\left[-\sin\theta_v|\nu_1\rangle e^{i\delta m^2 t/4k} + \cos\theta_v|\nu_2\rangle e^{-i\delta m^2 t/4k}\right]. \qquad (7.117)$$

Now, if we compare Eqs. (7.111) and (7.117) we see that they are special cases of a more general problem. Suppose we write our initial neutrino wave function as

$$|\nu(t=0)\rangle = a_e(t=0)|\nu_e\rangle + a_\mu(t=0)|\nu_\mu\rangle. \qquad (7.118)$$

Then Eqs. (7.111) and (7.117) tell us that the subsequent propagation is described by changes in $a_e(x)$ and $a_\mu(x)$ according to

$$i\frac{d}{dx}\begin{pmatrix} a_e \\ a_\mu \end{pmatrix} = \frac{1}{4E}\begin{pmatrix} -\delta m^2\cos 2\theta_v & \delta m^2\sin 2\theta_v \\ \delta m^2\sin 2\theta_v & \delta m^2\cos 2\theta_v \end{pmatrix}\begin{pmatrix} a_e \\ a_\mu \end{pmatrix}. \qquad (7.119)$$

Note that the common phase has been ignored: it can be absorbed into the overall phase of the coefficients a_e and a_μ, and thus has no consequence. Also, we have equated $x = t$, that is, set $c = 1$.

The view of neutrino oscillations changed when Mikheyev and Smirnov [29] showed in 1985 that the density dependence of the neutrino effective mass, a phenomenon first discussed by Wolfenstein in 1978 [28], could greatly enhance oscillation probabilities: a ν_e is adiabatically transformed into a ν_μ as it traverses a critical density within the Sun. It became clear that the Sun was not only an excellent neutrino source, but also a natural regenerator for cleverly enhancing the effects of flavor mixing

$$i\frac{d}{dx}\begin{pmatrix} a_e \\ a_\mu \end{pmatrix} = \frac{1}{4E}\begin{pmatrix} 2E\sqrt{2}G_F\rho(x) - \delta m^2\cos 2\theta_v & \delta m^2\sin 2\theta_v \\ \delta m^2\sin 2\theta_v & -2E\sqrt{2}G_F\rho(x) + \delta m^2\cos 2\theta_v \end{pmatrix}\begin{pmatrix} a_e \\ a_\mu \end{pmatrix}, \qquad (7.120)$$

where G_F is the weak coupling constant and $\rho(x)$ the solar electron density. If $\rho(x) = 0$, this is exactly our previous result and can be trivially integrated to give the vacuum oscillation solutions given above. The new contribution to the diagonal elements, $2E\sqrt{2}G_F\rho(x)$, represents the effective contribution to the M_ν^2 matrix that arises from neutrino-electron scattering. The indices of refraction of electron and muon neutrinos differ because the former scatter by charged and neutral currents, while the latter have only neutral current interactions. The difference in the forward scattering amplitudes determines the density-dependent splitting of the diagonal elements of the matter equation, the generalization of Eq. (7.119).

It is helpful to rewrite this equation in a basis consisting of the light and heavy local mass eigenstates (i.e., the states that diagonalize the right-hand side of Eq. (7.120)),

$$\begin{aligned} |\nu_L(x)\rangle &= \cos\theta(x)|\nu_e\rangle - \sin\theta(x)|\nu_\mu\rangle \\ |\nu_H(x)\rangle &= \sin\theta(x)|\nu_e\rangle + \cos\theta(x)|\nu_\mu\rangle. \end{aligned} \qquad (7.121)$$

The local mixing angle is defined by

$$\sin 2\theta(x) = \frac{\sin 2\theta_v}{\sqrt{X^2(x) + \sin^2 2\theta_v}},$$

$$\cos 2\theta(x) = \frac{-X(x)}{\sqrt{X^2(x) + \sin^2 2\theta_v}}, \qquad (7.122)$$

where $X(x) = 2\sqrt{2}G_F\rho(x)E/\delta m^2 - \cos 2\theta_v$. Thus $\theta(x)$ ranges from θ_v to $\pi/2$ as the density $\rho(x)$ goes from 0 to ∞.

If we define

$$|\nu(x)\rangle = a_H(x)|\nu_H(x)\rangle + a_L(x)|\nu_L(x)\rangle, \qquad (7.123)$$

the neutrino propagation can be rewritten in terms of the local mass eigenstates

$$i\frac{d}{dx}\begin{pmatrix} a_H \\ a_L \end{pmatrix} = \begin{pmatrix} \lambda(x) & i\alpha(x) \\ -i\alpha(x) & -\lambda(x) \end{pmatrix}\begin{pmatrix} a_H \\ a_L \end{pmatrix}, \tag{7.124}$$

with the splitting of the local mass eigenstates determined by

$$2\lambda(x) = \frac{\delta m^2}{2E}\sqrt{X^2(x) + \sin^2 2\theta_v}, \tag{7.125}$$

and with mixing of these eigenstates governed by the density gradient

$$\alpha(x) = \left(\frac{E}{\delta m^2}\right)\frac{\sqrt{2}G_F \frac{d}{dx}\rho(x)\sin 2\theta_v}{X^2(x) + \sin^2 2\theta_v}. \tag{7.126}$$

The results above are quite interesting: the local mass eigenstates diagonalize the matrix if the density is constant. In such a limit, the problem is no more complicated than our original vacuum oscillation case, although our mixing angle is changed because of the matter effects. But if the density is not constant, the mass eigenstates evolve as the density changes. This is the crux of the MSW effect. Note that the splitting achieves its minimum value, $\frac{\delta m^2}{2E}\sin 2\theta_v$, at a critical density $\rho_c = \rho(x_c)$,

$$2\sqrt{2}EG_F\rho_c = \delta m^2\cos 2\theta_v, \tag{7.127}$$

that defines the point where the diagonal elements of the original flavor matrix cross.

Our local-mass-eigenstate form of the propagation equation can be trivially integrated if the splitting of the diagonal elements is large compared to the off-diagonal elements (see Eq. (7.124)),

$$\gamma(x) = \left|\frac{\lambda(x)}{\alpha(x)}\right| = \frac{\sin^2 2\theta_v}{\cos 2\theta_v}\frac{\delta m^2}{2E}\frac{1}{\left|\frac{1}{\rho_c}\frac{d\rho(x)}{dx}\right|}\frac{[X(x)^2 + \sin^2 2\theta_v]^{3/2}}{\sin^3 2\theta_v} \gg 1, \tag{7.128}$$

a condition that becomes particularly stringent near the crossing point,

$$\gamma_c = \gamma(x_c) = \frac{\sin^2 2\theta_v}{\cos 2\theta_v}\frac{\delta m^2}{2E}\frac{1}{\left|\frac{1}{\rho_c}\frac{d\rho(x)}{dx}\right|_{x=x_c}} \gg 1. \tag{7.129}$$

That is, adiabaticity depends on the density scale height at the crossing point. The resulting adiabatic electron neutrino survival probability [15], valid when $\gamma_c \gg 1$, is

$$P_{\nu_e}^{\text{adiab}} = \frac{1}{2} + \frac{1}{2}\cos 2\theta_v \cos 2\theta_i, \tag{7.130}$$

where $\theta_i = \theta(x_i)$ is the local mixing angle at the density where the neutrino was produced.

The physical picture behind this derivation is illustrated in Figure 7.14. One makes the usual assumption that, in vacuum, the ν_e is almost identical to the light mass eigenstate, $\nu_L(0)$, i.e., $m_1 < m_2$ and $\cos\theta_v \sim 1$. But as the density increases, the matter effects make the ν_e heavier than the ν_μ, with $\nu_e \to \nu_H(x)$ as $\rho(x)$ becomes large. The special property of the Sun is that it produces ν_es at high density that then propagate to the vacuum where they are measured. The adiabatic approximation tells us that if initially $\nu_e \sim \nu_H(x)$, the neutrino will remain on the heavy mass trajectory provided the density changes slowly. That is, if the solar density gradient is sufficiently gentle, the neutrino will emerge from the Sun as the heavy vacuum eigenstate, $\sim \nu_\mu$. This guarantees nearly complete conversion of ν_es into ν_μs.

But this does not explain the curious pattern of partial flux suppressions coming from the various solar neutrino experiments. The key to this is the behavior when $\gamma_c \lesssim 1$. Our expression for $\gamma(x)$ shows that the critical region for non-adiabatic behavior occurs in a narrow region (for small θ_v) surrounding the crossing point, and that this behavior is controlled by the density scale height. This suggests an analytic strategy for handling non-adiabatic crossings: one can replace the true solar density by a simpler (integrable!) two-parameter form that is constrained to reproduce the true density and its derivative at the crossing point x_c. Two convenient choices are the linear ($\rho(x) = a + bx$) and exponential ($\rho(x) = ae^{-bx}$) profiles. As the density derivative at x_c governs the non-adiabatic behavior, this procedure should provide an accurate description of the hopping probability between the local mass eigenstates when the neutrino traverses the crossing point. The initial and ending points x_i and x_f for the artificial profile are then chosen so that $\rho(x_i)$ is the density where the neutrino was produced in the solar core and $\rho(x_f) = 0$ (the solar surface). Since the adiabatic result ($P_{\nu_e}^{\text{adiab}}$) depends only on the local mixing angles at these points, this choice builds in that limit. But our original flavor-basis equation can then be integrated exactly for linear and exponential profiles, with the results given in terms of parabolic cylinder and Whittaker functions, respectively.

That result can be simplified further by observing that the non-adiabatic region is generally confined to a narrow region around x_c, away from the endpoints x_i and x_f. We can then extend the artificial profile to $x = \pm\infty$. As the neutrino propagates adiabatically in the unphysical region $x < x_i$, the exact solution in the physical region can be recovered by choosing the initial boundary conditions

$$a_L(-\infty) = -a_\mu(-\infty) = \cos\theta_i \exp\left[-i\int_{-\infty}^{x_i}\lambda(x)dx\right],$$

$$a_H(-\infty) = a_e(-\infty) = \sin\theta_i \exp\left[i\int_{-\infty}^{x_i}\lambda(x)dx\right]. \tag{7.131}$$

That is, $|\nu(-\infty)\rangle$ will then adiabatically evolve to $|\nu(x_i)\rangle = |\nu_e\rangle$ as x goes from $-\infty$ to x_i. The unphysical region $x > x_f$ can be handled similarly.

With some algebra a simple generalization of the adiabatic result emerges that is valid for all $\delta m^2/E$ and θ_v

$$P_{\nu_e} = \frac{1}{2} + \frac{1}{2}\cos 2\theta_v \cos 2\theta_i (1 - 2P_{\text{hop}}), \tag{7.132}$$

where P_{hop} is the *Landau-Zener probability* of hopping from the heavy mass trajectory to the light trajectory on traversing the crossing point. For the linear approximation to the density,

$$P_{\text{hop}}^{\text{lin}} = \exp\left[-\pi\gamma_c/2\right]. \tag{7.133}$$

As it must by our construction, P_{ν_e} reduces to $P_{\nu_e}^{\text{adiab}}$ for $\gamma_c \gg 1$. When the crossing becomes non-adiabatic (e.g., $\gamma_c \ll 1$), the hopping probability goes to 1, allowing the neutrino to exit the Sun on the light mass trajectory as a ν_e, i.e., no conversion occurs.

Thus there are two conditions for strong conversion of solar neutrinos: there must be a level crossing (that is, the solar core density must be sufficient to render $\nu_e \sim \nu_H(x_i)$ when it is first produced) and the crossing must be adiabatic. The first condition requires that $\delta m^2/E$ not be too large, and the second $\gamma_c \gtrsim 1$. The combination of these two constraints defines a triangle of interesting parameters in the $\frac{\delta m^2}{E} - \sin^2 2\theta_v$ plane, as Mikheyev and Smirnov found by numerically integration. A remarkable feature of this triangle is that strong $\nu_e \to \nu_\mu$ conversion can occur for very small mixing angles ($\sin^2 2\theta \sim 10^{-3}$), unlike the vacuum case.

The analysis of neutrino experiments have obtained mass differences of the order of $\delta m^2 \sim 10^{-5} - 10^{-6}\text{eV}^2$ and $\sin^2 2\theta_v$ as small as 6×10^{-3}, and as large as 0.6. These solutions can be distinguished by their characteristic distortions of the solar neutrino spectrum. The MSW mechanism provides a natural explanation for the pattern of observed solar neutrino fluxes. While it requires profound new physics, both massive neutrinos and neutrino mixing are expected in extended models. This phenomenon is known as the *Mikheyev-Smirnov-Wolfenstein (MSW) effect*. It provides an explanation for the observed reduction in the solar neutrino flux.

Cosmic rays collide with nuclei at the top of the atmosphere and produce secondary showers of hadrons, leptons, and neutrinos, such as the reaction $p + p \to p + n + \pi^+$. The pion decays as $\pi^+ \to e^+ + \nu_e$ or $\mu^+ + \nu_\mu$. The muon then decays by $\mu^+ \to e^+ + \nu_e + \bar{\nu}_\mu$ and the net result is a neutrino flavor ratio $(\nu_\mu + \bar{\nu}_\mu)/\nu_e = 2$. But experimentally one finds that this ratio is closer to one. This puzzle is also explained in terms of neutrino oscillations. The path-length is limited by the Earth's diameter and for a neutrino energy of the order of 1 GeV one finds that *atmospheric neutrinos are sensitive to* $\delta m^2 \gtrsim 10^{-4}$ eV^2. The Super-Kamiokande collaboration obtained the first strong evidence for neutrino oscillations with observations that muon-neutrinos produced in the upper atmosphere consistently changed into tau-neutrinos. It was observed that fewer neutrinos were detected coming through the Earth than coming directly from above the detector. The Sudbury Neutrino Observatory (SNO) obtained another evidence for solar neutrino oscillations by finding that only about 35% of the arriving solar neutrinos were ν_es, with the rest being ν_μs and μ_τs. The observed number of neutrinos agrees quite well with the SSM model for the Sun. The 2015 Nobel Prize for Physics was awarded to Takaaki Kajita from the Super-K Observatory and Arthur McDonald from Sudbury Neutrino Observatory for experiments on neutrino oscillations. There has been many estimates of the lower limit for neutrino masses from such type of experiments and the various neutrinos flavors have been estimated to have a total mass of 0.06 eV.

The neutrino masses are thought to suppress the growth of dense structures which are responsible for the formation of galaxy clusters. In addition to 3 neutrinos one also allows for the a sterile neutrino, which does not take part in weak interactions but can appear through flavor oscillations in the same way as the other neutrinos. The neutrino mixing should include all neutrinos. For a 3-neutrino mixing $|\nu_l\rangle = \sum_j U_{lj} |\nu_j\rangle$, where the 3×3 matrix U_{lj} is known as the Pontecorvo-Maki-Nakagawa-Sakata (PMNS) or Maki-Nakagawa-Sakata (MNS) mixing matrix. Adding the sterile neutrino to the mixing obviously extends it to a 4×4 matrix. Neutrino astrophysics is a thriving research field, with several nations supporting experiments which connect neutrinos to deep questions of scientific interest. One of the greatest scientific discoveries of the 21st century was the recent detection of gravitational waves, as predicted by Eintein's theory of General Relativity about 100 years ago. These and other possibilities will open an new era of astronomy with gravitational waves and neutrinos, two of the weakest interacting fields in nature, becoming valuable tools together with photons and cosmic rays to probe the structure and history of our Universe.

7.8   Nucleosynthesis in massive stars

The pp-chain is very effective to generate energy in low temperature stars such as our Sun. But as temperatures increase for more massive stars, the CNO cycle becomes more effective, provided the CNO-nuclei are already present (from earlier stellar generations). This is shown in Figure 7.15 where the energy output of the CNO cycle is compared to that of the pp-chain as a function of the temperature. The CNO energy output, assuming the CNO solar abundance, dominates for for $20 \leq T_6 \leq 130$, corresponding to Gamow peak energies of 30-110 keV.

Population II and population III are early generation of stars which obtained their energy mainly through the pp-chain. One still observes such stars in *globular clusters* and the center of a galaxy. They usually have masses smaller than the Sun, are older, cooler and have fewer heavy elements, i.e. low metallicity. In contrast, population I stars are later generation stars which formed from the debris of heavier stars containing heavy elements. Later generation stars, heavier than the Sun, achieve higher central temperatures due to higher gravity. In such a scenario, hydrogen is more efficient through reaction chains involving C, N, and O. Such nuclei have a non-negligible abundance of 1% and slightly more compared to Li, Be, B which as we have seen before, have a very low abundance.

The CN cycle, shown in Figure 7.16 (right) as cycle I, is the chain of reactions

$$^{12}\text{C}(\text{p},\gamma)^{13}\text{N}(e^+\nu_e)^{13}\text{C}(\text{p},\gamma)^{14}\text{N}(\text{p},\gamma)^{15}\text{O}(e^+\nu)^{15}\text{N}(\text{p},\alpha)^{12}\text{C}, \tag{7.134}$$

catalyzing an α-particle from four protons, $4p \rightarrow\, ^4\text{He} + 2e^+ + 2\nu_e$, and releasing $Q = 26.7$ MeV. The reactions in this cycle are relatively well-known experimentally. The $^{15}\text{N}(\text{p},\gamma)^{16}\text{O}$ reaction leads to a loss of catalytic nuclei from the CN cycle. But the catalytic material returns to the CN cycle by means of the CNO cycle II,

$$^{15}\text{N}(\text{p},\gamma)^{16}\text{O}(\text{p},\gamma)^{17}\text{F}(e^+\nu_e)^{17}\text{O}(\text{p},\alpha)^{14}\text{N}. \tag{7.135}$$

Two low energy neutrinos are emitted from the beta decays of ^{13}N and ^{15}O. The slowest reaction is $^{14}\text{N}(\text{p},\gamma)^{15}\text{O}$ in the CN cycle because it involves $Z = 7$ nuclei and therefore its Coulomb barrier

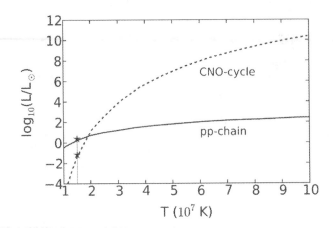

FIGURE 7.15   Temperature dependence of the energy production by the pp-chain and CNO cycles, assuming a CNO composition as that of the Sun. The passage by the Sun's core temperature is marked by "star" symbols.

is largest, and also because it is governed by electromagnetic forces, whereas the other reactions involving N isotope, $^{15}N(p,\alpha)^{12}C$, is governed by strong forces and as a consequence is faster. Thus the rate of energy production in the CN cycle is dictated by the $^{14}N(p,\gamma)^{15}O$ reaction. Sometimes the CNO cycle II is called the ON cycle and, together with the CN cycle, it constitutes the CNO cycle. The ON cycle is much slower than the CN cycle, by a factor of 1000, because the S-factor for $^{15}N(p,\alpha)^{12}C$ is about a 1000 times larger than that for $^{15}N(p,\gamma)^{16}O$. The energy released in the CNO cycle at $T_6 = 20$ is approximately given by $\epsilon \sim \epsilon_{CNO} X X_{CNO} T_6^{19.9}$, where X_{CNO} the mass fraction of oxygen/carbon/nitrogen (an average of the three) and $\epsilon_{CNO} \simeq 8.2 \times 10^{-31}$ Jm^3 kg^{-2} s^{-1}. Notice the much larger temperature dependence compared to the pp chain ($\epsilon \sim T_6^4$. That is why lower mass stars, with cooler core temperatures, generate most of their energy with pp chains, whereas in more massive stars with higher core temperatures the CNO cycle is more important.

7.8.1   Hot CNO and rp-process

The CNO cycle works at $T_6 \geq 20$ and is found in stars with the solar composition, and slightly more massive than the Sun, and burning hydrogen slowly. But CNO cycles also work at much larger temperatures ($T \sim 10^8 - 10^9$ K) in other stellar sites, such as (a) the accreting surface of a neutron star and (b) explosive *novae*, i.e., burning on the surface of a WD, or (c) in the exterior layers of the supernova shock heated material. Such stellar temperatures can induce the *hot CNO cycles* which operate on a timescale of only few seconds, as shown in Figure 7.17. In the hot CNO hydrogen burning is constrained by the β-decay lifetimes of the participating proton-rich nuclei, e.g., ^{14}O and ^{15}O. At $T \geq 5 \times 10^8$ K, material from the CNO cycle can leak and lead to the formation of heavier nuclei by means of the *rapid proton capture*, or rp-process.

The rp-process follows a path in the nuclear chart analogous to the *r-process* due to neutron capture. It converts CNO nuclei into proton-rich isotopes near the *the proton drip line*. Following a succession of fast proton captures, a mass number A is reached when another proton capture must wait until the occurrence of β^+-decay. The rp-process rate is hindered as the charge and mass numbers increase due to the increasing Coulomb barrier. Therefore, the rp-process does not really reach the proton drip line, being more effective by running near the beta-stability valley where the β^+-decay rates are of similar magnitude as the proton capture rates. Figure 7.17 shows schematically the paths for occurrence of the rp- and the r-processes in the nuclear chart.

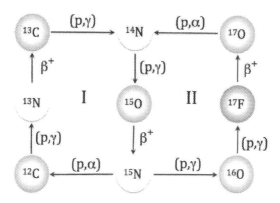

FIGURE 7.16   The ("cold") CNO cycle is composed of cycles I and II. Cycle I is also known as the CN cycle [19] and produces only about 1% of the solar energy, but leads to a significant flux of neutrinos from the Sun. Cycle II is a breakout off the CN cycle and is also known as the ON cycle. Cycle I operates 1000 times for each chain of reactions completing cycle II.

7.8.2   Triple-α capture

When the core temperature and density in a star reach $T_6 = 100 - 200$ and $\rho_c = 10^2 - 10^5$ gm cm^{-3}, it starts burning ^{4}He steadily. The most probable path to heavier elements is the fusion of three α particles into ^{12}C. The reaction occurs in two steps: (a) first by the fusion of two α particles into ^{8}Be, which is unstable to decay into two α particles, (b) followed by the fusion of ^{8}Be with another α-particle into ^{12}C as displayed in Figure 7.18. ^{8}Be has a very short lifetime of 10^{-16} s (with a decay width, $\Gamma = 6.8$ eV), nonetheless long compared to 10^{-19} s, the time that two α-particles stay close to each other. The Q-value of the reaction is $Q = -92.1$ keV. The reaction $\alpha + {}^8$Be $\rightarrow$ ^{12}C has therefore time to occur before ^{8}Be decays, for a sufficiently large density of ^{8}Be. At lower temperatures $T_6 = 10$, the α-particle energies are not high enough to produce the ^{8}Be(0_1^+) resonance and the process is non-resonant. The *Saha Equation* is an equation obtained by assuming thermal equilibrium for the process $a + b \leftrightarrow c + d$ and also for formation and decay of a nucleus. It requires the knowledge of the concentrations of the particles entering Eq. (7.6). For particle concentrations in the reaction $1 + 2 \leftrightarrow 3$

$$n_3 = \frac{n_1 n_2}{2} \left(\frac{2\pi}{\mu kT} \right)^{3/2} \hbar^3 \frac{(2J_3 + 1)}{(2J_1 + 1)(2J_2 + 1)} \exp\left(-\frac{E}{kT} \right), \qquad (7.136)$$

can be used to calculate the equilibrium concentration of ^{8}Be at $T_6 = 100$ and density $\rho = 10^5$ gm cm^{-3}. One obtains, using $n_3 = n_{8\text{Be}}$, $n_1 = n_2 = n_\alpha$ one gets n(^{8}Be)/n(^{4}He) $= 10^{-9}$. At such concentrations, the amount of ^{12}C produced is not enough to explain the observed abundance of ^{12}C, unless the reaction passes by a resonance. If one assumes an s-wave ($l = 0$) resonance in ^{12}C slightly above the threshold, the ^{8}Be $+ \alpha$ reaction is greatly enhanced. Both ^{8}Be and ^{4}He have $J^\pi = 0^+$, and an s-wave resonance would imply a 0^+ resonance state in ^{12}C. Fred Hoyle proposed that the excitation energy in ^{12}C had to be $E_r \sim 7.68$ MeV. In fact, this state was discovered experimentally only a few years later. The state became known as the *Hoyle state*. It has a total width of $\Gamma = 9$ eV, mostly due to α-decay. The γ-decay to the ground state of ^{12}C is not allowed by angular momentum conservations, as all states have $J^\pi = 0^+$. The γ-decay width is several thousand times smaller than the α-decay width, i.e, $\Gamma = \Gamma_\alpha + \Gamma_{rad} \sim \Gamma_\alpha$. Experimentally, it was determined that $\Gamma_{rad} = \Gamma_\gamma + \Gamma_{e^+e^-} = 3.7$ meV, dominated by the γ-decay width, $\Gamma_\gamma = 3.6$ meV.

The triple-α reaction rate is given by

$$r_{3\alpha} = \rho^2 N_A^2 Y_{{}^8\text{Be}} Y_\alpha \langle \sigma v \rangle_{{}^8\text{Be}+\alpha}. \qquad (7.137)$$

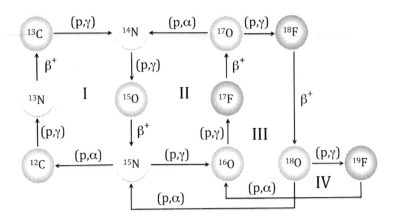

FIGURE 7.17 The CNO I, II, III and IV cycles. Cycle I, also known as CN cycle [19]. Cycle II is a breakout of the CN cycle and is sometimes called by ON cycle. Similarly the other breakout III and IV cycles may occur. The stable nuclei are represented in boxes with bottom-right gray shaded areas. The hot CNO cycles II and IV occur at high temperatures because of the additional cycles leaking from the CNO cycle through the $^{17}O(p, \gamma)^{18}F$ reaction.

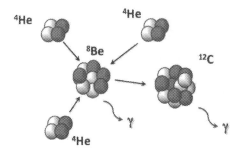

FIGURE 7.18 Schematic view of the sequential triple-alpha process.

The cross section can be parametrized by a Breit-Wigner form for a resonance reaction. For a narrow resonance with width $\Gamma \ll E_r$, the Maxwellian exponent $\exp(-E/kT)$ in Eq. (7.9) can be taken outside the integral, yielding

$$\langle \sigma v \rangle = \left(\frac{2\pi}{\mu kT} \right)^{3/2} \hbar^2 (\omega \gamma)_r \exp\left(-\frac{E_r}{kT} \right), \tag{7.138}$$

where $(\omega \gamma)_r$ is the *resonance strength*, given by

$$(\omega \gamma)_r = \frac{2J_r + 1}{(2J_j + 1)(2J_k + 1)} (1 + \delta_{jk}) \frac{\Gamma_r \Gamma_\gamma}{\Gamma}. \tag{7.139}$$

In the equation above, $\Gamma_r = \Gamma_\alpha$ dominates over Γ_γ so that $(\Gamma_\alpha \Gamma_\gamma / \Gamma) \sim \Gamma_\gamma$.

Using Saha's equilibrium Eq. (7.136) one finds the equilibrium concentration of ^{8}Be,

$$n(^8\text{Be}) = n_\alpha^2 (\omega \gamma)_r f \frac{h^3}{(2\pi \mu_{\alpha\alpha} kT)^{3/2}} \exp\left(-\frac{E}{kT} \right), \tag{7.140}$$

where a factor f was introduced to account for the enhancement of the cross section due to electron screening in stellar environments. The triple-alpha reaction rate is obtained by multiplying the equilibrium concentration of the resonant state in ^{12}C by the gamma-decay rate $\Gamma_\gamma/\hbar$ leading to the ground state of ^{12}C. That is,

$$r_{3\alpha} = \rho^2 N_A^2 Y_{^8\mathrm{Be}} Y_\alpha \hbar^2 \left(\frac{2\pi}{\mu_{\alpha^8\mathrm{Be}} kT} \right)^{3/2} (\omega\gamma)_r \, f \exp\left(-\frac{E'}{kT}\right). \tag{7.141}$$

Combining all equations in one,

$$r_{3\alpha \to {}^{12}\mathrm{C}} = \rho^3 N_A^3 \frac{Y_\alpha^3}{2} 3^{3/2} \left(\frac{2\pi\hbar^2}{M_\alpha kT} \right)^3 f\omega \frac{\Gamma_\alpha \Gamma_\gamma}{\Gamma\hbar} \exp\left(-\frac{Q}{kT}\right), \tag{7.142}$$

where ω now denotes only the spin-related part of the Eq. (7.139). The Q-value is the sum of $E'(^8\mathrm{Be}+\alpha) = E_r = 287$ keV and $E(\alpha+\alpha) = |Q| = 92$ keV, i.e., $Q_{3\alpha} = (M_{^{12}\mathrm{C}^*} - 3M_\alpha)c^2 = 380$ keV. The energy production rate is given by

$$\epsilon_{3\alpha} = \frac{r_{3\alpha} Q_{3\alpha}}{\rho} = 3.17 \times 10^{14} \frac{\rho^2 X_\alpha^3}{T_9^3} f \exp\left(-\frac{4.4}{T_9}\right) \text{ MeV g}^{-1}\text{ s}^{-1}, \tag{7.143}$$

leading to an estimate of energy production in red giants of 100 ergs/g/s at $T_8 \sim 1$. The triple-α reaction runs more efficiently a temperature value of $T_8 = 1$. Making a a Taylor series expansion around $T_8 = 1$ leads to $\epsilon_{3\alpha} \sim T_8^{41}$. Hence, a small temperature increase leads to a much larger reaction rate and energy production. AGB stars experience thermal pulses when the He burning shell is suddenly activated at high temperatures and densities and the helium burning becomes explosive. When this occurs, for a stellar core under degenerate conditions, an explosive condition known as *helium flash* develops. The luminosity from this process can reach values much larger than $10^{11} L_\odot$. It is like the luminosity of a supernovae. However, this energy does not reach the surface as it is absorbed by the expansion of the outer layers. As the flash continues, the core also looses its degeneracy and expands. Helium flash models are complicated and not very accurate. Maybe the ignition really occurs due to neutrino losses in the dense regions of the star. In dense regions at temperatures of $T > 10^8$ K, electron-electron collisions may produce neutrino anti-neutrino pairs, instead of photons. This process can cool the regions close to the center of the star. Modeling of helium flashes may also include chemical mixing in the core and asymmetries.

7.8.3 Red giants and AGB stars

The Hertzsprung-Russel diagram is shown in Figure 7.19. It is very popular among astronomers to catalogue stars in terms of their temperature and luminosity and to compare their location with those of the main sequence of stars in the diagram. After the hydrogen fuel in the core of a star is exhausted, the core contracts and its temperature rises. The outer layers of the star expand and cool down. The star luminosity greatly increases turning it to a *red giant*. When its core temperature reaches about 3×10^8 K, helium burning starts. The star's cooling stops and its luminosity is further increased. When helium burning completes the star has followed a path in the *Hertzprung-Russell* (HR) diagram first to the right, then to the left and then to the right again and up. Therefore, based on its path on the HR diagram, it has been named *asymptotic giant branch* (AGB) star. Stars with initial masses $< 9M_\odot$ reach the AGB phase in the final phases of their evolution. He burning turns from convective to radiative, but eventually switches off and the convective envelope penetrates the inner He inter-shell and brings to the surface the He burning products, a phenomenon known as the *dredge up*. This process can lead to carbon and elements heavier than iron generated by the s-process. Dredge up can occur a few hundreds of times depending on the stellar mass and the mass-loss rate as the star gravitationally contracts and heats up again. H burning resumes again until another flash occurs. AGB stars emits newly synthesized material into the interstellar medium in strong stellar winds, eroding its envelope within a million years. With the formation of ^{12}C via the triple-alpha capture, the following α-capture reaction can occur,

$$^{12}\mathrm{C} + \alpha \to {}^{16}\mathrm{O} + \gamma. \tag{7.144}$$

For large reaction rates for this process, all the carbon will end up into oxygen. But, after hydrogen, helium and oxygen, carbon is the most abundant element in the Universe, and even the

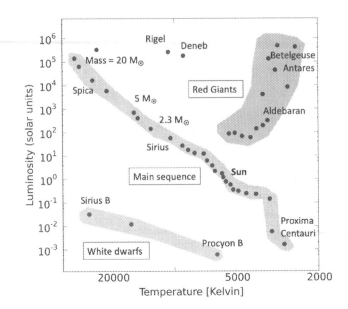

FIGURE 7.19 Hertzsprung-Russel diagram showing surface temperature, for stars born on the main sequence as a function of their absolute luminosity relative to the solar luminosity L and their temperature in Kelvin.

cosmic C/O ratio is of the order of 0.6. In fact, the main outcome of He burning in red giants are C and O. The reaction in Eq. (7.144) is very complicated one because of capture to resonances, non-resonant capture, and interfering sub-threshold resonances. One believes that the S-factor is about 0.3 MeV b. Using the same procedure as described in the triple-alpha reaction case, based on Saha's equation, one gets the reaction rate

$$\omega_{12C} = \left(\frac{n_\alpha}{7.5 \times 10^{26}/\text{cm}^3}\right)\left(2.2 \times 10^{13}/\text{s}\right) e^{-69/T_8^{1/3}} T_8^{-2/3}, \tag{7.145}$$

which has a very strong temperature dependence. For a red giant density of 10^4 g/cm^3 and an α fraction of 0.5, we get lifetimes for ^{12}C of 1.8×10^9 y for $T_8 = 1$ and 1.8×10^3 y for $T_8 = 2$.

At temperatures $T_9 = 0.1$ and above, the Gamow window is about $E_0 = 0.3$ MeV. The S-factor for this reaction of about 0.3 MeV barn is not enough to burn ^{12}C completely to ^{16}O, and one obtains the ratio C/O ~ 0.1. Direct measurements of this reaction are very difficult at the stellar energy of 300 keV region. It is often mentioned that if the cross section for this reaction is twice larger than the presently accepted value, a 25M$_\odot$ star will not produce enough ^{20}Ne since carbon burning would cease. The impact of this result is enormous as an oxygen-rich star probably collapses into a black hole whereas carbon-rich stars would collapse into neutron stars. The burning of ^{16}O via ^{16}O$(\alpha, \gamma)^{20}$Ne reaction is very slow during helium burning in red giant stars because of its very small cross section. The major ashes in red giants are therefore carbon and oxygen. The ^{16}O$(\alpha, \gamma)^{20}$Ne reaction is identified as the "end-point" of the reaction chain ^{4}He$(2\alpha, \gamma)^{12}$C$(\alpha, \gamma)^{16}$O$(\alpha, \gamma)^{20}$Ne. The reaction rate is very small because there are no resonances in the Gamow window, around $E_0 = 300$ keV.

7.9   Carbon, neon, oxygen, and silicon burning

When helium burning ceases the core of the stars becomes a mixture so C and O. Since H and He are the most abundant elements, C and O are formed in appreciable large amounts. Helium burning still continues in a layer surrounding the C and O rich core and hydrogen burning in another layer surrounding the helium burning shell. When helium burning ceases, there is not enough pressure to keep the star gravitationally stable and it begins to contract again. The temperature of the helium-poor core rises again but the contraction continues until the next burning fuel becomes effective or until electron degeneracy pressure stops the contraction.

7.9.1 Carbon burning

After helium burning, stars with masses $M \geq 8 - 10M_\odot$ have their CO-rich core contract until temperatures and densities reach $T \sim 5 \times 10^8 K$ and $\rho = 3 \times 10^6$ gcm^{-3} and carbon starts burning. Energy is released and the contraction stops, and a quiescent carbon-burning continues. Two ^{12}C nuclei can fuse, leading to a compound nucleus of ^{24}Mg nucleus with an excitation energy of 14 MeV. The Gamow window at such temperatures is about 1 MeV, a region with many excited levels in ^{24}Mg. Thus, the free proton is converted into a neutron and the α-particle fuse with ^{12}C into ^{16}O. We have discussed the reaction ^{12}C(p,γ)^{13}N when we referred to the CNO chain.

In AGB stars some protons diffuse from the convective envelope into the He inter-shell at the end of each dredge up and create ^{13}C explaining the known abundances of the s-process elements at the surface of AGB stars. A thin layer known as the ^{13}C pocket is produced by means of the ^{12}C(p,γ)^{13}N(e$^+\nu_e$)^{13}C. When the temperature reaches $T_6 \sim 100$ the ^{13}C(α,n)^{16}O reaction starts generating the neutrons triggering the s-process. The cross-section is strongly influenced by a subthreshold resonant tail contribution.

Carbon burning produces nuclei such as ^{16}O, ^{20}Ne, ^{24}Mg and ^{28}Si with each pair of ^{12}C releasing on average about 13 MeV of energy. After the completion of the carbon burning phase, reactions such as ^{12}C+^{16}O and ^{12}C+^{20}Ne will occur with reaction rates smaller than for ^{12}C+^{12}C due to the larger Coulomb barriers. The star will lose most of tis energy during the carbon-burning and later stages, by neutrino emission, a process that is very sensitive to the core temperature. The neutrino luminosity soon becomes larger than the photon luminosity at the surface of the star. The timescale for energy production in the star by neutrino emission is very short compared to the timescale for gravitational cooling, given by $t_{th} = GM_s^2/R_s L_s$, where M_s, R_s, and L_s are the mass, radii, and the star surface luminosity, respectively. Because of the slow photon diffusion, there is not enough time for this energy to propagate to he surface, and its temperature does not change considerably during and beyond the carbon burning phase.

7.9.2 Neon and oxygen burning

During the carbon burning stage mainly neon, sodium and magnesium are synthesized, with small traces of aluminum and silicon due to the capture of α, p, and n released during that stage. After carbon is exhausted, the core contracts and its temperature raises again. When $T_9 \sim 1$, the high energy photons present in the tail of the Planck distribution begin to disintegrate ^{20}Ne through the reaction ^{20}Ne(γ,α)^{16}O. Since α-decay processes release α-particles at nearly the same energy as of an emitted nucleon, by time-inversion, one concludes that the most common photodisintegration reactions are $(\gamma,\text{n}), (\gamma,\text{p})$, and (γ,α). The photonuclear rate passing by an excited state E_x is

$$r(\gamma,\alpha) = \left[\exp\left(-\frac{E_x}{kT}\right) \frac{2J_R + 1}{2J_0 + 1} \frac{\Gamma_\gamma}{\Gamma} \right] \frac{\Gamma_\alpha}{\hbar}, \qquad (7.146)$$

where the term in brackets is the probability of finding the nucleus in the excited state E_x with spin J_R (J_0 is the ground state spin) and the decay rate to alpha particle emission is given by $\Gamma_\alpha/\hbar$. Using $E_x = E_R + Q$,

$$r(\gamma,\alpha) = \frac{\exp(-Q/kT)}{\hbar(2J_0 + 1)}(2J_R + 1)\frac{\Gamma_\alpha \Gamma_\gamma}{\Gamma} \exp(-E_R/kT). \qquad (7.147)$$

The 5.63 MeV level in ^{20}Ne dominates the photodissociation process for $T_9 \geq 1$. When $T_9 \sim 1.5$, the photodissociation rate is greater than the alpha capture rate on ^{16}O leading to ^{20}Ne, which is the reverse reaction. The released α-particle then also begins to react with ^{20}Ne by means of ^{20}Ne(α,γ)^{24}Mg. The net result is the conversion of two ^{20}Ne nuclei in the form ^{20}Ne(^{20}Ne,^{16}O)^{24}Mg with $Q = 4.58$ MeV. The neon burning stage is very quick and yields a core with a mixture of ^{4}He, ^{16}O and ^{24}Mg nuclei. At the end of Ne burning, the core contracts and the temperature reaches

$T_9 \sim 2$. Then ^{16}O begins to react by means of ^{16}O(^{16}O,α)^{28}Si and ^{16}O(^{16}O,γ)^{32}Si each occurring approximately 50% of the time. The oxygen burning phase also produces, Ar, Ca, and traces of Cl, K, $\cdots$, up to Sc.

7.9.3   Silicon burning

At $T_9 \sim 3$, the produced ^{28}Si produced during the oxygen burning phase begins to burn in the Si burning phase. During the neon burning phase, not only heavier nuclei are produced, but photons are sufficiently energetic to dissociate neon, before the temperature is high enough to ignite reactions involving oxygen. Thus α-particles are produced in the neon burning phase, a trend which continues in the silicon burning phase. In general, photodissociation of the nuclear products of neon and oxygen burning continues when the temperature surpasses $T_9 \geq 3$ because of the high energy photons in the tail of Planck's distribution. In fact, photodissociation destroys nuclei with small binding energies and leads to many nuclear reactions involving protons, neutrons, and α-particles with nuclei in the mass range $A = 28 - 65$ such as

$$\gamma + \ {}^{28}_{14}\text{Si} \longrightarrow \ {}^{24}_{12}\text{Mg} + {}^{4}_{2}\text{He}, \qquad {}^{4}_{2}\text{He} + {}^{28}_{14}\text{Si} \longrightarrow {}^{32}_{16}\text{S} + \gamma, \text{ etc.} \qquad (7.148)$$

The large number of free neutrons leads to several (n,γ)-reactions, i.e., to *radiative neutron capture*. Elements with the largest binding energies in the iron mass region are produced abundantly. ^{56}Fe has the maximum value for binding energy per nucleon, leading to a full stop of the fusion reactions around the iron-group.

   The compound nuclear levels in the nuclei with $A = 28 - 65$ formed during silicon burning are dense and overlapping. At the high temperatures of $T_9 = 3 - 5$, a quasi-equilibrium is reached of reactions occurring backward and forward between the nuclei in this mass region. Among these there are a few slow reactions known as *bottlenecks*. As the the nuclear fuel is consumed, the thermal energy decreases by neutrinos escaping from the star, and many nuclear reactions stop occurring leading to a *freeze-out* of the nuclear burning process.

7.10   Slow and rapid nucleon capture

Due to the large Coulomb barriers, the abundance of heavy elements above the iron peak would drop very much if they were to be synthesized by reactions during the silicon burning phase and would require larger temperatures. Higher temperatures also increase photon energies in the Planck distribution, and eventually the photodisintegrations of nuclei will dominate. Therefore, a statistical chemical equilibrium will be reached during silicon burning and a maximum abundance of nuclei will be reached at the highest binding energies. This effect can only be circumvented with high densities, like in the rp-process on accreting neutron stars, where the high densities still favor capture reactions, acting against photodisintegrations. However, the study of reaction chains used in stellar evolution show that large neutron fluxes can be produced at the stellar core. Electron capture and beta decay reactions influence the reaction flow by changing the value of Y_e (the number of electrons per nucleon) changing the stellar core density and entropy not only in the silicon burning phase, but also in earlier oxygen burning phases. Fusion reactions involving charged particles cannot explain the formation of nuclei with $A > 100$. Such heavy nuclei are formed by the successive capture of slow neutrons followed by β^--decay. This process can explain the maxima of the observed elemental abundance for $N = 50, 82, 126$ which are due to the small capture cross sections for nuclei with magic numbers, leading to a peak of nuclear isotopes with these neutron values.

   The speed of the process to produce heavy elements is determined by the timescales of the nuclear transformations encountered along the process path. In case of an only small neutron density available neutron-captures are slower than beta-decays, in the other extreme of very high neutron densities beta-decays are slower than neutron captures. Therefore, the s-process nucleosynthesis runs very close to the valley of β-stability. On the other hand, rapid neutron capture, or *r-processes* can only occur when $\tau_n \ll \tau_\beta$. It requires extremely neutron-rich environments, because the capture timescale is inversely proportional to the neutron density. The r-process path includes very neutron rich and unstable nuclei, far from the valley of stability. Nuclear half-lives are modified in the stellar environment and excited states can also be thermally populated.

7.10.1   The s-process

The s-process path occurs along the valley of stability and avoids some nuclei because, when N and Z are even, (N, Z) and $(N + 2, Z − 2)$ are stable to β decay, whereas the neighboring odd-odd nuclei $(N + 1, Z − 1)$ are unstable. These odd-odd nuclei have an unpaired proton and also an unpaired neutron, and have therefore a rather large ground state energy. The production of the $(N + 2, Z − 2)$ nucleus is therefore suppressed in the s-process. The observation of such isotopes with non-negligible abundances is a hint that other processes than the s-process might occur. The reaction chain along the s-process process is of the form

$$\frac{dn_A(t)}{dt} = n_n(t)n_{A-1}(t)\langle\sigma v\rangle_{A-1} - n_n(t)n_A(t)\langle\sigma v\rangle_A - \lambda_\beta(t)n_A(t), \qquad (7.149)$$

where $n_n(t)$ is the neutron density and λ_β is the β decay rate. The first term accounts for neutron capture to form nucleus A, the second and the third are related to its destruction by neutron-induced reactions and by β-decay. A simple estimate can be applied for the s-process by assuming a slow β-decay rate, (i.e., the last term above) and also by assuming that $\langle\sigma v\rangle = \sigma_A\langle v\rangle$. Then one finds

$$\frac{dn_A(t)}{dt} = \langle v\rangle n_n(t)\left(\sigma_{A-1}n_{A-1} - \sigma_A n_A\right), \qquad (7.150)$$

and, at equilibrium, $dn/dt = 0$, yielding

$$\frac{n_A}{n_{A-1}} = \frac{\sigma_{A-1}}{\sigma_A}. \qquad (7.151)$$

The abundance for mass A is inversely proportional to the neutron cross section to A. If the cross section is small, mass A will be produced abundantly. A similar argumentation also works if β decay is faster than neutron capture because the small capture cross sections at the closed shells will also result in mass peaks in that region, compatible with the observations. Equilibrium will also occur more quickly in the plateaus between the magic numbers because mass will pile up around the closed shells before they can be surpassed. Calculations suggest that a probable site for the s-process is the helium-burning shell of a red giant, with sufficiently high temperatures to produce neutrons by means of ^{22}Ne$(\alpha, n)^{25}$Mg, where the ^{22}Ne nucleus is produced during the helium burning of the ^{14}N ashes in the CNO cycle. This occurs via the reaction chain ^{14}N$(\alpha, \gamma)^{18}$F$(\beta^+\nu)^{18}$O$(\alpha, \gamma)^{22}$Ne. Both ^{22}Ne$(\alpha,n)^{25}$Mg and the competing ^{22}Ne$(\alpha, \gamma)^{26}$Mg reactions are poorly known. The s-process is ineffective beyond ^{209}Bi because it leads to a decay chain that ends up with α emission when the neutron is captured in this isotope. That is the end product of the s-process and *transuranic elements* must be produced in some other way.

7.10.2   The r-process

Nuclei with mass up to about $A = 90$ above the iron group are mainly produced by the s-process in massive stars. For $A \geq 100$, the contribution of the s-process is very small in massive stars and one believes that most of the s-process contribution in this mass range occurs in AGB stars. The competing process, i.e., the r-process, is very fast and lasts for a few seconds in dense neutron environments with neutron densities of $n_n^{r \text{ process}} \sim 10^{20} - 10^{25}$ cm^{-3}, much larger than those for the s-process, $n_n^{s \text{ process}} \sim 10^8$ cm^{-3}. During compression of electrons in a core-collapse supernova, β^- decay process in nuclei are Pauli-blocked due to the Fermi energy of the environment electrons being larger than the energy in the beta decay. However, electrons can still be captured by nuclei, yielding a higher neutronized matter and a large flux of neutrons and higher temperatures. Neutron capture on heavy nuclei is faster than β^- decays when the matter expands and cools down again. The r-process produces highly unstable neutron rich nuclei, and its path runs close to the *neutron-drip line* (see Figure 7.20). More than 90% of elements such as europium (Eu), gold (Au), and platinum (Pt) in the solar system are believed to be synthesized in the r-process.

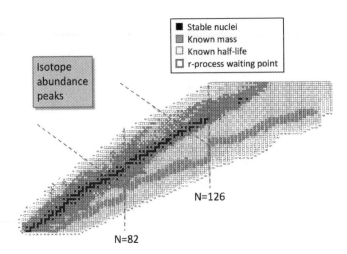

FIGURE 7.20 Schematic view of the r-process (red squares) in the nuclear chart. The process occurs close to the neutron drip drip line. The waiting point nuclei at the neutron magic numbers are responsible for peaks in the abundance of isotopes as a function of the nuclear mass.

If the radiative capture reaction $A + n \rightarrow (A + 1) + \gamma$ is *exothermic*, and assuming a resonant reaction in a high-level density nucleus, and further neglecting spins, Eq. (7.138) leads to

$$\langle \sigma v \rangle_{(n,\gamma)} = \left(\frac{2\pi}{\mu kT} \right)^{3/2} \frac{\Gamma_n \Gamma_\gamma}{\Gamma} e^{-E_R/kT}. \tag{7.152}$$

But the resonance energy is very close to zero, i.e., $E_R \sim 0$ and the rate is simply given by

$$r_{(n,\gamma)} \sim n_n n_A \left(\frac{2\pi}{\mu kT} \right)^{3/2} \frac{\Gamma_n \Gamma_\gamma}{\Gamma}. \tag{7.153}$$

The high-energy tail of the Planck distribution for thermal photons (with $\hbar = c = 1$) is given by $n(E_\gamma) \sim (E_\gamma^2 / n_\gamma \pi^2) e^{-E_\gamma/kT}$, where $n_\gamma \sim (\pi/13)(kT)^3$, is the approximate value of the total photon density obtained by integrating Planck's distribution over all energies. For a resonant reaction, and using the asymptotic form of the Maxwell-Boltzmann distribution in the reaction rate integral, we get

$$r_{(\gamma,n)} \sim 2n_{A-1} \frac{\Gamma_\gamma \Gamma_n}{\Gamma} e^{-E_R/kT}. \tag{7.154}$$

In equilibrium, i.e., when the (n,γ) and (γ, n) rates are the same, and using $n_A \sim n_{A-1}$, leads to

$$n_n \sim \frac{2}{(\hbar c)^3} \left(\frac{\mu c^2 kT}{2\pi} \right)^{3/2} e^{-S_n/kT},$$

where we assumed that E_R is the same as neutron separation energy, S_n. If we assume that the r-process works under the conditions of $n_n \sim 3 \times 10^{23}/\text{cm}^3$ and $T_9 \sim 1$, we obtain that $S_n \sim 2.4$ MeV. Hence, neutrons are bound by about $30kT$, a small value compared to typical binding energies of 8 MeV in a normal nucleus. This means that contour line of constant neutron separation energy is close to the neutron drip-line.

Assuming that one knows under what conditions the r-process operates, the goal is to reproduce the relative abundance of elements observed in our Galaxy, shown in Figure 7.21. For $A < 100$ it decreases exponentially with A, while for $A > 100$ it is nearly constant, except for the peaks around the magic numbers $Z = 50$ and $N = 50, 82,$ and 126. At the shell closure gaps $N = 82, 126$, the neutron number stays fixed because the nuclei cannot overcome the gap energy. They have to wait

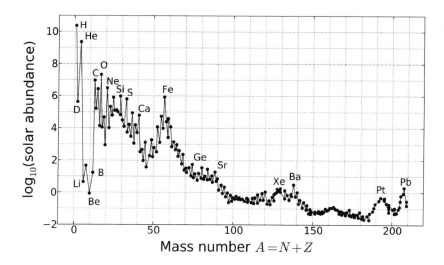

FIGURE 7.21 Isotopic mass dependence of the relative abundances of elements in the solar system. The abundances have been normalized so that the abundance of silicon is 10^6. The data have been collected from numerous sources.

for the several β-decays to overcome the shell closures. Moreover, the β decays are slow at the shell closures. That is why the closed neutron shells are called *waiting points*. The abundance of an isotope is inversely proportional to the β decay time, opposite to what occurs in the s-process, where it is proportional to the time scale of neutron capture (i.e., inversely proportional to the neutron-capture cross section). Therefore, mass accumulates at the waiting points, yielding the peaks shown in Figure 7.21. The abundance peaks near $A = 130$ and $A = 195$ are a signature of an environment of rapid neutron capture near closed nuclear shells.

When the r-process ends, the nuclei β-decay to the valley of stability. Neutron spallation moves the peaks to to a lower value in A. But the peaks are also shifted to lower values of N because β-decay transforms neutrons into protons during the r-process. At the end of the r-process very heavy nuclei ($A \sim 270$) might be produced until β-delayed and neutron-induced fission occurs leading the nuclei back to $A \sim A_{max}/2$. For the r-process to exist one needs very high densities and temperatures and reactions occurring at very short timescales, e.g., $n_n \sim 10^{20}$ cm^{-3}, $T \sim 10^9$K, and $t \sim 1$ s. But these are explosive conditions, which could occur (a) in the neutronized region above the proto-neutron star in Type II supernova, (b) or in neutron-rich jets from supernovae or from neutron star mergers, (c) during the inhomogeneous Big Bang, (d) in He/C zones in Type II supernovae, (e) in red giant He flashes, and (f) in neutrino spallation on neutrons in the He zone.

One possible scenario for the occurrence of the r-process are neutron stars mergers. In fact, recent calculations suggest that heavy elements with $A > 110$ are difficult to be produced in core-collapse supernovae due to a not sufficiently neutron-rich environment. Neutron star mergers take a long time to merge (about 100 Myr) at a low rate. They occur in neutron star binary systems that lose their energy very slowly due to gravitational radiation.

Long-lived radioactive nuclei are often utilized as *cosmo-chronometers*. They are useful for investigating the nucleosynthesis process history of the cosmic evolution before the formation of the solar system, an idea proposed by Rutherford about 70 years ago. Only a handful of cosmo-chronometers with half-live values useful for dating the cosmic age, $1 - 100$ Gyr, are known, e.g., ^{40}K, ^{176}Lu, ^{187}Re, ^{232}Th and ^{238}U. The pair of nuclei ^{187}Re-^{187}Os, was also proposed as cosmo-chronometer to investigate the galactic history of the r-process. ^{187}Re predominantly decays by β^- after the freezeout of the r-process. The s-process nuclei ^{186,187}Os are not produced in the r-process but ^{186}Os is produced in the s-process and its observed abundance normalizes the overall ^{186}Os production. The half-life for ^{187}Re to decay into ^{187}Os is 4.35×10^{10} yr, longer than the age of the Universe. After subtraction of the contributions of s-process, the epoch from an r-process nucleosynthesis event can be determined by the presently observed abundances of ^{187}Re and ^{187}Os.

7.11   White dwarfs and neutron stars

If the thermonuclear processes in massive stars achieve the production of iron, there are the following possibilities for the star evolution.

(a) For stars with masses $< 1.2\ M_\odot$ the internal pressure of the degenerated electron gas (i.e., when the electrons occupy all states allowed by the Pauli principle) does not allow the star compression due to the gravitational attraction to continue indefinitely. For a free electron gas at temperature $T = 0$ (lowest energy state), the electrons occupy all energy states up to the Fermi energy. The total density of the star can be calculated adding up the individual electronic energies. Since each phase-space cell $d^3p \cdot V$ (where V is the volume occupied by the electrons) contains $d^3p \cdot V/(2\pi\hbar)^3$ states, we get

$$\frac{E}{V} = 2 \int_0^{p_F} \frac{d^3p}{(2\pi\hbar)^3} E(p) = 2 \int_0^{p_F} \frac{d^3p}{(2\pi\hbar)^3} \sqrt{p^2c^2 + m_e^2 c^4} = n_0 m_e c^2 x^3 \epsilon(x),$$

$$\epsilon(x) = \frac{3}{8x^3}\left\{ x(1 + 2x^2)(1 + x^2)^{1/2} - \log[x + (1 + x)^{1/2}] \right\}, \tag{7.155}$$

where the factor 2 is due to the electron spin, and

$$x = \frac{p_F c}{m_e c^2} = \left(\frac{n}{n_0}\right)^{1/3} = \left(\frac{\rho}{\rho_0}\right)^{1/3}, \tag{7.156}$$

where

$$n_0 = \frac{m e^3 c^3}{\hbar^3} \quad \text{and} \quad \rho_0 = \frac{m_N n_0}{Y_e} = 9.79 \times 10^5\ Y_e^{-1}\ \text{g/cm}^3. \tag{7.157}$$

In the above relations p_F is the Fermi momentum of the electrons, $m_e(m_N)$ is the electron (nucleon) mass, n is the density of electrons, and ρ is the mass density in the star. Y_e is the number of electrons per nucleon.

The variable x characterizes the electron density in terms of

$$n_0 = 5.89 \times 10^{29} \text{cm}^{-3}. \tag{7.158}$$

At this density the Fermi momentum is equal to the inverse of the Compton wavelength of the electron.

Using traditional methods of thermodynamics, the pressure is related to the energy variation by

$$P = -\frac{\partial E}{\partial V} = -\frac{\partial E}{\partial x}\frac{\partial x}{\partial V} = -\frac{\partial E}{\partial x}\left(-\frac{x}{3V}\right) = \frac{1}{3} n_0 m_e c^2 x^4 \frac{d\epsilon}{dx}. \tag{7.159}$$

This model allows us to calculate the pressure in the electron gas in a very simple form. Since the pressure increases with the electron density, which increases with the decreasing volume of the star, we expect that the gravitational collapse stops when the electronic pressure equals the gravitational pressure. When this occurs the star cools slowly and its luminosity decreases. The star becomes a *white dwarf* and in some cases its diameter can become smaller than that of the Moon.

(b) For stars with masses in the interval 1.2 - 1.6 $M_\odot$, the electron pressure is not sufficient to balance the gravitational attraction. The density increases to 2×10^{14}g.cm^{-3} and the matter "neutronizes". This occurs via the electron capture by the nuclei (inverse beta decay), transforming protons into neutrons. The final product is a *neutron star*, with a small radius (see Figure 7.22). For example, if it were possible to form a neutron star from the Sun it would have a radius given by

$$\left(\frac{M_\odot}{\frac{4\pi}{3}\rho}\right)^{1/3} = \left(\frac{2 \times 10^{33}\ \text{g}}{\frac{4\pi}{3} \times 2 \times 10^{14}\ \text{g cm}^{-3}}\right)^{1/3} \simeq 14\ \text{km}.$$

The process of transformation of iron nuclei into neutron matter occurs as following: for densities of the order of 1.15×10^9 g.cm^{-3} the Fermi energy of the electron gas is larger than the upper

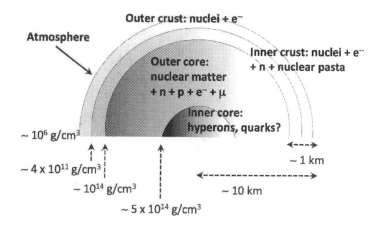

FIGURE 7.22 The structure of a neutron star. The outer regions of a neutron star may consist of thin layers of various elements that were produced by nuclear reactions during the star's lifetime. These outer layers are thought to have a rigid crystalline structure because of the intense gravitational field of the neutron star. The composition of the inner core is unknown.

energy of the energy spectrum for the β-decay of the isotope $^{56}_{25}$Mn. The decay of this isotope can be inverted and two neutron-rich isotopes of $^{56}_{25}$Mn are formed, i.e.,

$$^{56}_{26}\text{Fe} + \text{e}^- \longrightarrow \, ^{56}_{25}\text{Mn} + \nu_e. \tag{7.160}$$

These nuclei transform in $^{56}_{24}$Cr by means of the reaction

$$^{56}_{25}\text{Mn} + \text{e}^- \longrightarrow \, ^{56}_{24}\text{Cr} + \nu_e. \tag{7.161}$$

With the increasing of the pressure more isotopes can be formed, until neutrons start being emitted:

$$^A_Z\text{X} + \text{e}^- \longrightarrow \, ^{A-1}_{Z-1}\text{X} + \text{n} + \nu_e. \tag{7.162}$$

For $^{56}_{26}$Fe this reaction network starts to occur at an energy of 22 MeV, which corresponds to a density of 4×10^{11} g·cm^{-3}. With increasing density, the number of free neutrons increases and, when the density reaches 2×10^{14} g·cm^{-3}, the density of free neutrons is 100 times larger than the density of the remaining electrons.

A *pulsar* is a rapidly rotating neutron star. Like a black hole, it is an endpoint to stellar evolution. The "pulses" of high-energy radiation we see from a pulsar are due to a misalignment of the neutron star's rotation axis and its magnetic poles (see Figure 7.23). Neutron stars for which we see such pulses are called "pulsars". They have a mass 40% larger than the Sun and their radius is just 20 kilometers. This means that a cubic centimeter of this matter has 100 million tons! The neutron stars are in the limit of the density that matter can have: the following step is a black hole. Today one knows more than 600 pulsars and they are formidable astrophysics laboratories, since: (a) their density is comparable to that of an atomic nucleus; (b) their mass and size give place to gravitational fields only smaller than those of the black holes, but easier to measure; (c) the fastest of the pulsars have 600 turns about its axis in one second. Thus, its surface rotates by 36,000 kilometers a second; (d) neutrons stars have the more intense magnetic fields than any other known object in the Universe, million of times stronger than those produced in any terrestrial laboratory; (e) in some cases the regularity of their pulsations is the same or larger than the precision of the atomic clocks, the better that we have.

General observations

Neutron-star masses can be deduced from observations of supernova explosions and from binary stellar systems. For example, Newtonian mechanics (i.e., *Kepler's third law*) relates the mass of a

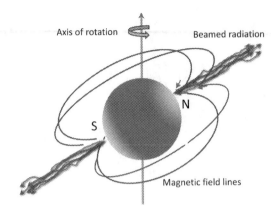

FIGURE 7.23 A rapidly rotating neutron star, or pulsar. At the magnetic poles, particles can escape and give rise to radio emission. If the magnetic axis is misaligned with the rotation axis of the neutron star (as shown), the star's rotation sweeps the beams over the observer as it rotates like a lighthouse, and one sees regular, sharp pulses of light (optical, radio, X-ray, etc.).

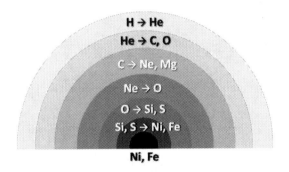

FIGURE 7.24 The "onion" structure of a $20M_\odot$ star just before a supernova explosion.

neutron star M_{NS} and its companion mass M_C in a binary system though

$$\frac{(M_C \sin\theta)^3}{(M_{NS} + M_C)^2} = \frac{T_{NS} v_\theta^3}{2\pi G}, \tag{7.163}$$

where T_{NS} is the period of the orbit, v_θ is its orbital velocity projected along the line of sight, and θ is the angle of inclination of the orbit. But this equation is not enough to determine M_{NS} as we need to know the mass of the companion, too. General relativity predicts an advance of the periastron of the orbit, $d\omega/dt$, given by

$$\frac{d\omega}{dt} = 3\left(\frac{2\pi}{T_{NS}}\right)^{5/3} T_\odot^{2/3} \frac{(M_{NS} + M_C)^{2/3}}{1 - e}, \tag{7.164}$$

where and $T_\odot = GM/c^3 = 4.9255 \times 10^{-6}$ s. The observation of this quantity and comparison to this equation, together with the range parameter $R = T_\odot M_C$ associated with the *Shapiro time delay* of the pulsar signal as it propagates through the gravitational field of the companion star, and the Keplerian Eq. (7.163), allows one to determine the neutron star mass M_{NS}. Other so-called post-Keplerian parameters such as the the orbital decay due to the emission of quadrupole gravitational radiation, or the Shapiro delay shaper parameter, can be used together with Eq. (7.163) to determine the mass of a neutron star. Typical masses obtained with this procedure

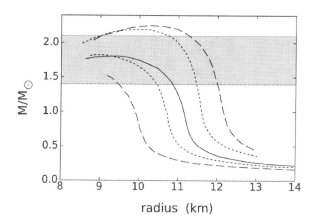

FIGURE 7.25 Mass of a neutron star (in units of solar masses) versus radius in kilometers calculated (solid, dashed and dotted curves) with a few realistic EoS. The horizontal band displays the limits of the observed masses.

range from $M_{NS} = 1.44 M_\odot$ from observations of binary pulsars systems to $M_{NS} = 2.01 M_\odot$ from millisecond pulsars in binary systems formed by a neutron star and a white dwarf.

The radius of a neutron star is more difficult to determine because they are very small. But measurements of the X-ray flux Φ stemming from a neutron star in a binary system at a distance d from us can be used, assuming that the radiation stems from a blackbody at temperature T, leading to the effective radius

$$R_{eff} = \sqrt{\frac{\Phi d^2}{\sigma T^4}}. \tag{7.165}$$

This can be used together with a relativistic correction to get the neutron star radius R from

$$R = R_{eff}\sqrt{1 - \frac{2GM_{NS}}{Rc^2}}. \tag{7.166}$$

The radii of neutron stars found with this method range within $9 - 14$ km.

Neutron stars can rotate very fast due to the conservation of angular momentum when they were created as leftovers of supernova explosions. In a binary system the matter absorption from the companion star can increase its rotation. The angular speed can reach several hundred times per second and turn it into an oblate form. It slows down because its rotating magnetic field radiates energy into free space and its shape becomes more spherical. The slow-down rate is nearly constant and extremely small, of the order of $-(d\Omega/dt)/\Omega = 10^{-10} - 10^{-21}$ s/rotation. Sometimes a neutron star suddenly rotates faster, a phenomenon know as a "glitch", thought to be associated with a "starquake" when there is a rupture in their stiff crust. As a consequence, the equatorial radius contracts and angular momentum conservation leads to an increase of rotation. Glitches could also be due to vortices in the superfluid core transiting from a metastable state to a lower-energy state.

The hitherto compiled knowledge about neutron stars seems to indicate that they contain a dense core surrounded by a much thinner crust with mass $\lesssim M_\odot/100$ and a thickness of $\lesssim 1$ km (see Figure 7.22). The crust, divided into outer and inner crust, consists of neutron-rich nuclei coexisting with strongly degenerate electrons and for densities $\rho \gtrsim 4 \times 10^{11}$ g cm^{-3} free neutrons drip from nuclei. The density increases going from crust to the core and at their interface the density is $\approx \rho_0/2$, where $\rho_0 = 2.8 \times 10^{14}$ g cm^{-3} is the saturation density of nuclear matter. The outer core is probably composed of neutrons, protons and electrons, whereas the inner core, found in massive and more compact neutron stars might contain pion, kaon or hyperon condensates, or quark matter. The Equation of State (EoS), i.e., how pressure depends on the density in the core of a neutron star, is usually separated in an EoS for the outer core ($\rho \lesssim 2\rho_0$) and another for the inner core ($\rho \gtrsim 2\rho_0$).

Structure equations

The equations of hydrostatic equilibrium for a neutron star are modified to account for special and general relativity corrections. The so-called Tolman-Oppenheimer-Volkoff (TOV) equation is the state-of-the-art equation for the structure of a spherically symmetric neutron star in static equilibrium, given by

$$\frac{dp(r)}{dr} = -\frac{G}{r^2} \left[\rho(r) + \frac{p(r)}{c^2} \right] \left[m(r) + 4\pi r^3 \frac{p(r)}{c^2} \right] \left[1 - \frac{2Gm(r)}{c^2 r} \right]^{-1}, \tag{7.167}$$

where $m(r)$ is the total mass within radius r,

$$m(r) = 4\pi \int_0^r r'^2 \rho^2(r') dr'. \tag{7.168}$$

The second terms within the square brackets in Eq. (7.167) stem from special and general relativity corrections to order $1/c^2$, and in their absence one has the equations of hydrostatic equilibrium following Newton's gravitation theory. To obtain how mass increases with the radial distance, the two equations above need to be supplemented by an EoS linking the pressure to density, $p(\rho)$. The relativistic corrections factors have all the effect of enhancing the gravity at a distance r from the center.

The TOV equation is integrated enforcing the boundary conditions $M(0) = 0$, $\rho(R) = 0$ and $p(R) = 0$, where R denotes the neutron star radius. It is usually integrated radially outward for p, ρ and m until it reaches $p \simeq 0$ at a star radius R. A maximum value of the neutron star mass is obtained for a specific value of the central density. Typical results are plotted as in Figure 7.25 with the star mass M given in terms of the central density ρ_c, or in terms of the radius R. An EoS based on non-interacting neutron gas yields the maximum mass of a neutron star as $\sim 0.7 M_\odot$, whereas a stiffer equation of state yields $\sim 3 M_\odot$. The horizontal band displays the limits of the observed masses. At very high pressures, Eq. (7.167) is quadratic in the pressure and if a star develops a too high central pressure, it will quickly develop an instability and will not be able to support itself against gravitational implosion.

7.11.1 The equation of state of neutron stars

Neutron stars are almost exclusively made of neutrons with a small fraction, $\sim 1/100$, of electrons and protons. The neutron is transformed into a proton and an electron via the weak decay process $n \rightarrow p + e^- + \bar{\nu}_e$, liberating an energy of $\Delta E = m_n - m_p - m_e = 0.778$ MeV which is carried away by the electron and the neutrino[*]. In weak decay equilibrium, as many neutrons decay as electrons are captured in $p + e^- \rightarrow n + \nu_e$, which can be expressed in terms of the chemical potentials for each particle as $\mu_n = \mu_p + \mu_e$. The chemical potential is the energy required to add one particle of a given species to the system. The Pauli principle impedes decays when low-energy levels for the proton, electron, or the neutron are already occupied. Since the matter is neutral, the Fermi momenta of protons and electrons are the same, $k_{F,p} = k_{F,e}$.

A large number of experimental data on stable nuclei has obtained important results for symmetric nuclear matter ($Z = N$) such as an equilibrium number density $\rho_0 = 0.16$ nucleons/fm^3, and a binding energy per nucleon at saturation of $E/A = -16$ MeV, where $\epsilon(\rho)$ is the energy density. The saturation density ρ_0 corresponds to a Fermi momentum of $k_F = 263$ MeV/c, small compared with $m_N = 939$ MeV/c^2 and thus justifying a non-relativistic treatment. A Taylor expansion of the energy per particle for asymmetric nuclear matter ($Z \neq N$) can be done,

$$\frac{E}{A}(\rho, \delta) = \frac{E}{A}(\rho_0, 0) + \frac{1}{2} K_0 x^2 + \frac{1}{6} Q_0 x^3 + S(\rho)\delta^2 + \cdots, \tag{7.169}$$

---

[*]In this and in the next equations we use $\hbar = c = 1$.

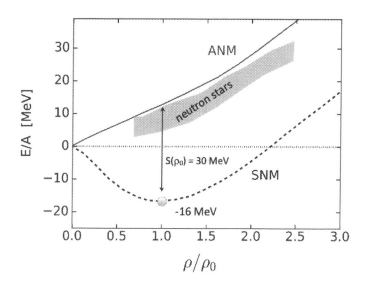

FIGURE 7.26 Microscopic (e.g, Hartree-Fock, Relativistic Mean Field, etc.) calculations of the nuclear matter (SNM) equation of state (EoS) compared to the result obtained for pure (ANM) neutron matter. The two curves are roughly separated by the bulk symmetry energy factor $S(\rho_0) \equiv L$. The EoS for neutron stars is likely to be somewhat different than the ANM curve shown and lie somewhere within the hypothetical region shown in the figure.

here $x = (\rho - \rho_0)/3\rho_0$, K_0 is known as the incompressibility parameter, Q_0 the so-called skewness, and $S(\rho)$ is the symmetry energy which measures the contribution to the E/A from the difference between N and Z, parametrized as $\delta = (N - Z)/A$.

For symmetric nuclear matter, the proton and neutron densities are equal, $\rho_n = \rho_p$, and the total nucleon density is $\rho = \rho_p + \rho_n = 2\rho_n$. The energy per nucleon is related to the energy density, $\epsilon(\rho)$, by means of $\epsilon(\rho) = \rho E(\rho)/A$, which includes the nucleon rest mass, m_N. The density dependent function $E(\rho)/A - m_N$ has a minimum at $\rho = \rho_0$ with a value $E(\rho_0)/A = -16$ MeV, obtained with

$$\frac{d}{d\rho}\left(\frac{E(\rho)}{A}\right) = \frac{d}{d\rho}\left(\frac{\epsilon(\rho)}{\rho}\right) = 0 \quad \text{at } \rho = \rho_0. \tag{7.170}$$

The EoS of homogeneous nuclear matter is the relation of the pressure and density,

$$p(\rho, \delta) = \rho^2 \frac{d[\epsilon(\rho, \delta)/\rho]}{d\rho}, \tag{7.171}$$

with $\delta = (\rho_n - \rho_p)/\rho$.

The value of incompressibility of nuclear matter, K_0, or the curvature of the energy per particle, or the derivative of the pressure at $\rho = \rho_0$, is

$$K_0 = 9\frac{dp(\rho)}{d\rho} = 9\rho^2 \left.\frac{\partial^2 (E/A)}{\partial \rho^2}\right|_{\rho_0} = 9 \left[\rho^2 \frac{d^2}{d\rho^2}\left(\frac{\epsilon}{\rho}\right)\right]_{\rho_0}. \tag{7.172}$$

It has been extracted from the analysis of excitations of isoscalar giant monopole resonances in heavy ion collisions.

The symmetry energy $S(\rho)$ can also be expanded around $x = 0$

$$S = \frac{1}{2}\left.\frac{\partial^2 E}{\partial \delta^2}\right|_{\delta=0} = J + Lx + \frac{1}{2}K_{sym}x^2 + \cdots \tag{7.173}$$

where $J = S(\rho_0)$ is known as the bulk symmetry energy, L determines the slope of the symmetry energy, and K_{sym} is its curvature at the saturation density $\rho = \rho_0$. The slope parameter is given by

$$L = 3\rho_0 \left. \frac{dS(\rho)}{d\rho} \right|_{\rho_0} . \tag{7.174}$$

Experimental values of masses, excitation energies and other nuclear properties have been compared to microscopic models yielding $J \approx 30$ MeV for the symmetry volume term and $L \approx 40 - 80$ MeV for the slope parameter.

In Figure 7.26 we see a microscopic calculation of the nuclear matter (SNM) equation of state (EoS) compared to the result pure (ANM) neutron matter. The two curves are roughly separated by the bulk symmetry energy factor $S(\rho_0)$. The EoS for neutron stars is likely to be somewhat different than the ANM curve shown and lie somewhere within the hypothetical region shown in the figure.

7.12   Novae and Supernovae

It has long been observed that, occasionally, a new star appears in the sky, increases its brightness to a maximum value, and decays afterwards until its visual disappearance. Such stars were called by *novae*. Among the novae some stars present an exceptional variation in their brightness and are called by *supernovae*.

7.12.1   Novae

Novae (or "new") stars increase their brightness by a factor of a million before their visual disappearance. The process lasts for a few days only, reaching peak luminosities at $L = 10^4 - 10^5 L_\odot$ and ejected energies of 10^{45} ergs, and takes a few months to decrease. More than half of all stars are believed to be part of a binary system, consisting of a WD star orbiting around an orange/red dwarf or a red giant. The fuel originating the outbursts by the WD are gases falling from the larger star. The novae events leave the participating stars almost intact and the phenomenon can recur within $10^4 - 10^5$ yr. *Supernovae*, on the other hand, are one-time events leading to the total destruction of the star. Novae occur by a few dozen a year in our galaxy. Observations show that nova ejecta are rich in He, C, N, O, Ne, and Mg, but they only contribute by 1/50 as much intergalactic material as supernovae, and only by 1/200 as much as red giant and supergiant stars. In novae, the matter accretes from the binary companion and accumulates at the surface of the WD, building up a hydrogen layer, below which is the main components of the WD, mainly of carbon and oxygen. The accreting material leads to an increase in pressure and temperature and the WD surface grows, eventually becoming hot enough to burn hydrogen into helium. When the mass accreted reaches $M_\odot/100,000$, a nuclear explosion starts at the base of the accreted material. The surface layer is ejected at ~ 1000 km/s or greater. The WD is left intact underneath the explosion and the cycle can repeat again as long as the companion star can provide fresh hydrogen-rich matter. In classical novae, the triggering reaction is ^{12}C(p,γ)^{13}N leading to the reactions ^{13}N(β^+)^{13}C(p,γ)^{14}N with is part of the CNO cycle. As the temperature increases and $\tau_{\mathrm{p},\gamma}[^{13}N] < \tau_{\beta+}[^{13}N]$ the reactions ^{13}N(p,γ)^{14}O and ^{16}O(p,γ)^{17}F, and ^{14}N(p,γ)^{15}O occur. The reactions ^{18}F(p,α)^{15}O, ^{25}Al(p,γ)^{26}Si, ^{30}P(p,γ)^{31}S have been identified as major sources of nuclear uncertainties. The endpoint for classical nova nucleosynthesis, as predicted by theory and obtained in observations, is around the Ca isotopes.

7.12.2   X-ray bursters

X-ray bursts are also a common phenomenon besides novae and supernovae. In X-ray binary systems, the compact object is not a WD but either a *neutron star* (NS) or a *black hole*. The larger gravitational field of the compact stars gives rise to large velocities of the accreting gas rich in hydrogen and helium (Figure 7.27). The falling material collides with the already accreted material, leading to the formation of an accretion disk and releasing a humongous amount energy within a short time ($\sim 10^{38}$ erg s^{-1} within 1 to 10 s). The energy is mostly released as X-ray photons. The

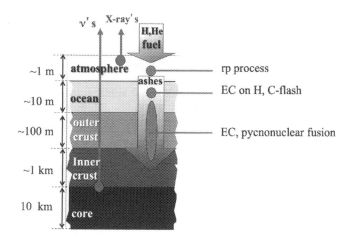

FIGURE 7.27 The accreted matter on a neutron star leads to X-ray bursts and a runaway reaction chain involving H, He, and C capture, electron capture (EC) and also pycnonuclear reactions in the deeper crust of the NS.

accreted material forms a dense layer of degenerate electron gas at the surface of the star, due to its strong gravitational field. Changes in temperature do not yield appreciable changes in pressure in a degenerate electron gas, but after enough material accumulates on the surface of the star, thermal instabilities occur triggering nuclear fusion reactions, further increasing its temperature, now greater than 10^9 K, and finally enabling a runaway thermonuclear explosion. Nucleosynthesis starts with the hot CNO cycle and quickly becomes an *rp-process*, i.e. a succession of rapid proton captures before the nuclei are able to β-decay by e^+ emission.

X-ray bursts can also be recurrent (from hours to days) because they are not powerful enough to disrupt either the binary system orbit or the stars. The recurring bursts occur at irregular periods, from few hours to several months. Modeling of X-ray bursts and the subproducts of the nuclear reaction networks involve rp-process, the 3α-reaction, and αp-process (i.e., (α,p) combined with (p,γ) reactions). In some of these modeling several hundreds of nuclei are included in the reaction network. It is not easy to identify the reactions of most relevance, although some reactions such as ^{65}As$(p,\gamma)^{66}$Se, ^{61}Ga$(p,\gamma)^{62}$Ge, ^{12}C$(\alpha,\gamma)^{16}$O and ^{96}Ag$(p,\gamma)^{97}$Cd have been identified as key reactions. These reactions as well as the masses of several nuclei would have a strong influence on the predicted yields.

7.12.3 Supernovae

Schematically a pre-supernova has the onion structure presented in Figure 7.24. Starting from the center of the star, we first find a core of iron, the remnant of silicon burning. After that we pass by successive regions where ^{28}Si, ^{16}O, ^{12}C, ^{4}He, and ^{1}H form the dominant fraction. In the interfaces, the nuclear burning continues to happen. *Type I Supernovae* (SNI) are thermonuclear stellar explosions triggered by material accreted into a massive carbon-oxygen WD in a binary system. They play an important part in astronomy as "*standard candles*", compared to which luminosities of other stars are used to infer cosmic distances, because all are thought to blaze with equal brightness at their peaks. Hundreds of nuclear isotopes and thousands of nuclear reactions participate in the process, at densities of up to $\rho \sim 10^{10}$ g/cm^3 and temperatures up to $T \sim 10^{10}$ K. The explosion can lead to turbulent instabilities and energy deflagration during the expansion of a SNI. Detailed 3D hydrodynamic modeling is necessary and accurate nuclear physics input must be provided. *Type II Supernovae* (SNII) have hydrogen in their spectra and their light curves peak around 10^{42} erg s^{-1}. In contrast, SNI do not show hydrogen in their spectra. If their spectra contain some silicon lines, they are called *Type Ia*. If the spectra contain helium lines they are called *Type Ib*. If they do not contain helium lines, they are called *Type Ic*. The luminosities of Type Ia supernovae can reach a peak value of about 2×10^{43} erg s^{-1}.

The silicon burning exhausts the nuclear fuel. As we mentioned previously, the gravitational collapse of the iron core cannot be hold by means of pressure heat from nuclear reactions. However, Chandrasekhar [16] showed that a total collapse can be avoided by the electronic pressure. In this situation, the core is stabilized due to the pressure of the degenerated electron gas, $P(r)$, and the inward gravitational pressure with the EoS given by

$$\frac{dP}{d\rho} = Y_e \frac{m_e}{m_N} \frac{x^2}{3\sqrt{1+x^2}}. \tag{7.175}$$

where x and Y_e are defined following Eq. (7.155). This model is appropriate for a non-rotating white dwarf. With the boundary conditions $m(r = 0) = 0$ and $\rho(r = 0) = \rho_c$ (the central density), the stellar structure equations can be solved easily. For a given Y_e, the model is totally determined by ρ_c. The total mass of a white dwarf (of the order of a solar mass, $M_\odot = 1.98 \times 10^{33}$ g), increases with ρ_c. Nonetheless, and perhaps the most important, it cannot exceed the finite value of

$$M \leq M_{Ch} \simeq 1.45(2Y_e)^2 M_\odot, \tag{7.176}$$

which is know as the *Chandrasekhar mass* [16]. Applying these results to the nucleus of a star with any mass, we get from Eq. (7.175) that stars with mass $M > M_{Ch}$ cannot be stable against the gravitational collapse by the pressure of the degenerate electron gas. The collapse occurs inevitably for a massive star, since the silicon burning adds more and more material to the stellar core.

Type II supernovae develop from stars with mass larger than $10M_\odot$, burning hydrogen in their core, and contracting when hydrogen is exhausted, until densities and temperatures are so high that the $3\alpha \to {}^{12}$C reaction ignites. He burning follows, producing nuclear ashes which are also burned and an ensuing cycle of fuel exhaustion, contraction, followed by ignition of the ashes from the previous burning cycle repeats, finally bringing the star to the ignite an explosive burning of ^{28}Si up to Fe nuclei. This evolution is fast for heavy stars, with a $25M_\odot$ star going through all of the cycles within 7 My, and the final Si explosion stage only taking a few days. The start will then posses a "*onion-like*" structure, as shown in Figure 7.27. Starting from the center, there is an iron core, remnant of silicon burning, followed by successive thick layers of ^{28}Si, ^{16}O, ^{12}C, ^{4}He, and ^{1}H. At the interfaces between these regions, nuclear burning continues. Silicon burning exhausts the nuclear fuel and the heat from nuclear reactions cannot hold the gravitational contraction of the iron core. A collapse occurs inevitably. But stars with $M > 20 - 30M_\odot$ might not explode as a supernova and instead collapse into black holes. At the beginning of the collapse of a massive star the temperature and density are of the order of $T \sim 10^{10}$ K and $\rho \sim 3 \times 10^9$ g/cm^3. The core is made of ^{56}Fe and of electrons.

There are two possibilities, both accelerating the collapse:

(a) At conditions present in the collapse the strong reactions and the electromagnetic reactions between the nuclei are in equilibrium with their inverse, i.e.,

$$\gamma + {}^{56}_{26}\text{Fe} \Longleftrightarrow 13({}^{4}\text{He}) + 4\text{n} - 124 \text{ MeV}. \tag{7.177}$$

For example, with $\rho = 3 \times 10^9$ g/cm^3 and $T = 11 \times 10^9$ K, half of ^{56}Fe is dissociated. This dissociation takes energy from the core and causes pressure loss. The collapse is thus accelerated.

(b) If the mass of the core exceeds M_{Ch}, electrons are captured by the nuclei to avoid the violation of the Pauli principle:

$$e^- + (Z, A) \longrightarrow (Z - 1, A) + \nu_e. \tag{7.178}$$

The neutrinos can escape the core, taking away energy. This is again accompanied by a pressure loss due to the decrease of the free electrons (this also decreases M_{Ch}). The collapse is again accelerated.

The gravitational contraction increases the temperature and density of the core. An important change in the physics of the collapse occurs when the density reaches $\rho_{\text{trap}} \simeq 4 \times 10^{11}$ g/cm^3. The neutrinos become essentially confined to the core, since their diffusion time in the core is larger than the collapse time. After the neutrino confinement no energy is taken out of the core.

Also, all reactions are in equilibrium, including the capture process (7.178). The degeneracy of the neutrino Fermi gas avoids a complete neutronization, directing the reaction (7.178) to the left. As a consequence, Y_e remains large during the collapse ($Y_e \approx 0.3 - 0.4$). To equilibrate the charge, the number of protons must also be large. To reach $Z/A = Y_e \approx 0.3 - 0.4$, the protons must be inside heavy nuclei which will therefore survive the collapse.

Two consequence follows:

(a) The pressure is given by the degenerate electron gas that controls the whole collapse; the collapse is thus adiabatic, with the important consequence that the collapse of the most internal part of the core is *homologous*, i.e., the position $r(t)$ and the velocity v(t) of a given element of mass of the core are related by

$$r(t) = \alpha(t)r_0; \qquad v(t) = \frac{\dot{\alpha}}{\alpha}r(t), \tag{7.179}$$

where r_0 is the initial position.

(b) Since the nuclei remain in the core of the star, the collapse has a reasonably large order and the entropy remains small during the collapse ($S \approx 1.5\ k$ per nucleon, where k is the Boltzmann constant).

The collapse continues homologously until nuclear densities of the order of $\rho_N \approx 10^{14}$ g/cm^3 are reached, when the matter can be thought as approximately a degenerate Fermi gas of nucleons. Since the nuclear matter has a finite compressibility, the homologous core decelerate and starts to increase again as a response to the increase of the nuclear matter. This eventually leads to a *shock wave* which propagates to the external core (i.e., the iron core outside the homologous core) which, during the collapse time, continued to contract reaching the supersonic velocity. The collapse break followed by the shock wave is the mechanism which breads the supernova explosion. Nonetheless, several ingredients of this scenario are still unknown, including the equation of state of the nuclear matter. The compressibility influences the available energy for the shock wave, which must be of the order of 10^{51} erg.

The exact mechanism for the explosion of a supernova is still controversial.

(a) In the *direct mechanism*, the shock wave is not only strong enough to stop the collapse, but also to explode the exterior stellar shells.

(b) An enhancement in electron capture rates occurs on the free protons left over by the shock wave. This together with a sudden reduction of the neutrino opacity ($\sigma^{\nu}_{coherent} \sim \rho^2$), greatly enhances neutrino emission (*neutrino eruption*), leading to more energy loss, a process which can lead to the shock to stall. The neutrinos will escape and take away energy from the core, it looses pressure and the collapse, is again accelerated. The chemical equilibrium condition for the reaction $e^- + p \rightarrow \nu_e + n$ is $\mu_e + \mu_p = \mu_n + \langle E_\nu \rangle$, with μ_i being the chemical potential of particle i. But since the neutrinos escape and the electron Fermi energy increases as the density increases, the reaction above breaks equilibrium and leads to a rise of neutronization of the matter. The neutrinos also carry off energy and lepton number. All these features lead to a very fast collapse, as attested by accurate numerical simulations. Neutrinos of the all three species are generated by the production of pairs in the hot environment. A new shock wave can be generated by the outward diffusion of neutrinos, what indeed carries the most part of the energy liberated in the gravitational collapse of the core ($\approx 10^{53}$ erg). If about 1% of the energy of the neutrinos is converted into kinetic energy due to the coherent neutrino-nucleus scattering, a new shock wave arises. This will be strong enough to explode the star. This process is know as the *retarded mechanism* for supernova explosion.

The second scenario above deserves a few more words. For an iron core of mass $\sim 1.2 - 1.5 M_\odot$ models predict that the collapse happens at about 0.6 of the free fall time for matter under gravitation, given by $t_{fall} = (3/8\pi G\rho)^{1/2}$. The iron core has a density of $\sim 10^8$ g cm^{-3} at its edge and $\sim 3 \times 10^9$ g cm^{-3} at its center when the collapse starts. Thus, $t_{fall} = 0.13\ \rho_8^{-1/2}$ s, where ρ_X is the density in Xg.cm$^{-1/3}$. The neutrino mean free path for scattering on matter is $\lambda_\nu = 1/\langle n\sigma_\nu \rangle$, and for a typical neutrino energy of 20 MeV it is only $\lambda_\nu \sim 0.5\rho_{12}^{-1}$ km. The time it takes for the neutrinos to diffuse to a radial distance R is given by $t_{diff} = R^2/3\lambda_\nu c$. The neutrino scattering off matter occurs through both coherent neutral current and charged current processes. The neutral current neutrino scattering cross section is determined by the total nuclear weak charge, proportional to N^2, where N is the neutron number. The neutrino random walks in the core, leading to its diffusion up to to the core surface. On its way, neutrinos can induce nucleosynthesis, which can be enhanced with neutrino oscillations. For a core with mass $1.5 M_\odot$ and and a neutrino energy approximately equal to the approximate local Fermi energy, $E_F \sim 35 \rho_{12}^{1/3}$ MeV, one finds $t_{diff} = 5 \times 10^{-2} \rho_{12}$ s. If we compare the diffusion time scale with the free fall time scale, we get $t_{diff}/t_{fall} = 40 \, \rho_{12}^{3/2}$ s. Hence, for density larger than 10^{11} g cm^{-3} the neutrinos become fully confined in the core. More energy is released by the continued gravitational collapse and the star's still existing lepton number are trapped within the core. The fact that t_{diff} much larger than t_{fall} reinforces the relevance of the neutrino diffusion process. Accurate calculations show that neutrino diffusion occurs on a time scale of about 2 s. All three species of neutrinos are generated by pair production in the hot medium. A new shock wave can be reinstated by the outward push due to neutrino diffusion which carry most part of the gravitational energy liberated in the collapse of the core ($\approx 10^{53}$ erg). For a new shock wave to develop one needs only $\sim 1\%$ of the energy contained the neutrinos, if this energy is converted into kinetic energy of nuclear matter by means of coherent neutrino-nucleus scattering. The first shock wave stalls at a radius of $200 - 300$ km. The shock wave revival probably occurs at about 0.5 s later.

The daughter nucleus in Eq. (7.178) may beta decay and the original nucleus can be restored, with the creation of an electron-neutrino and electron-antineutrino ($\bar{\nu}_e$) pair and the emission of the neutrinos from the core. This process can repeat, draining energy from the core by the escaping neutrinos. The two-step process is given by

$$
\begin{aligned}
(N, Z) + e^- &\longrightarrow (N+1, Z-1) + \nu_e \quad \text{(electron capture)}, \\
(N+1, Z-1) &\longrightarrow (N, Z) + e^- + \bar{\nu}_e \quad \beta\text{-decay}.
\end{aligned} \tag{7.180}
$$

For a core of $M \sim 1.5 M_\odot$ and a 10 km radius, an estimate of its binding energy is $GM^2/2R \sim 10^{53}$ ergs, which is roughly the trapped energy later radiated by the neutrinos. The electron captures on nuclei occur at small momentum transfer, and they are dominated by Gamow-Teller (GT) transitions; i.e. the response of nuclei to spin-isospin ($\sigma\tau$) operators, in which a proton is changed into a neutron. The electron chemical potential μ_e grows with density like $\rho^{1/3}$ and when it is is of the same order as the reaction Q-value, the electron capture rates are very sensitive to the detailed GT distribution of the involved nuclei. Experiments are very difficult to carry out in order to obtain such nuclear matrix elements to obtain the weak-interaction rates for $A \sim 50 - 65$, which are relevant at such densities.

To know which of the above mechanism is responsible for the supernova explosion one needs to know the rate of electron capture, the nuclear compressibility, and the way neutrinos are transported. The iron core, remnant of the explosion (the homologous core and part of the external core) will not explode and will become either a neutron star, and possibly later a *pulsar* (rotating neutron star), or a *black-hole*, as in the case of more massive stars, with $M \geq 25 - 35 M_\odot$.

Supernova models together with recent observations of low metallicity stars have confirmed that the ejecta of SNII are characterized by elevated ratios of α-elements (a convenient designation for the observation that some even-Z elements such as O, Mg, Si, S, Ca, and Ti) relative to iron: $[(\alpha\text{-elements})/\text{Fe}] \simeq +0.5$. Therefore, the contributions of SNIa to ^{56}Fe comprise $\simeq 2/3$ of the observed galactic iron. Since SNIa produce $\simeq 0.6 M_\odot$ per event while SNII produce only $\simeq 0.1 M_\odot$ per event, than SNII have been about 3 times more numerous when averaged over the galactic history.

7.12.4   Supernova radioactivity and light-curve

During a supernova explosion nucleosynthesis proceeds in the silicon core leading to several radioactive isotopes, including ^{56}Ni, ^{57}Ni and ^{44}Ti, which have short half-lives. The decays of these isotopes can be directly observed. They are are characterized by either the half-life, $t_{1/2}$, or by the decay time $\tau = t_{1/2}/\ln 2$. An example is ^{26}Al, which has a 720 kyr lifetime to decay into ^{26}Mg. The decay from the ^{26}Al 5^+ ground state goes to the first two excited states of ^{26}Mg, both of them 2^+ states at 2.938 and 1.809 MeV above its ground state. 97% of the time the later state is populated, and decays by emitting a 1.809 MeV γ. The 2.938 MeV state decays to the 1.809 MeV level, and therefore a small quantity of 1.129 MeV γ's are created. The primary site for ^{26}Al production is believed to be Type II and IIb supernovae. For ^{56}Ni, $\tau = 8.8$ days, due to the electron capture

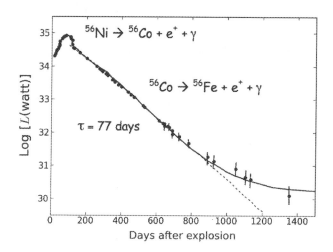

FIGURE 7.28 Observed luminosity of supernova SN1987 A as a function of time (data from [30]). The dashed line is the exponential decay of ^{56}Co. The solid curve is based on theoretical predictions.

process ^{56}Ni$(e^-, \gamma)^{56}$Co, with γ-rays energies of the order of 0.16 - 0.8 MeV. However, ^{56}Co is not stable either and decays according to ^{56}Co$(e^-, \gamma)^{56}$Fe or ^{56}Co $\rightarrow$ ^{56}Fe $+ e^+ + \nu_e$, with a branching ratio for the first decay is 80% and 20% for the second, yielding gamma-ray and positron sharing the energy emission by 96% and 4%, respectively. The most intense gamma-ray lines occur at 0.847 MeV and 1.237 MeV and the mean positron energy is 0.66 MeV.

Astronomical observations are in agreement with the model of an explosion of a 1.5 $M_\odot$ WD, providing a proof of the standard hypotheses of Type Ia supernovae explosion. ^{57}Ni also decays by electron capture, ^{57}Ni$(e^-, \gamma)^{57}$Co, within a very short time, $\tau = 52$ hours. Another interesting decay is ^{57}Co$(e^-, \gamma)^{57}$Fe, with $\tau = 390$ days. Produced in a nuclear statistical equilibrium at high densities of $(10^4 - 10^{10}$ g/cm$^3)$ and temperatures $(4 - 10) \times 10^9$ K, ^{44}Ti has a half-life of 59 yr. It decays first to ^{44}Sc by means of ^{44}Ti$(e^-, \gamma)^{44}$Sc, followed by a very quick decay time, $\tau = 5.4$ h, to ^{44}Sc$(e^-, \gamma)^{44}$Ca or ^{44}Sc $\rightarrow$ ^{44}Ca $+ e^+ + \nu_e$. All these radioactive decays yield either γ-rays or positrons. The γ-rays Compton-scatter off electrons, loosing about half of their energy to electrons in each collision. Initially, the γ-ray energies are in the MeV range, much larger than the atomic electron binding energies, and thus both bound and free electrons contribute to the scattering. The γ-ray energies decrease until the photoelectric absorption cross section is larger than the Compton cross section, at $E_\gamma \sim 10-100$ keV, mostly on iron. When the shock stops the radiation leaks out on a diffusion time scale, $t_{diff} \simeq 3R^2 \rho \kappa / \pi^2 c$, where κ is now the opacity to radiation. The expansion time scale is $t_{exp} \simeq R/v$ and the opacity due to Thompson scattering is $\kappa = 0.4$ cm^2/g. The two time scales are comparable, $t_{diff}/t_{exp} \sim 1$ for a typical mass of 10 $M_\odot$, when the supernova has expanded to $R_{peak} \sim 4 \times 10^{15}$ cm. Only after $t_{peak} = R_{peak}/v \sim 40$ days, the radiation can leak out faster than the ejecta can expand.

The first information indicating that a core-collapse supernova event has occurred is the burst of neutrinos. A few hours later electromagnetic radiation is released initially as a ultra-violet flash. The expanding supernova only becomes visible at the optical wavelengths, with a rise and fall of the light curve which is a result of an increasing surface area and a slow temperature decrease. The peak in the light curve appears when the temperature of the outer layers begin to decrease. A change in the opacity of the outer layer of the exploded star is created by the shock wave, which heats it up to about 100,000 K, leading to hydrogen ionization, which has a high opacity. The radiation from the star's interior cannot escape, and we thus only observe photons from the surface of the star. The outer parts of the star only cool off after a few weeks to about $4,000 - 6,000$ K, at which point the ionized hydrogen start recombining into neutral hydrogen. Being transparent at most wavelengths, neutral hydrogen forces the opacity to change appreciably at the photosphere of the star. Then photons from the hotter, inner parts of the hydrogen envelope start to escape. After full hydrogen recombination in the hydrogen envelope, the light curves are dimmer and generated by a radioactive tail of the ^{56}Co to ^{56}Fe conversions. ^{56}Ni decays by electron capture in 6.1 d, ^{56}Co also by electron capture in 77.3 d, but ^{56}Fe is stable. Figure 7.28 shows the light curve from the SN1887A supernova. It could be beautifully related to 0.85 and 1.24 MeV γ-lines from ^{56}Co decay. A lot ($\sim 0.1 M_\odot$) of ^{56}Ni and ^{56}Fe were formed in the first moments after the supernova explosion. The data are also consistent with predictions of a total neutrino count of 10^{58} and total energy of 10^{46} J (solid curve in the figure). Observations of ^{44}Ti gamma-ray emission lines from SN1987A have been able to reveal an asymmetric supernovae explosion, open a new and challenging field for supernova explosion models. This is expected because some neutron stars are observed with velocities up to 500 Km/s or greater. But the mechanism leading to this recoil velocity during the supernovae explosion remains unknown.

7.13   Exercises

1. (a) What is the most probable kinetic energy of an hydrogen atom at the interior of the Sun ($T = 1.5 \times 10^7$ K)? (b) What fraction of these particles would have kinetic energy in excess of 100 keV?

2. Suppose the iron core of a supernova star has a mass of 1.4 $M_\odot$ (the Sun's mass is $M_\odot = 1.99 \times 10^{30}$ kg) and a radius of 100 km and that it collapses to a uniform sphere of neutrons of radius 10 km. Assume that the virial theorem

$$2 \langle T \rangle + \langle V \rangle = 0$$

   holds, where $\langle T \rangle$ is the average of the internal kinetic energy and $\langle V \rangle$ the average of the gravitational potential energy. $E = \langle T \rangle + \langle V \rangle$ is the total mechanical energy of the system. Calculate the energy consumed in neutronization and the number of electron neutrinos produced. Given that the remaining energy is radiated as neutrino-antineutrino pairs of all kinds of average energy 12 + 12 MeV, calculate the total number of neutrinos radiated.

3. About 3 s after the onset of the Big Bang, the neutron-proton ratio became frozen when the temperature was still as high as 10^{10} K ($kT \simeq 0.8$ MeV). About 250 s later, fusion reactions took place converting neutrons and protons into ^{4}He nuclei. Show that the resulting ratio of the masses of hydrogen and helium in the Universe was close to 3. The neutron half-life = 10.24 min and the neutron-proton mass difference is 1.29 MeV.

4. Given that the supernova of Exercise 3 is at a distance of 163,000 light years, calculate the total number of neutrinos of all types arriving at each square meter at the Earth. Also estimate the number of reactions

$$\bar{\nu}_e + \mathrm{p} \longrightarrow \mathrm{n} + e^+$$

   that will occur in 1000 tones of water. Assume that the cross section is given by

$$\sigma = \frac{4 p_e E_e G_F^2}{\pi \hbar^4 c^3},$$

   where p_e and E_e are the positron momentum and energy respectively and G_F is the Fermi coupling constant. Assume that only one-sixth of the neutrinos are electron neutrinos.

5. Evaluate the radius of a neutron star with mass of 1.2 $M_\odot$.

6. The rate of energy delivered by the Sun to the Earth is known as the solar constant and equals 1.4×10^6 erg cm^{-2} s^{-1}. Knowing that the distance between the Sun and the Earth is of 1.5×10^8 km, give a lower estimate of the rate at which the Sun is loosing mass to supply the radiated energy.

7. Calculate the energy radiated during the contraction of the primordial gas into Sun (the Sun diameter is 1.4×10^6 km). What energy would be released if the solar diameter suddenly shrink by 10%?

8. Given that the Sun was originally composed of 71% hydrogen by weight and assuming it has generated energy at its present rate (3.86×10^{36} W) for about 5×10^9 years by converting hydrogen into helium, estimate the time it will take to burn 10% of its remaining hydrogen. Take the energy release per helium nucleus created to be 26 MeV.

9. The CNO cycle that may contribute to energy production in stars similar to the Sun begins with the reaction $p + {}^{12}C \longrightarrow {}^{13}N + \gamma$. Assuming the temperature near the center of the Sun to be 15×10^6 K, find the peak energy and width of the reaction rate.

10. Assume that we know the bound state wave function for the relative motion of particles a and b. Initially, the particles are thought to be in a quasi-stationary excited state of the compound nucleus. There is an exponential decrease of the wave function through the potential barrier which turns into an outgoing wave at infinity. We define the decay rate of that state (for particle emission) as

$$\lambda = \frac{1}{\tau} = \text{probability/sec for a decay through a large spherical shell.}$$

(a) If the wave function is written as

$$\Psi_{nlm} = \frac{u_l(r)}{r} Y_{lm}(\theta, \phi),$$

show that $\lambda = v |u_l(\infty)|^2$.

We define the penetration factor for particles of relative angular momentum l as

$$P_l = \frac{|u_l(\infty)|^2}{|u_l(R)|^2},$$

where $r = R$ is a position where the nuclear potential is close to zero (see Figure 7.29).

Thus the decay rate can be written as $\lambda = vP_l |u_l(R)|^2$. For a uniform probability density inside the compound nucleus

$$|u_l(R)|^2 \, dr = \frac{4\pi R^2 dr}{4\pi R^3/3} = \frac{3}{R} dr.$$

One defines the reduced width θ_l by means of

$$|u_l(R)|^2 = \theta_l^2 \frac{3}{R}.$$

For realistic nuclear states

$$0.01 < \theta_l^2 < 1,$$

and θ_l^2 gives a measure of the degree to which a quasi-stationary nuclear state can be described by a relative motion of a and b in a potential.

(b) Show that the partial width of that state is given by

$$\Gamma_l = \frac{3\hbar v}{R} P_l \theta_l^2.$$

(c) Show that for charged particles, with

$$V_l(r) = \begin{cases} l(l+1)\,\hbar^2/2\mu r^2 + Z_1 Z_2 e^2/r, & r > R \\ l(l+1)\,\hbar^2/2\mu r^2 + V_N, & r < R, \end{cases}$$

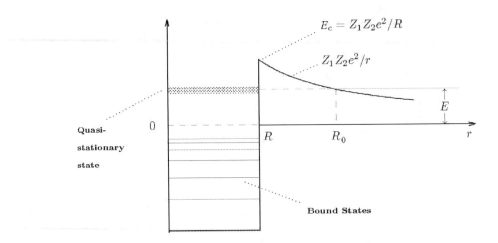

FIGURE 7.29   Potential barrier for charged particles.

we get

$$P_l = \frac{1}{F_l^2(R) + G_l^2(R)},$$

where F_l and G_l are the regular and irregular Coulomb functions.

(d) Show that using WKB wave functions for $u_l(r)$, one gets

$$P_l = \left[\frac{V_l(R) - E}{E}\right]^{1/2} \exp\left\{-\frac{2\sqrt{2\mu}}{\hbar}\int_R^{R_0} [V_l(r) - E]^{1/2}\, dr\right\}, \qquad (7.181)$$

with $l(l+1) \longrightarrow (l + 1/2)^2$.

11. A numerical computation of the above equation shows that a good approximation consists in replacing $V_l(R) - E \cong E_c = Z_1 Z_2 e^2/R$ in the factor preceding the exponential. This approximation is justified since at astrophysical energies $E \ll E_c$.

(a) Calling the exponent in Eq. (7.181) by W_l show that, to lowest order in E/E_c,

$$W_0 = 2\pi \frac{Z_1 Z_2 e^2}{\hbar v}\left[1 - \frac{4}{\pi}\left(\frac{E}{E_c}\right)^{1/2} + \frac{2}{3\pi}\left(\frac{E}{E_c}\right)^{3/2} - \cdots\right]. \qquad (7.182)$$

(b)

Write the expression for the lifetime, Γ_0, of a state with $l = 0$ in terms of the reduced width θ_0^2 and W_0. Show that, to first order in E/E_c,

$$\Gamma_0 \propto \exp\left(-bE^{-1/2}\right).$$

12. For $l \neq 0$,

$$W_l = \frac{2\sqrt{2\mu}}{\hbar}\int_R^{R_0}\left[E_c\frac{R}{r} + E_l\left(\frac{R}{r}\right)^2 - E\right]^{1/2} dr.$$

where $E_l = (l + 1/2)^2 \hbar^2/2\mu R^2$. For astrophysically relevant cases, $R/R_0 \lesssim 10^{-3}$ and the ratio R/r is quite small for most range of the integration. As a consequence the second term in the square-root bracket never dominates, and the integrand may be expanded, whereupon the leading terms becomes

$$W_l \simeq \frac{2\sqrt{2\mu}}{\hbar}\int_R^{R_0}\left(E_c\frac{R}{r} - E\right)^{1/2} dr + \frac{\sqrt{2\mu}}{\hbar}\int_R^{R_0}\frac{E_l R^{3/2}}{E_c^{1/2}}\frac{dr}{r^{3/2}}. \qquad (7.183)$$

The first term is just equal to W_0, whereas the second term reflects the additional effects of the centrifugal barrier.

(a) Show that (7.183) becomes

$$W_l = W_0 + 2 \left[\frac{(l+1/2)^2 \, E_l}{E_c} \right]^{1/2} \left[1 - \left(\frac{E}{E_c} \right)^{1/2} \right].$$

(b) Neglecting the correction in $(E/E_c)^{1/2}$ in the above equation, show that

$$P_l \approx \left(\frac{E_c}{E} \right)^{1/2} \exp \left[-2\pi \frac{Z_1 Z_2 e^2}{\hbar v} + 4 \left(\frac{2\mu R^2 E_c}{\hbar^2} \right)^{1/2} - 2\,(l+1/2)^2 \left(\frac{\hbar^2}{2\mu R^2 E_c} \right)^{1/2} \right],$$

where the correction of order $(E/E_c)^{3/2}$ in W_0 has also been dropped.

(c) Show that

$$\Gamma_l = 6\theta_l^2 \left(\frac{\hbar^2 E_c}{2\mu R^2} \right)^{1/2} \exp \left(-W_l \right). \tag{7.184}$$

(d)

The reaction $^{12}\text{C}\,(\text{p},\gamma)\,^{13}\text{N}$ has a peak at 424 keV center-of-mass energy, corresponding to a $J^\pi = \frac{1}{2}^+$ resonance. The resonance has a full width at half maximum $\Gamma = 40$ keV. This width is essentially the proton width, since the only other channel is Γ_γ, which is much smaller than Γ_p. What is the value of the dimensionless reduced width θ_l^2 for that state?

13. (a) Show that close to a resonance the astrophysical S-factor is given by

$$S\,(E) = \frac{\pi \hbar^2}{2\mu} \frac{g \Gamma_\alpha\,(E)\,\Gamma_\beta\,(E)}{(E - E_r)^2 + \Gamma^2/4} \exp \left(2\pi \frac{Z_1 Z_2 e^2}{\hbar v} \right),$$

where $g = (2J+1)/(2J\alpha+1)(2J\beta+1)$ is the statistical spin factor. The decay width for the entrance and decay channels, $\Gamma_\alpha\,(E)$ and $\Gamma_\beta\,(E)$ respectively, also depend on the energy for reactions of astrophysical interest.

(b) In some situations, Γ_β is approximately constant, i.e., when the final channel is a γ- or α- decay. In this case, use the Eq. (7.184) for $\Gamma_\alpha^{(l)}\,(E)$ and write an expression for $S\,(E)$ in terms of $l, E, E_c, \Gamma_\beta^{(l)}$, and Γ.

14. In stellar interiors, and also in laboratory experiments, the nuclear fusion cross sections for bare naked nuclei is modified due to the presence of electrons. The electron shielding around the nuclei is equivalent to a constant (negative) potential U_e that is usually much smaller than the energy E.

(a) Show that this potential modifies the fusion cross section so that

$$\sigma_{screened}\,(E) = \exp \left(\pi \eta \frac{U_e}{E} \right) \sigma_{bare}\,(E).$$

Find in the literature the experimental values of U_e (screening by electrons in the target) for 5 reactions of astrophysical interest. Comment about the comparison with theoretical values.

15. Obtain the results of Eq. (7.18) from Eq. (7.17). Follow the instructions preceding the equation.

16. Show that, expanding the argument of the exponential in Eq. (7.17) around the Gamow energy, E_0, i.e.

$$f(E) = \frac{E_0}{kT} + \frac{b}{\sqrt{E}} \sim f(E_0) + (E - E_0) \frac{df}{dE_0} + \frac{1}{2}(E - E_0)^2 \frac{d^2 f}{dE_0^2} + \cdots$$

and since $f'(E_0)$ vanishes by definition of E_0,

$$f(E) = f(E_0) + f''(E_0)\frac{1}{2}(E - E_0)^2 + \cdots ,$$

one gets

$$\langle \sigma v \rangle = \frac{16}{9\sqrt{3}} \frac{1}{\mu} \frac{1}{2\pi\alpha Z_1 Z_2} S(E_0) \exp\left(-\frac{3E_o}{kT}\right)\left(\frac{3E_0}{kT}\right)^2 .$$

17. Show that for a resonant (p, γ) cross section reaction, e.g. $^{12}\mathrm{C}(p,\gamma)^{13}\mathrm{N}$ at 424 keV, one finds

$$\langle \sigma v \rangle_{\text{resonant}} = \left(\frac{2\pi}{\mu kT}\right)^{3/2} \frac{\Gamma_p \Gamma_\gamma}{\Gamma} e^{-E_r/kT} .$$

18. Show that for non-resonant reactions in the Sun, Eq. (7.8) becomes

$$r_{12} = \frac{n_1 n_2}{1 + \delta_{12}} 7.21 \cdot 10^{-19} \mathrm{cm}^3/\mathrm{s} \frac{1}{A Z_1 Z_2}$$
$$\times \left[\frac{S(E_0)}{\mathrm{keV \cdot barns}} (Z_1^2 Z_2^2 A)^{2/3} \left(\frac{42.5}{T_6^{1/3}}\right)^2 \exp\left(\frac{(Z_1^2 Z_2^2 A)^{1/3} 42.5}{T_6^{1/3}}\right) \right], \qquad (7.185)$$

where $A = A_1 A_2 / (A_1 + A_2)$.

This tells us that small $Z_1 Z_2$ is favored, and that rates are expected to rise as $e^{-1/T^{1/3}}$. In the above, T_6 is the temperature in units of 10^6K. The factor δ_{12} is to prevent double counting for fusion of the same nuclear species.

19. A "network" calculation of the ppI cycle in the Sun has the main contributing reactions

$$\mathrm{p} + \mathrm{p} \to \mathrm{d} + e^+ + \nu_e, \qquad r_{pp} \sim \lambda_{pp} \frac{n_p^2}{2}$$
$$\mathrm{d} + \mathrm{p} \to \,^3\mathrm{He} + \gamma, \qquad r_{pd} \sim \lambda_{pd} n_p n_d$$
$$^3\mathrm{He} +\,^3\mathrm{He} \to \,^4\mathrm{He} + \mathrm{p} + \mathrm{p}, \qquad r_{33} \sim \lambda_{33} \frac{n_{3He}^2}{2} .$$

Here r represents a rate and $\lambda = \langle \sigma v \rangle$. From one calculated S-factor (for pp) and two that are measured, one can calculate the production of He once the composition and temperature is specified.

One feature of interest in this simple network is that d and ^{3}He both act as "catalysts": they are produced and then consumed in the burning. In a steady state process, this implies they must reach some equilibrium abundance where the production rate equals the destruction rate. That is, the general rate equation

$$\frac{dn_d}{dt} = \lambda_{pp} \frac{n_p^2}{2} - \lambda_{pd} n_p n_d$$

is satisfied at equilibrium by replacing the LHS by zero. Thus,

$$\left(\frac{n_p}{n_d}\right)_{\text{equil}} = \frac{2\lambda_{pd}}{\lambda_{pp}} .$$

(a) Using the S-factors

$$S_{12}(0) = 2.5 \cdot 10^{-4} \text{ kev b}, \qquad S_{11}(0) = 4.07 \cdot 10^{-22} \text{ kev b}$$

and the rate formula (7.185) find an equation for $\left(n_p/n_d\right)_{\text{equil}}$ as a function of $T_7^{1/3}$.

The result shows that this ratio is a decreasing function of T_7: the higher the temperature, the lower the equilibrium abundance of deuterium. Therefore in the region of the Sun where the ppI cycle is operating, the deuterium abundance is lowest in the Sun's center.

(b) Plugging in the solar core temperature show that

$$\left(\frac{n_d}{n_p}\right) = 3.6 \cdot 10^{-18}.$$

There isn't much deuterium about: using $n_p \sim 3 \cdot 10^{25}/\text{cm}^3$ one finds $n_d \sim 10^8/\text{cm}^3$.

(c) Using the value of $S_{11}(0)$ given above and Eq. (7.185) show that

$$r_{pp} \sim 0.6 \times 10^8/\text{cm}^3/\text{sec}$$

and that the typical life time of a deuterium nucleus at the core is

$$\tau_d \sim 1 \text{ sec.}$$

That is, deuterium is burned instantaneously and thus reaches equilibrium very, very quickly.

20. The result above allows us to write the analogous equation for ^{3}He as

$$\frac{dn_3}{dt} = \lambda_{pp}\frac{n_p^2}{2} - 2\lambda_{33}\frac{n_3^2}{2},$$

where the factor of two in the term on the right comes because the ^{3}He+^{3}He reaction destroys two ^{3}He nuclei. Thus at equilibrium

$$\left(\frac{n_3}{n_p}\right)_{\text{equil}} = \sqrt{\frac{\lambda_{pp}}{2\lambda_{33}}}.$$

Using

$$S_{33}(0) = 5.15 \cdot 10^3 \text{ keV b,}$$

do again the rate algebra to find

$$\left(\frac{n_3}{n_p}\right)_{\text{equil}} = (1.33 \cdot 10^{-13}) \exp\left(\frac{20.65}{T_7^{1/3}}\right) = \left(\begin{array}{cc} 9.08 \cdot 10^{-6} & , \quad T_7 = 1.5 \\ 1.24 \cdot 10^{-4} & , \quad T_7 = 1.0 \end{array}\right).$$

This ratio is clearly a sharply decreasing function of T_7 and thus a sharply increasing function of r. That is, a sharp gradient in ^{3}He is established in the Sun.

21. The non-existence of a bound nucleus with $A = 8$ was one of the major puzzles in nuclear astrophysics. How could heavier elements than $A = 8$ be formed? Using typical values of concentration of α particles in the core of a heavy star, $n_\alpha \sim 1.5 \cdot 10^{28}/\text{cm}^3$ (corresponding to $\rho_\alpha \sim 10^5 \text{ g/cm}^3$) and $T_8 \sim 1$ one obtains

$$\frac{n(^8\text{Be})}{n(\alpha)} \sim 3.2 \times 10^{-10}.$$

Salpeter suggested that this concentration would then allow $\alpha + {}^8\text{Be} (\alpha + \alpha) \to {}^{12}\text{C}$ to take place. Hoyle then argued that this reaction would not be fast enough to produce significant burning unless it was also resonant. Now the mass of $^8\text{Be} + \alpha$ is 7.366 MeV, and each nucleus has $J^\pi = 0^+$. Thus s-wave capture would require a 0^+ resonance in ^{12}C at ~ 7.4 MeV. No such state was then known, but a search by Cook, Fowler, Lauritsen, and Lauritsen revealed a 0^+ level at 7.644 MeV, with decay channels $^8\text{Be}+\alpha$ and γ decay to the 2^+ 4.433 level in ^{12}C. The parameters are:

$$\Gamma_\alpha \sim 8.9\text{eV},$$

$$\Gamma_\gamma \sim 3.6 \cdot 10^{-3}\text{eV}.$$

(a) Show that

$$r_{48} = n_\alpha^3 T_8^{-3} \exp\left(-\frac{42.9}{T_8}\right) (6.3 \cdot 10^{-54}\text{cm}^6/\text{sec}).$$

If we denote by $\omega_{3\alpha}$ the decay rate of an α in the plasma, then

$$\omega_{3\alpha} = 3n_\alpha^2 T_8^{-3} \exp\left(-\frac{42.9}{T_8}\right)\left(6.3 \cdot 10^{-54} \mathrm{cm}^6/\mathrm{sec}\right)$$

$$= \left(\frac{n_\alpha}{1.5 \cdot 10^{28}/\mathrm{cm}^3}\right)^2 \left(4.3 \cdot 10^3/\mathrm{sec}\right) T_8^{-3} \exp\left(-\frac{42.9}{T_8}\right).$$

(b) Since the energy release per reaction is 7.27 MeV show that the energy produced per gram, ϵ, is

$$\epsilon = \left(2.5 \cdot 10^{21}\mathrm{erg/g\ sec}\right)\left(\frac{n_\alpha}{1.5 \cdot 10^{28}/\mathrm{cm}^3}\right)^2 T_8^{-3} \exp\left(-\frac{42.9}{T_8}\right).$$

References

1. Yao W M et al. 2006 *J. Phys.* **G 33** 1
2. Clayton D D 1984 *Principles of Stellar Evolution and Nucleosynthesis* (University of Chicago Press, Chicago)
3. Adelberger E, et al. 2011 *Rev. Mod. Phys.* **83** 195
4. Friedlander M W 1990 *Cosmic Rays* (Harvard Univ Press)
5. Press W H et al. 2007 *Numerical Recipes* (Cambridge University Press)
6. Weinberg S 1993 *The First Three Minutes: A Modern View of the Origin of the Universe* (Basic Books)
7. Kolb E,Turner M 1994 *The Early Universe* (Westview Press)
8. Peacock J A 1999 *Cosmological Physics* (Cambridge University Press)
9. Liddle A R, Lyth D 2000 *Cosmological Inflation and Large-Scale Structure* (Cambridge University Press)
10. Dodelson S 2003 *Modern Cosmology* (Academic Press)
11. Weinberg S 2008 *Cosmology* (Oxford University Press)
12. Steigman G 2007 *Annu. Rev. Nucl. Part. Sci.* **57** 463
13. Bertulani C A, Kajino T 2016 *Prog. Part. Nucl. Phys.* **89** 56
14. Bahcall J N 1989 *Neutrino Astrophysics* (Cambridge: Cambridge University Press)
15. Bethe H 1986 *Phys. Rev. Lett.* **56** 1305
16. Chandrasekhar S 1931 *Astrophys. J.* **74**, 81; Chandrasekhar S 1984 *Rev. Mod. Phys.* **56, 137**
17. Kippenhahn R, Weigert A 1991 *Stellar Structure and Evolution* (Springer-Verlag)
18. Fowler W A, Caughlan G E and Zimmermann B A 1967 *Ann. Rev. Astron. Astrophys.* **5** 525
19. Bethe H A 1937 *Rev. Mod. Phys.* **9** 69
20. Rolfs C and Rodney W S 1988 *Cauldrons in the Cosmos* (Chicago: University of Chicago Press)
21. Langanke K and Barnes C A 1996 *Advances in Nuclear Physics*, Negele J W and Vogt E (eds) **22** 173 (New York: Plenum Press)
22. Burbidge E M et al. 1957 *Rev. Mod. Phys.* **29** 547
23. Schatz H et al. 1988 *Phys. Rep.* **294** 167
24. Fisher P, Kayser B, McFarland K S 1999 *Annual Review of Nuclear and Particle Science* **49** 481
25. Bertulani C A 2003 *Comput. Phys. Commun.* **156** 123
26. Hammer H W, König S and van Kolck U 2020 *Rev. Mod. Phys.* **92** 025004
27. Christenson J H, Cronin J W, Fitch V L and Turlay R 1964 *Phys. Rev. Lett.* **13** 138
28. Wolfenstein L 1978 *Phys. Rev.* **D17** 2369
29. Mikheyev S P and Smirnov A 1985 *Sov. J. Nucl. Phys.* **42** 913
30. Suntzeff N, et al. 1992 *Ast. J. Lett.* **384** L33

8

High-energy collisions

8.1 Introduction

The most popular techniques to study nuclear scattering at high energies are (a) the *Glauber theory* and (b) *transport equations*. The name Glauber theory is used in a very wide context. In the original lectures by Glauber [1], use was made of the eikonal wave functions to describe the nuclear scattering at high energies. Thus, in principle, this approach was purely quantum mechanical. However, many of his results can be interpreted in terms of classical mechanics of billiard balls. In some respects, the resulting equations can neither be considered as purely quantum mechanical nor as purely classical, i.e. they are *semiclassical*. In this chapter we will discuss the Glauber theory in different ways, both classically and quantum mechanically, depending on the purpose of the calculation. Although some of the results were not new to Glauber, nor they were derived by Glauber himself, the name "Glauber" will be used here in a general context, e.g. when we use probability concepts based on nucleon mean free path and forward scattering or when quantum mechanical methods based on the eikonal wave functions are used.

Transport equations are also based on classical concepts, and quantum mechanics is sometimes put by hand on the equations. However, as with the Glauber theory, many features of the transport equations can be derived from a quantum mechanical limit, e.g. the time-dependent Hartree-Fock theory. In this chapter we show that most equations used in the description of nuclear collisions at high energies can also be deduced from very simple probability arguments.

8.2   Nucleons as billiard balls

Let us assume, as shown in Figure 8.1, that a nucleus-nucleus collision occurs at an impact parameter b. We can define the probability of having a nucleon-nucleon collision within the transverse element area $d\mathbf{b}$ as $t(\mathbf{b})\,d\mathbf{b}$, where $t(\mathbf{b})$ is known as the thickness function. It is defined in a normalized way, i.e.,

$$\int t(\mathbf{b})\,d\mathbf{b} = 1. \tag{8.1}$$

For unpolarized projectiles $t(\mathbf{b}) = t(b)$. In most practical situations one can use $t(\mathbf{b}) \simeq \delta(\mathbf{b})$, which simplifies the calculations considerably.

Since the total transverse area for nucleon-nucleon collisions is given by σ_{NN}, the probability of having an inelastic nucleon-nucleon collision is given by $t(\mathbf{b})\sigma_{NN}$.

The probability of finding a nucleon in $d\mathbf{b}_B dz_B$ is given by $\rho(\mathbf{b}_B, z_B)d\mathbf{b}_B dz_B$, where the nuclear density is normalized to unity:

$$\int \rho(\mathbf{b}_B, z_B)\,d\mathbf{b}_B dz_B = 1. \tag{8.2}$$

Using these definitions, it is easy to verify that the probability dP of occurrence of a nucleon-nucleon collision is given by

$$dP = \rho(\mathbf{b}_B, z_B)\,d\mathbf{b}_B dz_B \times \rho(\mathbf{b}_A, z_A)\,d\mathbf{b}_A dz_A \times t(\mathbf{b} - \mathbf{b}_A - \mathbf{b}_B). \tag{8.3}$$

Thus, as in the case of free nucleon-nucleon collisions, we define $T(\mathbf{b})\sigma_{NN}$ as the probability of occurrence of a nucleon-nucleon collision in nucleus-nucleus collisions at impact parameter $\mathbf{b}$. This is obtained by multiplying dP by σ_{NN} and integrating it over all the projectile and target volumes, i.e.,

$$T(\mathbf{b})\sigma_{NN} = \int \rho(\mathbf{b}_B, z_B)\,d\mathbf{b}_B dz_B \rho(\mathbf{b}_A, z_A)\,d\mathbf{b}_A dz_A t(\mathbf{b} - \mathbf{b}_A - \mathbf{b}_B)\sigma_{NN}. \tag{8.4}$$

The thickness function for nucleus-nucleus collisions, $T(\mathbf{b})$, can thus be related to the corresponding thickness function for nucleon-nucleon collisions as

$$T(\mathbf{b}) = \int \rho(\mathbf{b}_B, z_B)\,d\mathbf{b}_B dz_B \rho(\mathbf{b}_A, z_A)\,d\mathbf{b}_A dz_A t(\mathbf{b} - \mathbf{b}_A - \mathbf{b}_B). \tag{8.5}$$

We notice that our definition immediately implies that $T(\mathbf{b})$ is also normalized to unity:

$$\int T(\mathbf{b})\,d\mathbf{b} = 1. \tag{8.6}$$

We can also define the individual thickness functions for each nucleus. That is, for nucleus A,

$$T_A(\mathbf{b}_A) = \int \rho(\mathbf{b}_A, z_A)\,dz_A, \tag{8.7}$$

and similarly for the nucleus B. In terms of these definitions

$$T(\mathbf{b}) = \int d\mathbf{b}_A d\mathbf{b}_B T_A(\mathbf{b}_A) T_B(\mathbf{b}_B) t(\mathbf{b} - \mathbf{b}_A - \mathbf{b}_B). \tag{8.8}$$

Now we are able to describe more specific aspects of nucleus-nucleus collisions in terms of nucleon-nucleon collisions. For example, we may want to calculate the probability of occurrence of n nucleon-nucleon collisions in a nucleus-nucleus collision at impact parameter $\mathbf{b}$. If for simplicity we call A (B) the number of nucleons in nucleus A (B), this probability is given by

$$P(n, \mathbf{b}) = \binom{AB}{n} [T(\mathbf{b})\sigma_{NN}]^n [1 - T(\mathbf{b})\sigma_{NN}]^{AB-n}. \tag{8.9}$$

The first term is the number of combinations for finding n collisions out of AB possible nucleon-nucleon encounters. The second term is the probability of having exact n collisions, while the last

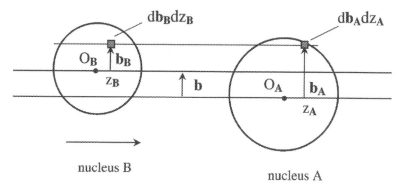

FIGURE 8.1   Geometry of nucleus-nucleus collisions at high energies.

term is the probability of having $AB - n$ misses. The total probability, or differential cross section, is given by

$$\frac{d\sigma}{d\mathbf{b}} = \sum_{n=1}^{AB} P(n, \mathbf{b}) = 1 - [1 - T(\mathbf{b})\sigma_{NN}]^{AB}, \tag{8.10}$$

and the total nucleus-nucleus cross section is given by

$$\sigma^{AB} = \int d\mathbf{b}\left\{1 - [1 - T(\mathbf{b})\sigma_{NN}]^{AB}\right\}. \tag{8.11}$$

The quantity

$$|S(\mathbf{b})|^2 = [1 - T(\mathbf{b})\sigma_{NN}]^{AB} \tag{8.12}$$

is the square of the scattering matrix, for reasons that will become clear when we discuss the eikonal approximation.

In the optical limit, where a nucleon of projectile undergoes only one collision in the target nucleus

$$|S(\mathbf{b})|^2 \simeq \exp(-ABT(\mathbf{b})\sigma_{NN}). \tag{8.13}$$

For nucleus-nucleus collisions one may ask what is the average number of nucleon-nucleon collisions at a given impact parameter $\mathbf{b}$. One has

$$\langle n(\mathbf{b})\rangle = \sum_{n=0}^{AB} nP(n, \mathbf{b}) = \sum_{n=0}^{AB} n\binom{AB}{n}[T(\mathbf{b})\sigma_{NN}]^n [1 - T(\mathbf{b})\sigma_{NN}]^{AB-n}$$

$$= \alpha\frac{\partial}{\partial\alpha} \sum_{n=0}^{AB} \binom{AB}{n}[\alpha\sigma_{NN}]^n [1 - T(\mathbf{b})\sigma_{NN}]^{AB-n}\Big|_{\alpha=T(\mathbf{b})}$$

$$= \alpha\frac{\partial}{\partial\alpha} [1 - T(\mathbf{b})\sigma_{NN} + \alpha\sigma_{NN}]^{AB}\Big|_{\alpha=T(\mathbf{b})}$$

$$= \left\{\alpha AB\sigma_{NN}[1 - T(\mathbf{b})\sigma_{NN} + \alpha\sigma_{NN}]^{AB-1}\right\}\Big|_{\alpha=T(\mathbf{b})}, \tag{8.14}$$

or

$$\langle n(\mathbf{b})\rangle = ABT(\mathbf{b})\sigma_{NN}. \tag{8.15}$$

One can also calculate the standard deviation in the number of nucleon-nucleon collisions. First, we need to calculate $\langle n^2(\mathbf{b})\rangle$. One can use the same trick as in the derivation above, replacing the sum over $n^2 P(n, \mathbf{b})$ by the application of twice the operator $\alpha\partial/\partial\alpha$. The net result is

$$\langle n^2(\mathbf{b})\rangle = ABT(\mathbf{b})\sigma_{NN} + AB(AB - 1)[T(\mathbf{b})\sigma_{NN}]^2. \tag{8.16}$$

Using (8.14) and (8.15) we find for the standard deviation

$$\langle n^2(\mathbf{b}) - \langle n(\mathbf{b})\rangle^2\rangle \equiv \langle n^2\rangle - \langle n\rangle^2 = ABT(\mathbf{b})\sigma_{NN}[1 - T(\mathbf{b})]. \tag{8.17}$$

It is worthwhile to apply this formalism to nucleon-nucleus collisions. In this case, we can set $B = 1$ in Eq. (8.9) and

$$
P(n, \mathbf{b}) = \binom{A}{n} [T(\mathbf{b}) \sigma_{NN}]^n [1 - T(\mathbf{b}) \sigma_{NN}]^{A-n} = \frac{A!}{n! (A-n)!} [T(\mathbf{b}) \sigma_{NN}]^n [1 - T(\mathbf{b}) \sigma_{NN}]^{A-n}
$$
$$
\simeq \frac{A^n}{n!} [T(\mathbf{b}) \sigma_{NN}]^n \exp[-T(\mathbf{b}) \sigma_{NN} (A - n)] \simeq \frac{[AT(\mathbf{b}) \sigma_{NN}]^n}{n!} \exp[-AT(\mathbf{b}) \sigma_{NN}],
$$
$$
(8.18)
$$

valid for $A \gg n$. Thus, for nucleon-nucleus collisions,

$$
P(n, \mathbf{b}) = \frac{[\langle n(\mathbf{b}) \rangle]^n}{n!} \exp[-\langle n(\mathbf{b}) \rangle], \tag{8.19}
$$

where

$$
\langle n(\mathbf{b}) \rangle = A \sigma_{NN} \int \rho(\mathbf{b}, z) \, dz \tag{8.20}
$$

is the average number of nucleon-nucleon collisions. Thus, the probability of having n nucleon-nucleon collisions in a nucleon-nucleus collision is given in terms of the Poisson distribution of the average number of nn-collisions.

We can also ask a more refined question: what is the probability of n nucleons in A colliding with m nucleons in B? To answer this let us write the probability of finding a nucleon in a tube of cross section σ_{NN} at $\mathbf{b}_B$. It is given by $T_B(\mathbf{b}_B) \sigma_{NN}$. The probability of finding n nucleons in the same tube is

$$
\binom{B}{n} [T_B(\mathbf{b}_B) \sigma_{NN}]^n [1 - T_B(\mathbf{b}_B) \sigma_{NN}]^{B-n}. \tag{8.21}
$$

Following the same steps as in Eq. (8.14) we get for the average number of collisions in this tube,

$$
\langle n(\mathbf{b}_B) \rangle = B T_B(\mathbf{b}) \sigma_{NN}. \tag{8.22}
$$

Thus, the probability of having n nucleons of B in a tube of cross section σ_{NN} colliding with m nucleons of A in a similar tube is given by

$$
P(n, m, \mathbf{b}_A, \mathbf{b}_B) = \int t(\mathbf{b} - \mathbf{b}_A - \mathbf{b}_B) \, d\mathbf{b}_A \cdot \binom{B}{n} [T_B(\mathbf{b}_B) \sigma_{NN}]^n
$$
$$
\times \quad [1 - T_B(\mathbf{b}_B) \sigma_{NN}]^{B-n} \binom{A}{n} [T_A(\mathbf{b}_A) \sigma_{NN}]^m [1 - T_A(\mathbf{b}_A) \sigma_{NN}]^{A-m}.
$$
$$
(8.23)
$$

Using $t(\mathbf{b}) \simeq \delta(\mathbf{b})$ we get

$$
P(n, m, \mathbf{b}_A, \mathbf{b}_B) = \binom{A}{n} \binom{B}{n} [T_B(\mathbf{b}_B) \sigma_{NN}]^n [1 - T_B(\mathbf{b}_B) \sigma_{NN}]^{B-n}
$$
$$
\times [T_A(|\mathbf{b} - \mathbf{b}_B|) \sigma_{NN}]^m [1 - T_A(|\mathbf{b} - \mathbf{b}_B|) \sigma_{NN}]^{A-m}.
$$
$$
(8.24)
$$

The *abrasion-ablation* model is based on this equation. It can be extended to account for the isospin dependence of the nucleon-nucleon collisions in a trivial way. In this case m is interpreted as the number of holes created in the nucleon orbitals in the target. Many of these equations can also be derived quantum-mechanically using the eikonal approximation. The derivation presented in this section is much simpler and only requires the use of probability concepts.

In the coordinate space $T(\mathbf{b})$ is derived as

$$
T(\mathbf{b}) = \int \rho_A^z(\mathbf{b}_A) d\mathbf{b}_A \rho_B^z(\mathbf{b}_B) d\mathbf{b}_B t(\mathbf{b} - \mathbf{b_A} + \mathbf{b_B}), \tag{8.25}
$$

where ρ_i^z are the integral of the density along the z-direction (see Eqs. (8.7) and (8.8)). It is a four dimensional integration: two over $\mathbf{b_A}$ and two over $\mathbf{b_B}$. It is convenient to write it in momentum space as

$$T(\mathbf{b}) = \frac{1}{(2\pi)^2} \int \rho_A^z(\mathbf{b}_A) d\mathbf{b}_A \rho_B^z(\mathbf{b}_B) d\mathbf{b}_B \exp\left[-i q.(\mathbf{b} - \mathbf{b_A} + \mathbf{b_B})\right] f_{NN}(q) d^2 q. \qquad (8.26)$$

Here $f_{NN}(q)$ is the q dependence of NN scattering amplitude given by

$$t(\mathbf{b}) = \frac{1}{(2\pi)^2} \int e^{-i\mathbf{q}.\mathbf{b}} f_{NN}(q) d^2 q. \qquad (8.27)$$

Thus,

$$T(b) = \frac{1}{(2\pi)^2} \int \exp(-i\mathbf{q}.\mathbf{b}) \rho_A^z(\mathbf{b_A}) \exp(i\mathbf{q}.\mathbf{b_A}) d\mathbf{b}_A \rho_B^z(\mathbf{b_B}) \exp(-i\mathbf{q}.\mathbf{b_B}) d\mathbf{b}_B \times f_{NN}(q) d^2 q$$

$$= \frac{1}{(2\pi)^2} \int e_A^{-i\mathbf{q}.\mathbf{b}} \widetilde{\rho}_A(\mathbf{q}) \widetilde{\rho}_B(-\mathbf{q}) f_{NN}(q) d^2 q = \frac{1}{2\pi} \int J_0(qb) \widetilde{\rho}_A(q) \widetilde{\rho}_B(-q) f_{NN}(q) q dq. \qquad (8.28)$$

Here $\widetilde{\rho}_A(q)$ and $\widetilde{\rho}_B(-q)$ are the Fourier transforms of the nuclear densities, assumed to be spherical. The function $f_{NN}(q)$ is the Fourier transform of the profile function $t(\mathbf{b})$. The profile function for the NN scattering can be taken as delta function, assuming the nucleons are point particles. In general it is taken as a Gaussian function of width r_0 as

$$t(\mathbf{b}) = \frac{\exp(-b^2/r_0^2)}{\pi r_0^2}. \qquad (8.29)$$

Thus,

$$f_{NN}(q) = \int e^{i\mathbf{q}.\mathbf{b}} t(\mathbf{b}) d\mathbf{b} = \frac{1}{\pi r_0^2} \int e^{i\mathbf{q}.\mathbf{b}} \exp(-b^2/r_0^2) d\mathbf{b} = \exp(-r_0^2 q^2/4). \qquad (8.30)$$

Here, $r_0^2 = 0.439$ fm^2 is the range parameter.

Figure 8.2 shows the free total nucleon-nucleon cross sections [2]. A very useful chi-square fit of the experimental data, yields the expressions obtained in Ref. [3],

$$\sigma_{pp} = \begin{cases} 19.6 + 4253/E - 375/\sqrt{E} + 3.86 \times 10^{-2} E \\ \qquad \text{(for } E < 280 \text{ MeV)} \\[2mm] 32.7 - 5.52 \times 10^{-2} E + 3.53 \times 10^{-7} E^3 - 2.97 \times 10^{-10} E^4 \\ \qquad \text{(for } 280 \text{ MeV} \leq E < 840 \text{ MeV)} \\[2mm] 50.9 - 3.8 \times 10^{-3} E + 2.78 \times 10^{-7} E^2 + 1.92 \times 10^{-15} E^4 \\ \qquad \text{(for } 840 \text{ MeV} \leq E \leq 5 \text{ GeV)}, \end{cases} \qquad (8.31)$$

for proton-proton collisions, and

$$\sigma_{np} = \begin{cases} 89.4 - 2025/\sqrt{E} + 19108/E - 43535/E^2 \\ \qquad \text{(for } E < 300 \text{ MeV)} \\[2mm] 14.2 + 5436/E + 3.72 \times 10^{-5} E^2 - 7.55 \times 10^{-9} E^3 \\ \qquad \text{(for } 300 \text{ MeV} \leq E < 700 \text{ MeV)} \\[2mm] 33.9 + 6.1 \times 10^{-3} E - 1.55 \times 10^{-6} E^2 + 1.3 \cdot 10^{-10} E^3 \\ \qquad \text{(for } 700 \text{ MeV} \leq E \leq 5 \text{ GeV)}, \end{cases} \qquad (8.32)$$

for proton-neutron collisions, where E is the laboratory energy. The fits are represented by dotted and solid curves in Figure 8.2. For nucleus-nucleus collisions $(1 + 2)$ one often uses an *isospin average* of the cross sections

$$\langle \sigma_{NN} \rangle = \frac{Z_1 Z_2 + N_1 N_2}{A_1 A_2} \sigma_{pp} + \frac{Z_1 N_2 + Z_2 N_1}{A_1 A_2} \sigma_{pn}. \qquad (8.33)$$

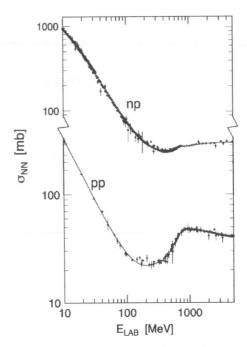

FIGURE 8.2 Least square fit (solid curves) to the nucleon-nucleon cross section described by Eqs. (8.31,8.32). The experimental data are from Ref. [2].

8.3   Applications of the semiclassical model

8.3.1   Reaction cross sections

We first assume that the nuclear densities are of Gaussian shape given by

$$\rho_i(r) = \rho_i(0) \exp(-r^2/a_i^2) \qquad (i = 1, 2). \tag{8.34}$$

Here the parameters $\rho_i(0)$ and a_i are adjusted to reproduce the experimentally determined surface texture of the nucleus. Instead of the normalization of Eq. (8.2), we will now use

$$\int \rho(\mathbf{r}) \, d^3r = A, \tag{8.35}$$

so that the factors AB in the previous equations will be included in the density distributions. The z-integrated density will be

$$\rho_i^z(r) = \rho_i(0) \sqrt{\pi} a_i \exp(-b^2/a_i^2). \tag{8.36}$$

This one can write in the momentum representation as

$$\widetilde{\rho}_i(q) = \rho_i(0)(\sqrt{\pi} a_i)^3 \exp(-q^2 a_i^2/4). \tag{8.37}$$

The overlap integral $T(b)$ can be written as (P denotes projectile and T the target nucleus)

$$T(b) = \frac{1}{(2\pi)^2} \int e_P^{-i\mathbf{q}\cdot\mathbf{b}} \widetilde{\rho}(\mathbf{q}) \widetilde{\rho}_T(-\mathbf{q}) f_{NN}(q) d^2q \tag{8.38}$$

$$= \frac{1}{(2\pi)^2} \pi^3 \rho_P(0) \rho_T(0) a_P^3 a_T^3 \int e^{-i\mathbf{q}\cdot\mathbf{b}} \exp(-a^2 q^2/4) d^2q, \tag{8.39}$$

where $a^2 = a_P^2 + a_T^2 + r_0^2$. Performing q integration we get

$$T(b) = \frac{1}{(2\pi)^2} \pi^3 \rho_P(0) \rho_T(0) a_P^3 a_T^3 \left[\frac{4\pi}{a^2} \exp(-b^2/a^2) \right] = \pi^2 \frac{\rho_P(0) \rho_T(0) a_P^3 a_T^3}{a_P^2 + a_T^2 + r_0^2} \exp\left(-\frac{b^2}{a_P^2 + a_T^2 + r_0^2} \right). \tag{8.40}$$

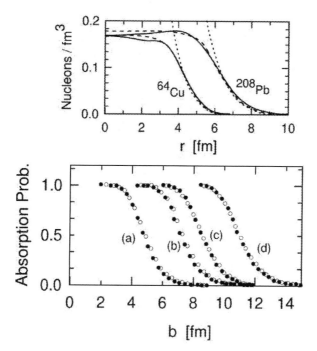

FIGURE 8.3 Upper figure: Nucleon density distribution for ^{64}Cu and ^{208}Pb. Solid lines represent the experimental data. Dashed lines are the Fermi distributions, and dashed lines are the Gaussian distributions which fit the surface region. Lower figure: absorption probability $1 - T(b)$ for (a) ^{64}Cu+^{14}N, ^{64}Cu+^{40}Ar, ^{208}Pb+^{14}N, and ^{208}Pb+^{40}Ar. Solid circles are calculations using Fermi (or experimental) distributions, open circles are the calculations using Gaussian densities, following Karol [4].

As shown in the last section, the *reaction cross section* in the optical limit of the Glauber theory is given by Eqs. (8.11-8.13). $1-|S(b)|^2$ is know as the *transparency function,* or absorption probability.

In the optical limit the integral over impact parameter in Eq. (8.11) can also be done analytically by using the identity

$$\int_0^x \frac{1 - e^{-u}}{u} du = E_1(x) + \ln x + \gamma, \tag{8.41}$$

where

$$E_1(x) = \int_x^\infty \frac{e^{-u}}{u} du \tag{8.42}$$

is the exponential integral, and $\gamma = 0.5772\ldots$ is the *Euler's constant.*

The final expression for σ_R is [4]*

$$\sigma_R = \pi \left(a_p^2 + a_T^2\right) \left[E_1(x) + \ln x + \gamma\right], \tag{8.43}$$

where

$$x = \frac{\pi^2 \sigma_{NN} \rho_T(0) \rho_P(0) a_T^3 a_P^3}{a_T^2 + a_P^2 + r_0^2}. \tag{8.44}$$

Except for the very light nuclei (as α-particles and carbon nuclei) the Gaussian parametrization is not a good one. A Fermi function is more adequate in most cases. However, as observed by Karol [4] the absorption probability term $1 - |S(b)|^2$ which enters in Eq. (8.11) is very little dependent

---

*Misprints appearing in Ref. [4] were corrected.

on the smaller values of b and consequently on the values of $\rho_{P(T)}(r)$ for small $r's$. Only the surface form of the density is relevant. Thus, one can fit the surface part of the densities by Gaussians and use the Eq. (8.43) for σ_R with the appropriate fitting parameters $\rho_P(0)$, $\rho_T(0)$, α_T and α_P. If the density distributions are described by a Fermi function

$$\rho(r) = \rho_0 \left\{ 1 + \exp\left[\frac{(r-c)}{a}\right] \right\}^{-1}, \tag{8.45}$$

it can be shown [4] that $T(b)$ is well reproduced with Eq. (8.38) if the parameters in the Gaussian distributions are given by

$$t = (4\ln 3)\, a = (4.39444\cdots)\, a, \qquad \alpha^2 = \frac{4ct + t^2}{k}, \qquad \rho(0) = \frac{1}{2}\rho_0 e^{c^2/\alpha^2}, \tag{8.46}$$

where

$$\rho_0 = \frac{3A}{4\pi c^3 \left[1 + (\pi^2 a^2/c^2)\right]}, \qquad k = 4\ln 5 = 6.43775\cdots \tag{8.47}$$

where A is the mass number.

The upper part of Figure 8.3 shows the nucleon density distributions for ^{64}Cu and ^{208}Pb. The solid curves are Fermi distributions with $R = 4.34$ fm and $t = 2.15$ fm for Cu, and $c = 6.32$ fm and $t = 2.73$ fm for Pb, respectively. The dotted curves correspond to the "surface normalized" Gaussian distributions (Eqs. (8.46,8.47)), which cross the Fermi distributions at the 50% and 10% central density values. In the lower part of Figure 8.3 one sees the absorption probabilities $1 - |S(b)|^2$ as a function of impact parameter for collisions of Cu and Pb targets with N and Ar projectiles at $E_{\text{lab}} = 2.1$ GeV/nucleon. The open circles are evaluated using the surface-normalized Gaussian distributions while the closed circles used Fermi distributions. One can see that the approximation proposed by Karol [4] works extremely well and is therefore very useful for the calculation of total reaction cross sections in high-energy collisions.

For light nuclei a Gaussian distribution gives a good description of the full densities and we can use

$$\rho(r) = \frac{A}{(a\sqrt{\pi})^3} e^{-r^2/a^2}, \tag{8.48}$$

where A is the mass number, and a is related to the root mean square radius R_{rms} by

$$a = \sqrt{\frac{2}{3}} R_{\text{rms}}. \tag{8.49}$$

The reaction cross sections can be calculated using the matter distributions of the target, or by means of the approximation (8.43).

The two parameter Fermi density is given by Eq. (8.45) where

$$\rho_0 = 3/\left[4\pi c^3 (1 + \pi^2 d^2/c^2)\right],$$

and the momentum density can be derived as

$$\widetilde{\rho}(q) = \frac{8\pi\rho_0}{q^3} \frac{ze^{-z}}{1 - e^{-2z}} \left(\sin x \frac{z(1 + e^{-2z})}{1 - e^{-2z}} - x\cos x \right), \tag{8.50}$$

where $z = \pi dq$ and $x = cq$. Here d is the diffuseness and c, the half value radius in terms of rms radius R for the 2 point Fermi distribution is calculated by $c = (5/3R^2 - 7/3\pi^2 d^2)^{1/2}$.

Good parametrizations of the ground state nuclear densities are given by Gaussian distributions for light nuclei (e.g., $\alpha's$ and ^{12}C) or by Fermi functions for heavy nuclei (e.g. ^{40}C and ^{208}Pb).

For a better description slight modifications of the Gaussian or Fermi distributions might be needed. In Table 8.1 we give examples of ground state nuclear densities for some nuclei with the 3 point Fermi (3pF) and the 3 point Gaussian (3pG) density distributions.

$$\rho(r) = \rho(0)\left(1 + \frac{r^2 d}{R^2}\right)\left\{1 + \exp\left[(r-c)/a\right]\right\}^{-1} \qquad \text{(3pF)}$$

$$= \rho(0)\left(1 + \frac{r^2 d}{R^2}\right)\left\{1 + \exp\left[(r^2 - c^2)/a^2\right]\right\}^{-1} \qquad \text{(3pG)}. \tag{8.51}$$

Nucleus	Model	c (fm)	a (fm)	d (fm)
^{4}He	Gaussian	1.37		
^{12}C	3pF	2.335	0.522	−0.149
^{16}O	3pF	2.608	0.513	−0.051
^{20}Ne	2pF	2.740	0.569	0
^{28}Si	3pF	3.300	0.545	− 0.18
^{40}Ca	3pF	3.725	0.591	− 0.169
^{42}Ca	3pF	3.627	0.594	−0.102
^{58}Ni	3pF	4.309	0.517	−0.131
^{90}Zr	3pG	4.522	2.522	0.245
^{208}Pb	2pF	6.624	0.549	0

TABLE 8.1   Data for parameters used in Gaussian, Modified Fermi (MF), and Modified Gaussian (MG) fits of the density matter distribution of some nuclei.

In the nucleus-nucleus collision at low energies it is the surface region of the nucleus that contributes to the scattering amplitude in a non-trivial manner. The basic assumption of the Glauber model is the description of the relative motion of the two nuclei in terms of straight line trajectory. The density overlaps are evaluated along straight lines associated with each impact parameter b. The modification in the straight line trajectory due to the Coulomb field at lower bombarding energies can not be ignored, especially in the case of heavily charged systems at relatively low bombarding energies.

For low energy heavy ion reactions the straight line trajectory is assumed at the distance of closest approach b' calculated under the influence of the Coulomb potentials for each impact parameter b as given by,

$$b' = \left(\eta + \sqrt{\eta^2 + b^2 k^2} \right) / k, \tag{8.52}$$

which is a solution of the equation

$$E - \frac{Z_1 Z_2 e^2}{r} - E \frac{b^2}{r^2} = 0, \tag{8.53}$$

where E (k) is the bombarding energy (momentum) and $\eta = Z_1 Z_2 e^2 / \hbar v$ is the dimensionless Sommerfeld parameter. The equation accounts for Coulomb repulsion and the centrifugal force at the distance of closest approach $b' = r$.

The Coulomb recoil modified reaction cross section will be modified from Eq. (8.11) to

$$\sigma_R^{AB} = 2\pi \int b \, db \left\{ 1 - \left[1 - T(b') \sigma_{NN} \right]^{AB} \right\}. \tag{8.54}$$

8.3.2   Isotope yield in high-energy collisions

Using the probability approach to high-energy scattering, described in the last section, we can develop a simple model to calculate the isotopic yield in high-energy collisions of heavy nuclei. This is know as the *abrasion-ablation model* [5, 6].

According to the probability concepts introduced in the last section, the differential primary yield can be written as the product of a density of states $\omega(E, Z_f, A_f)$ and an integral over impact parameter,

$$\frac{d\sigma}{dE} (E, Z_f, A_f) = \omega(E, Z_f, A_f) \int d^2 b [1 - P_\pi(b)]^{Z_P - Z_f} P_\pi(b)^{Z_f} [1 - P_\nu(b)]^{N_P - N_f} P_\nu(b)^{N_f}. \tag{8.55}$$

The integral gives the cross section for each primary fragment state as the sum over impact parameters of the probability that Z_f projectile protons and $N_f = A_f - Z_f$ projectile neutrons do not scatter, while the remaining ones do. The distinction between protons and neutrons generalizes the expression (8.24) and permits one to account for the differences in their densities.

We can use Eq. (8.18) and write the probability that a projectile proton does not collide with the target as

$$P_\pi(b) = \int d^2s dz \rho_\pi^P(z,\mathbf{s}) \exp\left[-\sigma_{pp} Z_T \int dz \rho_\pi^T(z,\mathbf{b}-\mathbf{s}) - \sigma_{pn} N_T \int dz \rho_\nu^T(z,\mathbf{b}-\mathbf{s})\right], \quad (8.56)$$

where ρ_π^P and ρ_ν^T are the projectile and target single-particle proton and neutron densities while σ_{pp} and σ_{pn} are the total (minus Coulomb) proton-proton and proton-neutron scattering cross sections, respectively. The neutron probability can be expressed likewise as

$$P_\nu(b) = \int d^2s dz \rho_\nu^P(z,\mathbf{s}) \exp\left[-\sigma_{pn} Z_T \int dz \rho_\pi^T(z,\mathbf{b}-\mathbf{s}) - \sigma_{pp} N_T \int dz \rho_\nu^T(z,\mathbf{b}-\mathbf{s})\right], \quad (8.57)$$

where we have identified the total neutron-neutron scattering cross section σ_{nn} with the proton-proton one. Note that Eqs. (8.55), (8.56) and (8.57) are slightly different than those that we have derived in last section. The difference is the replacement of the probability $1 - T(\mathbf{b})\sigma_{NN}$ by $\exp[-T(\mathbf{b})\sigma_{NN}]$, in the same spirit of the approximation (8.13). It is assumed that if a projectile nucleon collides with a target nucleon it will be removed from the projectile.

For each primary fragment, the density of states, $\omega(E, Z_f, A_f)$, is obtained by counting all combinations of projectile holes consistent with the fragment's charge and neutron numbers. Each hole is a state left vacant by an abraded nucleon. One can use the projectile single-particle energies in the calculation since the mean field rearrangements that would modify them do not have time to occur until long after the abrasion stage has passed. The distribution is shifted down in energy so that the lowest level is at zero excitation energy (This ground level is obtained by removing the nucleons from the highest energy levels in the projectile.). The total number of states is given by

$$\mathcal{N}(Z_f, A_f) = \int dE \omega(E, Z_f, A_f) = \begin{pmatrix} Z_P \\ Z_f \end{pmatrix} \begin{pmatrix} N_P \\ N_f \end{pmatrix}. \quad (8.58)$$

Thus, the energy-integrated primary cross section contains the combinatorial factors $\begin{pmatrix} Z_P \\ Z_f \end{pmatrix}$ and $\begin{pmatrix} N_P \\ N_f \end{pmatrix}$.

A collision between two nuclei of zero isotopic spin, in which the proton and neutron distributions for each nucleus are taken to be identical, results in energy-integrated primary cross sections that are symmetric in charge and neutron number about the point $Z_f = N_f = A_f/2$. This need not be the case when the distributions are different or when one of the nuclei is not of zero isotopic spin. The calculations based on Eqs. (8.55-8.58), show that the asymmetry of the primary yield from a $Z = N$ projectile on a $Z \neq N$ target is small. The target dependence of the primary projectile yield is thus an almost purely geometrical one.

We show in Figure 8.4 results for the average excitation energy obtained from the densities of states as a function of the primary fragment mass A_f for the system ^{16}O + Pb and for [6]. The average was performed over all isotopes with a given A_f in order to remove the slight isotopic dependence of the individual averages. The distributions' variance is displayed as an error bar on each point. Also shown, as a solid line in Figure 8.4, is the excitation energy that would be obtained using a surface energy estimate based on the liquid drop model. The latter yields about half the average energy of the hole distribution but remains within its variance for all but the largest mass losses. As can be seen in Figure 8.4, the average excitation energy of fragments that undergo little abrasion remains below the particle emission threshold, although their energy distribution extends above it. In these cases, use of the energy distribution rather than an average value is essential for describing the decay.

One can resort to several statistical models to calculate the particle evaporation during the ablation stage as a function of the primary fragment charge Z_f, mass A_f, and excitation energy ε. A Hauser-Feshbach or Weisskopf-Ewing evaporation formalism can be used (see Chapter 4). The result of the evaporation calculation can be expressed as the probability, $P(Z, A; E, Z_f, A_f)$, of yielding a residue of charge Z and mass A, given a primary compound nucleus of charge Z_f

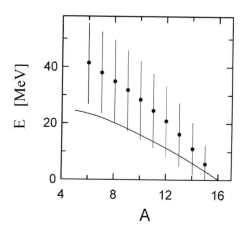

FIGURE 8.4   The average primary fragment excitation energy obtained from the densities of states is shown as a function of the fragment mass by the solid points. The variances of the distributions are displayed as error bars. The solid line shows the excitation energy obtained using the surface energy estimate based on a liquid drop model.

mass A_f and excitation energy ε. In terms of this quantity and the differential primary yield $d\sigma\left(E, Z_f,\ A_f\right)/d\varepsilon$, one can calculate the observed secondary yield, $\sigma\left(Z, A\right)$, as

$$\sigma\left(Z, A\right) = \sum_{Z_f, A_f} \int dE P\left(Z, A; E, Z_f, A_f\right) \frac{d\sigma}{dE}\left(E, Z_f, A_f\right), \qquad (8.59)$$

which can also include the decay to fission fragments for large mass fissionable projectiles.

The microscopic calculation of the absorption probabilities and cross sections permits the use of more realistic collision geometries. A natural step in this direction is to next replace the average single-particle projectile densities in Eqs. (8.56) and (8.57) by the probability distributions of the individual projectile orbitals. This allows one to take into account differences in the abrasion probabilities or the different orbitals. Thus generalized, the expression for the differential primary yield, $d\sigma$ $\left(E, Z_f,\ A_f\right)/dE$, becomes a sum over all the possible combinations of orbital transmission and absorption factors that result in a given primary charge Z_f and mass A_f. The population of primary fragment states will no longer be evenly distributed but will depend on the absorption probabilities of the single-particle states on which they are based. As the nucleons that are less bound are the more superficial ones and, thus, also the ones more likely to be absorbed, the average primary fragment excitation energy will be lower in this more realistic model.

The cross sections for mass distribution of each fragment calculated according to the evaporation model after ablasion and electromagnetic (EM) excitation of giant resonances are displayed in Figure 8.5 for the ^{238}U $+$ ^{208}Pb reaction at 1 GeV/nucleon [7]. We will describe the EM excitation of giant resonances later in this chapter. The diamonds (circles) show the results without (with) inclusion of fission channels. The squares include large impact parameters with only EM excitation. It is clear that fission products after the abrasion originate from primary fragments with mass $A \gtrsim$ 170. Not shown in the Figure are cross sections for primary fragments with mass $A = 235 - 237$ of about $196 - 726$ mb, contributing mostly to fission fragments. About 23% of the primary fragments yield fission products after ablation. These are seen as a clear bump in the plot, peaked around mass $A = 110$. On the other hand, the fission yields originating from EM excitation of the ^{238}U projectile is responsible for the double hump structure in the figure, characteristic of fission, peaked around masses $A = 100$ and $A = 140$. Fission products correspond to about 18% of the EM excitation of ^{238}U at large impact parameters.

Figure 8.6 displays the isotopic distribution of niobium (upper panel) and ruthenium (lower panel) fragments for the ^{238}U $+$ ^{208}Pb reaction at 1 GeV/nucleon [7]. Data (open circles) are from Ref. [8]. The solid (dashed) lines correspond to fragmentation calculations with (without)

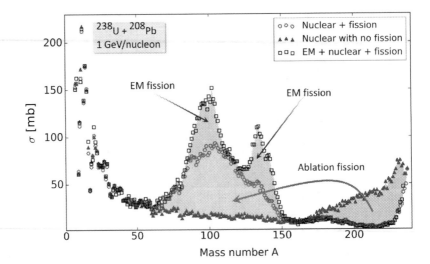

FIGURE 8.5   Mass distribution of projectile fragments for the ^{238}U + ^{208}Pb reaction at 1 GeV/nucleon. The diamonds correspond to the yields obtained with the abrasion-ablation model without fission, while the circles include fission decays. The squares include the electromagnetic excitation (EM) of giant resonances leading to particle evaporation and fission products.

inclusion of final state EM excitation of abraded fragments. The arrows point to the region (shaded area) of increasing contribution of fragments decaying by fission. The overall agreement with the experimental data is reasonable and differences in particular cases are estimated to be within a factor of two to ten. Clear trends are: (a) Evidently, the addition of final state EM excitation of primary fragments leads to an enhancement of the average yields across the isotopic chains. (b) The yields of lighter fragments become larger when the final state EM excitations are accounted for.

The model presented in this section shows that we can understand the main features of the isotopic fragmentation yield in heavy ion collisions at high energies in terms of simple Glauber calculations and with statistical decay models for the spectators and participants.

8.4   The eikonal wave function

In the last sections we have shown that a probability approach to heavy ion scattering in high-energy collisions gives reasonable results for the isotopic yields. The probability approach is based on the survival probabilities in forward nucleon-nucleon scattering. No reference to the quantum behavior of the particles was needed to derive the results. In this and next sections we will show that the probability approach is a subset of a more general description of nuclear scattering at high energies. This general description, known as *Glauber theory* of nuclear scattering [1], is based on the *eikonal wave function*, which is an approximation valid for forward scattering of high-energy particles. In this and the following sections we will deduce the eikonal wave function and the cross sections for elastic and inelastic scattering of highly energetic particles. Some of the equations have classical probabilistic interpretation, as in the previous sections.

The free-particle wave function

$$\psi \sim e^{i\mathbf{k}\cdot\mathbf{r}} \tag{8.60}$$

becomes "distorted" in the presence of a potential $V(\mathbf{r})$. As we have seen in Chapter 2 the distorted wave can be calculated numerically by solving the Schrödinger equation for each partial wave, i.e.,

$$\left[\frac{d^2}{dr^2} + k_l^2(r)\right] \chi_l(r) = 0, \tag{8.61}$$

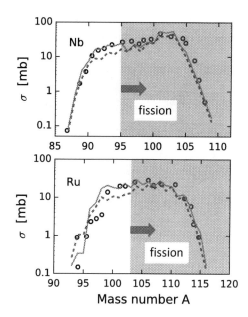

FIGURE 8.6 Isotopic distribution of niobium (upper panel) and ruthenium (lower panel) fragments. Data (open circles) are from Ref. [8]. The solid (dashed) lines correspond to calculations with (without) inclusion of final state electromagnetic excitation of abraded fragments. The arrows point to the region (shaded area) of increasing contribution of fragments decaying by fission.

where

$$k_l(r) = \left\{ \frac{2\mu}{\hbar^2} \left[E - V(r) - \frac{l(l+1)\hbar^2}{2\mu r^2} \right] \right\}^{1/2}. \tag{8.62}$$

with the condition that asymptotically $\psi(\mathbf{r})$ behaves as (8.60).

The solution of (8.61) involves a great numerical effort at large bombarding energies E. Fortunately, at large energies E a very useful approximation is valid when the excitation energies ΔE are much smaller than E and the nuclei (or nucleons) move in forward directions, i.e., $\theta \ll 1$.

Calling $\mathbf{r} = (z, \mathbf{b})$, where z is the coordinate along the beam direction, we can assume that

$$\psi(\mathbf{r}) = e^{ikz}\phi(z, \mathbf{b}), \tag{8.63}$$

where ϕ is a slowly varying function of z and b, so that

$$\left|\nabla^2\phi\right| \ll k\left|\nabla\phi\right|. \tag{8.64}$$

In cylindrical coordinates the Schrödinger equation

$$-\frac{\hbar^2}{2\mu}\nabla^2\psi(\mathbf{r}) + V(\mathbf{r})\psi(\mathbf{r}) = E\psi(\mathbf{r})$$

becomes

$$2ike^{ikz}\frac{\partial\phi}{\partial z} + e^{ikz}\frac{\partial^2\phi}{\partial z^2} + e^{ikz}\nabla_b^2\phi - \frac{2m}{\hbar^2}Ve^{ikz}\phi = 0,$$

or, neglecting the 2nd and 3rd terms because of (8.64),

$$\frac{\partial\phi}{\partial z} = -\frac{i}{\hbar v}V(\mathbf{r})\phi, \tag{8.65}$$

whose solution is

$$\phi = \exp\left\{-\frac{i}{\hbar v}\int_{-\infty}^{z}V(\mathbf{b}, z')dz'\right\}. \tag{8.66}$$

That is,

$$\psi(\mathbf{r}) = \exp\left\{ikz + i\chi(\mathbf{b}, z)\right\}, \tag{8.67}$$

where

$$\chi(\mathbf{b}, z) = -\frac{1}{\hbar v}\int_{-\infty}^{z} V(\mathbf{b}, z')dz' \tag{8.68}$$

is the *eikonal phase*. Given $V(\mathbf{r})$ one needs a single integral to determine the wave function: a great simplification of the problem.

The eikonal approximation, in the same form as given by Eqs. (8.67), can be obtained from the Klein-Gordon equation with a (scalar) potential V. The proof follows a similar sequence of steps as described above.

8.5   Elastic scattering

As shown in Section 3.3 the scattering amplitude for a potential $V(\mathbf{r})$ is given by

$$f_{\mathrm{el}}\left(\theta\right) = -2\pi^2\left(\frac{2\mu}{\hbar^2}\right)\int d\mathbf{r}\, e^{i\mathbf{k}\cdot\mathbf{r}'}V(\mathbf{r}')\psi_{\mathbf{k}}^{+}(\mathbf{r}'). \tag{8.69}$$

Using the eikonal wave function (8.67) for $\psi^{+}(\mathbf{r})$ in Eq. (8.69), we obtain

$$f_{\mathrm{el}}(\theta) = -\frac{\mu}{2\pi\hbar^2}\int d^2 b\, e^{i(\mathbf{k}-\mathbf{k}')\cdot\mathbf{b}}\int_{-\infty}^{\infty}dz\, e^{i(\mathbf{k}-\mathbf{k}')\cdot\hat{\mathbf{k}}z}$$

$$\times V(\mathbf{r})\exp\left\{-\frac{i}{\hbar v}\int_{-\infty}^{z} V(\mathbf{r}')dz'\right\}. \tag{8.70}$$

But, for $\theta \ll 1$, $(\mathbf{k}-\mathbf{k}')\cdot\mathbf{k} \cong 0$, and

$$\int_{-\infty}^{\infty} dz V(\mathbf{r})\exp\left\{-\frac{i}{\hbar v}\int_{-\infty}^{z} V(\mathbf{r}')dz'\right\} = i\hbar v\exp\left\{-\frac{i}{\hbar v}\int_{-\infty}^{z} V(\mathbf{r}')dz'\right\}\Big|_{-\infty}^{\infty} = i\hbar v\left\{e^{i\chi(\mathbf{b})} - 1\right\},$$

where

$$\chi(\mathbf{b}) \equiv \chi(\mathbf{b}, \infty) = -\frac{1}{\hbar v}\int_{-\infty}^{\infty} V(\mathbf{b}, z')dz' \tag{8.71}$$

is the total eikonal phase.

We will use the momentum transfer defined by

$$\mathbf{q} \equiv \Delta\mathbf{k} = \mathbf{k} - \mathbf{k}', \qquad q = 2k\sin\theta/2, \tag{8.72}$$

where we assumed that $|\Delta\mathbf{k}|/k \ll 1$. Then

$$f(\theta) = -\frac{ik}{2\pi}\int d^2 b\, e^{i\mathbf{q}\cdot\mathbf{b}}\left[e^{i\chi(\mathbf{b})} - 1\right]. \tag{8.73}$$

If the potential is spherically symmetric, $\chi(b)$ is a function of the absolute value of b only and using

$$\int_0^{2\pi} d\phi\, e^{iqb\cos\phi} = 2\pi J_0(qb),$$

we obtain

$$f_{\mathrm{el}}(\theta) = -ik\int db\, b J_0(qb)\left[e^{i\chi(b)} - 1\right]. \tag{8.74}$$

Thus, in the eikonal approximation the elastic scattering amplitude $f(\theta)$ is obtained from two simple integrals, Eqs. (8.71) and (8.74).

In an extreme version of the eikonal approximation the phase-shift function is assumed to satisfy the relation

$$e^{i\chi(b)} = \Theta(b - R) \equiv \begin{cases} 0 & b < R \\ 1 & b \geq R, \end{cases} \tag{8.75}$$

where R is a cut-off radius, and is of the order of the sum of the radii of the two colliding nuclei (in realistic cases $e^{i\chi(b)}$ has a smooth cut-off distribution). In this approximation the incoming flux corresponding to the collision with $b < R$ is completely absorbed, while the collision with $b > R$ receives no effect (note that the Coulomb interaction is ignored). Substitution of Eq. (8.75) into Eq. (8.74) leads to the scattering amplitude of $f_{\text{el}}(\theta) \propto J_1(qR)/qR$. The differential cross section $|f_{\text{el}}(\theta)|^2$ thus vanishes at the zeros of the Bessel function $J_1(qR)$, which occur at $qR \approx 3.83, 7.02, \cdots$. With increasing A, R in general increases, thus the first zero appears at smaller values of q.

8.6 Coulomb amplitude and Coulomb eikonal phase

In general, the scattering potential is given by

$$V(\mathbf{r}) = U_N^{\text{opt}}(r) + U_C(r), \tag{8.76}$$

where U_N^{opt} is the nuclear optical potential and $U_C(r) = Z_1 Z_2 e^2/r$ is the Coulomb potential between the nuclei.

Since U_N^{opt} (generally complex) is well localized in space, the eikonal phase for the nuclear part of (8.76) is obtained by a well convergent integral Eq. (8.71). However, the integral diverges logarithmically for the Coulomb potential. This is due to the use of the approximation (8.64) which is not valid for the (long range) Coulomb potential. But this does not pose a real problem since an analytical formula can be given for the Coulomb eikonal phase which reproduces the exact Coulomb amplitude, Eq. (2.149). The total eikonal phase is

$$\chi(b) = -\frac{1}{\hbar v} \int_{-\infty}^{\infty} U_N^{\text{opt}}(b, z')dz' + \chi_C(b), \tag{8.77}$$

where

$$\chi_C(b) = \frac{2Z_1 Z_2 e^2}{\hbar v} \ln(kb). \tag{8.78}$$

We now show that the Coulomb phase, as given by the above formula, reproduces the Coulomb amplitude in the eikonal approximation. We have for pure Coulomb scattering

$$f_C(\theta) = ik \int_0^{\infty} J_0(qb) \left[e^{i\chi_C(b)} - 1 \right] b\,db = \frac{ik}{q^2} \int_0^{\infty} J_0(x) \left[\left(\frac{kx}{q} \right)^{i2\eta} - 1 \right] x\,dx$$

$$= \frac{ik}{q^2} \left\{ \left(\frac{k}{q} \right)^{i2\eta} \int_0^{\infty} J_0(x) x^{i2\eta+1} dx - \int_0^{\infty} J_0(x) x\,dx \right\},$$

where

$$\eta = \frac{Z_1 Z_2 e^2}{\hbar v}; \qquad k = \frac{mv}{\hbar}; \qquad q = 2k \sin(\theta/2). \tag{8.79}$$

Integrating by parts and using $\int x J_0(x)dx = x J_1(x)$, one gets

$$f(\theta) = \frac{ik}{q^2} \left(\frac{k}{q} \right)^{i2\eta} x^{i2\eta+1} J_1(x) \Big|_0^{\infty} - i2\eta \left(\frac{k}{q} \right)^{i2\eta} \int_0^{\infty} x^{i2\eta} J_1(x)dx - x J_1(x) \Big|_0^{\infty}$$

$$= \frac{ik}{q^2} \left\{ x J_1(x) \left[e^{i2\eta \ln(kx/q)} - 1 \right] \Big|_{\infty}^{0} - 2i\eta \left(\frac{k}{q} \right)^{i2\eta} 2^{i2\eta} \frac{\Gamma(1+i\eta)}{\Gamma(1-i\eta)} \right\}.$$

The first term in the equation above can be neglected and we get

$$f_C(\theta) = \frac{2k\eta}{q^2} \left(\frac{2k}{q} \right)^{i2\eta} \frac{\Gamma(1+i\eta)}{\Gamma(1-i\eta)}.$$

But

$$\Gamma(1 \pm i\eta) = \pm i\eta \Gamma(\pm i\eta) = \pm i\eta \left| \Gamma(i\eta) \right| \begin{cases} e^{i\sigma_0} \\ e^{-i(\sigma_0+\pi)}, \end{cases}$$

where

$$\sigma_0 = \arg \Gamma\left(1 + i\eta\right). \tag{8.80}$$

Therefore,

$$f_C(\theta) = -\frac{Z_1 Z_2 e^2}{2\mu v^2 \sin^2 \frac{\theta}{2}} \exp\left\{-i\eta \ln\left(\sin^2 \frac{\theta}{2}\right) + i2\sigma_0\right\}, \tag{8.81}$$

which is the exact Coulomb amplitude, Eq. (2.149).

The phase (8.80) can also be written as

$$\sigma_0 = -\eta\gamma + \sum_{k=0}^{\infty}\left(\frac{\eta}{k+1} - \arctan\frac{\eta}{k+1}\right), \tag{8.82}$$

where $\gamma = 0.57721\ldots$ is the Euler's constant.

For numerical evaluation it is appropriate to rewrite Eq. (8.74) as

$$f(\theta) = ik \int_0^\infty J_0(qb) e^{i\chi_C(b)} \left[1 - e^{i\chi_N(b)}\right] b\,db + f_C(\theta), \tag{8.83}$$

which can be easily obtained by adding and subtracting f_C to (8.74) and combining terms. This is because $1 - \exp\left[i\chi_N(b)\right]$ drops to zero rapidly for $b > R_1 + R_2$ since U_N^{opt} goes to zero there. In Eq. (8.83) we use (8.81) for $f_C(\theta)$. $\chi_N(b)$ is given by the first term of Eq. (8.77) and $\chi^C(b)$ is given by (8.78).

A few modifications of the Coulomb eikonal phase are needed to account for the extended nature of the nuclear charge distributions. For light nuclei, one can assume Gaussian nuclear densities, and the Coulomb phase is given by

$$\chi_C(b) = 2\frac{Z_1 Z_2 e^2}{\hbar v}\left[\ln(kb) + \frac{1}{2}E_1\left(\frac{b^2}{R_G^2}\right)\right], \tag{8.84}$$

with $R_G^{(i)}$ equal to the size parameter of Gaussian matter densities of nucleus 1 and nucleus 2, respectively, $R_G^2 = [R_G^{(1)}]^2 + [R_G^{(2)}]^2$, and

$$E_1(x) = \int_x^\infty \frac{e^{-t}}{t}dt. \tag{8.85}$$

The first term in Eq. (8.84) is the contribution to the Coulomb phase of a point-like charge distribution. The second term in Eq. (8.84) is a correction due to the extended Gaussian charge distribution. It eliminates the divergence of the Coulomb phase at $b = 0$, so that

$$\psi_C(0) = 2\frac{Z_1 Z_2 e^2}{\hbar v}\left[\ln(kR_G) - \gamma\right], \tag{8.86}$$

where γ is the Euler's constant.

For heavy nuclei a "black-sphere" absorption model is more appropriate. Assuming an absorption radius R_0, the Coulomb phase is given by

$$\chi_C(b) = 2\frac{Z_1 Z_2 e^2}{\hbar v}\left\{\Theta(b - R_0)\ln(kb) + \Theta(R_0 - b)\left[\ln(kR_0)\right.\right.$$
$$\left.\left. + \ln\left[1 + (1 - b^2/R_0^2)^{1/2}\right] - (1 - b^2/R_0^2)^{1/2} - \frac{1}{3}(1 - b^2/R_0^2)^{3/2}\right]\right\}. \tag{8.87}$$

Again, the first term inside the curly brackets is the Coulomb eikonal phase for pointlike charge distributions. The second term accounts for the finite extension of the charge distributions.

8.7   Total reaction cross sections

According to the optical theorem (see Section 2.5) the total scattering cross section is given by

$$\sigma_{\text{tot}} = \frac{4\pi}{k}\text{Im}f(\theta = 0°).$$

Using (8.73) we obtain

$$\sigma_{\text{tot}} = 2\int \left[1 - \text{Re}e^{i\chi(\mathbf{b})}\right]d^2b. \tag{8.88}$$

We can also verify Eq. (8.88) by an integration of the scattering amplitude over angles, using the eikonal amplitude $f(\theta)$, Eq. (8.73),

$$\int |f(\theta)|^2 d\Omega_k = \left(\frac{k}{2\pi}\right)^2 \int e^{i(\mathbf{k}-\mathbf{k}')\cdot(\mathbf{b}-\mathbf{b}')} \left\{e^{i\chi(\mathbf{b})} - 1\right\}\left\{e^{-i\chi(\mathbf{b}')} - 1\right\} d^2b d^2b' d\Omega_{k'}. \tag{8.89}$$

But

$$d\Omega_{k'} = \sin\theta d\theta d\phi = \frac{k^2\sin\theta d\theta d\phi}{k^2} \simeq \frac{(k\theta d\theta)(k d\phi)}{k^2} \qquad \text{valid for } \theta \ll 1$$

$$\simeq \frac{kd\phi dk}{k^2} = \frac{d^2k'}{k^2}. \tag{8.90}$$

Furthermore, using

$$\int e^{(\mathbf{k}-\mathbf{k}')\cdot(\mathbf{b}-\mathbf{b}')} d^2k' = (2\pi)^2\delta^{(2)}(\mathbf{b}, -\mathbf{b}') \tag{8.91}$$

where $\delta^{(2)}(\mathbf{b} - \mathbf{b}')$ is a two-dimensional delta-function, we find

$$\sigma_{\text{scatt}} = \int \left|e^{i\chi(\mathbf{b})} - 1\right|^2 d^2b. \tag{8.92}$$

For $\chi(\mathbf{b})$ real (real optical potential), Eq. (8.92) reduces to Eq. (8.88).

However, when $\chi(\mathbf{b})$ is complex (optical potential has an imaginary part), Eq. (8.92) is not equal to Eq. (8.88). The difference is equal to the reaction cross section. That is,

$$\sigma_R = \sigma_{\text{tot}} - \sigma_{\text{scatt}} = \int \left[1 - \left|e^{i\chi(\mathbf{b})}\right|^2\right]d^2b. \tag{8.93}$$

8.8   Scattering of particles with spin

In the scattering of a spin-$\frac{1}{2}$ particle by a nucleus, e.g., proton-nucleus scattering, the optical potential contains a spin-orbit term, i.e.,

$$U_N^{\text{opt}} = U_N(r) + U_S(r)\boldsymbol{\sigma}\cdot\mathbf{L}, \tag{8.94}$$

where

$$\mathbf{L} = \frac{1}{\hbar}(\mathbf{r}\times\mathbf{p}), \qquad \mathbf{s} = \frac{1}{2}\boldsymbol{\sigma}, \tag{8.95}$$

and the Dirac matrices $\boldsymbol{\sigma} = (\sigma_1, \sigma_2, \sigma_3)$ obey the commutation rule

$$\sigma_i\sigma_j + \sigma_j\sigma_i = 2\delta_{ij}. \tag{8.96}$$

As in the spinless case, the distorted wave function can be written as

$$\psi_{\mathbf{r}}(\mathbf{r}) \cong e^{i\mathbf{k}\cdot\mathbf{r}}\varphi(\mathbf{r})u_i(\mathbf{k}), \tag{8.97}$$

where $u_i(\mathbf{k})$ is a spinor wave function. One obtains

$$\varphi(\mathbf{r}) \cong \exp\left\{-\frac{i}{\hbar v}\int_{-\infty}^{z} dz' \left\{U_N(b, z') + U_S(b, z')\boldsymbol{\sigma}\cdot(\mathbf{b}\times\hat{\mathbf{k}})k\right\}\right\}, \tag{8.98}$$

since $\mathbf{z} = z\hat{\mathbf{k}}$.

The eikonal amplitude is

$$f(\theta) = \frac{k}{2\pi i} \int e^{i\mathbf{q}\cdot\mathbf{b}} \left\{ e^{i\chi_N(b)+i\chi_S(b)\boldsymbol{\sigma}\cdot(\mathbf{b}\times\hat{\mathbf{k}})k} - 1 \right\} d^2b, \tag{8.99}$$

where

$$\chi_N(b) = -\frac{1}{\hbar v} \int_{-\infty}^{\infty} U_N(b,z)dz, \quad \chi_S(b) = -\frac{1}{\hbar v} \int_{-\infty}^{\infty} U_S(b,z)dz. \tag{8.100}$$

Using Eq. (8.96), we can write

$$\exp\left\{i\chi_S(b)\boldsymbol{\sigma}\cdot(\mathbf{b}\times\hat{\mathbf{k}})k\right\} = \sum_n \frac{1}{n!} \left[i\chi_S(b)\boldsymbol{\sigma}\cdot(\mathbf{b}\times\hat{\mathbf{k}})k\right]^n$$

$$= \sum_{n=\text{odd}} \frac{1}{n!}(kb)^n \left[i\chi_S(b)\right]^n \boldsymbol{\sigma}\cdot(\hat{\mathbf{b}}\times\hat{\mathbf{k}}) + \sum_{n=\text{even}} \frac{1}{n!}(kb)^n \left[i\chi_S(b)\right]^n$$

$$= i\boldsymbol{\sigma}\cdot(\hat{\mathbf{b}}\times\hat{\mathbf{k}})\sin\left[kb\chi_S(b)\right] + \cos\left[kb\chi_S(b)\right]. \tag{8.101}$$

Including the Coulomb amplitude as in Eq. (8.83) we obtain

$$f(\theta) = \frac{ik}{2\pi} \int d^2b\, e^{i\mathbf{q}\cdot\mathbf{b}} e^{i\chi_C(b)} \left[1 - e^{i\chi_N(b)+i\chi_S(b)\boldsymbol{\sigma}\cdot(\mathbf{b}\times\mathbf{k})}\right] + f_C(\theta), \tag{8.102}$$

which, by using Eq. (8.101) becomes

$$f(\theta) = \frac{ik}{2\pi} \int d^2b\, e^{i\mathbf{q}\cdot\mathbf{b}} \left\{\Gamma_0(b) + i\boldsymbol{\sigma}\cdot(\hat{\mathbf{b}}\times\hat{\mathbf{k}})\Gamma_1(b)\right\} + f_C(\theta), \tag{8.103}$$

where the *profile functions* $\Gamma_0(b)$ and $\Gamma_1(b)$ are defined as

$$\Gamma_0(b) = e^{i\chi_C(b)}\left\{1 - e^{i\chi_N(b)}\cos\left[kb\chi_S(b)\right]\right\},$$

$$\Gamma_1(b) = -e^{i\chi_C(b)+i\chi_N(b)}\sin\left[kb\chi_S(b)\right]. \tag{8.104}$$

It is more convenient to rewrite Eq. (8.103) as

$$f(\theta) = F(\theta) + (\boldsymbol{\sigma}\cdot\hat{\mathbf{n}})G(\theta), \tag{8.105}$$

where $\hat{\mathbf{n}} = \dfrac{\mathbf{k}\times\mathbf{k}'}{|\mathbf{k}\times\mathbf{k}'|}$. The azimuthal integrals in (8.103) can be easily performed,[*]

$$F(\theta) = f_C(\theta) + ik \int_0^{\infty} J_0(qb)\Gamma_0(b)b\,db,$$

$$G(\theta) = -ik \int_0^{\infty} J_1(qb)\Gamma_1(b)b\,db. \tag{8.106}$$

Since $J_1(x = 0) = 0$, $G(\theta = 0) = 0$, which is a consequence of the conservation of angular momentum + spin.

For unpolarized beams

$$\frac{d\sigma_{\text{el}}}{d\Omega} = \frac{1}{2}\sum_{\text{spins}} |f(\theta)|^2 = |F(\theta)|^2 + |G(\theta)|^2. \tag{8.107}$$

---

[*]$\boldsymbol{\sigma}\cdot(\mathbf{b}\times\hat{\mathbf{k}}) = (\boldsymbol{\sigma}\cdot\hat{\mathbf{n}})b_{||}$, where $b_{||}$ is the component of $\mathbf{b}$ parallel to $\mathbf{k}'$. So, $\boldsymbol{\sigma}\cdot(\mathbf{b}\times\hat{\mathbf{k}}) = (\boldsymbol{\sigma}\cdot\hat{\mathbf{n}})b\cos\theta$. The component of $\mathbf{b}$ perpendicular to $\mathbf{k}\times\mathbf{k}'$ defines another term, proportional to $(\boldsymbol{\sigma}\cdot\boldsymbol{\xi})$, which is unnecessary by symmetry arguments.

E_{lab} [MeV]	σ_{pp} [fm^2]	α_{pp}	β_{pp} [fm^2]	σ_{pn} [fm^2]	α_{pn}	β_{pn} [fm^2]
40	7.0	1.328	0.385	21.8	0.493	0.539
60	4.7	1.626	0.341	13.6	0.719	0.410
80	3.69	1.783	0.307	9.89	0.864	0.344
100	3.16	1.808	0.268	7.87	0.933	0.293
120	2.85	1.754	0.231	6.63	0.94	0.248
140	2.65	1.644	0.195	5.82	0.902	0.210
160	2.52	1.509	0.164	5.26	0.856	0.181
180	2.43	1.365	0.138	4.85	0.77	0.154
200	2.36	1.221	0.117	4.54	0.701	0.135
240	2.28	0.944	0.086	4.13	0.541	0.106
300	2.42	0.626	0.067	3.7	0.326	0.081
425	2.7	0.47	0.078	3.32	0.25	0.0702
550	3.44	0.32	0.11	3.5	-0.24	0.0859
650	4.13	0.16	0.148	3.74	-0.35	0.112
700	4.43	0.1	0.16	3.77	-0.38	0.12
800	4.59	0.06	0.185	3.88	-0.2	0.12
1000	4.63	-0.09	0.193	3.88	-0.46	0.151
2000	4.67	0.	0.12	3.88	-0.50	0.151

TABLE 8.2 Parameters [11] for the nucleon-nucleon amplitude, as given by Eq. (8.109).

8.9 The optical limit of Glauber theory

In its simplest version, neglecting the spin-orbit and surface terms, the optical potential for proton-nucleus collisions can be obtained as a folding of the nucleon-nucleon interaction $v(\mathbf{r})$ and the nuclear density $\rho(\mathbf{r})$. We can also relate the nucleon nucleon potential v to the t-matrix $t = \langle \mathbf{k}'|v(\mathbf{r}|\mathbf{k})$. This is needed essentially only at small angle scattering $\mathbf{k}' \simeq \mathbf{k}$ for large bombarding energies. Then the optical potential can be written as

$$U_{opt}(\mathbf{r}) = \langle t_{pn} \rangle \rho_n(\mathbf{r}) + \langle t_{pp} \rangle \rho_p(\mathbf{r}), \qquad (8.108)$$

where ρ_n (ρ_p) are the neutron (proton) ground state densities and $\langle t_{pi} \rangle$ is the (isospin averaged, as in Eq. (8.33)) transition matrix element for nucleon-nucleon scattering at forward directions,

$$t_{pi}(\mathbf{q} \simeq 0) = -\frac{2\pi\hbar^2}{\mu} f_{pi}(\mathbf{q} \simeq 0) = -\frac{\hbar \mathrm{v}}{2} \sigma_{pi}(\alpha_{pi} + i), \qquad (8.109)$$

where σ_{pi} is the free proton-nucleon cross section and α_{pi} is the ratio between the imaginary and the real part of the proton-nucleon scattering amplitude. The basic assumption here is that the scattering is given solely in terms of the forward proton-nucleon scattering amplitude and the local one-body density [9]. The potential generated in this way is often called by *"tρ"* approximation.

For nucleus-nucleus collisions, the extension of this method leads to an optical potential of the form

$$U_{opt}(\mathbf{r}) = \int \langle t_{NN}(\mathbf{q} \simeq 0) \rangle \rho_1(\mathbf{r} - \mathbf{r}')\rho_2(\mathbf{r}')d^3r', \qquad (8.110)$$

where $\mathbf{r}$ is the distance between the center-of-mass of the nuclei. The potential generated in this way is often called by *"tρρ"* approximation.

The formula (8.109) can be improved to account for the scattering angle dependence of the nucleon-nucleon amplitudes. A good parametrization [10] for the nucleon-nucleon scattering amplitude is given by

$$f_{NN}(\mathbf{q} \simeq 0) = \frac{k_{NN}}{4\pi}\sigma_{NN}(i + \alpha_{NN})e^{-\xi_{NN}\mathbf{q}^2}. \qquad (8.111)$$

The parameters of the nucleon-nucleon cross scattering amplitudes for $E_{lab} \geq 40$ MeV/nucleon are shown in Table (8.2), extracted from Ref. [11]. An isospin average of these quantities can be done in most practical cases, while in other situations they can be used to separate cross sections sensitiveness to the proton and neutron densities separately.

The nuclear part of the eikonal scattering phase can also be written

$$\chi_N(\mathbf{b}) = \int \int d\mathbf{r}d\mathbf{r}' \rho_1(\mathbf{r})\Gamma_{NN}(|\mathbf{b} - \mathbf{s} - \mathbf{s}'|)\rho_2(\mathbf{r}'), \qquad (8.112)$$

where the *profile function* $\Gamma_{NN}(\mathbf{b})$ is defined in terms of the two-dimensional Fourier transform of the elementary scattering amplitude

$$\Gamma_{NN}(\mathbf{b}) = \frac{1}{2\pi i k_{NN}} \int \exp\left(-i\mathbf{q}.\mathbf{b}\right) f_{NN}(\mathbf{q}) d\mathbf{q}, \tag{8.113}$$

and $\mathbf{s}$, $\mathbf{s}'$ are the projections of the coordinate vectors $\mathbf{r}$, $\mathbf{r}'$ of the nuclear densities on the plane perpendicular to the z-axis (beam-axis). For spherically symmetric ground-state densities Eq. (8.112) reduces to the expression

$$\chi_N(b) = \frac{1}{k_{NN}} \int_0^\infty dq q \tilde{\rho}_A(q) f_{NN}(q) \tilde{\rho}_B(q) J_0(qb), \tag{8.114}$$

where $\tilde{\rho}_i(q)$ are the Fourier transforms of the ground state densities.

The optical potential can also be obtained by using an inversion method of the eikonal phases. These phases might be chosen to fit the experimental data. In this approach, one uses the Abel transform [1]

$$U(r) = \frac{\hbar v}{\pi r} \frac{d}{dr} \int_r^\infty \frac{\chi(b)}{(b^2 - r^2)^{1/2}} b \, db. \tag{8.115}$$

This procedure leads to effective potentials which on the tail, where the process takes place, are very close to those obtained with phenomenological potentials.

From Table 8.2 we observe that at very high energies ($\gtrsim 500$ MeV/nucleon) the real part of the "$t\rho\rho$" potential tends to zero, i.e.,

$$U_N^{\text{opt}}(\mathbf{R}) \simeq -i\sigma_{NN} \frac{\hbar v}{2} \int \rho_1(\mathbf{r})\rho_2(\mathbf{R} + \mathbf{r}) d^3r \qquad [\gtrsim 500 \text{ MeV}]. \tag{8.116}$$

But at E_{lab}/nucleon $\gtrsim 1$ GeV the real part of the potential becomes repulsive. In any case, we can write

$$\sigma_R \simeq \int [1 - |S(\mathbf{b})|^2] d^2 b, \tag{8.117}$$

where

$$|S(\mathbf{b})|^2 = \exp\left[-\sigma_{NN} \int_{-\infty}^\infty dz \int \rho_1(\mathbf{r})\rho_2(\mathbf{R} + \mathbf{r}) d^3r\right]. \tag{8.118}$$

We thus see that the "t-$\rho\rho$" approximation recovers the optical limit of the Glauber approximation for large bombarding energies. It is worthwhile to mention that Eq. (8.118) has again a simple classical interpretation. The mean free path for a nucleon-nucleon collision is given by

$$\lambda_{NN}(\mathbf{R}) = \left[\sigma_{NN} \int \rho_1(\mathbf{r})\rho_2(\mathbf{R} + \mathbf{r}) d^3r\right]^{-1}. \tag{8.119}$$

Thus, the probability that the nuclei "survive" without a nucleon-nucleon collision is given by

$$P_{surv} = \exp\left[-\int \frac{dz}{\lambda_{NN}(\mathbf{R})}\right], \tag{8.120}$$

which is the sum of all survival probabilities $e^{-dz/\lambda}$ after the nuclei move through each path element dz along the trajectory (assumed to be a straight-line). This probabilistic interpretation is consistent with the concepts introduced in this section, which justifies calling the results obtained using eikonal wave functions by *semiclassical models*.

8.10 Medium effects of nucleon-nucleon scattering

Most practical studies of the medium corrections of nucleon-nucleon scattering are done by considering the effective two-nucleon interaction in infinite nuclear matter, or G-matrix, as a solution of

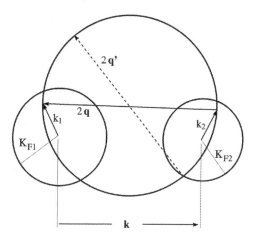

FIGURE 8.7 A nucleon from inside a Fermi sphere scatters off a nucleon from another Fermi sphere.

the Bethe-Goldstone equation [12]

$$\langle \mathbf{k}|G(\mathbf{P},\rho_1,\rho_2)|\mathbf{k}_0\rangle = \langle \mathbf{k}|v_{NN}|\mathbf{k}_0\rangle - \int \frac{d^3k'}{(2\pi)^3} \frac{\langle \mathbf{k}|v_{NN}|\mathbf{k}'\rangle Q(\mathbf{k}',\mathbf{P},\rho_1,\rho_2)\langle \mathbf{k}'|G(\mathbf{P},\rho_1,\rho_2)|\mathbf{k}_0\rangle}{E(\mathbf{P},\mathbf{k}') - E_0 - i\epsilon},$$

(8.121)

with $\mathbf{k}_0$, $\mathbf{k}$, and $\mathbf{k}'$ the initial, final, and intermediate relative momenta of the NN pair, $\mathbf{k} = (\mathbf{k}_1 - \mathbf{k}_2)/2$ and $\mathbf{P} = (\mathbf{k}_1 + \mathbf{k}_2)/2$. If energy and momentum is conserved in the binary collision, $\mathbf{P}$ is conserved in magnitude and direction, and the magnitude of $\mathbf{k}$ is conserved. v_{NN} is the nucleon-nucleon potential. E is the energy of the two-nucleon system, and E_0 is the same quantity on-shell. Thus $E(\mathbf{P},\mathbf{k}) = e(\mathbf{P} + \mathbf{k}) + e(\mathbf{P} - \mathbf{k})$, with e the single-particle energy in nuclear matter. It is also implicit in Eq. (8.121) that the final momenta $\mathbf{k}$ of the NN-pair also lie outside the range of occupied states.

Eq. (8.121) is density-dependent due to the presence of the Pauli projection operator Q, defined by

$$Q(\mathbf{k},\mathbf{P},\rho_1,\rho_2) = \begin{cases} 1, & \text{if } k_{1,2} > k_{F1,F2} \\ 0, & \text{otherwise.} \end{cases},$$

(8.122)

with $k_{1,2}$ the magnitude of the momenta of each nucleon. Q prevents scattering into occupied intermediate states. The Fermi momenta $k_{F1,F2}$ are related to the proton and neutron densities by means of the *local density approximation*, $k_{Fi} = (3\pi^2\rho_i/2)^{1/3}$. For finite nuclei, one usually replaces the local densities to obtain the local Fermi momenta. This is obviously a rough approximation, but very practical and extensively used in the literature.

Only by means of several approximations, Eq. (8.121) can be related to nucleon-nucleon cross sections. If one neglects the medium modifications of the nucleon-mass, and scattering though intermediate states, the medium modification of the NN cross sections can be accounted for by the geometrical factor Q only, i.e.

$$\sigma_{NN}(k,\rho_1,\rho_2) = \int \frac{d\sigma_{NN}^{free}}{d\Omega} Q(k,\rho_1,\rho_2)d\Omega,$$

(8.123)

where Q is now a simplified geometrical condition on the available scattering angles for the scattering of the NN-pair to unoccupied final states.

An usual approximation for the Pauli blocking is to assume that the effect of the Q operator is equivalent to a restricted angular integration in the domain (for symmetric nuclear matter)

$$\frac{k_F^2 - P^2 - k^2}{2Pk} \le \cos\theta \le \frac{P^2 + k^2 - k_F^2}{2Pk}.$$

(8.124)

The integral in Eq. (8.123) becomes zero if the upper limit is negative (as determined by the condition in Eq. (8.122)), whereas the full integration range is used if the upper limit is greater than 1. (Notice that the average angle θ in Eq. (8.124), namely the angle between the directions of $\mathbf{k}$ and $\mathbf{P}$, is also the colatitude of $\mathbf{k}$ in a coordinate system where the z-axis is along $\mathbf{P}$ and, thus, in such a reference frame it coincides with the scattering angle to be integrated over in Eq. (8.123)). The method of using Eq. (8.123), with (8.124), is not correct and misses an important part of the Pauli blocking geometry, as we show next.

In nuclear matter, the energy of a single particle with momentum p, which appears in the energy-denominator of Eq. (8.121), is also density-dependent and defined by

$$e(p) = T(p) + U(p), \tag{8.125}$$

where $T(p)$ is the kinetic energy and $U(p)$ is a single-particle potential generated by the average interaction of all the nucleons in the Fermi sea. For nucleons below and above the Fermi level one defines

$$U(p) = \langle p|U|p\rangle = \text{Re} \sum_{q \leq k_F} \langle pq|\text{G}|pq - qp\rangle, \tag{8.126}$$

with $|p\rangle$ and $|q\rangle$ single particle momentum, spin, and isospin states.

Thus, the propagator in Eq. (8.121) depends on $U(p)$ through the single-particle energy ϵ in Eq. (8.125). Consequently, the determination of G depends on the choice of $U(p)$. Since the potential $U(p)$ must be determined from the reaction matrix through Eq. (8.126), which depends on G, a solution for both quantities must be reached in a self-consistent way. From the self-consistently produced G-matrix, scattering parameters (positive energies) or the bound state properties (negative energies) can be predicted.

A proper geometric description of the Pauli operator Q was first studied in Ref. [13] where an analytical expression was obtained for the scattering of a nucleon on a nucleus described by a nucleon Fermi gas. By using the local density approximation, their procedure has been widely used to describe Pauli-blocking in nucleon-nucleus scattering. Much later, in Ref. [14], an expression was obtained for the geometrical Q operator for nucleus-nucleus scattering, both treated as Fermi gases. Besides reproducing the results when one of the gases reduces to a single nucleon, it can also be used to describe asymmetric nuclear matter, involving two Fermi spheres, one for the proton and another for the neutron. In contrast to Eq. (8.123), the generalized procedure also allows for the treatment NN-scattering with the relative momentum vector lying outside the symmetry axis of the two Fermi gas system.

The in-medium nucleon-nucleon cross section corrected by Pauli-blocking can be defined as

$$\sigma_{NN}(k, K_{F1}, K_{F2}) = \int \frac{d^3k_1 d^3k_2}{(4\pi K_{F1}^3/3)(4\pi K_{F2}^3/3)} \frac{2q}{k} \sigma_{NN}^{free}(q) \frac{\Omega_{Pauli}}{4\pi}, \tag{8.127}$$

where $\mathbf{k}$ is the relative momentum per nucleon of the nucleus-nucleus collision (Figure 8.7), and $\sigma_{NN}^{free}(q)$ is the free nucleon-nucleon cross section for the relative momentum $2\mathbf{q} = \mathbf{k}_1 - \mathbf{k}_2 - \mathbf{k}$, of a given pair of colliding nucleons. Clearly, Pauli-blocking enters through the restriction that $|\mathbf{k}_1'|$ and $|\mathbf{k}_2'|$ lie outside the Fermi spheres. From energy and momentum conservation in the collision, $\mathbf{q}'$ is a vector which can only rotate around a circle with center at $\mathbf{p} = (\mathbf{k}_1 + \mathbf{k}_2 + \mathbf{k})/2$. These conditions yield an allowed scattering solid angle given by

$$\Omega_{Pauli} = 4\pi - 2(\Omega_a + \Omega_b - \bar{\Omega}), \tag{8.128}$$

where Ω_a and Ω_b specify the excluded solid angles for each nucleon, and $\bar{\Omega}$ represents the intersection angle of Ω_a and Ω_b (Figure 8.8).

The solid angles Ω_a and Ω_b are easily determined. They are given by

$$\Omega_a = 2\pi(1 - \cos\theta_a), \qquad \Omega_b = 2\pi(1 - \cos\theta_b), \tag{8.129}$$

where $\mathbf{q}$ and $\mathbf{p}$ were defined above, $\mathbf{b} = \mathbf{k} - \mathbf{p}$, and

$$\cos\theta_a = (p^2 + q^2 - K_{F1}^2)/2pq, \qquad \cos\theta_b = (b^2 + q^2 - K_{F2}^2)/2bq. \tag{8.130}$$

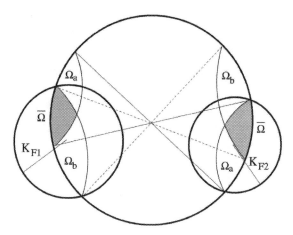

FIGURE 8.8 Due to the Pauli principle the scattering angle of the nucleons is restricted to lie outside the Fermi spheres.

The evaluation of $\bar{\Omega}$ is tedious but can be done analytically [14]. To summarize, there are two possibilities:

$$(1) \quad \bar{\Omega} = \Omega_i(\theta, \theta_a, \theta_b) + \Omega_i(\pi - \theta, \theta_a, \theta_b) , \quad \text{for } \theta + \theta_a + \theta_b > \pi, \tag{8.131}$$

$$(2) \quad \bar{\Omega} = \Omega_i(\theta, \theta_a, \theta_b) , \quad \text{for } \theta + \theta_a + \theta_b \leq \pi, \tag{8.132}$$

where θ is given by

$$\cos \theta = (k^2 - p^2 - b^2)/2pb. \tag{8.133}$$

The solid angle Ω_i has the following values

(a) $\quad \Omega_i = 0 , \quad \text{for } \theta \geq \theta_a + \theta_b,$ \hfill (8.134)

$$(b) \quad \Omega_i = 2 \left[\cos^{-1} \left(\frac{\cos \theta_b - \cos \theta \cos \theta_a}{\sin \theta_a (\cos^2 \theta_a + \cos^2 \theta_b - 2 \cos \theta \cos \theta_a \cos \theta_b)^{1/2}} \right) \right.$$

$$+ \cos^{-1} \left(\frac{\cos \theta_a - \cos \theta \cos \theta_b}{\sin \theta_b (\cos^2 \theta_a + \cos^2 \theta_b - 2 \cos \theta \cos \theta_a \cos \theta_b)^{1/2}} \right)$$

$$\left. - \cos \theta_a \cos^{-1} \left(\frac{\cos \theta_b - \cos \theta \cos \theta_a}{\sin \theta \sin \theta_a} \right) - \cos \theta_b \cos^{-1} \left(\frac{\cos \theta_a - \cos \theta \cos \theta_b}{\sin \theta \sin \theta_b} \right) \right]$$

$$\text{for } |\theta_b - \theta_a| \leq \theta \leq \theta_a + \theta_b, \tag{8.135}$$

(c) $\quad \Omega_i = \Omega_b \quad \text{for } \theta_b \leq \theta_a, \ \theta \leq |\theta_b - \theta_a| ,$ \hfill (8.136)

(d) $\quad \Omega_i = \Omega_a \quad \text{for } \theta_a \leq \theta_b, \ \theta \leq |\theta_b - \theta_a|.$ \hfill (8.137)

The integrals over $\mathbf{k}_1$ and $\mathbf{k}_2$ in Eq. (8.127) reduce to a five-fold integral due to cylindrical symmetry. Two approximations can be done which greatly simplify the problem: (a) on average, the symmetric situation in which $K_{F1} = K_{F2} \equiv K_F$, $q = k/2$, $p = k/2$, and $b = k/2$, is favored, (b) the free nucleon-nucleon cross section can be taken outside of the integral in Eq. (8.127). The assumption (a) implies that $\Omega_a = \Omega_b = \bar{\Omega}$, which can be checked using Eq. (8.137). One gets from (8.128) the simple expression

$$\Omega_{Pauli} = 4\pi - 2\Omega_a = 4\pi \left(1 - 2\frac{K_F^2}{k^2} \right). \tag{8.138}$$

Furthermore, the assumption (b) implies that

$$\sigma_{NN}(k, K_F) = \sigma_{NN}^{free}(k) \frac{\Omega_{Pauli}}{4\pi} = \sigma_{NN}^{free}(k) \left(1 - 2\frac{K_F^2}{k^2} \right). \tag{8.139}$$

The above equation shows that the in-medium nucleon-nucleon cross section is about $1/2$ of its free value for $k = 2K_F$, i.e., for $E/A \simeq 150$ MeV, in agreement with the numerical results of Ref. [9]. Since the effect of Pauli blocking at these energies is very large it is important to calculate the in-medium nucleon-nucleon scattering cross section according to Eq. (8.127), including the energy dependence of the free nucleon-nucleon cross sections.

The connection with the nuclear densities is accomplished through the local density approximation including a surface correction by means of

$$K_F^2 = \left[\frac{3\pi}{4} \rho(r) \right]^{2/3} + \frac{5}{2} \xi \left(\nabla \rho / \rho \right)^2 , \tag{8.140}$$

where $\rho(\mathbf{r})$ is the sum of the nucleon densities of each colliding nucleus at the position $\mathbf{r}$.

The second term is small and amounts to a surface correction, with ξ of the order of 0.1. Inserting Eq. (8.140) into Eq. (8.139), and using $E = \hbar^2 k^2 / 2m_N$, one gets (with $\bar{\rho} = \rho/\rho_0$)

$$\sigma_{NN}(E, \rho) = \sigma_{NN}^{free}(E) \left(1 + \alpha' \bar{\rho}^{2/3} \right) \quad \text{with} \quad \alpha' = -\frac{48.4}{E \text{ (MeV)}}, \tag{8.141}$$

where the second term of Eq. (8.140) has been neglected. This equation shows that the local density approximation leads to a density dependence proportional to $\bar{\rho}^{2/3}$. The Pauli principle yields a $1/E$ dependence on the bombarding energy. This behavior arises from a larger phase space available for nucleon-nucleon scattering with increasing energy.

The nucleon-nucleon cross section at $E/A \lesssim 200$ MeV decreases with E approximately as $1/E$. We thus expect that, in nucleus-nucleus collisions, this energy dependence is flattened by the Pauli correction, i.e., the in-medium nucleon-nucleon cross section is less dependent of E, for $E \lesssim 200$, than the free cross section. For higher values of E the Pauli blocking is less important and the free and in-medium nucleon-nucleon cross sections are approximately equal. These conclusions are in agreement with the experimental data for nucleus-nucleus reaction cross sections. Notice that, for $E/A = 100 - 200$ MeV, and $\rho \simeq \rho_0$, Eq. (8.141) yields a coefficient α' between -0.2 and -0.5.

A practical expression was obtained in Ref. [3] by fitting the numerical results obtained for the five-fold integral of Eq. (8.127) assuming an isotropic nucleon-nucleon cross section. This is a rough approximation because the anisotropy of the free NN cross section is markedly manifest at large energies. In the isotropic case, one obtains [3]

$$\sigma_{NN}(E, \rho_1, \rho_2) = \sigma_{NN}^{free}(E) \frac{1}{1 + 1.892 \left(\frac{|\rho_1 - \rho_2|}{\tilde{\rho}\rho_0} \right)^{2.75}} \begin{cases} 1 - \dfrac{37.02\tilde{\rho}^{2/3}}{E}, & \text{if } E > 46.27\tilde{\rho}^{2/3} \\[2mm] \dfrac{E}{231.38\tilde{\rho}^{2/3}}, & \text{if } E \leq 46.27\tilde{\rho}^{2/3}, \end{cases} \tag{8.142}$$

where E is the laboratory energy in MeV, $\tilde{\rho} = (\rho_1 + \rho_2)/\rho_0$, with $\rho_0 = 0.17$ fm^{-3}, and $\rho_i(\mathbf{r})$ is the local density at position $\mathbf{r}$ within nucleus i.

8.10.1   Comparison to elastic scattering data

In Figure 8.9 we show the data on the elastic scattering of ^{17}O projectiles with an energy of $E_{\text{lab}} = 84$ MeV/nucleon bombarding ^{208}Pb targets. Data are from Ref. [15]. The calculation was done using Eq. (8.74) together with the "$t\rho\rho$" approximation with the parameters for σ_{NN}, α_{NN}, ξ_{NN} as given in Table (8.2) and ground state densities from electron scattering experiments. The nucleon-nucleon cross sections σ_{NN}, α_{NN}, ξ_{NN} were averaged over isospin and over Pauli blocking, according to Eq. (8.127). We see that the approximation works extremely well (solid line). It should be said however that this system is not very sensitive to the optical potential since it is dominated by Coulomb scattering. Note that the vertical scale is a ratio of the elastic scattering cross section and the Rutherford cross section. At $\theta \sim 3°$ the cross section deviates from the Rutherford cross section and decreases rapidly. This is due to the strong absorption at small impact parameters. Any potential which is strongly absorptive (large imaginary part) at small impact parameters and rapidly decreases to zero at the strongly absorption radius will reproduce well the data.

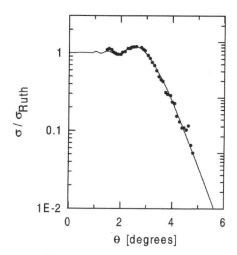

FIGURE 8.9 Elastic scattering of ^{17}O projectiles with an energy of $E_{\text{lab}} = 84$ MeV/nucleon bombarding ^{208}Pb targets.

A better test of the theory is provided by more "transparent" systems as, e.g., ^{12}C $+^{12}$ C. In Figure 8.9 we show the elastic scattering data of ^{12}C$+^{12}$C at 84 MeV/nucleon. The nucleon-nucleon scattering parameters were again calculated as before. The scattering is not dominated by Coulomb scattering as in the previous case. It is now much more sensitive to the optical potential chosen. The dashed curve is the one obtained with the "$t\rho\rho$" approximation. We see that the agreement is quite good at forward angles, but it fails at large angles. However, this is not a failure of the eikonal approximation but of a good optical potential, which in this case is not provided by the "$t\rho\rho$" approximation. To show this point we also plot in Figure 8.10 the result of an eikonal calculation with the same optical potential used in Ref. [15]. In fact, at these energies and for not a too large scattering angle ($\theta \lesssim 30°$) the eikonal approximation works very well.

It is instructive to decompose $f(\theta)$ into *"near"* and *"far" side* components. This is accomplished by first writing the Bessel function J_0 in Eq. (8.74) as

$$J_0(qb) = \frac{1}{2} \left[H_0^{(1)}(qb) + H_0^{(2)}(qb) \right], \qquad (8.143)$$

where $H_0^{1(2)}(qb)$ is the Hankel function of order zero and first (second) type. Asymptotically, these functions behave as running waves. With that the amplitude $f(\theta)$ can be written as $f(\theta) = f_{\text{near}}(\theta) + f_{\text{far}}(\theta)$, where $f_{\text{near}}(\theta)$ [or $f_{\text{far}}(\theta)$] is given by Eq. (8.74) with $J_0(qb)$ replaced by $\frac{1}{2}H_0^{(2)}(qb)$ [or $\frac{1}{2}H_0^{(1)}(qb)$]. The function $H^{(2)}(qb)$ is more sensitive to large values of b than $H^{(1)}(qb)$ does.

This fact is mainly due to the Coulomb interaction. In the limit when $\chi_C(qb)$ is negligible and $\chi_N(qb)$ is pure imaginary (no refraction) it is easy to see that the following relation holds (from the properties of the $H_0^{1(2)}$ functions)

$$f_{\text{near}}(\theta) = -f_{\text{far}}^*(\theta). \qquad (8.144)$$

The above results in an angular distribution, $f(\theta)$, that exhibits simple black-disk Fraunhofer diffraction patterns since the near and far amplitudes are equal in magnitude and interfere, as shown in Figure 8.11.

Back to Figure 8.9 we observe a small bump in the ratio-to-Rutherford cross section before the angular distribution goes down in magnitude. This is called the *nuclear rainbow* effect, discussed in Chapter 1. It is a situation characterized by the dominance of the far side component over the near side. In other words, as the impact parameter decreases the influence of the Coulomb interaction also diminishes and the nuclear force pushes the wave strongly (refracts strongly) to the other side

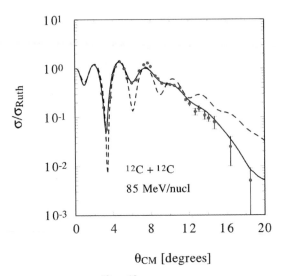

FIGURE 8.10   Elastic scattering data of ^{12}C+^{12}C at 84 MeV/nucleon.

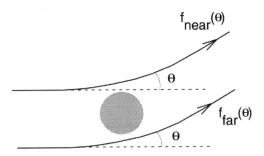

FIGURE 8.11   Near and far decomposition of the scattering amplitude.

of the nucleus interfering there with the other part of the wave. At very small angles one always encounters the opposite situation, namely, $f_{near}/f_{far} \gg 1$, owing to the influence, on the angular region, of Coulomb repulsion which affects mostly f_{near}.

Figure 8.12 shows the differential cross section, $d\sigma/dt$, versus the Mandelstam variable $t = -2p^2(1 - \cos\theta)$, for p^{4}He, p^{6}He and p^{8}He elastic scattering at energies $E_p = 628$ and 721 MeV, respectively. The calculated cross sections are based on the Glauber theory for elastic scattering using nucleon-nucleon scattering data as input [16]. The analysis of these data are useful to determine the nuclear matter distribution in ^{6}He and ^{8}He.

Dirac phenomenology

A very successful phenomenological approach was developed in Ref. [17] based on an optical potential consisting of two parts: $U_0(r)$, transforming like the time-like component of a Lorentz four-vector; and a second potential, $U_S(r)$, is a Lorentz scalar. U_0 and U_S are regarded as effective interactions arising from nucleons interacting via the meson exchange, folded with proton and neutron densities. They depend on the masses and coupling constants can be of the neutral vector ω and scalar σ bosons of the one boson exchange (OBE) model of the two-nucleon interaction. The Dirac equation for the scattering wave for the proton-nucleus system is

$$\left\{ \boldsymbol{\alpha} \cdot \mathbf{p} + \beta \left[m + U_S(r) \right] + \left[U_0(r) + V_C(r) \right] \right\} \Psi(\mathbf{r}) = E\Psi(\mathbf{r}), \tag{8.145}$$

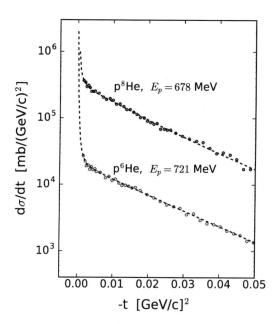

FIGURE 8.12 Differential cross sections, $d\sigma/dt$, versus the Mandelstam variable $t = -2p^2(1-\cos\theta)$, for p^{4}He, p^{6}He and p^{8}He elastic scattering at energies $E_p = 628$ and 721 MeV, respectively. The calculated cross sections are based on the Glauber theory for elastic scattering using nucleon-nucleon scattering data as input [16]. The analysis of these data are useful to determine the nuclear matter distribution in ^{6}He and ^{8}He.

where, m is the proton mass, E the nucleon total energy in the c.m. frame, and α_k ($k = 1, 2, 3$) and β are Dirac matrices. One problem with this method is the difficulty to disentangle the contributions of $\rho_0(r)$ from $\rho_S(r)$, including medium effects on the meson couplings and masses.

An effective Schrödinger equation can be obtained from this procedure, which still carries the essence of relativistic corrections and has been shown to be very successful to describe proton-nucleus scattering at high energies. The Dirac equation (8.145) can be rewritten as two coupled equations for the upper (Ψ_u) and lower (Ψ_l) components of the Dirac wave function, $\Psi(\mathbf{r})$. Keeping only the two upper components of the wave function, using the definition $\Psi_u = \sqrt{B}\phi$, where

$$B(r) = \frac{m + U_S + E - U_0 - V_C}{m + E},$$

one obtains the coupled equations

$$\left[p^2 + 2E\left(U_{cent} + U_{SO}\boldsymbol{\sigma} \cdot \mathbf{L}\right)\right]\phi(\mathbf{r}) = \left[(E - V_C)^2 - m^2\right]\phi(\mathbf{r}), \tag{8.146}$$

with

$$U_{cent} = \frac{1}{2E}\left[2EU_0 + 2mU_S - U_0^2 + U_S^2 - 2V_cU_0 + U_D\right], \tag{8.147}$$

where the Darwin potential is given by

$$U_D(r) = -\frac{1}{2r^2 B}\frac{\partial}{\partial r}\left(r^2\frac{\partial B}{\partial r}\right) + \frac{3}{4B^2}\left(\frac{\partial B}{\partial r}\right)^2, \tag{8.148}$$

and the spin-orbit potential by

$$U_{SO}(r) = -\frac{1}{2EBr}\frac{\partial B}{\partial r}.$$

The potentials U_0 and U_S are treated exactly in the same way as the non-relativistic optical potentials, usually parametrized in terms of a sum of real and imaginary Woods-Saxon potentials.

However, as an inheritance of their relativistic nature, their depths are quite different than those in low-energy scattering. Typically, e.g., for p+^{40}Ca at $E_p = 200$ MeV one has Re$U_0 \sim -350$ MeV, Im$U_0 \sim -100$ MeV, Re$U_S \sim -550$ MeV, and Im$U_S \sim -100$ MeV [17]. There are therefore big cancelations in Eq. (8.147), leading a central depth of U_{cent} compatible with the non-relativistic models. But relativity also introduces modifications in U_{cent} rendering them different than a simple Woods-Saxon form. Moreover, the approach is more consistent than in the non-relativistic case, as there is a prescription on how to get the spin-orbit potential out of U_0 and U_S. Hence in the Dirac approach, the spin-orbit potential appears naturally when one reduce the Dirac equation to a Schrödinger-like second-order differential equation, while in the non-relativistic Schrödinger approach, one has to insert the spin-orbit potential by hand.

Figure 8.13 shows the elastic p-^{40}Ca cross sections at 181 MeV. The curves are the results of the relativistic optical model analysis [17]. The agreement with the data is excellent. Many other examples can be found in the literature. At high energies the experimental data displays an exponential smooth decrease with angle due to the rather transparent charge distribution in the light nucleus.

Elastic scattering data is a simple way to access sizes, density profiles, and other geometric features of nuclei. For example, the beautiful exponential decrease of the cross sections with the nuclear diffuseness is clearly seen in elastic scattering data at large energies. In contrast, inelastic scattering requires many other pieces of information about the intrinsic nuclear properties and are sensitive to the models used to describe nuclear excitation. Often, the coupling to many excitation channels has to be considered. This contrasts to the nice features of elastic scattering as a probe of the nuclear geometry and density profiles.

8.11 Glauber theory of multiple scattering

The conditions of validity of the eikonal approximation are that the momentum and energy transfers in high-energy collisions are much smaller than the bombarding energy. However, the eikonal-Born approximation described in last section assumes that the transition $|i> \to |f>$ occurs in one step. Glauber [1] has shown how to treat the general problem of multi-step collisions using the eikonal approximation. The derivation is quite similar to the one leading to Eq. (8.74) and we only report the results here. The inelastic amplitudes (note that $f^{inel} = -\frac{\mu}{2\pi\hbar^2}T_{if}$)

$$f^{inel}(\theta) = \frac{k}{2\pi i} \int d^2b\, e^{i\mathbf{q}\cdot\mathbf{b}} \int d^3r'\, \psi_f^*(\mathbf{r}')\left[e^{i\chi(\mathbf{b}-\mathbf{s})} - 1\right]\psi_i(\mathbf{r}'), \tag{8.149}$$

where $\psi_i(\psi_f)$ denote the initial (final) internal wave functions of the projectile, or target (or both) and

$$\mathbf{s} = \mathbf{r}' - \hat{\mathbf{k}}(\hat{\mathbf{k}}\cdot\mathbf{r}') \tag{8.150}$$

is the component of $\mathbf{r}'$ perpendicular to the propagation direction $\hat{\mathbf{k}}$.

The advantage of using the Glauber amplitude (8.149) is to treat multiple collisions between the constituents of the nuclei. Then the amplitude is calculated from a single fundamental interaction, namely the nucleon-nucleon potential. Fortunately, this interaction is not needed and in most situations the only inputs needed are the nucleon-nucleon cross sections and nuclear ground-state densities, which are well known.

For an explicitly many-particle system we replace the single particle wave function $\psi(\mathbf{r})$ in (8.149) by the many-particle wave function

$$\psi(\mathbf{r}) \to \psi(\mathbf{r}_1, \mathbf{r}_2, \cdots, \mathbf{r}_n)$$

and the single-particle phase-shift function $\chi(\mathbf{b})$ by the phase-shift of the projectile wave function due to multiple collisions with the target nucleons. That is,

$$\chi(\mathbf{b}-\mathbf{s}) \to \sum_{j=1}^{n} \chi_j(\mathbf{b}-\mathbf{s}_j).$$

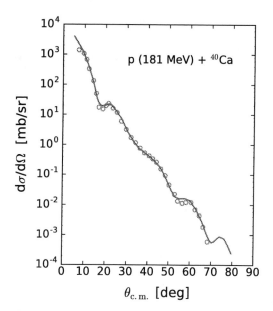

FIGURE 8.13 Elastic p-^{40}Ca cross sections at 181 MeV. The curves are the results of the relativistic Dirac phenomenology optical model analysis [17].

Thus, (8.149) becomes

$$f^{\text{inel}}(\theta) = \frac{k}{2\pi i} \int e^{i\mathbf{q}\cdot\mathbf{b}} d^2b \int \psi_f^*(\mathbf{r}_1,\ldots,\mathbf{r}_n) \left[e^{i\sum_{j=1}^n \chi_j(\mathbf{b}-\mathbf{s}_j)} - 1 \right] \psi_i(\mathbf{r}_1,\ldots,\mathbf{r}_n) \prod_j d^3r_j. \quad (8.151)$$

which is the well-known Glauber formula for high-energy collisions between composite particles. A lengthy manipulation of this formula with some simplifying assumptions, leads to Eq. (8.10). However, we note that Eq. (8.10), or the optical limit of the Glauber theory (e.g., the $t\rho\rho$ approximation) does not include *"shadowing"* effects, which are inherent in the Glauber formula (8.151). The simplest shadowing effect occurs in the scattering of a system composed by two particles, e.g., the deuteron. Applying Eq. (8.151) to it, one gets a term due to the scattering of the proton, another term for the scattering of the neutron, and a third term due to their interference. This last term is called the shadowing term.

Let us consider the effect of multiple scattering in the case of elastic scattering. The optical phase-shift function $\chi_{\text{el}}(\mathbf{b})$ is in this case included in a *profile function* Γ defined as

$$\Gamma(\mathbf{b}) = e^{i\chi_{\text{elast}}(\mathbf{b})} = \langle \psi_0^P \psi_0^T | \prod_{i\in P}\prod_{j\in T}[1 - \Gamma(\mathbf{b} + \mathbf{s}_i^P - \mathbf{s}_j^T)]|\psi_0^P \psi_0^T\rangle, \quad (8.152)$$

where $\mathbf{b}$ is the impact parameter, and $\psi_0^P (\psi_0^T)$ is the intrinsic projectile (target) ground state wave function. In Eq. (8.152) one integrates over all (independent) intrinsic coordinates $\mathbf{r}_i^P$ and $\mathbf{r}_j^T$. The two-dimensional vectors, $\mathbf{s}_i^P$ and $\mathbf{s}_j^T$, are the projections of $\mathbf{r}_i^P$ and $\mathbf{r}_j^T$ onto the xy-plane which is perpendicular to the incident direction of the projectile.

Usually Γ is not calculated from the nucleon–nucleon interaction, as we described in the previous section, but parametrized in a convenient form (like Gaussians) so as to fit the empirical nucleon–nucleon scattering amplitude using

$$\Gamma(\mathbf{b}) = \frac{\sigma_\tau}{4\pi\beta_\tau}(1 - i\alpha_\tau)\exp\left(-\frac{\mathbf{b}^2}{2\beta_\tau}\right), \quad (8.153)$$

where the parameters depend on the isospin of the nucleons (τ=pp, nn, pn), and are related to similar parameters appearing in Eq. (8.111). Other operators, like spin-orbit and tensor, may also be necessary (especially if the spins of the target and projectile are nonzero), but those are most often neglected.

As seen in Eq. (8.152), the optical phase-shift function is defined by a many-body multiple-scattering operator and obviously its calculation is very involved. One often uses some simplification like the *optical limit* (OL), which is just the first-order approximation

$$e^{i\chi_{OL}(\mathbf{b})} = \exp\left\{-\int\int d^3r_P d^3r_T \rho_P(\mathbf{r}_P)\rho_T(\mathbf{r}_T)\Gamma(\mathbf{s}_P - \mathbf{s}_T + \mathbf{b})\right\}. \tag{8.154}$$

A method of calculating the complete Glauber amplitude (8.152) uses wave functions in an expansion of the multiple-scattering operator and evaluates each term analytically provided that the profile function Γ is expressed by Gaussians. However, this method is limited by two things: one is that the number of terms in the expansion becomes very large for heavier nuclei; another is that if Γ is not a Gaussian then the analytic integration for correlated wave functions is difficult and has to resort to statistical methods such as *Monte Carlo integration*.

8.12   Relativistic Coulomb excitation

Inelastic scattering of heavy ions in intermediate energy collisions is an important tool to investigate the structure of stable and unstable nuclei. The angular distribution of the inelastically scattered fragments are particularly useful to identify unambiguously the multipolarity of the interaction, and consequently the spin and parities of the excited states. In order to describe correctly the angular distribution, absorption and diffraction effects have to be included properly.

Defining $\mathbf{r}$ as the separation between the center of mass of the two nuclei and $\mathbf{r}'$ as the intrinsic coordinate of the target nucleus, the inelastic scattering amplitude to first-order is given by

$$f(\theta) = \frac{ik}{2\pi\hbar v}\int d^3r d^3r' \left\langle \Phi_{\mathbf{k}'}^{(-)}(\mathbf{r})\phi_f(\mathbf{r}')\left|V_{int}(\mathbf{r},\mathbf{r}')\right|\Phi_{\mathbf{k}}^{(+)}(\mathbf{r})\phi_i(\mathbf{r}')\right\rangle, \tag{8.155}$$

where $\Phi_{\mathbf{k}'}^{(-)}(\mathbf{r})$ and $\Phi_{\mathbf{k}}^{(+)}(\mathbf{r})$ are the incoming and outgoing distorted waves, respectively, for the scattering of the center of mass of the nuclei, and $\phi(\mathbf{r}')$ is the intrinsic nuclear wave function of the target nucleus.

At intermediate energies, $\Delta E/E_{lab} \ll 1$, and forward angles, $\theta \ll 1$, we can use eikonal wave functions for the distorted waves; i.e.,

$$\Phi_{\mathbf{k}'}^{(-)*}(\mathbf{r})\Phi_{\mathbf{k}}^{(+)}(\mathbf{r}) = \exp\left\{-i\mathbf{q}.\mathbf{r} + i\chi(b)\right\}, \tag{8.156}$$

where $\chi(b)$ is the eikonal phase, Eq. (8.77), including the nuclear and Coulomb phase. In Eq. (8.155) the interaction potential, assumed to be purely Coulomb, is given by the retarded potential [18]

$$V_{int}(\mathbf{r},\mathbf{r}') = \frac{v^\mu}{c^2}j_\mu(\mathbf{r}')\frac{e^{i\kappa|\mathbf{r}-\mathbf{r}'|}}{|\mathbf{r}-\mathbf{r}'|}, \tag{8.157}$$

where $v^\mu = (c,\mathbf{v})$, with $\mathbf{v}$ equal to the projectile velocity, $\kappa = \omega/c$, and $j_\mu(\mathbf{r}')$ is the charge four-current for the intrinsic excitation of the target nucleus by an energy of $\hbar\omega$. Inserting Eqs. (8.156) and (8.157) in Eq. (8.155) one finds [19, 20]

$$f(\theta) = i\frac{Z_1 ek}{\gamma\hbar v}\sum_{\pi\lambda m} i^m \left(\frac{\omega}{c}\right)^\lambda \sqrt{2\lambda+1}e^{-im\phi}\Omega_m(q)G_{\pi\lambda m}\left(\frac{c}{v}\right)\langle I_f M_f|\mathcal{M}(\pi\lambda,-m)|I_i M_I\rangle \tag{8.158}$$

where $\pi\lambda (= E1, E2, \cdots, M1, M2, \cdots)$ denotes the multipolarity, $G_{\pi\lambda m}$ are the *Winther-Alder functions* [21]. The function $\Omega_m(q)$ is given by [20]

$$\Omega_m(q) = \int_0^\infty dbb J_m(qb)K_m\left(\frac{\omega b}{\gamma v}\right)\exp\left\{i\chi(b)\right\}, \tag{8.159}$$

where $q = 2k\sin(\theta/2)$ is the momentum transfer, θ and ϕ are the polar and azimuthal scattering angles, respectively.

For the $E1$, $E2$ and $M1$ multipolarity, the functions $G_{\pi\lambda m}(c/v)$ are given by [21]

$$G_{E11}(x) = -G_{E1-1}(x) = (1/3)\,x\sqrt{8\pi}; \quad G_{E10}(x) = -i\,(4/3)\,\sqrt{\pi(x^2-1)};$$
$$G_{M11}(x) = G_{M1-1}(x) = -i\,(1/3)\,\sqrt{8\pi}; \quad G_{M10}(x) = 0;$$
$$G_{E22}(x) = G_{E2-2}(x) = -\,(2/5)\,x\sqrt{\pi(x^2-1)/6};$$
$$G_{E21}(x) = -G_{E2-1}(x) = i\,(2/5)\,\sqrt{\pi/6}(2x^2-1);$$
$$G_{E20}(x) = (2/5)\,x\sqrt{\pi(x^2-1)}\,. \tag{8.160}$$

Using the Wigner-Eckart theorem, one can calculate the inelastic differential cross section from (8.158), using techniques similar to those discussed in Chapter 6. One obtains

$$\frac{d^2\sigma_C}{d\Omega dE_\gamma}\,(E_\gamma) = \frac{1}{E_\gamma}\sum_{\pi\lambda}\frac{dn_{\pi\lambda}}{d\Omega}\sigma_\gamma^{\pi\lambda}\,(E_\gamma)\,, \tag{8.161}$$

where $\sigma_\gamma^{\pi\lambda}\,(E_\gamma)$ is the photonuclear cross section for the absorption of a real photon with energy E_γ by nucleus 2, and $dn_{\pi\lambda}/d\Omega$ is the *virtual photon number*, which is given by [20]

$$\frac{dn_{\pi\lambda}}{d\Omega} = Z_1^2\alpha\left(\frac{\omega k}{\gamma v}\right)^2\frac{\lambda\,[(2\lambda+1)!!]^2}{(2\pi)^3(\lambda+1)}\sum_m |G_{\pi\lambda m}|^2|\Omega_m(q)|^2, \tag{8.162}$$

where $\alpha = e^2/\hbar c$.

The total cross section for Coulomb excitation can be obtained from Eqs. (8.161) and (8.162), using the approximation $d\Omega \simeq 2\pi qdq/k^2$, valid for small scattering angles and small energy losses. Using the closure relation for the Bessel functions, one obtains

$$\frac{d\sigma_C}{dE_\gamma}\,(E_\gamma) = \frac{1}{E_\gamma}\sum_{\pi\lambda}n_{\pi\lambda}\,(E_\gamma)\,\sigma_\gamma^{\pi\lambda}\,(E_\gamma)\,, \tag{8.163}$$

where the total number of virtual photon with energy $\hbar\omega$ is given by

$$n_{\pi\lambda}(\omega) = Z_1^2\alpha\frac{\lambda\,[(2\lambda+1)!!]^2}{(2\pi)^3(\lambda+1)}\sum_m |G_{\pi\lambda m}|^2 g_m(\omega), \tag{8.164}$$

and

$$g_m(\omega) = 2\pi\left(\frac{\omega}{\gamma v}\right)^2\int dbbK_m^2\left(\frac{\omega b}{\gamma v}\right)\exp\{-2\chi_I(b)\}\,, \tag{8.165}$$

where $\chi_I(b)$ is the imaginary part of the eikonal phase $\chi(b)$.

For very light heavy ion partners, the distortion of the scattering wave functions caused by the nuclear field is not important. This distortion is manifested in the diffraction peaks of the angular distributions, characteristic of strong absorption processes. If $Z_1Z_2\alpha \gg 1$, one can neglect the diffraction peaks in the inelastic scattering cross sections and a purely Coulomb excitation process emerges. One can gain insight into the excitation mechanism by looking at how the classical limit of the excitation amplitudes emerges from the general result of Eq. (8.162).

In Figure 8.14 we show (with $E_\gamma = \hbar\omega$) the virtual photons for the $E1$ multipolarity, "as seen" by a projectile passing by a lead target at impact parameters $b > 12.3$ fm, at three projectile energies. As the bombarding energy increases, virtual photons with larger energies become available for the reaction. The number of states accessed in the excitation process is concomitantly increased, which means that high-lying states such as giant resonances, $E_{GR} \sim 10 - 20$ MeV, can be excited by the Coulomb field.

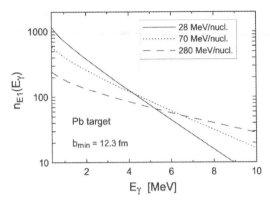

FIGURE 8.14 Number of virtual photons for the E1 multipolarity, as "seen" by a projectile flying by a lead target at impact parameters $b > 12.3$ fm, at three projectile energies.

8.12.1 Classical limit of the Coulomb excitation amplitudes

If we assume that Coulomb scattering is dominant and neglect the nuclear eikonal phase, we get

$$\Omega_m(q) \simeq \int_0^\infty db\, b\, J_m(qb) K_m\left(\frac{\omega b}{\gamma v}\right) \exp\{i\chi_C(b)\}. \tag{8.166}$$

This integral can be done analytically by rewriting it as

$$\Omega_m(q) = \int_0^\infty db\, b^{1+i2\eta}\, J_m(qb) K_m\left(\frac{\omega b}{\gamma v}\right), \tag{8.167}$$

where we used the simple form $\chi_C(b) = 2\eta \ln(kb)$, with $\eta = Z_1 Z_2 e^2/\hbar v$. The analytical result is

$$\Omega_m(q) = 2^{2i\eta} \frac{1}{m!} \Gamma(1+m+i\eta)\Gamma(1+i\eta)\Lambda^m \left(\frac{\gamma v}{\omega}\right)^{2+2i\eta} F\left(1+m+i\eta; 1+i\eta; 1+m; -\Lambda^2\right), \tag{8.168}$$

where

$$\Lambda = \frac{q\gamma v}{\omega}, \tag{8.169}$$

and F is the hypergeometric function.

The connection with the classical results may be obtained by using the low momentum transfer limit

$$J_m(qb) \simeq \sqrt{\frac{2}{\pi q b}} \cos\left(qb - \frac{\pi m}{2} - \frac{\pi}{4}\right) = \frac{1}{\sqrt{2\pi q b}}\left[e^{iqb} e^{-i\pi(m+1/2)/2} + e^{-iqb} e^{i\pi(m+1/2)/2}\right], \tag{8.170}$$

and using the stationary phase method, i.e.,

$$\int G(x) e^{i\phi(x)} dx \simeq \left(\frac{2\pi i}{\phi''(x_0)}\right)^{1/2} G(x_0) e^{i\phi(x_0)}, \tag{8.171}$$

where

$$\frac{d\phi}{dx}(x_0) = 0 \quad \text{and} \quad \phi''(x_0) = \frac{d^2\phi}{dx^2}(x_0). \tag{8.172}$$

This result is valid for a slowly varying function $G(x)$.

Only the second term in brackets of Eq. (8.170) will have a positive ($b = b_0 > 0$) stationary point. Thus,

$$\Omega_m(q) \simeq \frac{1}{\sqrt{2\pi q}} \left(\frac{2\pi i}{\phi''(b_0)}\right)^{1/2} \sqrt{b_0}\, K_m\left(\frac{\omega b_0}{\gamma v}\right) \exp\left\{i\phi(b_0) + i\frac{\pi(m+1/2)}{2}\right\}, \tag{8.173}$$

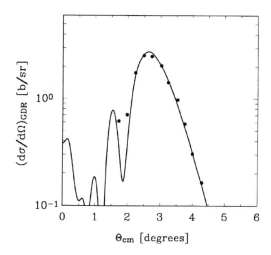

FIGURE 8.15 Cross section for the GDR excitation of ^{208}Pb for the ^{17}O+^{208}Pb reaction at 84A MeV, as a function of the center-of-mass scattering angle. Data are from Ref. [22].

where

$$\phi(b) = -qb + 2\eta \ln(kb). \tag{8.174}$$

The condition $\phi'(b_0) = 0$ implies

$$b_0 = \frac{2\eta}{q} = \frac{a_0}{\sin(\theta/2)}, \tag{8.175}$$

where $a_0 = Z_1 Z_2 e^2/\mu v^2$ is half the distance of closest approach in a classical head-on collision.

We observe that the relation (8.175) is the same [with $\cot(\theta/2) \sim \sin^{-1}(\theta/2)$] as that between impact parameter and deflection angle of a particle following a classical Rutherford trajectory. Also,

$$\phi''(b_0) = -\frac{2\eta}{b_0^2} = -\frac{q^2}{2\eta}, \tag{8.176}$$

which implies that in the classical limit

$$|\Omega_m(q)|^2_{s.c.} = \frac{4\eta^2}{q^4} K_m^2 \left(\frac{2\omega\eta}{\gamma vq}\right) = \frac{1}{k^2} \left(\frac{d\sigma}{d\Omega}\right)_{Ruth} K_m^2 \left(\frac{\omega a_0}{\gamma v \sin(\theta/2)}\right). \tag{8.177}$$

Using the above results, Eq. (8.162) becomes

$$\frac{dn_{\pi\lambda}}{d\Omega} = \left(\frac{d\sigma}{d\Omega}\right)_{Ruth} Z_1^2 \alpha \left(\frac{\omega}{\gamma v}\right)^2 \frac{\lambda\left[(2\lambda+1)!!\right]^2}{(2\pi)^3(\lambda+1)} \sum_m |G_{\pi\lambda m}|^2 K_m^2 \left(\frac{\omega a_0}{\gamma v \sin(\theta/2)}\right). \tag{8.178}$$

If strong absorption is not relevant, the above formula can be used to calculate the equivalent photon numbers. The stationary value given by Eq. (8.175) means that the important values of b which contribute to $\Omega_m(q)$ are those close to the classical impact parameter. Dropping the index 0 from Eq. (8.175), we can also rewrite Eq. (8.178) as

$$\frac{dn_{\pi\lambda}}{2\pi b db} = Z_1^2 \alpha \left(\frac{\omega}{\gamma v}\right)^2 \frac{\lambda\left[(2\lambda+1)!!\right]^2}{(2\pi)^3(\lambda+1)} \sum_m |G_{\pi\lambda m}|^2 K_m^2 \left(\frac{\omega b}{\gamma v}\right), \tag{8.179}$$

which is equal to the classical expression given in Ref. [19].

For very forward scattering angles, such that $\Lambda \ll 1$, a further approximation can be made by setting the hypergeometric function in Eq. (8.168) equal to unity, and one obtains

$$\Omega_m(q) = 2^{2i\eta} \frac{1}{m!} \Gamma(1+m+i\eta)\Gamma(1+i\eta)\Lambda^m \left(\frac{\gamma v}{\omega}\right)^{2+2i\eta}. \tag{8.180}$$

The main value of m in this case will be $m = 0$, for which one gets

$$\Omega_0(q) \simeq 2^{2i\eta}\Gamma(1+i\eta)\Gamma(1+i\eta)\left(\frac{\gamma v}{\omega}\right)^{2+2i\eta} = -\eta^2 2^{2i\eta}\Gamma(i\eta)\Gamma(i\eta)\left(\frac{\gamma v}{\omega}\right)^{2+2i\eta}, \tag{8.181}$$

and

$$|\Omega_0(q)|^2 = \eta^4\left(\frac{\gamma v}{\omega}\right)^4 \frac{\pi^2}{\eta^2 \sinh^2(\pi\eta)}, \tag{8.182}$$

which, for $\eta \gg 1$, results in

$$|\Omega_0(q)|^2 = 4\pi^2\eta^2\left(\frac{\gamma v}{\omega}\right)^4 e^{-2\pi\eta}. \tag{8.183}$$

This result shows that in the absence of strong absorption and for $\eta \gg 1$, Coulomb excitation is strongly suppressed at $\theta = 0$. This also follows from semiclassical arguments, since $\theta \to 0$ means large impact parameters, $b \gg 1$, for which the action of the Coulomb field is weak.

8.12.2   Coulomb excitation of giant dipole resonances

The photo-nuclear cross sections σ_γ^{GDR} can be parametrized by a Lorentzian shape

$$\sigma_\gamma^{GDR}(E) = \sigma_0 \frac{E^2\Gamma^2}{(E^2 - E_{GDR}^2)^2 + E^2\Gamma^2}, \tag{8.184}$$

where $E_{GDR} = 31.2A_P^{-1/3} + 20.6A_P^{-1/6}$ is a fit to the mass dependence of the centroid of the experimentally observed GDR. It is a mixture of the mass dependence predicted by the hydrodynamical Goldhaber-Teller and Steinwedel-Jensen models [35, 24]. The parameter σ_0 can be chosen to yield the Thomas-Reiche-Kuhn (TRK) sum rule

$$\int dE\sigma_\gamma^{GDR}(E) = 60\frac{NZ}{A} \text{ MeV mb}, \tag{8.185}$$

which is a nearly model independent results for the nuclear response to a dipole operator.

The width Γ of the GDR is more complicated to explain. It has a strong dependence on the nuclear shell structure. Experimental systematics provides widths ranging within $4 - 5$ MeV for a closed shell nucleus and can grow to 8 MeV for a nucleus between closed shells. One can also use a simple phenomenological parameterization of the GDR width in the form, $\Gamma_{GDR} = 2.51 \times 10^{-2}E_{GDR}^{1.91}$ MeV, with E_{GDR} in units of MeV. The centroid of the ISGQR can be taken as $E_{ISGQR} = 62/A_P^{1/3}$ MeV.

We consider a projectile of ^{17}O with bombarding energy of $E_{lab} = 84A$ MeV exciting a giant dipole resonance (GDR) on the target nucleus ^{208}Pb. One can use a standard optical potential and a Lorentzian parameterization for the photoabsorption cross section of ^{208}Pb. Inserting this form into Eq. (8.163) and doing the calculations implicit in Eq. (8.162) for $dn_{E1}/d\Omega$, one obtains the angular distribution which is compared to the data in Figure 8.15. The agreement with the data is excellent, provided one adjusts the overall normalization to a value corresponding to 93% of the energy weighted sum rule (EWSR) in the energy interval $7 - 18.9$ MeV (see section 6.10). Taking into account the $\pm 10\%$ uncertainty in the absolute cross sections quoted in Ref. [22], this is consistent with photoabsorption cross section in that energy range.

To unravel the effects of relativistic corrections, one can repeat the previous calculations unplugging the factor $\gamma = (1 - v^2/c^2)^{-1}$ which appears in the expressions (8.164) and (8.165) and using the non-relativistic limit of the functions G_{E1m} of Eq. (8.160). These modifications eliminate the relativistic corrections on the interaction potential. The result of this calculation is shown in Figure 8.16 (dotted curve). For comparison, we also show the result of a full calculation, keeping the relativistic corrections (dashed curve). One observes that the two results have approximately the same pattern, except that the non-relativistic result is slightly smaller than the relativistic one. In fact, if one repeats the calculation for the excitation of GDR using the non-relativistic limit of Eqs. (8.164) and (8.165), one finds that the best fit to the data is obtained by exhausting 113% of the EWSR. This value is very close to the 110% obtained in Ref. [22].

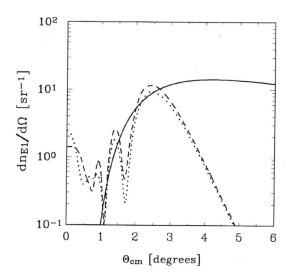

FIGURE 8.16 Virtual photon numbers for the electric dipole multipolarity generated by 84A MeV ^{17}O projectiles incident on ^{208}Pb, as a function of the center-of-mass scattering angle. The solid curve is a classical calculation. The dashed and dotted curves are eikonal calculations with and without relativistic corrections, respectively.

In Figure 8.16 we also show the result of a classical calculation (solid curve) for the GDR excitation in lead, using Eq. (8.178) for the virtual photon numbers. One observes that the classical curve is not able to fit the experimental data. This is mainly because diffraction effects and strong absorption are not included. But the classical calculation displays the region of relevance for Coulomb excitation. At small angles the scattering is dominated by large impact parameters, for which the Coulomb field is weak. Therefore the Coulomb excitation is small and the classical approximation fails. It also fails in describing the large angle data (dark-side of the rainbow angle), since absorption is not treated properly. One sees that there is a "window" in the inelastic scattering data near $\theta = 2 - 3°$ in which the classical and full calculations give approximately the same cross section.

In Figure 8.17 we show a similar calculation, but for the excitation of the GDR in Pb for the collision ^{208}Pb + ^{208}Pb at 640A MeV. The dashed line is the result of a classical calculation. Here we see that a purely classical calculation, is able to reproduce the quantum results up to a maximum scattering angle θ_m, at which strong absorption sets in. This justifies the use of classical calculations for heavy systems, even to calculate angular distributions. The cross sections increase rapidly with increasing scattering angle, up to an approximately constant value as the maximum Coulomb scattering angle is neared. This is explained as follows. Very forward angles correspond to large impact parameter collisions in which case $\omega b/\gamma v > 1$, the virtual photon numbers in Eq. (8.164) drop quickly to zero, and the excitation of giant resonances in the nuclei is not achieved. As the impact parameter decreases, increasing the scattering angle, $\omega b/\gamma v \lesssim 1$ and excitation occurs.

As discussed above, the classical result works for large Z nuclei, or small excitation energies, and for relativistic energies where the approximation of Eq. (8.166) is justified. However, angular distributions are not useful at relativistic energies since the scattering is concentrated at extremely forward angles. The quantity of interest in this case is the total inelastic cross section. If we use a *sharp-cutoff model (or black-sphere model)* for the strong absorption, so that $\chi_I(b) = \infty$ for $b < b_{min}$ and 0 otherwise, then Eqs. (8.164) and (8.165) yield the same result as an integration of the classical expression, Eq. (8.179), from b_{min} to ∞, as shown in Ref. [19].

One of the most dramatic findings in the application of relativistic Coulomb excitation was the discovery of the Double Giant Dipole Resonance (DGDR). The excitation of the DGDR can be

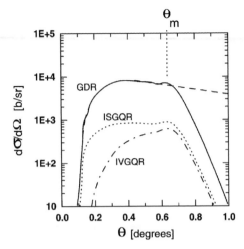

FIGURE 8.17   Differential cross section for the excitation of a giant dipole resonance in Pb+Pb collisions at 640 MeV/nucleon, as a function of the center of mass scattering angle.

described as a two-step mechanism induced by the Coulomb field [19]. In Figure 8.18 we show the result of one of these experiments [25], which detected neutron decay channels of giant resonances excited with relativistic projectiles. The excitation spectrum of relativistic ^{136}Xe projectiles incident on Pb are compared with the spectrum using C targets. The comparison proves that nuclear contribution to the excitation is small for large-Z targets.

It was earlier on recognized [19] that the electromagnetic excitation of nuclei in relativistic heavy ion collisions lead to large cross sections because of the excitation of giant resonances. The decay of these resonances, including the DGDR, leads to several decay channels, such as xn evaporation and fission. This was demonstrated by comparison of experiment to theory in Ref. [26]. Therefore, EM dissociation leads to large interaction cross sections, which can easily measure in inverse kinematics. In Figure 8.19 we show the cross section for electromagnetic dissociation of ^{238}U [26]. The curves show theoretical calculations using two sets of experimental GDR parameters as an input. It is clear that, except for light targets, one should expect that EM dissociation will always be relevant in any nucleus-nucleus collision at relativistic energies and might be either used as a spectroscopic tool, or become a background for processes where the effects of the strong interaction are studied.

8.13   Nuclear excitation

As we have seen in Chapter 6 the deformed potential model is based on the surface peaked assumption for the transition density. The inelastic scattering amplitudes can be calculated according to Eqs. (6.229), (6.230) and (6.235). Using the eikonal approximation, $\phi_{\mathbf{k}}^{(-)*}(\mathbf{r})\phi_{\mathbf{k}_0}^{(+)}(\mathbf{r}) \cong e^{i\mathbf{q}\cdot\mathbf{r}+i\chi(b)}$, the integrals in Eqs. (6.229), (6.230) and (6.235) become

$$I_{\lambda\mu} = \int d^3r\, F_\lambda(r) Y_{\lambda\mu}(\hat{\mathbf{r}}) e^{i\mathbf{q}\cdot\mathbf{r}+i\chi(b)}, \tag{8.186}$$

where

$$F_\lambda(r) = \begin{cases} 3U_0 + r\dfrac{dU_0}{dr} & , \quad \lambda = 0 \\[2ex] \dfrac{dU_0}{dr} & , \quad \lambda > 0. \end{cases} \tag{8.187}$$

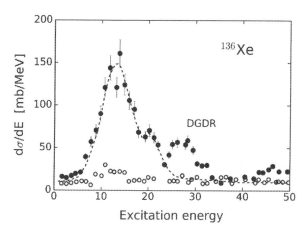

FIGURE 8.18 Experimental cross sections for the excitation of ^{136}Xe (700 MeV/nucleon) projectiles incident on lead (solid circles) and carbon targets (open circles). The dashed curve is a calculation including the excitation of isoscalar and isovector giant quadrupole resonances and the isovector giant dipole resonance (IVGDR). Altogether, these resonances compose the large bump in the spectrum. The double giant dipole resonance (IVGDR) is identified as the bump at double the energy of the IVGDR. Data are from Ref. [25].

We can rewrite Eq. (8.186) as

$$I_{\lambda\mu} = \sqrt{\pi(2\lambda+1)}\sqrt{\frac{(\lambda-\mu)!}{(\lambda+\mu)!}} i^\mu \int_0^\infty db \cdot bJ_\mu(q_t b)e^{i\chi(b)} \int_{-\infty}^\infty P_{\lambda\mu}\left(\frac{z}{\sqrt{b^2+z^2}}\right) F_\lambda(b,z)e^{iq_l z}dz,$$

(8.188)

with

$$q_l = k_0 - k_\lambda \cos\theta \cong k_0 - k_\lambda \cong \frac{\omega\lambda}{v} \quad \text{and} \quad q_t \cong 2\sqrt{k_0 k_\lambda}\sin\frac{\theta}{2},$$

(8.189)

where θ is the scattering angle and we used the mean wavenumber $\langle k \rangle = \sqrt{k_0 k_\lambda}$ to compute q_t. Thus, to compute the inelastic scattering amplitudes in the deformed potential model one needs to calculate two simple integrals. The scattering amplitudes will depend on the optical potential parameters and on deformation length δ_λ.

It is also instructive to consider the calculation of the scattering amplitude, or transition matrix in terms of the transition densities instead of transition potentials (e.g., the deformed potential model). We can write the transition matrix element as

$$T_{\lambda\mu} = \int d^3R \int d^3r \phi_{\mathbf{k}}^{(-)*}(\mathbf{R})U_{\text{int}}\left(|\mathbf{R}-\mathbf{r}|\right)\delta\rho_{\lambda\mu}(\mathbf{r})\phi_{\mathbf{k}_0}^{(+)}(\mathbf{R}),$$

(8.190)

where $\delta\rho_{\lambda\mu} = \psi_{\lambda\mu}^*\psi_0$ is the transition density. Instead of using the deformed potential model we can think of $U_{\text{int}}\left(|\mathbf{R}-\mathbf{r}|\right)$ as the potential between each nucleon and the projectile. That is, the transition $|0\rangle \rightarrow |\lambda\mu\rangle$ is caused by the (target-nucleon) - projectile interaction.

A link between the deformed potential model and the folding model is obtained by using the standard vibrational model presented in the Section 6.10. We can write

$$\delta\rho_\lambda = -\begin{cases} \dfrac{\delta_\lambda}{\sqrt{2\lambda+1}}\dfrac{d\rho_o}{dr} & \text{for} \quad \lambda \geq 1 \\ \alpha_0\left(3\rho_0 + r\dfrac{d\rho_0}{dr}\right) & \text{for} \quad \lambda = 0. \end{cases}$$

(8.191)

As in the deformed potential model, the scattering amplitude is determined by the optical potential parameters and the deformation parameters δ_λ and α_0. Isovector excitation are obtained by multiplying the above densities by $Q_\lambda^{(n)} + Q_\lambda^{(p)}$ (see Eq. (6.233) and the remark after Eq. (6.235)).

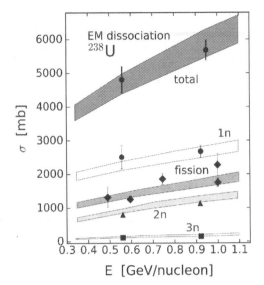

FIGURE 8.19   Cross section for electromagnetic dissociation of ^{238}U. The Coulomb fission cross sections (diamonds), measured with ^{208}Pb beams on ^{238}U-targets as well as the multiple neutron decay channels, xn, and fission cross sections obtained in inverse kinematics with ^{238}U beams. The data obtained with Au-targets were scaled to the data obtained with Pb targets. The curves show theoretical calculations using two sets of experimental GDR parameters as an input [26].

Another commonly used model for $\delta\rho_\lambda$ is the *Tassie Model* [23] which uses the form

$$\delta\rho_\lambda(r) = -\frac{\delta_\lambda}{\sqrt{2\lambda+1}} \left(\frac{r}{R_0}\right)^{\lambda-1} \frac{d\rho_0}{dr}, \qquad \lambda \geq 1. \tag{8.192}$$

where R_0 is the nuclear radius. For $\lambda = 0$, one can use (8.191). In general, both models yield basically the same transition density for heavy nuclei and low lying collective states. The Tassie model is a variant of the standard vibrational model (Section 6.10) and is more frequently used than the former (also known as the Bohr-Mottelson Model).

The *deformed potential model* yields the transition potentials given by

$$\delta\rho_0(r) = -\alpha_0 \left(3U + r\frac{d}{dr}\right)\rho_0(r), \tag{8.193}$$

for isoscalar monopole excitations,

$$\delta\rho_1^{IV}(r) = \beta_1^{IV} \left(\frac{d}{dr} + \frac{1}{3}R\frac{d^2}{dr^2}\right)\rho_0(r), \tag{8.194}$$

for isovector dipole excitations, and

$$\delta\rho_\lambda(r) = \beta_\lambda \frac{d}{dr}\rho_0(r), \tag{8.195}$$

for isoscalar giant quadrupole and higher multipoles. α_0 and β_i are the vibrational amplitudes. R is the nuclear radius at $1/2$ the central nuclear density and $\rho_0(r)$ are the ground state densities.

The corresponding transition potentials are

$$U_0(r) = -\alpha_0 \left(3 + r\frac{d}{dr}\right)U(r), \tag{8.196}$$

for isoscalar monopole,

$$U_1^{IV}(r) = \delta_1 \left(\frac{d}{dr} + \frac{1}{3}R\frac{d^2}{dr^2}\right)U(r), \tag{8.197}$$

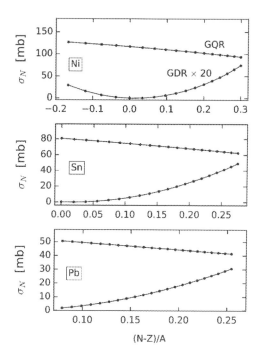

FIGURE 8.20 Nuclear excitation cross sections of ISGQR (GQR) and IVGDR (GDR) resonances in nickel, tin and lead projectiles bombarding carbon targets at 1 GeV/nucleon. The dependence on the asymmetry coefficient $(N - Z)/A$ of the projectile is shown. The upper curves in each frame display the excitation of ISGQR and the lower ones are calculations for the excitation of IVGDR multiplied by 20.

for isovector dipole, and

$$U_\lambda(r) = \delta_\lambda \frac{dU(r)}{dr}, \tag{8.198}$$

for $\lambda \geq 2$. In these equations, $\delta_i = \beta_i R$, are known as deformation lengths.

The deformation parameters entering the equations above are obtained from sum-rules, as discussed in the end of Section 6.10. But, for dipole isovector excitations the transition potential in is modified from Eq. (8.197) to account for the isospin dependence. It becomes

$$U_1^{IV}(r) = -\Lambda\left(\frac{N - Z}{A}\right)\left(\frac{dU}{dr} + \frac{1}{3}R\frac{d^2U}{dr^2}\right), \tag{8.199}$$

where the factor Λ depends on the difference between the proton and the neutron matter radii as

$$\Lambda\frac{2(N - Z)}{3A} = \frac{R_n - R_p}{\frac{1}{2}(R_n + R_p)} = \frac{\Delta r_{np}}{R}. \tag{8.200}$$

The strength of isovector excitations increases with the *neutron skin* Δr_{np} which is larger for neutron-rich nuclei.

As shown in Ref. [28], for the isoscalar giant dipole resonance, the transition density in the deformed model is expressed as

$$\delta\rho^{(IS)}(r) = -\frac{\beta_1^{(IS)}}{R\sqrt{3}}\left[3r^2\frac{d}{dr} + 10r - \frac{5}{3}\left\langle r^2\right\rangle\frac{d}{dr} - \epsilon\left(r\frac{d^2}{dr^2} + 4\frac{d}{dr}\right)\right]\rho_0(r), \tag{8.201}$$

where

$$\epsilon = \left(\frac{4}{E_2} + \frac{5}{E_0}\right)\frac{\hbar^2}{3mA}, \tag{8.202}$$

with A being the nuclear mass, E_2 and E_0 are the excitation energies of the giant isoscalar quadrupole and monopole resonances, respectively. In Eq. (8.201) $\langle\rangle$ means average value.

The collective coupling parameter for the isoscalar dipole resonance is

$$\beta_1^{(IS)} = \frac{6\pi\hbar^2 R^2}{m_N A E_x}\left(11\left\langle r^4\right\rangle - \frac{25}{3}\left\langle r^2\right\rangle^2 - 10\epsilon\left\langle r^2\right\rangle\right). \tag{8.203}$$

Analogously, the transition potential can be written, for isoscalar dipole excitations

$$U_1^{(IS)}(r) = -\frac{\delta_1^{(IS)}}{R\sqrt{3}}\left[3r^2\frac{d}{dr} + 10r - \frac{5}{3}\left\langle r^2\right\rangle\frac{d}{dr} - \epsilon\left(r\frac{d^2}{dr^2} + 4\frac{d}{dr}\right)\right]U(r), \tag{8.204}$$

Using the deformed potential model, we show in Figure 8.20 the cross sections for nuclear excitation of ISGQR (GQR) and IVGDR (GDR) resonances in nickel, tin and lead projectiles incident on carbon targets at 1 GeV/nucleon. The dependence on the asymmetry coefficient $\delta = (N-Z)/A$ of the projectile is displayed. The upper curves in each frame are calculations for the ISGQR and the lower curves are for the excitation of IVGDR multiplied by a factor of 20. For ^{208}Pb projectiles the cross sections are of the order of 43 mb (1.11) mb, for the ISGQR (IVGDR).

The IVGDR cross sections are negligible for $N = Z$ with a negligible neutron skin, as the deformation parameter δ_1 is directly proportional to the neutron skin $\Delta R = R_n - R_p$ (see last Section of Chapter 6). Light nickel isotopes exhibit a non-zero proton skin and a reversing trend of the IVGDR excitation cross section around $\delta = 0$ is observed. The cross sections for IVGDR resonances are more than a factor 20 smaller than for ISGQR resonances. Hence, they are unimportant for the purposes of extracting neutron skins at such bombarding energies. Nonetheless, the method has been used previously at lower energies, below 100 MeV/nucleon, by measuring differential cross sections that can display marked differences between angular distributions for $L = 1$ and $L = 2$. The ISGQR cross sections decrease along an isotopic chain as the neutron numbers increase. This can be understood as due to the decrease of the deformation parameter δ_2 with the increase of the ISGQR centroid energy with mass number, as inferred from Eq. (6.211).

Larger theoretical uncertainties in treating nuclear excitations in high-energy collisions exist as compared to Coulomb excitation. Whereas the Coulomb interaction is well known, optical potentials used in the deformed potential model, described in Section 6.11, are not so constrained. Little can be done to improve these models with the state of the art knowledge of high-energy nuclear reactions. The deformed potential (Section 6.11) and the Tassie model, Eq. (8.192), should be considered rough approximations for nuclear reactions at high energies.

The deformed potential model used in collisions with a proton target yields cross sections that similar to those shown in Figure 8.20. Use different targets does not display appreciable changes in the nuclear excitation of giant resonances. A noticeable variation of the Coulomb excitation is possible. By varying the bombarding energies in the range 100–1000 MeV/nucleon will not help either because the cross sections for nuclear excitation remain practically unchanged. Thus, the 50 mb to 100 mb of nuclear excitation cross sections mainly contributing to the one-neutron decay channel will be hard to control systematically without adding other observables to the angle integrated cross sections. But, as discussed previously [27], simulations show that nuclear excitation events can be separated using the angular distribution of neutrons.

8.14   Charge-exchange reactions

Charge exchange reactions, i.e. (p, n), (n, p) reactions, are an important tool in nuclear structure physics, providing a measure of the Gamow-Teller strength function in the nuclear excitation spectrum. On microscopic grounds charge exchange is accomplished through charged meson exchange, mainly π- and ρ-exchange. It is well known that neutron-proton scattering at backward angles results from small angle (low momentum transfer) charge-exchange, and is one of the main pieces of evidence for the pion exchange picture of the nuclear force. The width of the peak is roughly given by the exchanged pion momentum divided by the beam momentum.

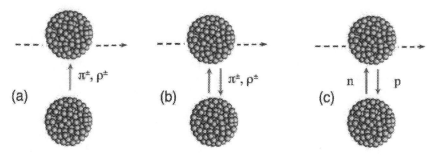

FIGURE 8.21 (a) One step charge exchange by means of $\pi^{\pm}$ and $\rho^{\pm}$ exchange. (b) Two step charge exchange by means of $\pi^{\pm}$ and $\rho^{\pm}$ exchange. (c) Two step charge exchange by means of neutron and proton exchange.

The DWBA scattering amplitude for inelastic processes in nucleus-nucleus collisions is given by

$$f_{aA \to bB}(\theta) = -\frac{m_0}{2\pi\hbar^2} \left\langle \chi_{\mathbf{k}'}^{(-)}(\mathbf{R})\phi_f(\mathbf{r}) \middle| V_{pt}(\mathbf{R},\mathbf{r}) \middle| \chi_{\mathbf{k}}^{(+)}(\mathbf{R})\phi_i(\mathbf{r}) \right\rangle, \qquad (8.205)$$

where $\chi_{\mathbf{k}',\mathbf{k}}^{(\pm)}$ are the (outgoing/incoming) distorted scattered waves for the c.m. motion of the two nuclei with reduced mass m_0 and $\phi_{i,f}(\mathbf{r})$ are the initial and final wave functions for the intrinsic nuclear motion, respectively. The c.m. of relative motion between the nuclei is described by the coordinate $\mathbf{R}$ and $\mathbf{r}$ is the intrinsic coordinate of the wave function for the nucleus of interest.

In Eq. 8.205, $V_{pt}(\mathbf{R},\mathbf{r})$ is the interaction potential between the two nuclei which leads to the inelastic process $aA \to bB$. It is taken as the interaction potential between a nucleon in nucleus t (A or B) at position $\mathbf{r}_t$, with a nucleon in nucleus p (a or b) at position $\mathbf{r}_p$. The meson exchange between these nucleons is responsible for the charge exchange process in nucleus-nucleus collisions at high energies ($E_{lab} > 50$ MeV/nucleon), for which the physical exchange of nucleons is negligible due to severe phase-space limitations. As shown in Figure 8.21 charge-exchange reactions can proceed via virtual pion and rho exchange in one or two step processes (a, b), whereas a physical nucleon exchange always requires at least a two step process (c). Therefore, pion and rho exchange is a more efficient way to induce charge exchange reactions with heavy ions, unless the reaction occurs at low energies of about 50 MeV/nucleon and lower. This fact also allows the reaction mechanism to be much simpler to handle because nucleon exchange is much harder to describe theoretically than pion or rho exchange.

Using eikonal wave functions, the DWBA amplitude becomes

$$f_{aA \to bB}(\theta) = -\frac{m_0}{2\pi\hbar^2} \sum_{p,t} \int d^3R \left\langle \phi^{(b)}(\mathbf{r}_p)\phi^{(B)}(\mathbf{r}_t) \middle| V_{pt}(\mathbf{r})S(b)\exp\left[-i\mathbf{q}\cdot\mathbf{R}\right] \middle| \phi^{(a)}(\mathbf{r}_p)\phi^{(A)}(\mathbf{r}_t) \right\rangle,$$

$$(8.206)$$

where the sum includes all pairs of nucleons, with one from nucleus p and another from nucleus t. The relative coordinates of the pair are related by $\mathbf{r} = \mathbf{R} + \mathbf{r}_p - \mathbf{r}_t$, $\mathbf{q} = \mathbf{k}' - \mathbf{k}$ is the momentum transfer, $\mathbf{k}$ and $\mathbf{k}'$ are the initial and final momentum in the center of mass, and $b = |\mathbf{R} \times \hat{\mathbf{k}}|$ is often interpreted as the impact parameter.

The nucleon-nucleon interaction was discussed in Section 4.2. If we make a simplistic approach and use the OBEP interaction with only pion and rho exchange, the following picture emerges. In momentum representation the pion+rho exchange potential is given by

$$V(\mathbf{q}) = -\frac{f_\pi^2}{m_\pi^2} \frac{(\boldsymbol{\sigma}_1 \cdot \mathbf{q})(\boldsymbol{\sigma}_2 \cdot \mathbf{q})}{m_\pi^2 + \mathbf{q}^2}(\boldsymbol{\tau}_1 \cdot \boldsymbol{\tau}_2) - \frac{f_\rho^2}{m_\rho^2} \frac{(\boldsymbol{\sigma}_1 \times \mathbf{q}) \cdot (\boldsymbol{\sigma}_2 \times \mathbf{q})}{m_\rho^2 + \mathbf{q}^2}(\boldsymbol{\tau}_1 \cdot \boldsymbol{\tau}_2), \qquad (8.207)$$

where the pion (rho) coupling constant is $f_\pi^2/4\pi = 0.08$ ($f_\rho^2/4\pi = 4.85$), $m_\pi c^2 = 145$ MeV, and $m_\rho c^2 = 770$ MeV.

The central part of the potential above has a zero-range component, which is a consequence of the point-like treatment of the meson-nucleon coupling. In reality the interaction extends over a finite region of space, so that the zero range force must be replaced by an extended source function. This can be done by adding a short-range interaction defined at $q = 0$ in terms of the *Landau-Migdal parameters* g'_π and g'_ρ. We will use $g'_\pi = 1/3$ and $g'_\rho = 2/3$, which amounts to remove exactly the zero-range interaction. Since the ρ-exchange interaction is of very short-range, its central part is appreciably modified by the ω-exchange force. The effect of this repulsive correlation is approximated by multiplying V^{cent}_ρ by a factor $\xi = 0.4$ and leaving V^{tens}_ρ unchanged since the tensor force is little affected by ω-exchange [29].

With these modifications the *pion+rho exchange potential* can be written as

$$V(\mathbf{q}) = \frac{1}{(2\pi)^{3/2}} \int d^3 r e^{i\mathbf{q}\cdot\mathbf{r}} V(\mathbf{r})$$

$$= V_\pi(\mathbf{q}) + V_\rho(\mathbf{q}) = [v(\mathbf{q})(\boldsymbol{\sigma}_1 \cdot \hat{\mathbf{q}})(\boldsymbol{\sigma}_2 \cdot \hat{\mathbf{q}}) + w(\mathbf{q})(\boldsymbol{\sigma}_1 \cdot \boldsymbol{\sigma}_2)](\boldsymbol{\tau}_1 \cdot \boldsymbol{\tau}_2), \tag{8.208}$$

where

$$v(\mathbf{q}) = v^{tens}_\pi(\mathbf{q}) + v^{tens}_\rho(\mathbf{q}), \tag{8.209}$$

and

$$w(\mathbf{q}) = w^{cent}_\pi(\mathbf{q}) + \xi w^{cent}_\rho(\mathbf{q}) + w^{tens}_\pi(\mathbf{q}) + w^{tens}_\rho(\mathbf{q}), \tag{8.210}$$

with

$$v^{tens}_\pi(\mathbf{q}) = -J_\pi \frac{\mathbf{q}^2}{m^2_\pi + \mathbf{q}^2}, \qquad v^{tens}_\rho(\mathbf{q}) = J_\rho \frac{\mathbf{q}^2}{m^2_\rho + \mathbf{q}^2}, \tag{8.211}$$

$$w^{cent}_\pi(\mathbf{q}) = -\frac{1}{3} J_\pi \left[\frac{\mathbf{q}^2}{m^2_\pi + \mathbf{q}^2} - 3g'_\pi \right], \qquad w^{cent}_\rho(\mathbf{q}) = -\frac{2}{3} J_\rho \left[\frac{\mathbf{q}^2}{m^2_\rho + \mathbf{q}^2} - \frac{3}{2} g'_\rho \right], \tag{8.212}$$

$$w^{tens}_\pi(\mathbf{q}) = \frac{1}{3} J_\pi \frac{\mathbf{q}^2}{m^2_\pi + \mathbf{q}^2}, \qquad w^{tens}_\rho(\mathbf{q}) = -\frac{1}{3} J_\rho \frac{\mathbf{q}^2}{m^2_\rho + \mathbf{q}^2}. \tag{8.213}$$

The values of the coupling constants J_π and J_ρ in nuclear units are given by

$$J_\pi = \frac{f^2_\pi}{m^2_\pi} \equiv f^2_\pi \frac{(\hbar c)^3}{(m_\pi c^2)^2} \simeq 400 \text{ MeV fm}^3$$

$$J_\rho = \frac{f^2_\rho}{m^2_\rho} \equiv f^2_\rho \frac{(\hbar c)^3}{(m_\rho c^2)^2} \simeq 790 \text{ MeV fm}^3. \tag{8.214}$$

Turning off the terms $w^{cent}_{\pi,\rho}$, or $v^{tens}_{\pi,\rho}$ and $w^{tens}_{\pi,\rho}$, allows us to study the contributions from the central and the tensor interaction, and from π- and ρ-exchange, respectively. Of course this approach is too simplistic. For example, we know that the nucleon-nucleon charge-exchange interaction is modified by medium effects (e.g., Pauli blocking, as we discussed earlier).

The important parts of the interaction for the charge-exchange reaction are the central and the tensor components. In general, the spin-orbit part of the interaction produces a negligible contribution to the cross sections. Figure 8.22 shows the phenomenological nucleon-nucleon potential (in momentum space) at forward angles and the separate contributions from the spin-isospin, $\sigma\tau$, and the isospin, τ, part of the interaction [30]. This potential includes medium corrections and has been fitted to explain a large number of nuclear reactions at intermediate energies ($E_{lab} \sim 100$ MeV/nucleon). One sees that, at $E \sim 100-300$ MeV, the $\sigma\tau$ contribution is larger than the τ one. This hints to a favored energy region for studies of the Gamow-Teller matrix elements needed for nuclear physics, astrophysics, and physics beyond the standard model of particle physics.

The (in-medium) nucleon-nucleon interaction may be written as

$$V_{pt}(\mathbf{r}) = V^C(r) + V^C_\sigma(r)(\boldsymbol{\sigma}_p \cdot \boldsymbol{\sigma}_t) + \left[V^C_\tau(r) + V^C_{\sigma\tau}(r)(\boldsymbol{\sigma}_p \cdot \boldsymbol{\sigma}_t) \right](\boldsymbol{\tau}_t \cdot \boldsymbol{\tau}_p)$$

$$+ \left[V^T(r) + V^T_\tau(r)(\boldsymbol{\tau}_t \cdot \boldsymbol{\tau}_p) \right] S_{pt}(\hat{\mathbf{r}}) + V^{LS}(r) \mathbf{l} \cdot (\boldsymbol{\sigma}_t + \boldsymbol{\sigma}_p), \tag{8.215}$$

where the tensor operator is given by $S_{pt}(\hat{\mathbf{r}}) = 3(\boldsymbol{\sigma}_p \cdot \hat{\mathbf{r}})(\boldsymbol{\sigma}_t \cdot \hat{\mathbf{r}}) - (\boldsymbol{\sigma}_p \cdot \boldsymbol{\sigma}_t)$, and the spin-orbit term contains the angular momentum operator $\mathbf{l} = (\mathbf{r}_t - \mathbf{r}_p) \times (\mathbf{p}_t - \mathbf{p}_p)/\hbar$, often yielding a negligible

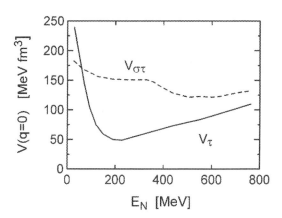

FIGURE 8.22 Nucleon-nucleon potential (in momentum space) at forward angles. The picture shows the separate contributions from the spin-isospin, $\sigma\tau$, and the isospin, τ, part of the interaction as a function of the laboratory energy.

contribution to the cross sections. The effective nucleon-nucleon potential has also to include an exchange operator to fully account for anti-symmetrization. Thus, the interaction (8.215) should be understood as $V_{pt}(\mathbf{r}) \to V_{pt}(\mathbf{r}) \left[1 + (-1)^l P^x\right]$, where P^x is the space exchange operator which changes $\mathbf{r} \to -\mathbf{r}$ on the right and $(-)^l$ (with $l =$ relative angular momentum in the nucleon-nucleon system) ensures anti-symmetrization.

Using a spherical basis ($\mu = 0, \pm 1$), the scalar products can be written as

$$\boldsymbol{\tau}_t \cdot \boldsymbol{\tau}_p = -\sum_\mu \sqrt{3} \langle 1\mu 1 - \mu|00\rangle \tau_t^{(\mu)} \tau_p^{(-\mu)} = \sum_\mu (-1)^\mu \tau_t^{(\mu)} \tau_p^{(-\mu)} \tag{8.216}$$

$$\boldsymbol{\sigma}_t \cdot \boldsymbol{\sigma}_p = -\sum_\mu \sqrt{3} \langle 1\mu 1 - \mu|00\rangle \sigma_t^{(\mu)} \sigma_p^{(-\mu)} = \sum_\mu (-1)^\mu \sigma_t^{(\mu)} \sigma_p^{(-\mu)}, \tag{8.217}$$

where

$$\tau^{(\pm 1)} = \mp \frac{1}{\sqrt{2}} \left(\tau_x \pm i\tau_y\right), \qquad \tau^{(0)} = \tau_z.$$

The $\boldsymbol{\tau}_t \cdot \boldsymbol{\tau}_p$ operator is responsible for isospin exchange via the combination of $\tau^\pm$ operators in Eq. (8.216). Likewise, the $\boldsymbol{\sigma}_t \cdot \boldsymbol{\sigma}_p$ is responsible for spin-flip interactions. Further, introducing $V^C(r) \equiv V_{00}^0(r), V_\tau^C(r) \equiv V_{01}^0(r), V_\sigma^C(r) \equiv V_{10}^0(r) V_{\sigma\tau}^C(r) \equiv V_{11}^0(r), V^T(r) \equiv V_{10}^2(r), V_\tau^T(r) \equiv V_{11}^2(r)$, Eq. (8.215) can be written in the following more compact form

$$V_{pt}(\mathbf{r}) = \sum_{(K=0,2)ST} V_{ST}^K(r) C_S^K Y_K(\widehat{\mathbf{r}}) \cdot \left[\boldsymbol{\sigma}_t^S \otimes \boldsymbol{\sigma}_p^S\right]^K \left[\boldsymbol{\tau}_t^T \cdot \boldsymbol{\tau}_p^T\right], \tag{8.218}$$

where $K = 0$ corresponds to the central force, $K = 2$ to the tensor force, and S, T to the spin and isospin labels of the force. The constants C_S^K have values $C_0^0 = (4\pi)^{1/2}$, $C_1^0 = -(12\pi)^{1/2}$, $C_0^2 = 0$, and $C_1^2 = (24\pi/5)^{1/2}$. When $S = 0$, $\boldsymbol{\sigma}^S$ is the unit operator, and when $S = 1$, it becomes the Pauli spin operator $\boldsymbol{\sigma}$. Likewise for the isospin operator $\boldsymbol{\tau}^T$. In Eq. 8.218 the following notation has been used

$$Y_K(\widehat{\mathbf{r}}) \cdot \left[\boldsymbol{\sigma}_t^S \otimes \boldsymbol{\sigma}_p^S\right]^K = \sum_\mu (-1)^\mu Y_{K,-\mu}(\widehat{\mathbf{r}}) \left[\boldsymbol{\sigma}_t^S \otimes \boldsymbol{\sigma}_p^S\right]^{K\mu}, \tag{8.219}$$

and

$$\left[\boldsymbol{\sigma}_t^S \otimes \boldsymbol{\sigma}_p^S\right]^{K\mu} = \sum_m \langle SmS(\mu - m)|K\mu\rangle \sigma_t^{S(m)} \sigma_p^{S(\mu-m)}. \tag{8.220}$$

Now, defining

$$\widetilde{V}_{ST}^K(p) = 4\pi \int V_{ST}^K(r) j_K(pr) r^2 dr, \tag{8.221}$$

it is straightforward to show that

$$f_{aA \to bB}(\theta) = -\frac{m_0}{(2\pi)^4 \hbar^2} \int d^3 p \, d^3 R \, S(b) \exp\left[-i(\mathbf{q}+\mathbf{p}) \cdot \mathbf{R}\right]$$

$$\times \left\langle \phi^{(b)}(\mathbf{r}_p) \phi^{(B)}(\mathbf{r}_t) \left| \exp\left[-i\mathbf{p} \cdot (\mathbf{r}_p - \mathbf{r}_t)\right] \widetilde{V}_{pt}(\mathbf{p}) \right| \phi^{(a)}(\mathbf{r}_p) \phi^{(A)}(\mathbf{r}_t) \right\rangle$$

$$= -\frac{m_0}{(2\pi)^4 \hbar^2} \sum_{(K=0,2)ST} i^K C_S^K \int d^3 p \, d^3 R \, S(b) \exp\left[-i(\mathbf{q}+\mathbf{p}) \cdot \mathbf{R}\right] \widetilde{V}_{ST}^K(p) Y_K(\widehat{\mathbf{p}})$$

$$\times \left\langle \phi^{(b)}(\mathbf{r}_p) \phi^{(B)}(\mathbf{r}_t) \left| \exp\left[-i\mathbf{p} \cdot (\mathbf{r}_p - \mathbf{r}_t)\right] \left[\sigma_t^S \otimes \sigma_p^S\right]^K \left[\boldsymbol{\tau}_t^T \cdot \boldsymbol{\tau}_p^T\right] \right| \phi^{(a)}(\mathbf{r}_p) \phi^{(A)}(\mathbf{r}_t) \right\rangle. \tag{8.222}$$

The differential cross section for charge exchange is obtained by an average of initial spins and sum over final spins, i.e.,

$$\frac{d\sigma}{d\Omega} = \frac{1}{(2J_a+1)(2J_A+1)} \sum_{\text{spins}} |f_{aA \to bB}(\theta)|^2. \tag{8.223}$$

Charge exchange induced reactions are often used to obtain values of Gamow-Teller, $B(GT)$, and Fermi, $B(F)$, matrix elements which cannot be extracted from β-decay experiments [31]. This method relies on the similarity in spin-isospin reaction operators in charge-exchange reactions and β-decay operators. Under certain approximations, the cross section for charge-exchange at small momentum transfer q can be shown theoretical to be closely proportional to $B(GT)$ and $B(F)$ [32],

$$\frac{d\sigma}{d\Omega}(\theta = 0°) = \left(\frac{\mu}{2\pi\hbar}\right)^2 \frac{k_f}{k_i} N_D |J_{\sigma\tau}|^2 \left[B(GT) + \mathcal{C}B(F)\right], \tag{8.224}$$

where μ is the reduced mass, $k_i(k_f)$ is the reactants relative momentum, N_D is a distortion factor (which accounts for initial and final state interactions), $J_{\sigma\tau}$ is the Fourier transform of the GT part of the effective nucleon-nucleon interaction, $\mathcal{C} = |J_\tau/J_{\sigma\tau}|^2$, and $B(\alpha = F, GT)$ is the reduced transition probability for non-spin-flip (τ_k is the isospin operator),

$$B(F) = \frac{1}{2J_i+1} |\langle f || \sum_k \tau_k^{(\pm)} ||i\rangle|^2, \tag{8.225}$$

and spin-flip (σ_k is the spin operator),

$$B(GT) = \frac{1}{2J_i+1} |\langle f || \sum_k \sigma_k \tau_k^{(\pm)} ||i\rangle|^2, \tag{8.226}$$

transitions. The condition that the momentum transfer is small, $q \sim 0$, is assumed to be valid for very small scattering angles, so that $\theta \ll 1/kR$, with R being the nuclear radius and k is the projectile wavenumber.

Eq. (8.224) can be derived easily from Eq. (8.206) if one can neglect the eikonal S-matrix, yielding the charge-exchange matrix element [32]

$$\mathcal{M}_{exch}(\mathbf{q}) = \left\langle \phi^{(b)}(\mathbf{r}_p) \phi^{(B)}(\mathbf{r}_t) \left| e^{-i\mathbf{q}\cdot\mathbf{r}_a} V_{pt}(\mathbf{q}) e^{i\mathbf{q}\cdot\mathbf{r}_b} \right| \phi^{(a)}(\mathbf{r}_p) \phi^{(A)}(\mathbf{r}_t) \right\rangle. \tag{8.227}$$

For forward scattering, low-momentum transfers, $\mathbf{q} \sim 0$, and small reaction q-values, the matrix element (8.227) becomes

$$\mathcal{M}_{exch}(\mathbf{q} \sim 0) \sim V_{exch}^{(0)}(\mathbf{q} \sim 0) \mathcal{M}_a(F, GT) \mathcal{M}_b(F, GT), \tag{8.228}$$

where $v_{exch}^{(0)}$ is the spinless part of the interaction, and

$$\mathcal{M}_{exch}(F, GT) = \left\langle \Phi^{(B)} ||(1 \text{ or } \sigma)\tau|| \Phi^{(A)} \right\rangle$$

are Fermi or Gamow-Teller (GT) matrix elements for the nuclear transition. The result above emerges by using plane waves for the nuclei.

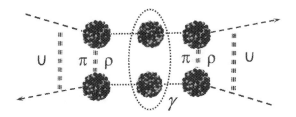

FIGURE 8.23   Schematic view of a double-charge exchange reaction, involving a two-step process induced by the nucleon-nucleon interaction. The potential U is responsible for the elastic scattering of the incoming and outgoing nuclei.

Double-charge-exchange and double-beta-decay

Double-charge exchange reactions, as shown schematically in Figure 8.23, could in principle be used to extract matrix elements for double beta decay in nuclei for a number of nuclei where such decays are energetically allowed. The reaction mechanism using a second-order matrix element involves the calculation of the amplitude

$$\mathcal{M}(\mathbf{k}, \mathbf{k}') = \sum_{\gamma, \mathbf{k}''} C_\gamma \left\langle \chi_{\mathbf{k}'}^{(-)} \left| V_{exch} \frac{1}{E_\mathbf{k} - \epsilon_{\gamma, \mathbf{k}''} - T - V_{exch}} V_{exch} \right| \chi_\mathbf{k}^+ \right\rangle, \qquad (8.229)$$

where $\chi_\mathbf{k}$ is the the distorted scattering wave due to an optical potential U in the initial and final channels, $\mathbf{k}$, $\mathbf{k}'$ are the initial and final momenta of the scattering nuclei, $\mathbf{k}''$ is the momentum of the intermediate state γ with energy $\epsilon_{\gamma, \mathbf{k}''}$ and T is the kinetic energy. C_γ are the spectroscopic amplitudes of the intermediate states. By using the Glauber scattering theory one can include the interaction U in all orders. Using the assumptions of forward scattering, and following the same approximations as used in Eq. (8.228) one can show again that a proportionality arises between double charge-exchange reactions and double beta-decay processes. The typical value of the cross section of a single step charge exchange relation is a few millibarns, while the a double charge exchange cross section is expected to be of the order of nanobarns [32].

Usually, double beta decay are ground state to ground state transitions. It can be accompanied by two neutrino emission, or by no emission of neutrinos. The latter process puts constraints on particle physics models beyond the standard one, such as the breaking of the lepton number conservation symmetry. In such case, the neutrino is a Majorana particle, e.g., its own anti-particle. To study neutrinoless double beta decay one needs to know the mass of the neutrino and the nuclear transition matrix element. Double beta decays emitting two neutrinos have been observed [33] but the observation of neutrinoless double beta decay remains elusive.

Usually, the Fermi type operator does not contribute appreciably to double beta-decay with neutrinos emitted, since the ground state of the final nucleus is not the double isobaric analog of the initial state. Therefore, the important transitions are those of double Gamow-Teller type. In the case of neutrinoless beta-decay one still expects that Gamow-Teller are larger than Fermi transitions. The problem still remains if one can control the contributions of the matrix elements for intermediate states entering Eq. (8.229). Perhaps, by measuring transitions to a large number of intermediate states in one step charge-exchange reactions, such as in (p,n) reactions, one in principle can determine the incoherent sum for the double charge-exchange transition. An obvious problem is that one is not sure if the same intermediate states excited in (p,n) and (n, p) experiments are involved in double-charge exchange. These intermediate states might also contribute very weakly in one-step reactions and very strongly in double charge-exchange, and vice-versa. Therefore, it seems that the best way to access information on the matrix elements needed for double beta-decay is to measure double charge exchange reactions directly.

In Figure 8.24 we show a correlation between calculated double charge-exchange (DCE) nuclear matrix elements (NME) for Gamow-Teller (GT) transitions and neutrino less double beta-decay ($0\nu\beta\beta$) obtained in Ref. [34]. The calculations have been done for $^{116}\mathrm{Cd} \rightarrow {}^{116}\mathrm{Sn}$, $^{128}\mathrm{Te} \rightarrow {}^{128}\mathrm{Xe}$,

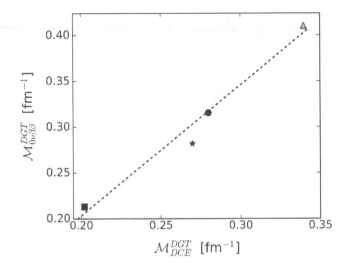

FIGURE 8.24 Correlation between calculated double charge-exchange (DCE) nuclear matrix elements (NME) for Gamow-Teller (GT) transitions and neutrino less double beta-decay ($0\nu\beta\beta$) [34]. The calculations have been done for ^{116}Cd $\rightarrow$ ^{116}Sn, ^{128}Te $\rightarrow$ ^{128}Xe, ^{82}Se $\rightarrow$ ^{82}Kr, and ^{76}Ge $\rightarrow$ ^{76}Se, respectively.

^{82}Se $\rightarrow$ ^{82}Kr, and ^{76}Ge $\rightarrow$ ^{76}Se, respectively. The linear correlation is explained in terms of a simple reaction theory and if it holds for cases of interest, it opens the possibility of constraining neutrinoless double beta-decay NMEs in terms of the experimental data on DCE at forward angles.

8.15 Exercises

1. Demonstrate Eq. (8.17).

2. Using Karol's formula (8.43), obtain the reaction cross sections for C + C , O + O, O + Pb and Pb + Pb at 100 MeV/nucleon.

3. Show that

$$\left\langle n^2\left(\mathbf{b}\right)\right\rangle = \sum_{n=0}^{AB} n^2 P\left(n, \mathbf{b}\right),$$

reduces to Eq. (8.16)

4. (a) Show that a nucleon-nucleus thickness function for the sharp-cutoff nuclear distribution of radius R and $t(\mathbf{b}) = \delta^{(2)}(\mathbf{b})$ is given by the formula:

$$T_A(b) = \frac{3\sqrt{R^2 - b^2}}{2\pi R^3}\Theta(R - b). \tag{8.230}$$

(b) Calculate the nucleon-nucleus thickness function for the Gaussian baryon distribution

$$\rho(\mathbf{b}_A, z_A) = \frac{1}{(2\pi)^{3/2}\sigma_A^3}\exp\left(-\frac{b_A^2 + z_A^2}{2\sigma_A^2}\right). \tag{8.231}$$

(c) Use the result of part (b) to show that for Gaussian nuclear distributions the nucleus-nucleus thickness function is a Gaussian characterized by the width $\sigma_{AB} = \sqrt{\sigma_A^2 + \sigma_B^2}$.

5. (a) How many participants sees a nucleon hitting a Au nucleus at zero impact parameter ($b = 0$?

   (b) How many binary collisions correspond to p + Au, $b = 0$ collision? Assume sharp sphere and $\sigma_{NN} = 40$ mb.

6. Consider the non-relativistic scattering of two particles with equal mass. The laboratory energy E_{lab} is related to the magnitude of the relative momentum k_{rel} which is the momentum that each particle has in the center-of-mass frame. If such relation is given by $E_{lab} = Ck_{rel}^2$, how is C related to other pertinent quantities? If the mass is $M_N = 939$ MeV, find the value of C in MeV fm^2?

7. Demonstrate the right-hand-side of Eq. (8.138).

8. Show that Eq. (8.164) follows from Eq. (8.162).

9. Find the total cross section for scattering off the square well potential

$$V(r) = \begin{cases} V_0, & \text{for} \quad r < R \\ 0, & \text{for} \quad r > R \end{cases}$$

in the Glauber and in the Born approximations. Compare the results for the cross section in the two approximations by plotting the results as functions of $\alpha = V_0 R/\hbar v$.

10. One can also derive the eikonal approximation for the scattering amplitude, Eq. (2.50b) in which the sum of partial waves l is replaced by an integral over the impact parameter b.

    (a) Show that if many partial waves enter the scattering, and if the scattering angle is very small, that the scattering amplitude can be written as

    $$f(\theta) \simeq \frac{1}{2ik} \int_0^\infty dbk \, (2kb) \, P_{kb}(\cos\theta) \left[e^{2i\delta(b)} - 1\right],$$

    where P_{kb} is the Legendre polynomial of order kb.

    (b) Show further that for very large k one obtains the Eq. (8.74) with $\delta(b) = \chi(b)$.

11. Using the properties of Hankel functions, prove Eq. (8.144).

12. Derive the eikonal approximation for the phase-shift, Eq. (8.71), from the WKB approximation.

13. One can consider Eq. (8.71) as an integral equation for $V(\mathbf{b})$. Assume spherical symmetry, $V(\mathbf{r}) = V(r)$, and use r as an integration variable in Eq. (8.71), so that it becomes

$$\chi(b) = -\frac{1}{\hbar v} \int_b^\infty \frac{V(r)}{\sqrt{r^2 - b^2}} r \, dr.$$

This is the *Abel integral equation*. Show that the solution is

$$V(r) = \frac{\hbar v}{2\pi} \int_r^\infty \frac{db}{\sqrt{b^2 - r^2}} \frac{d\chi}{db}. \tag{8.232}$$

Hint: note the result

$$\int_R^r \frac{b \, db}{\sqrt{(r^2 - b^2)(b^2 - R^2)}} = \frac{\pi}{2}.$$

Discuss Eq. (8.232) using a reasonable description of χ.

14. In addition to the Karol model, there are other ways to obtain fast solutions of Eq. (8.28). A Woods-Saxon density distribution can be well described by a folding of an Yukawa of range a and a hard sphere of size c. In this case, the Fourier transform $\widetilde{\rho}(q)$ can be calculated analytically:

$$\widetilde{\rho}(q) = \frac{4\pi\rho_0}{q^3} \left[\sin(qc) - qc\cos(qc) \right] \left[\frac{1}{1 + q^2 a^2} \right].$$

Make a plot of $\widetilde{\rho}(q)$ calculated numerically and using the above formula, as a function of q, for the Woods-Saxon parameterization of the densities of Al, Cu, Sn, Au and Pb (use the parameters from electron scattering experiments).

15. Prove the stationary phase method as stated in Eq. (8.171).

16. (a) Use the approximation (8.75) and obtain an analytical expression for Eq. (8.165).

(b) Insert your result in Eq. (8.164) and obtain analytic expressions for n_{E1}, n_{E2}, and n_{M1}. Compare this result with the expression for n_{E1} found in Ref. [19].

(c) Make a plot of $n_{E1}/Z_1^2\alpha$, $n_{E2}/Z_1^2\alpha$, and $n_{M1}/Z_1^2\alpha$ for $R = 10$ fm and $\gamma = 1$ and 10, as a function of the energy $\hbar\omega$. How do you understand the energy dependence of the virtual photon numbers?

17. (a) Show that charge-exchange reactions at forward angles $\theta \sim 0$ (momentum transfer $\mathbf{q} \sim 0$) are proportional to the Gamow-Teller matrix elements (as stated in Eq. (8.224)). The easiest path to obtain the proof is to assume a simple nucleon-nucleon interaction based on the pion and rho exchange model, as described by Eq. (8.208). Starting from it, show that Eq. (8.228) arises.

(b) How are the factors in Eq. (8.224) related to the coupling constants in the nucleon-nucleon interaction?

References

1. Glauber R J 1959 _Lecture notes on theoretical physics_ Vol. 1 (New York: Interscience)
2. Particle Data Group 2006 _J. Phys. G_ **33** 1
3. Bertulani C A, De Conti C 2010 _Phys. Rev._ **C81** 064603
4. Karol P J 1975 _Phys. Rev._ **C11** 1203
5. Hüfner J, Schäfer K and Schürmann B 1975 _Phys. Rev_ **C12** 1888
6. Carlson B V, Mastroleo R C and Hussein M S 1992 _Phys. Rev._ **C46** R30
7. Bertulani C A, Kuçuk Y and Lozeva R 2020 _Phys. Rev. Lett._ **124** 132301
8. T. Enqvist, et al. 1999 _Nucl. Phys._ **A658** 47
9. Hussein M S, Rego R A and Bertulani C A 1991 _Phys. Rep._ **201** 279
10. Ray L 1979 _Phys. Rev._ **C20** 1857
11. Aumann T and Bertulani C A 2020 _Progress in Particle and Nuclear Physics_ **112** 103753
12. Gomes L C, Walecka J D and Weisskopf V F 1958 _Ann. Phys._ (N.Y.) **3** 241
13. Ciementel E and Villi E 1955 _Nuovo Cimento_ **11** 176
14. Bertulani C A 1986 _Braz. J. Phys._ **16** 380
15. Buenerd M et al. 1982 _Phys. Rev._ **C26** 1299
16. Alkhazov G, et al. 2002 _Nuclear Physics_ **A712** 269
17. Arnold L G, et al. 1981 _Phys. Rev._ **C23** 1949
18. Eisenberg J M and Greiner W 1987 _Excitation Mechanisms of the Nuclei_ 1987 (Amsterdam: North-Holland)
19. Bertulani C A and Baur G 1988 _Phys. Rep._ **163** 299
20. Bertulani C A and Nathan A 1993 _Nucl. Phys._ **A554** 158
21. Winther A and Alder K 1979 _Nucl. Phys._ **A319** 518
22. Barrette J et al. 1988 _Phys. Lett._ **B209** 182

23. Tassie L J 1956 *Austrl. J. Phys.* **9** 407
24. Steinwedel H, Jensen J H D and Jensen P 1950 *Z. Naturf.* **5a** 343
25. Schmidt R, et al. 1993 *Phys. Rev. Lett.* **70** 1767
26. Aumann T, et al. 1996 *Nucl. Phys.* **A599 1** 321
27. Aumann T, Bertulani C A, Schindler F and Typel S 2017 *Phys. Rev. Lett.* **119** 262501
28. Harakeh M N and Dieperink A E L 1981 *Phys. Rev.* **C23** 2329
29. Anastasio M R and Brown G E 1977 *Nucl. Phys.* **A285** 516
30. Franey M A and Love W G 1985 *Phys. Rev.* **C31** 488
31. Taddeucci T N et al. 1987 *Nucl. Phys.* **A469** 125
32. Bertulani C A 1993 *Nuclear Physics* **A554** 493; Bertulani C A, Lotti P 1997 *Physics Letters* **B402** 237
33. Elliott S R, Hahn A A, and Moe M K 1987 *Phys. Rev. Lett.* **59** 2020
34. Santopinto E, García-Tecocoatzi H, Magaña Vsevolodovna R I, and Ferretti J 2018 *Phys. Rev.* **C98** 061601
35. Goldhaber M and Teller E 1948 *Phys. Rev.* **74** 1046

9

Relativistic collisions

9.1   Unpacking the nucleus

wave function As the bombarding energy in nucleus-nucleus collision increases the structure aspects of the nuclei become less relevant. Except for the bulk properties of the nuclei (size and number of nucleons), the physics involved is primarily due to the individual, and sometimes collective, hadronic collisions. At intermediate energies of $E_{lab} \sim 100 - 1000$ MeV/nucleon the nucleons and the products of their collisions can be described individually and their propagation can be described by semiclassical equations. One of such equations, and perhaps the most popular in such studies, is the so-called *Boltzmann-Uehling-Uhlenbeck* (BUU) equation. The BUU equation falls in the category of what one calls *quantum transport theories*. Hadronic transport theories have been quite successful in applications, describing a multitude of measured particle spectra. With a confidence stemming from the success of these predictions, one can gain through the transport theory a good insight into the history and mechanism of reactions. Transport theories are fairly flexible allowing one to include new particles as the energy domain changes and also to incorporate new collision processes if these become important. In this chapter we make a description of the BUU equation and show some of the results that can be obtained by its application to nucleus-nucleus collisions.

As the energy of nuclear collisions increases, one expects to probe the structure of the hadrons which compose them, i.e. the nucleons. In fact, one is not so interested in learning about the hadrons individually, but more on the collective effects resulting from simultaneous collisions between them. For example, one expects the formation of a *quark-gluon plasma*, a deconfined state of quarks and gluons, as shown schematically in Figure 9.1 . This is the major focus of relativistic heavy ion experiments at ultra high energies. A quark-hadron phase transition is predicted to have occurred at around ten micro-seconds after the Big Bang when the universe was at a temperature of approximately 150 to 200 MeV. This is the same transition as that from hadronic matter to a quark-gluon plasma, but in the reverse direction by cooling from a higher temperature as depicted in the phase diagram of Figure 9.2.

Also of interest in cosmology is general nucleosynthesis and the long-term effects of the possible existence of fluctuations in the quark-hadron phase transition. The region of high baryon densities

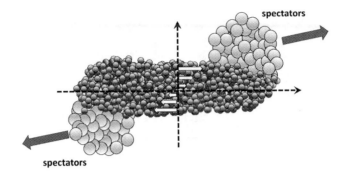

FIGURE 9.1 Schematic view of a near central collision between two nuclei at high energies. The incoherent and coherent nucleon-nucleon collisions produce numerous particles, and at extreme relativistic energies, it can de-confine the quarks and gluons (smaller circles) within the nucleons as predicted by Quantum-Chromodynamics. Identifying the properties of the final particles can lead to information about the nuclear Equation of State (EoS) in the hadronic and de-confined phase.

and very low temperatures is important for various aspects of stellar evolution. The nuclear matter equation of state governs neutron star collapse and supernova expansion dynamics. Collisions of relativistic heavy ions are expected to allow an investigation of the perturbative vacuum of *quantum chromodynamics* (QCD). We present the main concepts of hadronic structure, deconfinement of quarks, and the quark-gluon plasma towards the end of this chapter.

9.2 The Boltzmann-Uehling-Uhlenbeck equation

Let us call $dN(\mathbf{r}, \mathbf{p}, t)$ the number of particles with positions $\mathbf{r}$ and momenta $\mathbf{p}$ at time t. If dN is the number of particles in the volume element d^3r and whose momenta fall in the momentum element d^3p at time t, then the distribution function $f(\mathbf{r}, \mathbf{p}, t)$ is given by

$$dN = f(\mathbf{r}, \mathbf{p}, t)\, d^3r d^3p. \tag{9.1}$$

For a particle to be included in dN its position coordinates must lie between $\mathbf{r}_i$ and $\mathbf{r}_i + \Delta\mathbf{r}_i$, and its momentum must lie between $\mathbf{p}_i$ and $\mathbf{p}_i + \Delta\mathbf{p}_i$, where i runs from 1 to 3.

If there were no collisions, then a short time Δt later each particle would move from to $\mathbf{r} + \Delta\mathbf{r}$, and each particle momentum would change from to $\mathbf{p} + \mathbf{F}\Delta t$, where is $\mathbf{F}$ the external force on a particle at $\mathbf{r}$ with momentum $\mathbf{p}$. Therefore, any difference between $dN(\mathbf{r}, \mathbf{p}, t)$ and $dN(\mathbf{r} + \Delta\mathbf{r}, \mathbf{p} + \mathbf{F}\Delta t, t)$ is due to collisions, and we may set

$$[f(\mathbf{r} + \Delta\mathbf{r}, \mathbf{p} + \mathbf{F}\Delta t, t) - f(\mathbf{r}, \mathbf{p}, t)]\, d^3r d^3p = \left(\frac{\partial f}{\partial t}\right)_c d^3r' d^3p' \Delta t, \tag{9.2}$$

where $(\partial f/\partial t)_c$ is the time rate of change of f due to collisions. Expanding the first term on the left as a Taylor series about $f(\mathbf{r}, \mathbf{p}, t)$, we have (here repeated indices mean a summation, e.g., $a_i b_i = \mathbf{a} \cdot \mathbf{b} = a_1 b_1 + a_2 b_2 + a_3 b_3$)

$$f(\mathbf{r} + \Delta\mathbf{r}, \mathbf{p} + \mathbf{F}\Delta t, t) = f(\mathbf{r}, \mathbf{p}, t) + \left(\frac{\partial f}{\partial r_i}\frac{p_i}{m} + \frac{\partial f}{\partial p_i}F_i + \frac{\partial f}{\partial t}\right)\Delta t, \tag{9.3}$$

where m is the nucleon mass.

In the limit as $\Delta t \to 0$,

$$\frac{\partial f}{\partial r_i}\frac{p_i}{m} + \frac{\partial f}{\partial p_i}F_i + \frac{\partial f}{\partial t} = \left(\frac{\partial f}{\partial t}\right)_c, \tag{9.4}$$

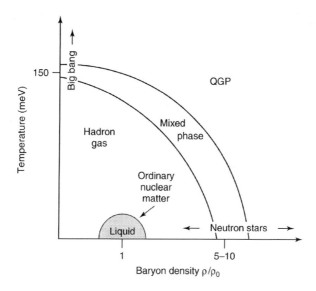

FIGURE 9.2 As water comes in different phases (solid, liquid, gas), so nuclear matter can come in its normal hadronic form and at sufficiently high temperature and density, in the form of a deconfined state of quarks and gluons. The diagram shows how nuclear matter should behave as a function of dentistry and temperature.

which is known at the *Boltzmann equation* for f. Note we have used the result that the Jacobian for the transformation $d^3r'd^3p' = |J| d^3rd^3p$ is unity, where J is the 6×6 element array,

$$J = \frac{\partial (x, y, z, p_x, p_y, p_z)}{\partial (x', y', z', p_x', p_y', p_z')} = 1. \tag{9.5}$$

This assumption is valid only if the collisions are elastic, i.e., if they conserve energy and momentum.

The system of nucleons are often free from external sources, so that one can drop off the term containing F_i in Eq. (9.4). However, to account for the effect of each particle interacting with all other, one introduces the concept of *mean-field*, $U(\mathbf{r}, \mathbf{p}, t)$. This mean-field exerts a force on each particle, given by $-\nabla_\mathbf{r} U(\mathbf{r}, \mathbf{p}, t)$. Re-deriving Eq. (9.4) in terms of a mean field yields

$$\frac{\partial f}{\partial t} + \left(\frac{\mathbf{P}}{m} + \nabla_\mathbf{p} U\right) \cdot \nabla_\mathbf{r} f - \nabla_\mathbf{r} U \cdot \nabla_\mathbf{r} f = \left(\frac{\partial f}{\partial t}\right)_c. \tag{9.6}$$

Note that the left hand side of this equation is simply the total time derivative of the distribution function, Df/Dt. In the absence of collisions, one obtains the *Vlasov equation*

$$\frac{Df}{Dt} = \frac{\partial f}{\partial t} + \left(\frac{\mathbf{P}}{m} + \nabla_\mathbf{p} U\right) \cdot \nabla_\mathbf{r} f - \nabla_\mathbf{r} U \cdot \nabla_\mathbf{r} f = 0. \tag{9.7}$$

Let us now assume that the system of nucleons form a dilute system of particles. Dilute means that the total volume of the gas particles is small compared to the volume available to the gas,

$$na^2 \ll 1, \tag{9.8}$$

where n is the number density of particles and a is the radius of a particle. Since the particles in a neutral gas do not have long-range forces like the particles in a plasma, they are assumed to interact only when they collide, i.e., when the separation between two particles is not much larger than $2a$. The term collision normally means the interaction between two such nearby particles. A particle moves in a straight line between collisions. The average distance travelled by a particle between two collisions is known as the mean free path. The mean free path depends on the cross section σ, and is given by

$$\lambda = \frac{1}{n\sigma}. \tag{9.9}$$

One consequence of the requirement that the gas be dilute is that $\lambda \gg a$. In other words the diluteness implies that the mean free path is much larger than the particle size so that a typical particle trajectory consists of long straight segments interrupted by almost discontinuous changes of direction when collisions occur. If the gas is dilute, the probability of three body collisions is much lower than for two body collisions and they can be neglected.

The Vlasov equation, $Df/Dt = 0$, says that f does not change as we move along the trajectory of a particle, provided collisions are neglected. Collisions can change f in two ways.

(1) Some particles originally having momentum $\mathbf{p}$ will have some different momentum after the collision. This causes a decrease in f.

(2) Some particles having other momentum may have the momentum $\mathbf{p}$ after a collision, increasing f.

The Boltzmann collisional term in the Boltzmann equation can be written as

$$\frac{Df}{Dt} d^3r d^3p = C_{in} - C_{out}, \tag{9.10}$$

where C_{in} and C_{out} are the rates at which particles enter and leave the infinitesimal volume $d^3r d^3p$ due to collisions.

Suppose two particle with initial velocities $\mathbf{v}_1$ and $\mathbf{v}_2$ have velocities $\mathbf{v}_1'$ and $\mathbf{v}_2'$ and after a collision. Since all particles have the same mass, conservation of momentum and energy require that

$$\mathbf{v}_1' + \mathbf{v}_2' = \mathbf{v}_1 + \mathbf{v}_2 \quad \text{and} \quad \frac{1}{2}\left|\mathbf{v}_1'\right|^2 + \frac{1}{2}\left|\mathbf{v}_2'\right|^2 = \frac{1}{2}\left|\mathbf{v}_1\right|^2 + \frac{1}{2}\left|\mathbf{v}_2\right|^2. \tag{9.11}$$

One would like to calculate the final velocities $\mathbf{v}_1'$ and $\mathbf{v}_2'$ from the initial velocities. Since we have six components we need six equations to solve for them. Four are provided by the conservation equations. A fifth condition comes from the fact that collisions are coplanar if the forces between particles are purely radial, i.e. $\mathbf{v}_1'$ will lie in the plane of $\mathbf{v}_1$ and $\mathbf{v}_2$, forcing $\mathbf{v}_2'$ to also lie in the same plane from conservation of momentum. We still need a sixth condition, which must come from the nature of the force between the particles. The short-range nature of the forces allow us to assume that the collision occurs at essentially only one value of $\mathbf{r}$ so we need not account for the changes of external forces. The unknown velocities are therefore specified once the impact parameter $\mathbf{b}$ and the azimuthal orientation ϕ of the collision is known. For an elastic collision the magnitude of the relative velocity is a collisional invariant:

$$\left|\mathbf{v}_1' - \mathbf{v}_2'\right| = \left|\mathbf{v}_1 - \mathbf{v}_2\right|, \tag{9.12}$$

which follows from kinetic energy conservation in the center of mass frame. Thus we may specify the remaining two pieces of information concerning the collision in terms of the change of the orientation of the relative velocity, i.e., in terms of two angles Θ and ϕ. In an elastic encounter the collision occurs in a single plane $\phi = const.$, turning the relative velocity $\mathbf{v}_1 - \mathbf{v}_2$ through an angle Θ without change in the magnitude of the relative velocity. For given intermolecular forces, the deflection, Θ, depends only on the impact parameter $\mathbf{b}$. The differential cross-section for the encounter $\sigma(\mathbf{v}_1, \mathbf{v}_2 | \mathbf{v}_1', \mathbf{v}_2')$ is defined so that $\sigma d\Omega = bdbd\phi$, where $d\Omega = 2\pi \sin\Theta d\Theta$.

Consider a beam of particles of number density n_1 and velocity $\mathbf{v}_1$ colliding with another beam of particles of density n_2 and velocity $\mathbf{v}_2$. A particle in the second beam experiences a flux $I = n_1 |\mathbf{v}_1 - \mathbf{v}_2|$ of particles from the first beam. We consider the number δn_c of collisions per unit time per unit volume which deflect particles from the second beam into a solid angle $d\Omega$,

$$\delta n_c = \sigma\left(\mathbf{v}_1, \mathbf{v}_2 | \mathbf{v}_1', \mathbf{v}_2'\right) n_1 |\mathbf{v}_1 - \mathbf{v}_2| n_2 d\Omega. \tag{9.13}$$

Consider the inverse collision where $(\mathbf{v}_1, \mathbf{v}_2) \longrightarrow (\mathbf{v}_1', \mathbf{v}_2')$. If the molecular processes are time-reversible, then we expect the reverse cross-section to equal the forward cross-section:

$$\sigma\left(\mathbf{v}_1', \mathbf{v}_2' | \mathbf{v}_1, \mathbf{v}_2\right) = \sigma\left(\mathbf{v}_1, \mathbf{v}_2 | \mathbf{v}_1', \mathbf{v}_2'\right). \tag{9.14}$$

It should be noted that this condition of time reversibility is by no means self-evident. Evidently, this is intimately related to the reciprocity theorem, Eq. (4.180). In the present case, the Eq. (9.14) is a consequence of the energy and momentum conservation in a binary collision, which leads to the equality of the phase space before and after the collision, i.e. $d^3 p_1 d^3 p_2 = d^3 p_1' d^3 p_2'$ (see Eq. (9.5)).

We now evaluate the term C_{out}. Consider the two streams of particles having the tips of their momentum vectors in $d^3 p_1$ and $d^3 p_2$. The first stream makes up a beam of number density $n_1 = f(\mathbf{r}, \mathbf{p}_1, t) d^3 p_1$, and velocity $\mathbf{v}_1$, whereas the second stream constitutes a beam of density $n_2 = f(\mathbf{r}, \mathbf{p}_2, t) d^3 p_2$ and velocity v_2. Substitution for n_1 and n_2 in the collision rate between the two beams is

$$\delta n_c = \sigma\left(\mathbf{v}_1, \mathbf{v}_2 | \mathbf{v}_1', \mathbf{v}_2'\right) |\mathbf{v}_1 - \mathbf{v}_2| f(\mathbf{r}, \mathbf{p}_1, t) f(\mathbf{r}, \mathbf{p}_2, t) d\Omega d^3 p_1 d^3 p_2. \tag{9.15}$$

Since C_{out} must be equal to the number of collisions per unit time with the volume $d^3 r_1 d^3 p_1$, C_{out} is obtained by multiplying δn_c by $d^3 r_1$ and then integrating over all solid angles, Ω, and collision partner momenta, p_2. Hence,

$$C_{out} = d^3 r_1 \int_{p_2} \int_\Omega \delta n_c = d^3 r_1 d^3 p_1 \int d^3 p_2 \int d\Omega \sigma\left(\mathbf{v}_1, \mathbf{v}_2 | \mathbf{v}_1', \mathbf{v}_2'\right) |\mathbf{v}_1 - \mathbf{v}_2| f(\mathbf{r}, \mathbf{p}_1, t) f(\mathbf{r}, \mathbf{p}_2, t). \tag{9.16}$$

To evaluate C_{in}, we consider the reverse collisions between particles in $d^3 p_1'$ and with momenta in $d^3 p_2'$ such that their velocities after collisions lie within $d^3 p_1$ and $d^3 p_2$, respectively. The number of such collision per unit volume per unit time is

$$\delta n_c' = \sigma\left(\mathbf{v}_1', \mathbf{v}_2' | \mathbf{v}_1, \mathbf{v}_2\right) |\mathbf{v}_1' - \mathbf{v}_2'| f(\mathbf{r}, \mathbf{p}_1', t) f(\mathbf{r}, \mathbf{p}_2', t) d\Omega d^3 p_1' d^3 p_2'. \tag{9.17}$$

Recall that the relative velocity of the particles is a collisional invariant, $|\mathbf{v}_1' - \mathbf{v}_2'| = |\mathbf{v}_1 - \mathbf{v}_2|$, and from *Liouville's theorem*, if the interaction can be described by a conservative Hamiltonian,

$$d^3 p_1' d^3 p_2' = d^3 p_1 d^3 p_2, \tag{9.18}$$

as already mentioned earlier.

Thus, assuming reversible collisions, the invariance of the relative velocity and the constant phase space volume,

$$\delta n_c' = \sigma\left(\mathbf{v}_1, \mathbf{v}_2 | \mathbf{v}_1', \mathbf{v}_2'\right) |\mathbf{v}_1 - \mathbf{v}_2| f(\mathbf{r}, \mathbf{p}_1', t) f(\mathbf{r}, \mathbf{p}_2', t) d\Omega d^3 p_1 d^3 p_2. \tag{9.19}$$

The term C_{in} is obtained by multiplying $\delta n_c'$ by $d^3 r_1$ and integrating over all solid angles, Ω, and collision partner momenta, $\mathbf{p}_2$, i.e.,

$$C_{out} = d^3 r_1 d^3 p_1 \int d^3 p_2 \int d\Omega \sigma\left(\mathbf{v}_1, \mathbf{v}_2 | \mathbf{v}_1', \mathbf{v}_2'\right) \times |\mathbf{v}_1 - \mathbf{v}_2| f(\mathbf{r}, \mathbf{p}_1', t) f(\mathbf{r}, \mathbf{p}_2', t). \tag{9.20}$$

Now that we have the rates at which particles leave and enter $d^3 r_1 d^3 p_1$ we can write the full Boltzmann equation as

$$\frac{\partial f}{\partial t} + \left(\frac{\mathbf{P}}{m} + \nabla_\mathbf{p} U\right) \cdot \nabla_\mathbf{r} f - \nabla_\mathbf{r} U \cdot \nabla_\mathbf{r} f = \int d^3 p_2 \int d\Omega \sigma\left(\mathbf{v}_1, \mathbf{v}_2 | \mathbf{v}_1', \mathbf{v}_2'\right) |\mathbf{v}_1 - \mathbf{v}_2|$$
$$\times \left[f(\mathbf{r}, \mathbf{p}_1', t) f(\mathbf{r}, \mathbf{p}_2', t) - f(\mathbf{r}, \mathbf{p}_1, t) f(\mathbf{r}, \mathbf{p}_2, t) \right]. \tag{9.21}$$

which is conveniently abbreviated as

$$\frac{\partial f}{\partial t} + \left(\frac{\mathbf{P}}{m} + \nabla_\mathbf{p} U\right) \cdot \nabla_\mathbf{r} f - \nabla_\mathbf{r} U \cdot \nabla_\mathbf{r} f = \int d^3 p_2 \int d\Omega \sigma(\Omega) |\mathbf{v}_1 - \mathbf{v}_2| \left[f_1' f_2' - f_1 f_2 \right]. \tag{9.22}$$

We have assumed that the differential cross-section is a function only of the scattering angle Ω between $\mathbf{p}_1$ and $\mathbf{p}_2$, since the differential cross-section for a simple spherically symmetric interaction potential can, due to symmetry, only be a function of the scattering angle. The complete (classical)

Boltzmann equation with the collision integral for binary collisions is a nonlinear integro-differential equation for the distribution function.

For a system of nucleons the classical Boltzmann equation can be modified to account for the Pauli principle. The principle states that no nucleon can scatter into a phase space already occupied by another nucleon. This amounts in modifying the term $f_1 f_2$ to $f_1 f_2 [1 - f_1'] [1 - f_2']$, where the $1 - f'$ terms yield zero if the final state is occupied ($f = 1$). Accordingly, $f_1' f_2'$ is modified to $f_1' f_2' [1 - f_1] [1 - f_2]$. Thus, for a nucleon system of particles, the appropriate Boltzmann equation for a nucleon-nucleon collisions (called the Boltzmann-Uehling-Uhlenbeck, or BUU, equation) is

$$\frac{\partial f}{\partial t} + \left(\frac{\mathbf{p}}{m} + \boldsymbol{\nabla}_{\mathbf{p}} U \right) \cdot \boldsymbol{\nabla}_{\mathbf{r}} f - \boldsymbol{\nabla}_{\mathbf{r}} U \cdot \boldsymbol{\nabla}_{\mathbf{r}} f = \int d^3 p_2 \int d\Omega \sigma_{NN} (\Omega) |\mathbf{v}_1 - \mathbf{v}_2|$$
$$\times \left\{ f_1' f_2' [1 - f_1] [1 - f_2] - f_1 f_2 [1 - f_1'] [1 - f_2'] \right\}. \tag{9.23}$$

The collision integral in Eq. (9.23), takes into account the nucleon scattering inside the nuclear medium. Its form can be justified on general physical grounds, but it can be also derived self-consistently from the quantum equations of motion of the one-body and two-body density .

Eq. (9.23) needs as basic ingredients the mean field U and the cross section σ_{NN}. Because these two quantities are related to each other, one should in principle derive them in a self-consistent microscopic approach, as the Brueckner theory. However, in practice the simulations are often done with a phenomenological mean field and free nuclear cross sections. The most commonly used mean field is of Skyrme-type, eventually with a momentum dependent part [1].

An important ingredient in the transport theory calculations is the compressibility K of nuclear matter, which refers to the second derivative of the compressional energy E with respect to the density, as discussed in Chapter 7:This is an important quantity, e.g., for nuclear astrophysics. In fact, the mechanism of supernova explosions described in Chapter 7 is strongly dependent on the value K. Supernova models might or not lead to explosions depending on the value of K. The central collisions of heavy nuclei are one of the few probes of this quantity in the laboratory. The dependence of the calculations on K follow from the dependence of the mean field potential U ($U \sim E/A+$ kinetic energy terms) on the particle density ρ. A typical parametrization for U is the *Skyrme parametrization*

$$U = a \frac{\rho}{\rho_0} + b \left(\frac{\rho}{\rho_0} \right)^\sigma .$$

Hydrodynamical equations are also frequently used in the analysis of heavy ion collisions at high energies. Viscous forces, due to friction acting on a fluid element because of viscosity in the fluid. These viscous forces are surface forces, like the forces due to pressure, but can act in any direction on the surface. In other words, viscous forces at a surface can have both normal and tangential (or shear) components. The net viscous force per unit volume for an incompressible Newtonian fluid is

$$\mathbf{f}_{\text{visc}} = \eta \nabla^2 \mathbf{v}, \tag{9.24}$$

where the right hand side is the Laplacian of the velocity vector times the viscosity coefficient η. Note that in hydrostatics, where the velocity is identically zero, there is no viscous force, regardless of the value of viscosity. The sum of all the forces on a volume element must equal the mass of the element times its acceleration. On a per unit volume basis, the equation of fluid motion is then

$$\rho \left[\frac{\partial \mathbf{v}}{\partial t} + (\mathbf{v} \cdot \boldsymbol{\nabla}) \mathbf{v} \right] = \mathbf{F}_{\text{ext}} - \nabla \mathbf{p} + \eta \nabla^2 \mathbf{v}, \tag{9.25}$$

where $\mathbf{p}$ stands for the pressure.

The above equation is the famous *Navier-Stokes equation*, valid for incompressible Newtonian flows. In some practical applications, the effects of viscosity can become negligible. The viscous terms in the Navier Stokes equation are then neglected, and the equation reduces to the *Euler equation*.

The Navier-Stokes equation is a second-order solution (the first-order solution is shown in exercise 1) of the Boltzmann transport equation [6]. Therefore, the equations of hydrodynamics are an approximation to the Boltzmann equation obtained by averaging that equation over an assumed distribution function.

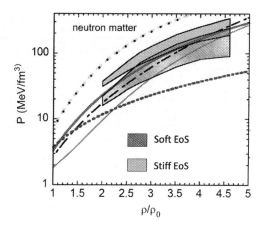

FIGURE 9.3 Pressure associated with neutron matter as a function of nucleon density. The various curves are different theoretical models. The shadow bands show the numerous possibilities for the EoS based on the experiment analyzes using a soft (lower shadow region) or stiff (upper shadow region) EoS.

The transport equations can be generalized to a covariant equations taking into account elements of relativity, although retardation and simultaneity are very difficult to handle. In such transport models the mean field U and the elementary cross sections σ are are correlated and one needs to use a self-consistent microscopic approach. In practice, the simulations are often done with a phenomenological mean field and free nuclear cross sections. Skyrme-type interactions are often adopted with a momentum dependent part [1]. As in the case of mean field calculations of nuclear densities, mentioned in previous sections of this review, this procedure allows one to deduce the compressibility K of nuclear matter, which refers to the second derivative of the compressional energy E with as well as the symmetry energy S related to the nuclear EoS. In Figure 9.3 (adapted from Ref. [9]), we show the pressure for neutron matter as a function of the density. The shadow bands represent the range of possible theoretical EoS based on soft and stiff mean-field potentials.

9.3 Wigner function

There are numerous ways to show that transport theories, e.g. the BUU equation, are classical limits of the quantum many-body problem. In this section we show one of these derivations. In particular, we show how the eikonal wave function, discussed in details in chapter 8, can be used for this purpose.

In classical mechanics the state of a particle is represented by a point in the phase space $(\mathbf{r}, \mathbf{p})$. In quantum mechanics the equivalent of this concept are the probability distributions $\rho(\mathbf{r})$ and $\rho(\mathbf{p})$. However, contrary to the classical statistical mechanics, in quantum mechanics it is impossible to determine the probability in the form $f(\mathbf{r}, \mathbf{p}, t)$ since, due to the uncertainty principle, it is impossible to measure simultaneously the position and momentum. A bridge between the classical and quantum mechanical case is obtained by introducing the *Wigner function* $W(\mathbf{r}, \mathbf{p}, t)$ defined by the relations [2]

$$\int d^3 p\, W(\mathbf{r}, \mathbf{p}, t) = |\psi(\mathbf{r}, t)|^2 = \rho(\mathbf{r}, t), \qquad (9.26)$$

$$\int d^3 r\, W(\mathbf{r}, \mathbf{p}, t) = \frac{1}{(2\pi\hbar)^3} \left|\tilde{\psi}(\mathbf{p}, t)\right|^2 = \rho(\mathbf{p}, t). \qquad (9.27)$$

Using the basic properties of the Fourier transforms it is easy to show that the following bilinear combinations of the wave function and its complex conjugate obey the relations above

$$W\left(\mathbf{r},\mathbf{p},t\right) = \frac{1}{\left(2\pi\hbar\right)^3}\int d^3 s\psi\left(\mathbf{r}+\frac{\mathbf{s}}{2},t\right)\exp\left(-i\frac{\mathbf{p}.\mathbf{s}}{\hbar}\right)\psi^*\left(\mathbf{r}-\frac{\mathbf{s}}{2},t\right) \tag{9.28}$$

$$= \frac{1}{\left(2\pi\hbar\right)^3}\int\frac{d^3 q}{\left(2\pi\hbar\right)^3}\widetilde{\psi}\left(\mathbf{p}+\frac{\mathbf{q}}{2},t\right)\exp\left(-i\frac{\mathbf{r}.\mathbf{q}}{\hbar}\right)\widetilde{\psi}^*\left(\mathbf{p}-\frac{\mathbf{q}}{2},t\right). \tag{9.29}$$

Taking the complex conjugate and changing the sign of the integration variable **s** one can show that the Wigner function is real. It is also normalized to unity:

$$\int d^3 r d^3 p W\left(\mathbf{r},\mathbf{p},t\right) = 1. \tag{9.30}$$

Nonetheless, the Wigner transform $W\left(\mathbf{r},\mathbf{p},t\right)$ can take negative values. For example let us take the function

$$\psi\left(\mathbf{r}\right) = Nx\exp\left(-\frac{r^2}{2l^2}\right), \tag{9.31}$$

where N is a normalization factor. The corresponding Wigner function is

$$W\left(\mathbf{r},\mathbf{p},t\right) = N'\left[\left(\frac{x}{l}\right)^2 + \left(p_x\frac{l}{\hbar}\right)^2 - 1\right]\exp\left[-\left(\frac{r}{l}\right)^2 - \left(\frac{pl}{\hbar}\right)^2\right], \tag{9.32}$$

where N' is a positive factor. The function above is negative inside an ellipse of semi-axis l and $\hbar l$ in the (x,p_x) plane. The center of the ellipse corresponds to zero values of the x components of momentum and position.

To investigate how the Wigner function varies with time, we differentiate (9.28) with respect to time and use the Schrödinger equation to show that

$$\frac{\partial}{\partial t}W\left(\mathbf{r},\mathbf{p},t\right) = \frac{1}{i\hbar}\int\frac{d^3 s}{\left(2\pi\hbar\right)^3}\exp\left(-i\frac{\mathbf{p}.\mathbf{s}}{\hbar}\right)\left\{-\frac{\hbar^2}{2m}\nabla^2\psi\left(\mathbf{r}+\frac{\mathbf{s}}{2},t\right)\psi^*\left(\mathbf{r}-\frac{\mathbf{s}}{2},t\right) + \right.$$

$$\left.\psi\left(\mathbf{r}+\frac{\mathbf{s}}{2},t\right)\frac{\hbar^2}{2m}\nabla^2\psi^*\left(\mathbf{r}-\frac{\mathbf{s}}{2},t\right) + \left[V\left(\mathbf{r}+\frac{\mathbf{s}}{2}\right) - V\left(\mathbf{r}-\frac{\mathbf{s}}{2}\right)\right]\psi\left(\mathbf{r}+\frac{\mathbf{s}}{2},t\right)\psi^*\left(\mathbf{r}-\frac{\mathbf{s}}{2},t\right)\right\}. \tag{9.33}$$

We now integrate by parts and use the relation $\phi\nabla^2\psi - \left(\nabla^2\phi\right)\psi = \nabla\cdot\left(\phi\nabla\psi - \psi\nabla\phi\right)$ to transform the Laplacian to the form $\left(-1/m\right)\mathbf{p}\cdot\nabla W\left(\mathbf{r},\mathbf{p},t\right)$. Expanding the potentials $V\left(\mathbf{r}\pm\frac{\mathbf{s}}{2}\right)$ in a Taylor series we obtain

$$\left[\frac{\partial}{\partial t} + \frac{1}{m}\mathbf{p}\cdot\nabla + \mathbf{F}\left(\mathbf{r}\right)\cdot\widetilde{\nabla}\right]W\left(\mathbf{r},\mathbf{p},t\right) = \sum_{k=1}^{\infty}\frac{\left(-\hbar^2/4\right)^k}{\left(2k+1\right)!}V\left(r\right)\left(\overleftarrow{\nabla}\cdot\widetilde{\nabla}\right)^{2k+1}W\left(\mathbf{r},\mathbf{p},t\right), \tag{9.34}$$

where $\widetilde{\nabla}\equiv\nabla_{\mathbf{p}}$ is the gradient operator in the momentum space and $\overleftarrow{\nabla}$ takes the derivative of the expression to the left.

In the classical limit ($\hbar\longrightarrow 0$), the above equation transforms into the *Liouville equation*:

$$\frac{\partial}{\partial t}W = -\left[\frac{1}{m}\mathbf{p}\cdot\nabla + \mathbf{F}\left(\mathbf{r}\right)\cdot\widetilde{\nabla}\right]W = \left[H,W\right]_P, \tag{9.35}$$

where $[\ ,\ ]_P$ denotes the *Poisson parentheses*. In this limit the time evolution of the Wigner function is determined by the solutions of the Newton equations, since the solutions of (9.35) have the form

$$W\left(\mathbf{r},\mathbf{p},t\right) = W\left[\mathbf{r}_0\left(\mathbf{r},\mathbf{p},t-t_0\right),\mathbf{p}_0\left(\mathbf{r},\mathbf{p},t-t_0\right),t_0\right], \tag{9.36}$$

where $\mathbf{r}_0\left(\mathbf{r},\mathbf{p},t-t_0\right)$ and $\mathbf{p}_0\left(\mathbf{r},\mathbf{p},t-t_0\right)$ are the initial conditions at t_0 so that the evolution according to the Newton equations leads to the point $(\mathbf{r},\mathbf{p})$ in the phase space at the instant t. The Wigner function, as the hydrodynamical variables, follows equations of motion that have classical limits. Thus, the Wigner function has applications in situations in which the corrections to the classical behavior are small. This happens only for potentials $V(\mathbf{r})$ which vary monotonously. This allows us to use only a few terms of the series (9.34).

The expectation values of the physical quantities connected to the particle are obtained integrating an appropriate function $A(\mathbf{r}, \mathbf{p})$, which represents the physical quantity, with a weight W over all phase space:

$$\langle A \rangle = \int d^3 r \int d^3 p\, W(\mathbf{r}, \mathbf{p}, t)\, A(\mathbf{r}, \mathbf{p}). \tag{9.37}$$

For the quantities which depend separately in $\mathbf{p}$ and $\mathbf{r}$, the obtained expectation values are those which we already know. For example we get for the energy

$$\langle E \rangle = \int d^3 r \int d^3 p\, W(\mathbf{r}, \mathbf{p}, t)\left[\frac{\mathbf{p}^2}{2m} + V(\mathbf{r})\right] = \int d^3 p\, \frac{\mathbf{p}^2}{2m}\rho_p(\mathbf{p}, t) + \int d^3 r\, V(\mathbf{r})\rho(\mathbf{r}, t). \tag{9.38}$$

In situations in which the dependence in $\mathbf{p}$ and $\mathbf{r}$ cannot be separated, the Eq. (9.37) becomes a postulate since it cannot be obtained from the properties of quantum mechanics.

Using the Cauchy-Schwartz inequalities* and the normalization condition of the wave function, one can show that

$$|W(\mathbf{r}, \mathbf{p}, t)|^2 \leq \int \frac{d^3 s}{(2\pi\hbar)^3}\left|\psi\left(\mathbf{r} + \frac{\mathbf{s}}{2}, t\right)\right|^2 \int \frac{d^3 s}{(2\pi\hbar)^3}\left|\psi\left(\mathbf{r} - \frac{\mathbf{s}}{2}, t\right)\right|^2 = \left(\frac{2}{\hbar}\right)^6. \tag{9.39}$$

From this relation, one obtains

$$|W(\mathbf{r}, \mathbf{p}, t)| \leq \left(\frac{2}{\hbar}\right)^3. \tag{9.40}$$

This inequality expresses the uncertainty principle in a clear way. The particle cannot be localized inside a phase space cell of a volume smaller than $(\hbar/2)^3$. In the classical case there is no upper limit for the values of the distribution function. Therefore we see that quantum mechanics can be formulated with the help of the Wigner function in a similar way as in statistical physics. Due to the complicated nature of the time evolution Eq. (9.34), this formulation is more appropriate to study the classical limit than to solve problems in quantum mechanics.

We now present a derivation of a transport equation using Wigner's function and multiple scattering together with the minimal set of assumptions needed in order to reach the simplest version of the transport equation, i.e., the analog of the classical Boltzmann equation. We address a system in which a particle passes through a medium of N particles experiencing optical distortion on the way. In addition we consider one direct collision between the projectile and the ith particle within the medium; generalizations to consider n such collisions, where $n \ll N$, are possible. The target is taken to be a totally uncorrelated set of particles,

$$|\Phi_0(\mathbf{r}_1, \ldots, \mathbf{r}_N)|^2 = \rho_1(\mathbf{r}_1) \ldots \rho_N(\mathbf{r}_N),, \tag{9.41}$$

where the complete lack of correlations implies that the densities for the individual particles are constants,

$$\rho_i(\mathbf{r}_i) = 1/\Omega,, \tag{9.42}$$

where Ω is the volume containing the particles.

We now assume that the projectile possesses sufficiently high energy that an eikonal form for its wave function is a good approximation (here we use for convenience $\hbar = c = 1$),

$$\phi_{\mathbf{k}}(\mathbf{r}; \mathbf{r}_i) = \exp\left[i\mathbf{k} \cdot \mathbf{r}\right]\exp\left[-\frac{i}{v}\int_{-\infty}^{z} \mathcal{U}(\mathbf{b}, \zeta)d\zeta\right]\exp\left[-\frac{i}{v}\int_{-\infty}^{z} V_{\text{int}}(\mathbf{b} - \mathbf{b}_i, \zeta - z_i)d\zeta\right]. \tag{9.43}$$

---

*The Cauchy-Schwartz theorem states that $|\langle\phi|\psi\rangle|^2 \leq |\langle\phi|\phi\rangle|\,|\langle\psi|\psi\rangle|$.

Here $\mathbf{k}$ is the momentum of the projectile, $\mathbf{v} = \mathbf{k}/m$ is its velocity with m its mass, $\mathcal{U}$ is the complex potential supplying the optical distortion, and V_{int} is the interaction between the projectile and the ith particle in the medium. The integrals are taken along the direction of $\mathbf{k}$, and the position vectors are taken as $\mathbf{r} = \{\mathbf{b}, z\}$, where $\mathbf{b}$ is the projection of $\mathbf{r}$ in a direction perpendicular to $\mathbf{k}$ and z is parallel to $\mathbf{k}$. Thus the ith particle in the target is located at $\mathbf{r}_i = \{\mathbf{b}_i, z_i\}$. The approximation of extreme high energy in Eq. (9.43) is to be expected in order to arrive at a transport equation of classical form. This eikonal form (chapter 8) satisfies the Schrödinger's equation at high energies with a potential given by the optical potential $\mathcal{U}$ plus the force of the direct scattering V_{int}.

Wigner's function [2] for this static problem is defined as

$$
W(\mathbf{r}; \mathbf{p}) \equiv \int d\mathbf{r}_1 \cdots d\mathbf{r}_N \, \Phi_0^*(\mathbf{r}_1, \ldots, \mathbf{r}_N) \int d\eta \exp\left[-i\mathbf{p} \cdot \eta\right] \phi_{\mathbf{k}}\left(\mathbf{r} + \frac{1}{2}\eta\right)
$$

$$
\times \phi_{\mathbf{k}}^*\left(\mathbf{r} - \frac{1}{2}\eta\right)\Phi_0(\mathbf{r}_1, \ldots, \mathbf{r}_N) = \int d\mathbf{r}_1 \cdots d\mathbf{r}_N |\Phi_0|^2 \int d\eta \exp\left[i(\mathbf{k} - \mathbf{p}) \cdot \eta\right]
$$

$$
\times \exp\left[-\frac{i}{v}\int_{-\infty}^{z + \frac{1}{2}\eta_\parallel} \mathcal{U}(\mathbf{b} + \frac{1}{2}\eta_\perp, \zeta)d\zeta\right] \times \exp\left[\frac{i}{v}\int_{-\infty}^{z - \frac{1}{2}\eta_\parallel} \mathcal{U}^*(\mathbf{b} - \frac{1}{2}\eta_\perp, \zeta)d\zeta\right]
$$

$$
\times \exp\left[-\frac{i}{v}\int_{-\infty}^{z + \frac{1}{2}\eta_\parallel - z_i} V_{int}(\mathbf{b} - \mathbf{b}_i + \frac{1}{2}\eta_\perp, \zeta)d\zeta\right] \exp\left[\frac{i}{v}\int_{-\infty}^{z - \frac{1}{2}\eta_\parallel - z_i} V_{int}^*(\mathbf{b} - \mathbf{b}_i - \frac{1}{2}\eta_\perp, \zeta)d\zeta\right],
$$

$$(9.44)$$

where $\eta_\parallel$ and $\eta_\perp$ are the components of η parallel and perpendicular to $\mathbf{k}$.

In order to generate a transport equation we consider the action of $v\partial/\partial z$ on the Wigner function of Eq. (9.44). We divide the calculation into two parts: first we examine the action of the derivative on the optical part of the expression, containing $\mathcal{U}$, to obtain the drift part of the transport equation, and then we let it act on the direct collision pieces, containing V_{int}, to produce the collision term. The first calculation is simple,

$$
v\frac{\partial}{\partial z}W(\mathbf{b}, z; \mathbf{p}) = -i \int d\mathbf{r}_1 \cdots d\mathbf{r}_N |\Phi_0|^2 \int d\eta \exp\left[i(\mathbf{k} - \mathbf{p}) \cdot \eta\right]
$$

$$
\times \left[\mathcal{U}(\mathbf{b} + \frac{1}{b} + \frac{1}{2}\eta_\perp, z + \frac{1}{2}\eta_\parallel) - \mathcal{U}^*(\mathbf{b} - \frac{1}{2}\eta_\perp, z - \frac{1}{2}\eta_\parallel)\right] \exp[\cdots] + \text{collision terms}
$$

$$
\approx -i \int d\mathbf{r}_1 \cdots d\mathbf{r}_N |\Phi_0|^2 \int d\eta \exp\left[i(\mathbf{k} - \mathbf{p}) \cdot \eta_\perp\right]
$$

$$
\times \left[2i \, \mathrm{Im}\mathcal{U}(\mathbf{b}, z) + \eta_\perp \cdot \boldsymbol{\nabla} \, \mathrm{Re}\mathcal{U}(\mathbf{b}, z)\right] \exp[\cdots] + \text{collision terms}
$$

$$
= \left[2 \, \mathrm{Im}\mathcal{U}(\mathbf{b}, z) + [\boldsymbol{\nabla} \, \mathrm{Re}\mathcal{U}(\mathbf{b}, z)] \cdot \boldsymbol{\nabla}_{\mathbf{p}}\right] W(\mathbf{r}, \mathbf{p}) + \text{collision terms}, \quad (9.45)
$$

where we have indicated only schematically the four exponentials of Eq. (9.44) that appear unchanged in each of the subsequent expressions. In Eq. (9.45) we have also used the usual approximation that the momentum $\mathbf{k}$ involved here is sufficiently high so that only small values of η enter in the integrals. Once again this is an approximation that is required to obtain the classical limit. Equation (9.45) can be recast in the form

$$
\left[v\frac{\partial}{\partial z} - 2 \, \mathrm{Im}\mathcal{U}(\mathbf{b}, z) - [\boldsymbol{\nabla} \, \mathrm{Re}\mathcal{U}] \cdot \boldsymbol{\nabla}_{\mathbf{p}}\right] W(\mathbf{r}, \mathbf{p}) = \text{collision terms}. \quad (9.46)
$$

We now turn to the collision term, omitting explicit reference to the contributions of the drift term that we have just calculated. We then have

$$
v\frac{\partial}{\partial z}W(\mathbf{b}, z; \mathbf{p}) = -v \int d\mathbf{r}_i \rho(\mathbf{r}_i) \int d\eta \exp[i(\mathbf{k} - \mathbf{p}) \cdot \eta]
$$

$$
\times \frac{\partial}{\partial z_i} \left\{ \exp\left[-\frac{i}{v} \int_{-\infty}^{z+\frac{1}{2}\eta_\parallel - z_i} V_{\mathrm{int}}(\mathbf{b} - \mathbf{b}_i + \frac{1}{2}\eta_\perp, \zeta)d\zeta \right] \right.
$$

$$
\times \exp\left[\frac{i}{v} \int_{-\infty}^{z-\frac{1}{2}\eta_\parallel - z_i} V_{\mathrm{int}}(\mathbf{b} - \mathbf{b}_i - \frac{1}{2}\eta_\perp, \zeta)d\zeta \right] \right\}
$$

$$
\times \exp\left[-\frac{i}{v} \int_{-\infty}^{z+\frac{1}{2}\eta_\parallel} \mathcal{U}(\mathbf{b} + \frac{1}{2}\eta_\perp, \zeta)d\zeta \right] \exp\left[\frac{i}{v} \int_{-\infty}^{z-\frac{1}{2}\eta_\parallel} \mathcal{U}^*(\mathbf{b} - \frac{1}{2}\eta_\perp, \zeta)d\zeta \right], \quad (9.47)
$$

where we have replaced the required derivative by z with one by z_i. The relevant limits of integration can be extended from $-\infty$ to ∞ if we assume that the longitudinal volume dimension $\Omega_\parallel$ is large relative to the domain of support of the scattering profiles. This is then leads to

$$
v\frac{\partial}{\partial z}W(\mathbf{b}, z; \mathbf{p}) = -v \int d\eta \exp[i(\mathbf{k} - \mathbf{p}) \cdot \eta] \int_\Omega d^2\mathbf{b}_i dz_i \frac{1}{\Omega}
$$

$$
\times \frac{\partial}{\partial z_i} \left\{ \exp\left[-\frac{i}{v} \int_{-\infty}^{z+\frac{1}{2}\eta_\parallel - z_i} V_{\mathrm{int}}(\mathbf{b} - \mathbf{b}_i + \frac{1}{2}\eta_\perp, \zeta)d\zeta \right] \right.
$$

$$
\times \exp\left[\frac{i}{v} \int_{-\infty}^{z-\frac{1}{2}\eta_\parallel - z_i} V_{\mathrm{int}}(\mathbf{b} - \mathbf{b}_i - \frac{1}{2}\eta_\perp, \zeta)d\zeta \right] \right\} \exp[\cdots] \quad (9.48)
$$

where we use Eq. (9.42) explicitly and again suppress the obvious "inert" exponentials, this time involving the optical potential. The integral over z_i leads to

$$
v\frac{\partial}{\partial z}W(\mathbf{b}, z; \mathbf{p}) = -v \int d\eta \exp[i(\mathbf{k} - \mathbf{p}) \cdot \eta] \int_{\Omega_\perp} d^2\mathbf{b}_i \frac{1}{\Omega}
$$

$$
\times \left\{ 1 - \exp\left[-\frac{i}{v} \int_{-\infty}^{\infty} V_{\mathrm{int}}(\mathbf{b} - \mathbf{b}_i + \frac{1}{2}\eta_\perp, \zeta)d\zeta \right] \exp\left[\frac{i}{v} \int_{-\infty}^{\infty} V_{\mathrm{int}}(\mathbf{b} - \mathbf{b}_i - \frac{1}{2}\eta_\perp, \zeta)d\zeta \right] \right\} \exp[\cdots]
$$

$$
= v \int d^2\eta_\perp \int_{-\infty}^{\infty} d\eta_\parallel \int \frac{d\mathbf{p}'}{(2\pi)^3} \exp\left[i(\mathbf{p}'_\perp - \mathbf{p}_\perp) \cdot \eta_\perp \right] \exp\left[i(p'_\parallel - p_\parallel)\eta_\parallel \right]
$$

$$
\int_{\Omega_\perp} d^2\mathbf{b}_i \frac{1}{\Omega} \left\{ [1 + \Gamma_i(\mathbf{b} - \mathbf{b}_i + \frac{1}{b_i} + \frac{1}{2}\eta_\perp)][1 + \Gamma_i^*(\mathbf{b} - \mathbf{b}_i - \frac{1}{2}\eta_\perp)] - 1 \right\} W(\mathbf{r}; \mathbf{p}'), \quad (9.49)
$$

where the integration over the transverse variable $\mathbf{b}_i$ ranges over the transverse cross section $\Omega_\perp$ of the volume Ω, and we have defined the usual profile functions,

$$
\Gamma_i(\mathbf{b}) \equiv \exp\left[-\frac{i}{v} \int_{-\infty}^{\infty} V_i(\mathbf{b}, \zeta)d\zeta \right] - 1. \quad (9.50)
$$

Hence,

$$
v\frac{\partial}{\partial z}W(\mathbf{b}, z; \mathbf{p}) = 2\pi v \int \frac{d\mathbf{p}'}{(2\pi)^3} \delta(p'_\parallel - p_\parallel) \int d^2\eta_\perp \int_{\Omega_\perp} d^2\mathbf{b}_i \frac{1}{\Omega}
$$

$$
\times \left\{ [1 + \Gamma_i(\mathbf{b} - \mathbf{b}_i + \frac{1}{b_i} + \frac{1}{2}\eta_\perp)][1 + \Gamma_i^*(\mathbf{b} - \mathbf{b}_i - \frac{1}{2}\eta_\perp)] - 1 \right\}
$$

$$
\times \exp\left[i(\mathbf{p}'_\perp - \mathbf{p}_\perp) \cdot \eta_\perp \right] W(\mathbf{r}; \mathbf{p}') = 2\pi v \int \frac{d\mathbf{p}'}{(2\pi)^3} \delta(p'_\parallel - p_\parallel) \int d^2\mathbf{B} d^2\mathbf{B}'
$$

$$
\times \frac{1}{\Omega} \left\{ [1 + \Gamma_i(\mathbf{B})][1 + \Gamma_i^*(\mathbf{B}')] - 1 \right\} \times \exp\left[i(\mathbf{p}'_\perp - \mathbf{p}_\perp) \cdot (\mathbf{B} - \mathbf{B}') \right] W(\mathbf{r}; \mathbf{p}'). \quad (9.51)
$$

The two-dimensional Fourier-transform integral over $\mathbf{B}$ and $\mathbf{B}'$ containing both profile functions immediately yields the scattering amplitudes at momentum transfer $\mathbf{p}'_\perp - \mathbf{p}_\perp$, assuming the transverse volume $\Omega_\perp$ is large relative to the domain of support of the scattering profile. Then

$$v\frac{\partial}{\partial z}W(\mathbf{b}, z; \mathbf{p}) = 2\pi v \int \frac{d\mathbf{p}'}{(2\pi)^3}\delta(p'_\parallel - p_\parallel)\left(\frac{2\pi}{k}\right)^2\frac{d\sigma_i(\mathbf{p}' \to \mathbf{p})}{d\Omega}W(\mathbf{r}; \mathbf{p}')\rho_i(\mathbf{r}), \qquad (9.52)$$

where we have suppressed terms with transverse delta functions which do not contribute. We finally arrive at

$$v\frac{\partial}{\partial z}W(\mathbf{b}, z; \mathbf{p}) = \frac{1}{mk}\int d^2\mathbf{p}'_\perp \frac{d\sigma_i(\mathbf{p}' \to \mathbf{p})}{d\Omega}W(\mathbf{r}; \mathbf{p}'_\perp, p_\parallel)\rho_i(\mathbf{r}) \qquad (9.53)$$

for the collision term.

Putting the drift and collision pieces together we arrive at the Boltzmann transport equation

$$\left[v\frac{\partial}{\partial z} - 2\,\mathrm{Im}\,\mathcal{U} - \frac{\partial\mathcal{U}}{\partial z}\frac{\partial}{\partial p_\parallel}\right]W(\mathbf{b}, z; \mathbf{p}'_\perp, p_\parallel) = \int \frac{d^2\mathbf{p}'_\perp}{mk}\frac{d\sigma_i(\mathbf{p}' \to \mathbf{p})}{d\Omega}W(\mathbf{b}, z; \mathbf{p}'_\perp, p_\parallel)\rho_i(\mathbf{r}), \quad (9.54)$$

which is the result arrived in the previous section.

Eq. (9.54) is the expected Boltzmann–Vlasov equation with the role of the drift potential played by the background optical potential and explicit scattering appearing on the right-hand side as a collision term [3]. This result is derived here in a relatively simple manner since we have been prepared to restrict ourselves from the very start to lowest-order features. In order to produce this classical transport equation the main assumptions have been (i) that the projectile enters the system at extremely high energy; (ii) that the scatterers within the bombarded system are completely without correlation; and (iii) that they are contained in a volume much larger than the support domain of the individual scatterings.

9.4 Numerical treatment of transport equations

The output of Eq. (9.23) is the distribution function $f(\mathbf{r}, \mathbf{p}, t)$, which allows one to calculate many properties of heavy-ion collisions. Let us quote *collective flows*, proton and neutron production rates, (sub-threshold and above threshold) pion and kaon yields, etc. Combining Eq. (9.23) with a phase-space coalescence model, one can also calculate such quantities as exclusive flows and intermediate fragment formations.

In order to numerically solve Eq. (9.23) one needs to go through the following general steps: initialization, mean field propagation, collisions and Pauli blocking. The solution of (9.23) is usually Monte Carlo simulated by using the test-particle method. According to this model the dynamics is traced by the one-body distribution function $f(\mathbf{r}, \mathbf{p}, t)$ expanded in terms of a set of generating functions centered on a finite number of points, Monte Carlo distributed in the whole phase-space. In this way the dynamics of nucleons is replaced by the dynamics of test-particles. Between two collisions a test-particle propagates following a classical trajectory determined by Newton-type equations. In order to have a good approximation of the exact continuous distribution function, $f(\mathbf{r}, \mathbf{p}, t)$, the number of test-particles per nucleon should be large enough. This requirement brings about, in the case of a large nucleus, a fast increase of the CPU time needed for running serial-code simulations.

Let us briefly discuss the general aspects of a typical numerical algorithm. In the first step one prepares nuclei in the ground state by discretizing the continuous distribution function as a sum of elementary functions. Here we describe them in terms of Gaussian functions both in coordinate and momentum space, with fixed widths σ_r and σ_p:

$$f(\mathbf{r}, \mathbf{p}, t) = \frac{1}{n_g (4\pi^2\sigma_r\sigma_p)^{3/2}} \sum_{i=1,n} \exp\left[-(\mathbf{r} - \mathbf{r}_i)^2/2\sigma_r^2\right]\exp\left[-(\mathbf{p} - \mathbf{p}_i)^2/2\sigma_p^2\right], \qquad (9.55)$$

where n_g is the number of generating functions per nucleon. This number should be quite large in order to have a good approximation of the exact continuous distribution function $f(\mathbf{r}, \mathbf{p}, t)$. The total number of test-particles ("Gaussians") is $N = n_g A$, where A is the total number of nucleons

θ [degrees]	90	45	40.4	15.4	15	10	5.7	2.1
η	0	0.8	1	2	2.03	2.44	3	4

TABLE 9.1 Relation between the scattering angle (in degrees) and the pseudorapidity.

in the nuclear system. The ground state is prepared by a Monte Carlo sampling of the phase space with a variational self- consistent procedure to reproduce the nuclear binding energy.

Once the initial phase-space configuration of the test particles in the two ground state nuclei is fixed, the two nuclei are translated and boosted to the center of mass frame where the calculation is performed. The evolution in time of the system is controlled by dividing the total reaction time in small time steps, dt (of the order of 0.5 fm/c). During a time step interval the test-particles are propagated freely in phase-space along the classical trajectories determined by the Newton equations, with the force term given by the derivative of the mean field. Actually, in the case of Gaussian generating functions (Eq. (9.55)), it can be shown [5] that the dynamics of test-particles ("the Gaussians") is given by Ehrenfest type equations, with the force term replaced by the convoluted derivatives of the mean field over the given Gaussian:

$$\frac{d\mathbf{r}_i}{dt} = \frac{\mathbf{p}_i}{m} + \langle \boldsymbol{\nabla}_p U\left(\mathbf{r},\mathbf{p}\right)\rangle_{\mathbf{r}_i,\mathbf{p}_i} \quad \text{and} \quad \frac{d\mathbf{p}_i}{dt} = \frac{\mathbf{r}_i}{m} + \langle \boldsymbol{\nabla}_r U\left(\mathbf{r},\mathbf{p}\right)\rangle_{\mathbf{r}_i,\mathbf{p}_i}. \tag{9.56}$$

At the end of each time-step the phase space is searched for allowed collisions. The algorithm for simulating the collision integral is based on the mean free path, λ. The procedure is as follow: for a given test particle one searches a possible scattering partner taken as the closest test-particle inside a sphere of a given radius. Then one estimates the averaged mean free path as $\lambda = (\rho\sigma)^{-1}$, where ρ is the local averaged density and is σ the cross section corresponding to the relative kinetic energy of the partners. Dividing the calculated mean free path by the relative velocity of the two test particles, one finds the averaged lifetime between two collisions, dt_{coll}. In terms of dt_{coll} the probability for scattering is

$$P = 1 - \exp\left(-dt/dt_{coll}\right). \tag{9.57}$$

If dt is chosen as to have $dt << dt_{coll}$ then P can be approximated by dt/dt_{coll}. After the probability P is calculated the decision for the scattering is made by the Monte Carlo method: the scattering is decided if P is greater than a generated random number smaller than one. As soon as a collision is decided, the final momenta of the two scattered test-particles are randomly generated, with the momentum-energy conservation constraints. The final decision for the scattering is taken only if the final scattering states are not Pauli-blocked. The Pauli-blocking factor is , $(1 - f_1)(1 - f_2)$ where f_1 and f_2 are the one-body distribution functions calculated in the phase-space points corresponding to the final states of the scattered test-particles. The decision about the Pauli-blocking is taken again by Monte Carlo method.Before proceeding a few words on the variables used in experimental analysis are necessary. The *rapidity* is a variable frequently used to describe the behavior of particles in inclusively measured reactions. It is defined by

$$y = \frac{1}{2}\ln\left(\frac{E + p_{\parallel}}{E - p_{\parallel}}\right) \tag{9.58}$$

which corresponds to

$$\tanh y = \frac{p_{\parallel}}{E}, \tag{9.59}$$

where y is the rapidity, $p_{\parallel}$ is the longitudinal momentum along the direction of the incident particle, E is the energy, both defined for a given particle. The accessible range of rapidities for a given interaction is determined by the available center-of-mass energy and all participating particles' rest masses. One usually gives the limit for the incident particle, elastically scattered at zero angle:

$$|y_{max}| = \ln\left[\frac{E + p}{m}\right] = \ln\left(\gamma + \gamma\beta\right), \tag{9.60}$$

where $\beta \equiv v$ is the velocity and all variables referring to the through-going particle given in the desired frame of reference (e.g. in the center of mass).

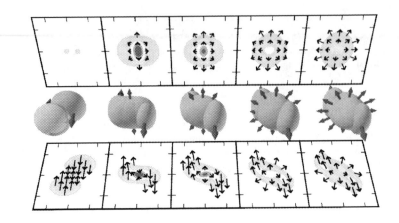

FIGURE 9.4   Matter distribution at several time stages of a nearly central nucleus-nucleus collision.

Note that $\partial y / \partial p_\parallel = 1/E$. A Lorentz boost β along the direction of the incident particle adds a constant, $\ln(\gamma + \gamma\beta)$, to the rapidity. Rapidity differences, therefore, are invariant to a Lorentz boost. Statistical particle distributions are flat in y for many physics production models. Frequently, the simpler variable *pseudorapidity* η is used instead of rapidity (and sloppy language mixes up the two variables).

The pseudorapidity is a handy variable to approximate the rapidity if the mass and momentum of a particle are not known. It is an angular variable defined by

$$\eta = -\ln\left[\tan\left(\frac{\theta}{2}\right)\right], \qquad (9.61)$$

whose inverse function is

$$\theta = 2\arctan\left(e^{-\eta}\right),$$

where θ is the angle between the particle being considered and the undeflected beam. η is the same as the rapidity y if one sets $\beta = 1$ (or $m = 0$). Statistical distributions plotted in pseudorapidity rather than rapidity undergo transformations that have to be estimated by using a kinematic model for the interaction.

Table 9.1 shows the relation between θ and η for some round values.

9.4.1   Head-on collisions

Next we shall present simulations carried out within a Boltzmann equation (BE) transport model with explicit nucleon, deuteron, $A = 3$ cluster, pion, and delta and N^* degrees of freedom [7]. All the particles X are represented in terms of their distribution functions f_X. The distribution functions satisfy equations of the form (9.23) with different collision integrals I. Pions within the model are produced in two steps. First a Δ or N^* resonance is produced in an NN collision, and then the resonance decays into a nucleon and pion. Clusters are produced in few-body processes that are inverse to cluster break-up.

The dynamics of the central high-energy reactions can be broken down into several stages. A 400 MeV/nucleon Au+Au system at $b = 0$ will serve to illustrate our points. Unless otherwise indicated the results of simulations correspond to a *stiff* equation of state with an incompressibility of $K = 380$ MeV.

If a nuclear system were very large, then it would be necessarily ruled by laws of hydrodynamics. In the initial state of a head-on reaction, a discontinuity exists in the velocity field at the contact area of two nuclei, with the velocities in the nuclei directed opposite. Within hydrodynamics, such an initial discontinuity would have to break at finite times into two shock fronts traveling in opposite directions into the projectile and target as seen in Figure 9.4.

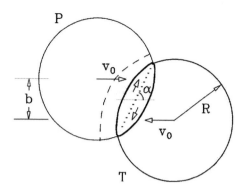

FIGURE 9.5   The initial discontinuity between the projectile and target velocities breaks at a finite b into two shock fronts propagating into the projectile P and target T (thick solid lines), and a weak tangential discontinuity in-between (dotted line). Shock-front position *within* projectile at a later time is indicated with a dashed line.

Let us now turn attention to the expansion. The hot matter in-between the shocks is exposed in transverse directions to vacuum and, as the pressure of the hot matter is finite and that of vacuum is zero, the matter begins to expand collectively into transverse directions. The features of this process may be understood at a qualitative level in terms of the self-similar cylindrically-symmetric hydrodynamic expansion. In the self-similar expansion, the velocity is proportional to the distance from symmetry axis, $v = (t)r$, which roughly holds in the direction of 90° in the Au + Au collision at 400 MeV/nucleon, at the center of the system. On solving, in a near-analytic manner, the hydrodynamic equations of motion for the expansion [7], one arrives at the expectation that the matter should accelerate collectively in the transverse direction for a time R/c_s, where c_s is the speed of sound. This time is such as necessary for a signal to propagate into the interior. It is generally consistent, $R/c_s \sim 20$ fm/c, with the time during which collective transverse energy, calculated from the local collective velocities

$$E_{coll}^{\perp} = \int d\mathbf{r}d\mathbf{p} f(\mathbf{r},\mathbf{p},t)\,m_N (v^{\perp})^2/2, \qquad (9.62)$$

rises in the simulation displayed in Figure 9.4.

After shocks reach the vacuum after a time $\sim (2R/v_0)(\rho_1 - \rho_0)/\rho_1 \sim 15$ fm/c, where v_0 is the initial velocity in the c.m. and ρ_1 is the density in the shocked region, an expansion along the beam axis sets in. However, as the expansion in transverse directions is already in progress and matter becomes decompressed, the expansion in longitudinal direction does not acquire same strength.

9.4.2   Semicentral reactions

In the initial state, an angle α of inclination of the plane of discontinuity in velocity, relative to the beam axis, is given nonrelativistically by

$$\cos \alpha \approx \frac{b}{2R}. \qquad (9.63)$$

While at $b = 0$ velocities are normal to the plane of discontinuity, the velocities have finite tangential components relative to that plane at finite b. Such components are continuous across the shocks which detach from the initial plane of discontinuity and, with tangential components being directed opposite for the matter from a projectile and from a target, at the center of the system a so-called tangential discontinuity develops, see the sketch in Figure 9.5. Density, pressure and entropy are continuous across that discontinuity. Only the tangential velocity component changes.

As only the normal velocity component drops to zero across shocks, the shocks are weaker at a finite b than in a head-on reaction. With the normal component being equal to $v_0 \sin \alpha$, the effective

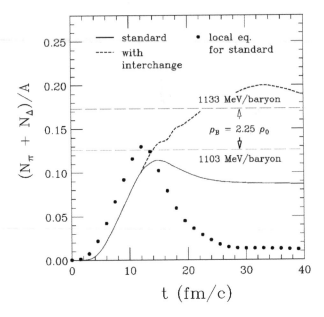

FIGURE 9.6   Time dependence of the number of pions and deltas normalized using the baryon number, in the $b = 0$ La + La reaction at 800 MeV/nucleon, from a standard calculation (solid line) and from a calculation with an interchange of particle positions (short-dashed line). The two horizontal long-dashed lines show the equilibrium number of pions and deltas for a baryon density $\rho = 2.25\rho_0$ and an energy per baryon of 1133 MeV and 1103 MeV, respectively. Dots show the number obtained by integrating over space the local equilibrium values of pion and delta density in a standard calculation.

c.m. kinetic energy for a shock at a finite b becomes

$$\frac{E_{\text{lab}}}{4} \cdot \sin^2 \alpha = \frac{E_{\text{lab}}}{4} \left(1 - \cos^2 \alpha\right) = \frac{E_{\text{lab}}}{4} \left(1 - \frac{b^2}{4R^2}\right), \tag{9.64}$$

rather than $E_{lab}/4$ in the nonrelativistic approximation. In consequence the density behind a shock at a finite b should be lower than at $b = 0$ and it should, in fact, coincide with the density in a $b = 0$ reaction at beam energy reduced by a factor $(1 - b^2/4R^2)$.

9.4.3   Effects on observables

Just as at $b = 0$ in a heavy system, an expansion develops at a finite b in-between the shock waves, in effect of an exposure of the hot matter to vacuum. Anisotropy in the expansion, with regard to the direction perpendicular to the reaction plane and the direction of the shock motion, results from the delay in the start of expansion in the second of these directions, as in the case $b = 0$. The anisotropy (squeeze-out) may be quantified with a ratio of the eigenvalue of the kinetic energy tensor associated with the direction out of the reaction plane, to the smaller of the eigenvalues associated with the directions within the plane,

$$R_{21} = \frac{\langle E_2 \rangle}{\langle E_1 \rangle}. \tag{9.65}$$

As $b \to 0$ in a heavy system (but not in light), the ratio R_{21} tends to $\langle E^\perp \rangle/2\langle E^\| \rangle$. The Au + Au data for the ratio [8] are compared to the results of calculations, corrected for the experimental inefficiencies, in Figure 9.7. At 250 MeV/nucleon and 400 MeV/nucleon a degree of agreement between the data and the calculations is observed. The inefficiencies reduce the sensitivity of the ratio to the nuclear compressibility.

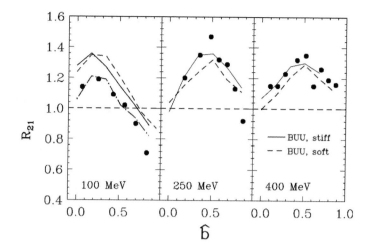

FIGURE 9.7 Ratio of the out-of-reaction-plane mean energy-component to the lower in-plane eigenvalue of the kinetic energy tensor, as a function of reduced impact parameter, in Au + Au collisions at beam energies indicated in the panels. Circles represent the measurements of Ref. [8]. Solid and dashed lines represent the results of the transport-model calculations with mean fields yielding a stiff and soft equation of state, respectively. T

With regard to the directions within the reaction plane, both the tangential motion behind the shock waves for finite impact parameters, and the anisotropy in expansion, contribute to the sideward deflection quantified usually in terms of the slope of mean momentum within the reaction plane at midrapidity, $F = d\langle p^x/m\rangle/dy$. If the motion associated with the weak discontinuity were not contributing to the sideward deflection, then the slope would have been limited from above [7] by $\chi/(2\sqrt{1+\chi})$ where $\chi = (\langle E^\perp\rangle/2 - \langle E^\parallel\rangle)/\langle E^\parallel\rangle$ and the energy components are from a $b = 0$ reaction.

As $b \to 0$, the expansion of matter between the shocks, favoring the transverse direction of motion, competes with the transparency effects in the corona, that favor the longitudinal direction. For light systems in simulations, such as Ca + Ca, the transparency effects prevail, leading to $\langle E^\perp\rangle/2\langle E^\parallel\rangle < 1$ at $b = 0$, but not so for heavy systems such as Au + Au. The experimental detection of $\langle E^\perp\rangle/2\langle E^\parallel\rangle > 1$, i.e., $\langle E^\perp\rangle/\langle E^\parallel\rangle > 2$, would demonstrate a violence of the nuclear hydrodynamic phenomena in heavy systems. On the other hand, if for any reason the in-medium NN cross sections were lower than the free-space cross-sections used in the simulations, then the transparency effects could prevail over the anisotropy in the expansion, even within heavy systems. A clear experimental answer requires a full 2π coverage either in the forward or in the backward c.m. hemisphere.

The transport simulations demonstrate various effects of the anisotropy in the collective expansion on observables from the collisions. However, the collective expansion should also affect observables at any one angle. Thus, in a globally equilibrated system, exponential spectra would be expected at any angle, with the same slope parameter for different particles, equal to the temperature,

$$\frac{dN_x}{dp^3} \propto \exp(-\frac{E_x}{T}). \tag{9.66}$$

In the presence of the collective expansion, on the other hand, the particle spectra should exhibit different slopes for particles with different mass, with particle spectrum becoming flatter as particle mass increases. This is due to an increased sensitivity to the collective motion with an increasing mass; for the average kinetic energy of a particle, one would, in fact, expect a linear rise with the mass number,

$$\langle E_x\rangle = \frac{3}{2}T + \frac{m_x\langle v^2\rangle}{2} = \frac{3}{2}T + A_x\frac{m_N\langle v^2\rangle}{2}, \tag{9.67}$$

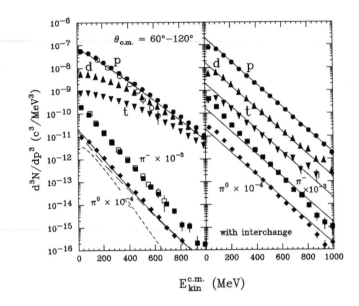

FIGURE 9.8 Momentum distribution of protons (circles), deuterons (triangles), helions (inverted triangles), negative (squares) and neutral (diamonds) pions from central 800 MeV/nucleon La + La reaction, in the vicinity of 90° in the c.m. Left panel shows the results of calculations at $b = 1$ fm (filled symbols) and data of Ref. [12] (open symbols). Right panel shows the results of calculations with particle positions interchanged during evolution. Solid lines in the left panel indicate results of the instantaneous freeze-out model, for protons and neutral pions. Long- and short-dashed lines for neutral pions indicate a contribution of free pions at freeze-out and a contribution from Δ decays, respectively. Straight parallel lines in the right panel serve to guide the eye.

where v is collective velocity and a uniform temperature T is assumed.

9.4.4 Mesons and expansion

Variation of the sum of pion and delta number in a central 800 MeV/nucleon La + La reaction is shown in Figure 9.6. This sum would be equal to the final pion number if interactions stopped at any one time instant. The sum maximizes at the time $t \sim 13.5$ fm/c when shocks reach the vacuum, and the sum decreases as system expands. The decrease contrasts with naive expectations; one would expect the sum to rise continuously tending towards some equilibrium value until the freeze-out. The number of pions and deltas expected under conditions of global equilibrium, given a representative baryon density and total kinetic energy per nucleon of 1133 MeV including mass at $t \sim 13.5$ fm/c, is indicated in the Figure by the upper of two horizontal long-dashed lines. The actual number turns down when away from that expectation.

Figure 9.8 shows, in its left panel, the momentum distributions of different particles emitted from the central La + La reactions at 800 MeV/nucleon, in the vicinity of 90° in the c.m. The spectra are not quite exponential and, generally, exhibit different slopes for particles with different mass. After correlations characteristic for collective motion are destroyed, though, the spectra turn out nearly exponential with the same slope for different particles, see right panel in Figure 9.8.

9.5   Structure of hadrons

The baryons (nucleons and hyperons) and the mesons, the properties of which we described in former sections, are the particles that obey the strong interaction. They receive, by that feature, the generic name of *hadrons*, which main attributes we have exposed in Table 9.2.

	B	S	t	t_z	s	m (MeV/c^2)
p	$+1$	0	$\frac{1}{2}$	$+\frac{1}{2}$	$\frac{1}{2}$	938.272
n	$+1$	0	$\frac{1}{2}$	$-\frac{1}{2}$	$\frac{1}{2}$	939.565
$\bar{\text{p}}$	-1	0	$\frac{1}{2}$	$-\frac{1}{2}$	$\frac{1}{2}$	938.272
$\bar{\text{n}}$	-1	0	$\frac{1}{2}$	$+\frac{1}{2}$	$\frac{1}{2}$	939.565
Λ	$+1$	-1	0	0	$\frac{1}{2}$	1115.68
Σ^+	$+1$	-1	1	$+1$	$\frac{1}{2}$	1189.4
Σ^0	$+1$	-1	1	0	$\frac{1}{2}$	1192.6
Σ^-	$+1$	-1	1	-1	$\frac{1}{2}$	1197.4
$\bar{\Lambda}$	-1	$+1$	0	0	$\frac{1}{2}$	1115.68
$\bar{\Sigma}^+$	-1	$+1$	1	-1	$\frac{1}{2}$	1189.4
$\bar{\Sigma}^0$	-1	$+1$	1	0	$\frac{1}{2}$	1192.6
$\bar{\Sigma}^-$	-1	$+1$	1	$+1$	$\frac{1}{2}$	1197.4
Ξ^0	$+1$	-2	$\frac{1}{2}$	$+\frac{1}{2}$	$\frac{1}{2}$	1315
Ξ^-	$+1$	-2	$\frac{1}{2}$	$-\frac{1}{2}$	$\frac{1}{2}$	1321
$\bar{\Xi}^0$	-1	$+2$	$\frac{1}{2}$	$-\frac{1}{2}$	$\frac{1}{2}$	1315
$\bar{\Xi}^-$	-1	$+2$	$\frac{1}{2}$	$+\frac{1}{2}$	$\frac{3}{2}$	1321
Ω^-	$+1$	-3	0	0	$\frac{3}{2}$	1672
π^0	0	0	1	0	0	134.976
π^+	0	0	1	$+1$	0	139.567
π^-	0	0	1	-1	0	139.567
K$^+$	0	$+1$	$\frac{1}{2}$	$+\frac{1}{2}$	0	493.7
K$^-$	0	-1	$\frac{1}{2}$	$-\frac{1}{2}$	0	493.7
K^0	0	$+1$	$\frac{1}{2}$	$-\frac{1}{2}$	0	497.7
$\bar{\text{K}}^0$	0	-1	$\frac{1}{2}$	$+\frac{1}{2}$	0	497.7

TABLE 9.2 Attributes of particles that interact strongly. The baryons are the particles with baryonic number $B \neq 0$; the mesons have $B = 0$. S is the strangeness, t the isotopic spin and t_z its projection; s is the particle spin and m its mass. Baryons have positive intrinsic parity, mesons have negative ones.

The quantum attributes of the particles in Table 9.2. are as following[*]. The direction of the third axis in charge space is chosen in such a way that $t_z = +\frac{1}{2}$ for the proton and $t_z = -\frac{1}{2}$ for the neutron. The pion has three charge states; thus it has isospin $t = 1$; the three pions form a charge *multiplet*, or isospin multiplet, with multiplicity $2t + 1 = 3$. The state $t_z = +1$ is attributed to the π^+, $t_z = 0$ to the π^0 and $t_z = -1$ to the π^-. This attribution is connected to the convention that was adopted for the nucleons and is necessary for the validity of Eq. (9.68) that follows.

The isospin modulus is an invariant quantity in a system governed by the strong interaction. In the electromagnetic interactions this quantity is not necessarily conserved and we shall have the opportunity to verify ahead that this is the only conservation law that has different behavior in relation to these two forces.

The number of nucleons before and after a reaction is always the same. This suggests the introduction of a new quantity, B, called *baryonic number*, that is always conserved in reactions. We attribute to the proton and to the neutron the baryonic number $B = 1$, and to the antiproton and antineutron $B = -1$. To the pions we ascribe $B = 0$ (in the same way that for electrons, neutrinos, muons and photons). In this way the conservation of baryonic number is extended to all reactions. This principle has its correspondence to the leptons, being defined in that case a *leptonic number*, which is also conserved in reactions.

From the isospin and baryonic number definition we can write the charge q, in units of e, as

$$q = t_z + \frac{B}{2}. \tag{9.68}$$

Since the antiparticle of a particle of charge q and baryonic number B has charge $-q$ and baryonic number $-B$, it must have also a third isospin component $-t_z$, where t_z is the isospin z-component of the corresponding particle.

---

[*]In the previous chapters we have used a different choice for the direction of this axis, namely, $t_z = -\frac{1}{2}$ for the proton and $t_z = +\frac{1}{2}$ for the neutron. This is just for convenience.

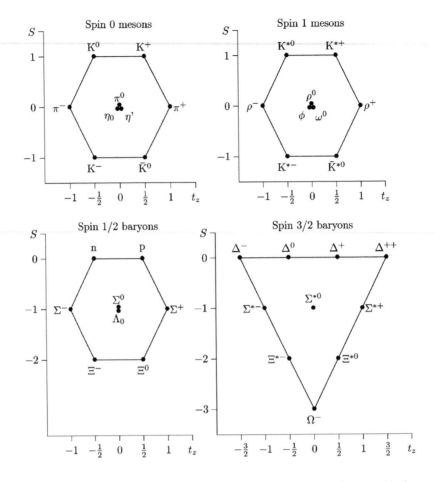

FIGURE 9.9   Strangeness versus t_z-component of the isospin for the several types of hadrons.

In 1953, Gell-Mann [13], and Nishijima and Nakano [14], have shown that the production of strange particles (e.g., K^0, Λ^0, Ξ^-, etc) could be explained by the introduction of a new quantum number, the *strangeness*, and postulating that the strangeness is conserved in the strong interactions. For example, two strange particles, but with opposite strangeness, could be produced by means of the strong interaction in a collision between a pion and a nucleon. The strangeness, however, is not conserved in the decay of a strange particle, and this decay is attributed to the weak interaction.

The strangeness, S, the baryonic number, B, the isospin z-component, t_z, and the charge, q, can be related by

$$q = t_z + \frac{B}{2} + \frac{S}{2}. \tag{9.69}$$

For $S = 0$, the above equation reduces to the relation (9.68) for nucleons and pions. Then, nucleons and pions have strangeness $S = 0$.

The great number of hadrons and their apparently complex distribution lead several investigators to question if these particles would not be complex structures composed by the union of simpler entities. Models were proposed for those structures and, after some unsuccessful attempts, a model independently created by M. Gell-Mann [15] and G. Zweig [16], in 1964, imposed itself and gained credibility along the time. The inspiration for that model came from the symmetries observed when one put mesons and baryons in plots of strangeness versus the t_z-component of isospin, as showed in Figure 9.9. The type of observed symmetry is a characteristic of the group called SU$_3$, where three basic elements can generate singlets (the mesons η' and ϕ), octets (the

Flavor	Charge	Spin	Strangeness
up	+2/3	1/2	0
down	-1/3	1/2	0
strange	-1/3	1/2	-1

TABLE 9.3   Quarks characteristic quantum numbers.

other eight mesons of the figures above and the eight spin 1/2 baryons) and decuplets (the spin 3/2 baryons). These three basic elements, initially conceived only as mathematical entities able to generate the necessary symmetries, ended acquiring the status of truly elementary particles, to which Gell-Mann coined the name *quarks*. To obtain hadronic properties, these three quarks, presented in *flavors up, down* and *strange* must have the characteristic values shown in Table 9.3.

The most striking fact is that, for the first time, the existence of particles with fractional charge (a fraction of the electron charge) were admitted. We can in this way construct a nucleon by composing three quarks:

$$\text{proton} \longrightarrow uud$$
$$\text{neutron} \longrightarrow udd$$

and is natural to attribute to quarks a baryonic number $B = 1/3$.

The pions, by their turn, are obtained by the junction of a quark and an antiquark:

$$\pi^+ \longrightarrow u\bar{d}$$
$$\pi^0 \longrightarrow d\bar{d}$$
$$\pi^- \longrightarrow d\bar{u}$$

where the properties of antiparticle for the quarks are obtained in the conventional way.

To reproduce the other baryons and mesons, the strange quarks have to play a role and an hyperon like the Σ^0, for example, has the constitution and a meson K^+

$$K^+ \longrightarrow u\bar{s}.$$

It is convenient to say at this point that a certain combination of quarks does not necessarily lead to only one particle. In the case of the combination above, we also have the possibility to build with it the hyperon

$$\Sigma^{*0} \longrightarrow uds.$$

The reason for this is that, besides other quantum numbers that will be discussed ahead, a combination of three fermions can give place to particles with different spin. If we consider null the quarks total orbital angular momentum, which is true for all particles that we discussed, the total spin of the three quarks can be 1/2 or 3/2. The hyperon Σ^0 corresponds to the first case and the hyperon Σ^{*0} to the last.

A first difficulty in the theory appears when we examine the particles

$$\Delta^{++} \longrightarrow uuu$$
$$\Delta^- \longrightarrow ddd$$
$$\Omega^- \longrightarrow sss.$$

Since the three quarks in each case are fermions with $l = 0$, it is immediate that at least two of them would be in the same quantum state, which violates Pauli principle. To overcome this difficulty, one has introduced a new quantum number, the *color*: the quarks, besides the flavors up, down or strange, also have a color, red (R), green (G) or blue (B), or an *anticolor*, $\bar{R}$, $\bar{G}$ or $\bar{B}$. It is clear that, in the same way as the flavor, the color has nothing to do with the usual notion that we have of that property. The introduction of this new quantum number solves the above difficulty, since now a baryon like the Δ^{++} is written

$$\Delta^{++} \longrightarrow u_R u_G u_B,$$

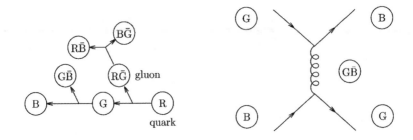

FIGURE 9.10 (a) Forces between quarks mediated by gluon exchange. (b) Diagram showing how a quark B changes to a quark G, and vice-versa, by the exchange of a gluon $G\bar{B}$.

the problems with the Pauli principle being eliminated. The adding of three new quantum numbers expands in excess the possibility of construction of hadrons but a new rule comes to play, limiting the possibilities of color combination: *all the possible states of hadrons are colorless*, where colorless in this context means absence of color or white color. The white is obtained when, in the building of a baryon, one adds three quarks, one of each color, and in this sense works the analogy with the common colors, since the addition of red, green and blue gives white. In building a meson, the absence of color results from the combination of a color and the respective anticolor. Another way to present this property is to understand the anticolor as the complementary color. In this case, the analogy with the common colors also works and the pair color-anticolor also results in white.

The concept of color is not only useful to solve the problem of Pauli principle obedience. It has a fundamental role in quark interaction processes. The accepted theory for this interaction establishes that the force between quarks works by the exchange of massless particles, with spin 1, called *gluons*. These gluons always carry one color and one different anticolor and in the mediation process they interchange the respective colors; one example is seen in Figure 9.10; one can also see in Figure 9.11 that the gluons themselves can emit gluons.

The fields around hadrons where exchange forces act by means of colors are denominated *color fields* and the gluons, the exchanged particles, turn to be the field particles of the strong interaction. In this task they come to replace the pions that, in the new scheme, are composite particles. The fact that the gluons have colors and can interact mutually makes the study of color fields (quantum chromodynamics) particularly complex.

Despite the success of the *quark model*, new difficulties arose and, in 1970, with the purpose to explain some decay times in disagreement with the predictions of the model, S. L. Glashow, J. Iliopoulos and L. Maiani [17] proposed the existence of a fourth quark whose flavor has received the designation of *charm* (c). This c quark has a charge of $+2/3$; it has strangeness zero but has a new quantum number, the charm C, with an attributed value $C = 1$. The proposition of quark c has received an indirect experimental confirmation in 1974, when two independent laboratories detected a new particle, called Ψ by the Stanford linear accelerator group (SLAC) and J by the Brookhaven National Laboratory team. The particle J/Ψ, as it is commonly designated, could be interpreted as a $c\bar{c}$ state, called *charmonium*, by analogy with the positronium e$\bar{\text{e}}$. The existence of particles with charm introduces some complication in the symmetries of Figure 9.9: an axis with the new quantum number is added and the new symmetries have to be searched in a three-dimensional space.

In 1977, a group of resonances in 10 GeV proton-proton reactions pointed to the existence of a new meson, that received the name Υ and that lead to the proposition of a new quark. This quark, b (from *bottom* or *beauty*), has charge $-1/3$ and a new quantum number, the *beauty* B^*. The quark b has $B^* = -1$.

Theoretical reasons make one believe that quarks exist in pairs and this lead us to a sixth flavor, that corresponds to quark t (from *top* or *true*), with charge $+2/3$. This quark was identified in experiments conducted at Fermilab in 1993.

The proposition of quarks, with its colors and flavors, have created a scheme in which a great number of experimental facts could be explained. But, the real existence of quarks as particles could

Quarks	Charge	Spin	Strangeness	Charm	Beauty	Truth
u	+2/3	1/2	0	0	0	0
d	−1/3	1/2	0	0	0	0
s	−1/3	1/2	−1	0	0	0
c	+2/3	1/2	0	1	0	0
b	−1/3	1/2	0	0	−1	0
t	+2/3	1/2	0	0	0	1

Leptons	Mass (MeV/c^2)	Charge	Spin	Half-life (s)
e^-	0.511	−1	1/2	∞
ν_e	0	0	1/2	∞
μ^-	105.66	−1	1/2	2.2×10^{-6}
ν_μ	0	0	1/2	∞
τ^-	1784	−1	1/2	3.4×10^{-13}
ν_τ	0	0	1/2	∞

Field particles	Mass (GeV/c^2)	Charge	Spin
Photon	0	0	1
W^+	81	1	1
Z^0	93	0	1
Gluons	0	0	1
Graviton	0	0	2

TABLE 9.4 Properties of the elementary particles - in the upper Table each quark can appear in three colors, R, G and B. Only one member of the pair particle-antiparticle appears in the table.

not yet be established. Although high-energy electron beams had detected an internal structure in nucleons with all the features of quarks [18], one could never pull out a quark from a hadron and study its properties separately. Models have been proposed to eliminate this possibility, *confining* quarks permanently to the hadrons. One consequence is that their masses can not be directly determined, since it depends on the binding energies, which are also unknown.

In any way, these difficulties are not enough to upset the model prestige and, with its acceptance, there is a substantial reduction in the number of elementary particles, that is, the point particles without an internal structure. These would be the quarks, the leptons and the field bosons. A sketch of the properties of these particles is shown in Table (9.4).

9.6 Quantum chromodynamics

Until now in this book the Schrödinger equation was considered as the fundamental equation governing the dynamics of quantum many-body systems (nuclei). It is well know however that quantum chromodynamics (QCD) is the fundamental theory for strongly interacting particles. In this section we make a brief description of QCD. The formalism is best described in terms of the Lagrangian formalism. Within this approach, the *Euler-Lagrangian equations* yield the equations of motion for the fundamental particles.

Strong interaction is indeed the strongest force of nature. It is responsible for over 80% of the baryon masses, and thus for most of the mass of everything on Earth. Strong interactions bind nucleons in nuclei which, being then dressed with electrons and bound into molecules by the much weaker electromagnetic force, give rise to the variety of the physical world.

Quantum Chromodynamics is the theory of strong interactions. The fundamental degrees of freedom of QCD, quarks and gluons, are already well established even though they cannot be observed as free particles, but only in color neutral bound states (*confinement*) . Today, QCD has firmly occupied its place as part of the *Standard Model of Particle Physics* (for a good introduction to the field, see Ref. [19]). However, understanding the physical world does not only mean understanding its fundamental constituents; it means mostly understanding how these constituents interact and bring into existence the entire variety of physical objects composing the universe. Here, we try to explain why high-energy nuclear physics offers us unique tools to study QCD.

QCD emerges when the naïve quark model is combined with local SU(3) gauge invariance. We will just summarize some of the main accomplishments in QCD. For an introduction to the concepts discussed here, we refer to, e.g., Ref. [20]. One can define a quark-state "vector" with three components,

$$q(x) = \begin{pmatrix} q^{\text{red}}(x) \\ q^{\text{green}}(x) \\ q^{\text{blue}}(x) \end{pmatrix}, \tag{9.70}$$

where $q^{\text{color}}(x)$ are field quantities which depend on the space-time coordinate $x = (t, \mathbf{r})$. The transition from quark model to QCD is made when one decides to treat color similarly to the electric charge in electrodynamics. The entire structure of electrodynamics emerges from the requirement of local gauge invariance, i.e. invariance with respect to the phase rotation of electron field, $\exp(i\alpha(x))$, where the phase α depends on the space–time coordinate. One can demand similar invariance for the quark fields, keeping in mind that while there is only one electric charge in quantum electrodynamics (QED), there are three color charges in QCD.

To implement this program, one requires the free quark Lagrangian,

$$\mathcal{L}_{\text{free}} = \sum_{q=u,d,s\ldots} \sum_{colors} \bar{q}(x) \left(i\gamma_\mu \frac{\partial}{\partial x_\mu} - m_q \right) q(x), \tag{9.71}$$

to be invariant under rotations of the quark fields in color space,

$$U: \qquad q^j(x) \quad \rightarrow \quad U_{jk}(x)q^k(x), \tag{9.72}$$

with $j, k \in \{1\ldots 3\}$ (we always sum over repeated indices). Since the theory we build in this way is invariant with respect to these "gauge" transformations, all physically meaningful quantities must be gauge invariant. In the Eq. (9.71), $\partial/\partial x_\mu = (\partial/\partial t, \nabla)$, and $\gamma_\mu = (\gamma_0, \gamma_i)$ are the (4×4) Dirac matrices, defined as

$$\gamma_i = \begin{pmatrix} 0 & \boldsymbol{\sigma}_i \\ -\boldsymbol{\sigma}_i & 0 \end{pmatrix}, \quad \gamma_0 = \begin{pmatrix} \mathbf{1} & 0 \\ 0 & -\mathbf{1} \end{pmatrix}. \tag{9.73}$$

The indices (i, j, k) can only have values 1, 2 e 3 (or, equivalently, x, y, z), and the index 0 means the time component of the matrix γ_μ. The matrices σ are the Pauli matrices

$$\sigma_1 = \begin{pmatrix} 0 & 1 \\ 1 & 0 \end{pmatrix}, \quad \sigma_2 = \begin{pmatrix} 0 & -i \\ i & 0 \end{pmatrix}, \quad \sigma_3 = \begin{pmatrix} 1 & 0 \\ 0 & -1 \end{pmatrix}. \tag{9.74}$$

The contraction of two four-vectors is defined as $A^\mu B_\mu = A^0 B^0 - \mathbf{A} \cdot \mathbf{B}$. In Eq. (9.71), $\bar{q}(x)$ is a matrix multiplication of the complex conjugate of the transpose of Eq. (9.70) and the Dirac matrix γ_0, i.e. $\bar{q}(x) = q^\dagger(x)\gamma_0$.

In electrodynamics, there is only one electric charge, and gauge transformation involves a single phase factor, $U = \exp(i\alpha(x))$. In QCD, one has three different colors, and U becomes a (complex valued) unitary 3×3 matrix, i.e. $U^\dagger U = UU^\dagger = 1$, with determinant Det $U = 1$. These matrices form the fundamental representation of the group $SU(3)$ where 3 is the number of colors, $N_c = 3$. The matrix U has $N_c^2 - 1 = 8$ independent elements and can therefore be parameterized in terms of the 8 generators T_{kj}^a, $a \in \{1\ldots 8\}$ of the fundamental representation of $SU(3)$,

$$U(x) = \exp\left(-i\phi_a(x)T^a\right). \tag{9.75}$$

By considering a transformation U that is infinitesimally close to the $\mathbf{1}$ element of the group, it is easy to prove that the matrices T^a must be Hermitian ($T^a = T^{a\dagger}$) and traceless (tr$T^a = 0$). The T^a's do not commute; instead one defines the $SU(3)$ structure constants f_{abc} by the commutator

$$\left[T^a, T^b\right] = if_{abc}T^c. \tag{9.76}$$

These commutator terms have no analog in QED which is based on the abelian gauge group $U(1)$. QCD is based on a non-abelian gauge group $SU(3)$ and is thus called a non-abelian gauge theory.

The generators T^a are normalized to

$$\text{tr}T^a T^b = \frac{1}{2}\delta_{ab}, \tag{9.77}$$

FIGURE 9.11 Due to the non-abelian nature of QCD, gluons carry color charge and can therefore interact with each others via these vertices.

where δ_{ab} is the Kronecker symbol. Useful information about the algebra of color matrices, and their explicit representations, can be found in many textbooks (see, e.g., [21]).

Since U is x-dependent, the free quark Lagrangian (9.71) is not invariant under the transformation (9.72). In order to preserve gauge invariance, one has to introduce, following the familiar case of electrodynamics, the gauge (or "gluon") field $A^\mu_{kj}(x)$ and replace the derivative in (9.71) with the so-called *covariant derivative*,

$$\partial^\mu q^j(x) \quad \rightarrow \quad D^\mu_{kj} q^j(x) \equiv \left\{\delta_{kj}\partial^\mu - iA^\mu_{kj}(x)\right\} q^j(x), \tag{9.78}$$

where ∂^μ si the four-dimensional derivative $\partial^\mu = (\partial/\partial t, \nabla)$.

Note that the gauge field $A^\mu_{kj}(x) = A^\mu_a T^a_{kj}(x)$ as well as the covariant derivative are 3×3 matrices in color space. Note also that Eq. (9.78) differs from the definition often given in textbooks, because we have absorbed the strong coupling constant in the field A^μ. With the replacement given by Eq. (9.78), all changes to the Lagrangian under gauge transformations cancel, provided A^μ transforms as

$$U: \qquad A^\mu(x) \rightarrow U(x)A^\mu(x)U^\dagger(x) + iU(x)\partial^\mu U^\dagger(x). \tag{9.79}$$

(From now on, we will often not write the color indices explicitly.)

The QCD Lagrangian then reads

$$\mathcal{L}_{\text{QCD}} = \sum_q \bar{q}(x)\left(i\gamma_\mu D^\mu - m_q\right)q(x) - \frac{1}{4g^2}\text{tr}G^{\mu\nu}(x)G_{\mu\nu}(x), \tag{9.80}$$

where the first term describes the dynamics of quarks and their couplings to gluons, while the second term describes the dynamics of the gluon field. The strong coupling constant g is the QCD analog of the elementary electric charge e in QED. The gluon field strength tensor is given by

$$G^{\mu\nu}(x) \equiv i\,[D^\mu, D^\nu] = \partial^\mu A^\nu(x) - \partial^\nu A^\mu(x) - i\,[A^\mu(x), A^\nu(x)]. \tag{9.81}$$

This can also be written in terms of the color components A^μ_a of the gauge field,

$$G^{\mu\nu}_a(x) = \partial^\mu A^\nu_a(x) - \partial^\nu A^\mu_a(x) + f_{abc}A^\mu_b(x)A^\nu_c(x). \tag{9.82}$$

For a more complete presentation, see modern textbooks like [21, 22, 23].

The crucial, as will become clear soon, difference between electrodynamics and QCD is the presence of the commutator on the *r.h.s.* of Eq. (9.81). This commutator gives rise to the gluon-gluon interactions shown in Figure 9.11 that make the QCD field equations non-linear: the color fields do not simply add like in electrodynamics. These non-linearities give rise to rich and non-trivial dynamics of strong interactions.

Let us now turn to the discussion of the dynamical properties of QCD. To understand the dynamics of a field theory, one necessarily has to understand how the coupling constant behaves as a function of distance. This behavior, in turn, is determined by the response of the vacuum to the presence of external charge. The vacuum is the ground state of the theory; however, quantum mechanics tells us that the "vacuum" is far from being empty – the uncertainty principle allows

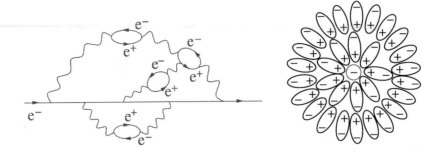

FIGURE 9.12   In QED, virtual electron-positron pairs popping out from the vacuum (left) screen the bare charge of the electron (right). The larger the distance, the more pairs are present to screen the bare charge and the electromagnetic coupling decreases. Conversely, the coupling is larger when probed at short distances.

particle-antiparticle pairs to be present in the vacuum for a period time inversely proportional to their energy. In QED, the electron-positron pairs have the effect of screening the electric charge, see Figure 9.12. Thus, the electromagnetic coupling constant increases toward shorter distances. The dependence of the charge on distance (running coupling constant) is given by [19]

$$e^2(r) = \frac{e^2(r_0)}{1 + \dfrac{2e^2(r_0)}{3\pi} \ln \dfrac{r}{r_0}}, \tag{9.83}$$

which can be obtained by resumming (logarithmically divergent, and regularized at the distance r_0) electron–positron loops dressing the virtual photon propagator.

The formula (9.83) has two surprising properties: first, at large distances r away from the charge which is localized at r_0, $r \gg r_0$, where one can neglect unity in the denominator, the "dressed" charge $e(r)$ becomes independent of the value of the "bare" charge $e(r_0)$ – it does not matter what the value of the charge at short distances is. Second, in the local limit $r_0 \to 0$, if we require the bare charge $e(r_0)$ be finite, the effective charge vanishes at any finite distance away from the bare charge! The screening of the charge in QED does not allow to reconcile the presence of interactions with the local limit of the theory. This is a fundamental problem of QED, which shows that i) either it is not a truly fundamental theory, or ii) Eq. (9.83), based on perturbation theory, in the strong coupling regime gets replaced by some other expression with a more accepTable behavior. The latter possibility is quite likely since at short distances the electric charge becomes very large and its interactions with electron–positron vacuum cannot be treated perturbatively.

Fortunately, because of the smallness of the physical coupling $\alpha_{em}(r) = e^2(r)/(4\pi) = 1/137$, this fundamental problem of the theory manifests itself only at very short distances $\sim \exp(-3/[8\alpha_{em}])$. Such short distances will probably always remain beyond the reach of experiment, and one can safely apply QED as a truly effective theory.

In QCD, as we are now going to discuss, the situation is qualitatively different, and corresponds to *anti*-screening – the charge is small at short distances and grows at larger distances. This property of the theory is called *asymptotic freedom* [24].

While the derivation of the running coupling is conventionally performed by using field theoretical perturbation theory, it is instructive to see how these results can be illustrated by using the methods of condensed matter physics. Indeed, let us consider the vacuum as a continuous medium with a dielectric constant ϵ. The dielectric constant is linked to the magnetic permeability μ and the speed of light c by the relation

$$\epsilon\mu = \frac{1}{c^2} = 1. \tag{9.84}$$

Thus, a screening medium ($\epsilon > 1$) will be diamagnetic ($\mu < 1$), and conversely a paramagnetic medium ($\mu > 1$) will exhibit antiscreening which leads to asymptotic freedom. In order to calculate the running coupling constant, one has to calculate the magnetic permeability of the vacuum. In QED one has [21, 22, 23]

$$\epsilon_{QED} = 1 + \frac{2e^2(r_0)}{3\pi} \ln \frac{r}{r_0} > 1. \tag{9.85}$$

So why is the QCD vacuum paramagnetic while the QED vacuum is diamagnetic? The energy density of a medium in the presence of an external magnetic field **B** is given by

$$u = -\frac{1}{2} 4\pi \chi \mathbf{B}^2, \tag{9.86}$$

where the magnetic susceptibility χ is defined by the relation

$$\mu = 1 + 4\pi\chi. \tag{9.87}$$

When electrons move in an external magnetic field, two competing effects determine the sign of magnetic susceptibility:

- The electrons in magnetic field move along quantized orbits, so-called Landau levels. The current originating from this movement produces a magnetic field with opposite direction to the external field. This is the diamagnetic response, $\chi < 0$.

- The electron spins align along the direction of the external **B**-field, leading to a paramagnetic response $(\chi > 0)$.

In QED, the diamagnetic effect is stronger, so the vacuum is screening the bare charges. In QCD, however, gluons carry color charge. Since they have a larger spin (spin 1) than quarks (or electrons), the paramagnetic effect dominates and the vacuum is anti-screening.

Based on the considerations given above, the energy density of the QCD vacuum in the presence of an external color-magnetic field can be calculated by using the standard formulas of quantum mechanics, see e.g. [25], by summing over Landau levels and taking account of the fact that gluons and quarks give contributions of different sign. Note that a summation over all Landau levels would lead to an infinite result for the energy density. In order to avoid this divergence, one has to introduce a cutoff Λ with dimension of mass. Only field modes with wavelength $\lambda \gtrsim 1/\Lambda$ are taken into account. The upper limit for λ is given by the radius of the largest Landau orbit, $r_0 \sim 1/\sqrt{gB}$, which is the only dimensionful scale in the problem; the summation thus is made over the wave lengths satisfying

$$\frac{1}{\sqrt{|gB|}} \gtrsim \lambda \gtrsim \frac{1}{\Lambda}. \tag{9.88}$$

The result is [26]

$$u_{vac}^{QCD} = -\frac{1}{2} B^2 \frac{11N_c - 2N_f}{48\pi^2} g^2 \ln \frac{\Lambda^2}{|gB|}, \tag{9.89}$$

where N_f is the number of quark flavors, and $N_c = 3$ is the number of flavors. Comparing this with eqs. (9.86) and (9.87), one can read off the magnetic permeability of the QCD vacuum,

$$\mu_{vac}^{QCD}(B) = 1 + \frac{11N_c - 2N_f}{48\pi^2} g^2 \ln \frac{\Lambda^2}{|gB|} > 1. \tag{9.90}$$

The first term in the denominator $(11N_c)$ is the gluon contribution to the magnetic permeability. This term dominates over the quark contribution $(2N_f)$ as long as the number of flavors N_f is less than 17 and is responsible for asymptotic freedom.

The dielectric constant as a function of distance r is then given by

$$\epsilon_{vac}^{QCD}(r) = \left. \frac{1}{\mu_{vac}^{QCD}(B)} \right|_{\sqrt{|gB|} \to 1/r}. \tag{9.91}$$

The replacement $\sqrt{|gB|} \to 1/r$ follows from the fact that ϵ and μ in Eq. (9.91) should be calculated from the same field modes: the dielectric constant $\epsilon(r)$ could be calculated by computing the vacuum

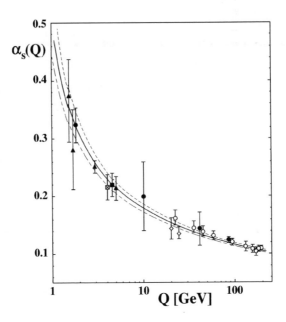

FIGURE 9.13   The running coupling constant $\alpha_s(Q^2)$ as a function of momentum transfer Q^2 determined from a variety of processes.

energy in the presence of two static colored test particles located at a distance r from each other. In this case, the maximum wavelength of field modes that can contribute is of order r so that

$$r \gtrsim \lambda \gtrsim \frac{1}{\Lambda}. \tag{9.92}$$

Combining eqs. (9.88) and (9.92), we identify $r = 1/\sqrt{|gB|}$ and find

$$\epsilon_{vac}^{QCD}(r) = \frac{1}{1 + \frac{11N_c - 2N_f}{24\pi^2} g^2 \ln(r\Lambda)} < 1. \tag{9.93}$$

With $\alpha_s(r_1)/\alpha_s(r_2) = \epsilon_{vac}^{QCD}(r_2)/\epsilon_{vac}^{QCD}(r_1)$ one finds to lowest order in α_s (the strong interaction coupling constant)

$$\alpha_s(r_1) = \frac{\alpha_s(r_2)}{1 + \frac{11N_c - 2N_f}{6\pi} \alpha_s(r_2) \ln\left(\frac{r_2}{r_1}\right)}. \tag{9.94}$$

Apparently, if $r_1 < r_2$ then $\alpha_s(r_1) < \alpha_s(r_2)$. The running of the coupling constant is shown in Figure 9.13, $Q \sim 1/r$ is the momentum transfer. The intuitive derivation given above illustrates the original field–theoretical result of [24].

At high-momentum transfer, corresponding to short distances, the coupling constant thus becomes small and one can apply perturbation theory, see Figure 9.13. There is a variety of processes that involve high-momentum scales, e.g., deep inelastic scattering, Drell-Yan dilepton production, $e^+ e^-$-annihilation into hadrons, production of heavy quarks/quarkonia, high p_T hadron production .... QCD correctly predicts the Q^2 dependence of these, so-called "hard" processes, which is a great success of the theory [19].

While asymptotic freedom implies that the theory becomes simple and treaTable at short distances, it also tells us that at large distances the coupling becomes very strong. In this regime we have no reason to believe in perturbation theory. In QED, as we have discussed above, the strong coupling regime starts at extremely short distances beyond the reach of current experiments – and this makes the "zero-charge" problem somewhat academic. In QCD, the entire physical world around us is defined by the properties of the theory in the strong coupling regime – and we have to construct accelerators to study it in the much more simple, "QED–like", weak coupling limit.

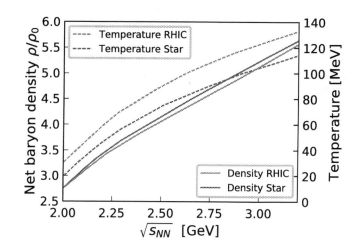

FIGURE 9.14 Largest values of baryon densities (solid lines) and temperatures (dashed lines) reached in relativistic heavy ion collisions (RHIC) and in neutron star mergers as a function of the center of mass beam energy $\sqrt{s_{NN}} = 2\gamma_{c.m.} m_N$. The densities and temperatures were calculated using a quark-hadron chiral parity doublet model for the EoS which depend on the beam energy. (Adapted from Ref. [28]).

We do not have to look far to find the striking differences between the properties of QCD at short and large distances: the elementary building blocks of QCD – the "fundamental" fields appearing in the Lagrangean (9.80), quarks and gluons, do not exist in the physical spectrum as asymptotic states. For some, still unknown to us, reason, all physical states with finite energy appear to be color–singlet combinations of quarks and gluons, which are thus always "confined" at rather short distances on the order of 1 fm. This prevents us, at least in principle, from using well–developed formal S-matrix approaches based on analyticity and unitarity to describe quark and gluon interactions.

Collisions of heavy ions are the best way to create high-energy density in a "macroscopic" (on the scale of a single hadron) volume. It thus could be possible to create and to study a new state of matter, the *Quark-Gluon Plasma* (QGP), in which quarks and gluons are no longer confined in hadrons, but can propagate freely. The search for QGP is one of the main motivations for the heavy ion research.

9.7 The quark-gluon plasma

The primary motivation for studying ultra-relativistic heavy ion collisions is to gain an understanding of the equation of state of nuclear, hadronic and partonic matter, commonly referred to as nuclear matter. Displayed in Figure 9.2 is a schematic phase diagram of nuclear matter. The behavior of nuclear matter as a function of temperature and density (or pressure), shown in Figure 9.2, is governed by its equation of state.

Conventional nuclear physics is concerned primarily with the lower left portion of the diagram at low temperatures and near normal nuclear matter density. Here normal nuclei exist and at low excitation a liquid-gas phase transition is expected to occur. This is the focus of experimental studies using low energy heavy ions, as we saw in Chapters 1-7. At somewhat higher excitation, nucleons are excited into baryonic resonance states, along with accompanying particle production and hadronic resonance formation. In relativistic heavy ion collisions, such excitation is expected to create hadronic resonance matter.

According to numerical simulations, in neutron star mergers, high temperatures $T \lesssim 100$ MeV can be reached [28]. In Figure 9.14 we show the largest values of baryon densities (solid lines)

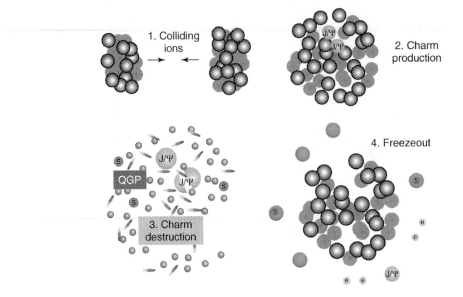

FIGURE 9.15   Formation and evolution of J/Ψ particles in relativistic heavy ion collisions. After the formation, the J/Ψs are dissociated in the plasma due to color screening. The end effect is a smaller number J/Ψs of than expected from pure hadron-hadron multiple collisions.

and temperatures (dashed lines) reached in relativistic heavy ion collisions and in neutron stars as a function of the center of mass beam energy $\sqrt{s_{NN}} = 2\gamma_{c.m.} m_N$. Beam energies in the range $\sqrt{s_{NN}} = 2.5 - 3$ GeV have been considered. The densities and temperatures were calculated using a quark-hadron chiral parity doublet model for the EoS which depend on the c.m. energy [29]. This EoS has different properties for different isospin content, and therefore the Figure shows that the temperatures in RHIC are larger and densities slightly smaller at the same relative velocities. In nucleus-nucleus collisions at relativistic energies the isospin per baryon is of the order of –0.1, whereas in NS it is about –0.38. NS also have a significantly different composition of strange particles compared to RHIC. The density compression in symmetric nuclear matter and in NS matter is very similar, but the temperature is quite different. This indicates that the additional degrees of freedom, such as leptons in beta equilibrium and non-conserved strangeness, decrease the temperature at a given compression. Therefore, the study of neutron star mergers requires the use of a consistent and realistic temperature dependent EoS, probably incorporating quark degrees of freedom.

We now briefly discuss the QGP signatures in nucleus-nucleus collisions. For more details see, e.g., Ref. [30]. One group of such signatures can be classified as *thermodynamic variables*. This class involves determination of the energy density ϵ, pressure P, and entropy density s of the interacting system as a function of the temperature T and the baryochemical potential μ_B. Experimental observables can be identified with these variables and thus their relative behavior can be determined. If a phase transition to QGP occurs, a rapid rise in the effective number of degrees of freedom, expressed by ϵ/T^4 or s/T^3, should be observed over a small range of T. The variables T, s, and ϵ, can be identified with the average transverse momentum $\langle p_T \rangle$, the hadron rapidity density dN/dy, and the transverse energy density dE_T/dy, respectively. The transverse energy produced in the interaction is

$$E_T = \sum_i E_i \sin \theta_i, \qquad (9.95)$$

where E_i and θ_i are the kinetic energies of the ejectiles and the emission angles.

Electromagnetic (EM) probes, such as photons and leptons (i.e. virtual photons), provide information on the various stages of the interaction without modification by final state interactions.

These probes may provide a measure of the thermal radiation from a QGP, if a region of photon energy, or equivalently lepton pair invariant mass, can be isolated for emission from a QGP relative to other processes. However, the yields for EM probes are small relative to background processes, which are primarily EM decays of hadrons and resonances. Lepton pairs from the QGP are expected to be identifiable in the 1-10 GeV invariant mass range. The widths and positions of the ρ, ω, and ϕ peaks in the lepton pair invariant mass spectrum are expected to be sensitive to medium-induced changes of the hadronic mass spectrum

The production of J/Ψ particles in a quark-gluon plasma is predicted to be suppressed in a QGP (see Figure 9.15). This is a result of the *Debye screening* of a $c\bar{c}$ pair, initially formed in the QGP by fusion of two incident gluons. Less tightly bound excited states of the $c\bar{c}$ system, such as Ψ' and χ_c , are more easily dissociated and will be suppressed even more than the J/Ψ.

A long-standing prediction for a signature of QGP formation is the enhancement of strange hadrons. The production of strange hadrons relative to nonstrange hadrons is suppressed in hadronic reactions. This suppression increases with increasing strangeness content of the hadron. In a QGP the strange quark content is rapidly saturated by $s\bar{s}$ pair production in gluon-gluon reactions, resulting in an enhancement in the production of strange hadrons. Thus, multi-strange baryons and strange antibaryons are predicted to be strongly enhanced when a QGP is formed.

The connection between energy loss of a quark and the color-dielectric polarizability of the medium can be established in analogy with the theory of electromagnetic energy loss. Although radiation is a very efficient energy loss mechanism for relativistic particles, it is strongly suppressed in a dense medium by the *Landau-Pomeranchuk effect* [30]. Adding the two contributions, the stopping power of a quark-gluon plasma is predicted to be higher than that of hadronic matter. A quark or *gluon jet* propagating through a dense medium will not only loose energy but will also be deflected. This effect destroys the coplanarity of the two jets from a hard parton-parton scattering with the incident beam axis. The angular deflection of the jets also results in an azimuthal asymmetry. The presence of a quark-gluon plasma is also predicted to enhance the emission of jet pairs with small azimuthal opening angles.

9.8   Ultra-peripheral nuclear collisions

In Section 8.12 we have shown how the electromagnetic field of a relativistic nucleus can lead to the excitation of giant resonances in another nucleus, thus providing useful insights of nuclear excitation of collective states and their decay dynamics. Ultra-peripheral collisions (UPC) with relativistic heavy ions are also used as a method to access information on (p,γ) and (n,γ) reactions or relevance for astrophysics, via the Coulomb dissociation method [31].

For ultra-relativistic nuclei, available in heavy ion colliders, such as the Large Hadron Collider (LHC) at the CERN laboratory, the electromagnetic fields generated in UPCs are so intense that elementary particles can be produced in one- and two-photon processes [31]. The cross section for photoproduction of a particle X, e.g, a J/Ψ particle, can be calculated with the expression

$$\sigma_X = \int dE \frac{n(E)}{E} \sigma_X^\gamma(E), \qquad (9.96)$$

where $n(E)$ is the virtual (or equivalent) photon number (see Section 8.12) and σ_X^γ is the photonuclear cross section for a real photon with energy E.

Photon-photon (or two-photon) processes are also used to study atomic, and particle physics processes, often involving electrodynamics in strong fields. One striking success of photon-photon processes in UPCs was the first time that anti-hydrogen atoms were produced at CERN in 1996 and at the Fermilab Tevatron in 1998. At the highest energy colliders, reactions like $\gamma\gamma \to X$ may be used to probe the quark and gluon content in parton distribution functions and spin structure of meson resonances. Production of meson or baryon pairs can also probe the internal structure of hadrons. At the LHC, electroweak processes such as $\gamma\gamma \to W^+W^-$ may be probed. The cross section for two-photon processes is [31]

$$\sigma_X = \int dE_1 dE_2 \frac{n(E_1)}{E_1} \frac{n(E_2)}{E_2} \sigma_X^{\gamma\gamma}(E_1, E_2), \qquad (9.97)$$

Particle X	Pb+Pb, LHC σ [mb]	# photons
ρ^0	5700	one
ω	510	one
ϕ	460	one
J/Ψ	36	one
$\Upsilon(1s)$	0.19	one
$f_0^{\bar{q}q}(980)$	25.8	two
$X(3915)$	7.34×10^{-3}	two
Muonic atom	0.16	two
Pionic atom	0.09	two

TABLE 9.5 Cross sections for particle production in Pb + Pb collisions at the LHC and collider energy of 5.6 TeV via one- and two-photon processes.

where $\sigma_X^{\gamma\gamma}$ is the photo nuclear cross section for the production of a state X by two photons with energies E_1 and E_2.

In Table 9.5 we list estimated cross sections for particle production in Pb + Pb collisions at the LHC collider at CERN for 5.6 TeV energy per nucleon pair in their center of mass. It has long been advertised that these cross sections are large enough to allow for their detection with present collider luminosities [31, 32]. UPC experiments have proven to be a relevant source of information on parton distribution functions by observing the production of vector mesons, such as the J/Ψ and Υ [33]. They are also useful to determine the $\gamma\gamma$ decay widths of several mesons for which these widths are unknown such as the $f_0^{\bar{q}q}(980)$ or to study the structure of exotic mesons, such as the $X(3872)$ which might have a charmonium-tetraquark, charmonium-molecule or tetraquark-molecule mixtures [34]. The investigation of muonic or pionic atoms is also possible, in which a muon (pion) pair is produced and the negative muon (pion) is captured in an orbit around the nucleus (last two columns of Table 9.5). Some cross sections such as the production of free e^+e^- and of light-by-light scattering ($\gamma^*\gamma^* \to \gamma\gamma$) are enormous, of several hundreds of kilobarns [31, 35].

9.9 Exercises

1. For a particle with momentum $\mathbf{p} = (p_z, \mathbf{p}_T)$ where z (T) denotes the longitudinal (transverse) direction, define

$$m_T^2 = E^2 - p_z^2 = p_T^2 + m^2.$$

   Show that m_T is invariant under Lorentz transformations.

2. Consider a relativistic heavy ion collision in which each nuclei can be treated as a disk with radius $R = 1.2A^{1/3}$ fm and assume that the two nuclei interact if the disks overlap, i.e., if the impact parameter is smaller than $2R$. What range of impact parameters does the most central 0-5 led on lead collisions correspond to?

3. For two particles 1 and 2 we define $s = (p_1 + p_2)^2 = (p_1 + p_2)_\mu (p_1 + p_2)^\mu$, where p_i is the particle four-momentum. Show that s is invariant under Lorentz transformations.

4. Use $s = m_1^2 + m_2^2 + 2E_1E_2 - 2\mathbf{p}_1 \cdot \mathbf{p}_2$ to obtain the center-of-mass momentum p_{cm} and the center-of-mass energy E_{cm} in terms of s and the particles masses.

5. Show that the quantity d^3p/E is Lorentz invariant. This is often used to construct invariant phase space for the scattering of interacting particles.

6. In a collider two beams of nuclei collider move in opposite directions and collide frontally. Show that if the maximum beam energy for proton-proton collisions is $\sqrt{s_{pp}}$, the maximal lead on lead beam energy is $\sqrt{s_{PbPb}} = Z\sqrt{s_{pp}}$, so that the beam energy per nucleon pair is $\sqrt{s_{NN}} = AZ\sqrt{s_{pp}}$.

7. In a situation of no mean field and in thermodynamical equilibrium one can show that the Boltzmann equation, Eq. (9.22), leads to the Maxwell-Boltzmann distribution, i.e.,

$$f(\mathbf{r}, \mathbf{p}) = \frac{n(\mathbf{r})}{(2\pi m_N kT)^{3/2}} \exp\left(-p^2/2m_N kT\right), \tag{9.98}$$

where $n(\mathbf{r})$ is the local density of nucleons, and T is the nuclear temperature.

The number of collisions per unit time in the nucleon gas is given by

$$Z = \frac{\sigma_{NN}}{m_N} \int d^3 p_2 \int d^3 p_1 \, |\mathbf{p}_1 - \mathbf{p}_2| \, f_1(\mathbf{r}, \mathbf{p}_1) \, f_2(\mathbf{r}, \mathbf{p}_2). \tag{9.99}$$

(a) Assuming $n(\mathbf{r}) = n = const$ find Z. (b) Using the free nucleon-nucleon cross section at $E = 40$ MeV ($\sim$ Fermi motion energy in nuclei) calculate the mean free path of nucleons in normal nuclear matter. Does the result make sense? Explain.

8. By explicitly calculating the derivatives entering the Jacobian, prove Eq. (9.5) when both energy and momentum is conserved.

9. Demonstrate Eq. (9.32).

10. What is the result for Eq. (9.40) when one uses a Gaussian and a harmonic oscillator function for $\psi(\mathbf{r}, t)$.

11. Using the definition of m_T from Exercise 1, show that

$$E = m_T \cosh y, \quad \text{and} \quad p_z = m_T \sinh y.$$

12. Prove that $y \sim \beta = v/c$ for $\beta \ll 1$.

13. Using elementary trigonometric identities derive the properties of the pseudo rapidity function:

$$\eta = \frac{1}{2} \ln\left(\frac{|\mathbf{p}| + p_z}{|\mathbf{p}| - p_z}\right) = \ln\left(\cot\frac{\theta}{2}\right) = -\ln\left(\tan\frac{\theta}{2}\right), \tag{9.100}$$

where $\sin\theta = 1/\cosh\eta$.

14. The rapidity of a particle in frame of reference O is y. Calculate the rapidity of the same particle in another frame of reference O' moving with respect to the first with a velocity v along the motion of the particle.

15. Show that to lowest order in $m^2/p_\perp^2$, where $p_\perp$ is the particle's transverse momentum, the relation between rapidity and pseudo-rapidity is

$$y \simeq \eta + \frac{m^2}{p_\perp^2} \cos\theta.$$

16. (a) Calculate the rapidities of the projectile nuclei in fixed-target experiments with the beam energy of 60 and 200 GeV per nucleon. (b) Prove that rapidities are additive under Lorentz boosts along the beam axis. (c) Show that for a high-energy particle one can measure independently its rapidity and longitudinal position.

17. Deduce the minimum laboratory energy needed to produce an antiproton in a proton-proton collision.

18. Calculate the compressibility of nuclear matter, Eq. (7.172), in the Fermi model (see Section 4.5).

19. Calculate the energy density of normal nuclear matter.

20. List what are the main motivations, both theoretical and experimental, for us to believe that QCD is the right theory of strong interactions.

21. Assume that the Lagrange density for free fermions is given by

$$\mathcal{L} = i\overline{\psi}\gamma^{\mu}\partial_{\mu}\psi - m\overline{\psi}\psi.$$

Take ψ and $\overline{\psi}$ as independent variables. Use the Euler-Lagrange equations to derive the Dirac equations for both the fermion field, ψ, and its adjoint $\overline{\psi}$.

22. Show that the Lagrangian density

$$\mathcal{L} = \frac{1}{2}\partial_{\mu}\phi\partial^{\mu}\phi - V(\phi)$$

generates the Klein-Gordon equation when $V(\phi) = m^2\phi^2/2$.

23. (a) Describe the differences between QED and QCD. (b) What fundamental feature of QCD led to the motivation to study relativistic heavy ion collisions? (c) What does it means when one says that QCD is a non-Abelian theory?

24. Write down the quark content of the mesons belonging to the spin-0 8-multiplet of $SU(3)_f$.

25. The J/Ψ particle decays into two charmed mesons (D^+, D^-). But the mass of $D^{\pm}$ is about $1.87~\text{GeV/c}^2$ and the mass of J/Ψ is $3.10~\text{GeV/c}^2$. Explain why the mass of J/Ψ is less than twice the mass of $D^{\pm}$.

26. Evaluate magnetic moments of all baryons in Table 9.2, assuming $m_u = m_d = m$, but $m_s \neq m$.

27. What are the main differences and what is studied in $e^+ + e^-$ and in p $+ A$ and $A + A$ relativistic collisions?

28. Browse the web and find plots that summarize the determination of α_s from different experiment types. Which measurement is the most precise? At which scale has it been performed? How was that measurement performed? Divide (when possible) the measurements at low energy from those at high energy. Is there a systematic behavior? Find a plot which shows the evolution of α_s with the scale and understand which data were used.

29. Calculate the energy density $\epsilon = \epsilon(T, \mu)$ of a weakly-interacting quark-gluon plasma consisting of massless partons, where μ is the chemical potential. Hint: Treat gluons and quarks as an ideal Bose-Einstein and Fermi-Dirac gas, respectively.

References

1. Gale C, Bertsch G and Das Gupta S 1987 *Phys. Rev.* **C35** 1666
2. Wigner E 1932 *Phys. Rev.* **40** 749
3. Eisenberg J M 1995 *Heavy Ion Phys.* **1** 53
4. Huefner J 1978 *Ann. Phys.* **115** 43
5. Gregoire C et al. 1987 *Nucl. Phys.* **A465** 317
6. Huang K 2008 *Statitical Mechanics* (Wiley)
7. Danielewicz P 1995 *Phys. Rev.* **C51** 716
8. Tsang M B et al. 1996 *Phys. Rev.* **C53** 1959
9. Danielewicz P, Lacey R and Lynch W G *Science* 2002 **298** 1592
10. Stock R et al. 1982 *Phys. Rev. Lett.* 49 1236; Harris J W et al. 1987 *Phys. Rev. Lett.* 58 463
11. Gustafsson H Å et al. 1988 *Mod. Phys. Lett.* **A3** 1323
12. Hayashi S et al. 1988 *Phys. Rev.* **C38** 1229
13. Gell-Mann M 1953 *Phys. Rev.* **92** 833
14. Nakano T and Nishijima K 1953 *Prog. Theor. Phys.* **10** 581

15. Gell-Mann M 1964 *Phys. Lett.* **8** 214

16. Zweig G 1964 CERN Report N° 8182/TH401 (unpublished)

17. Glashow S L, Iliopoulos J and Maiani L 1970 *Phys. Rev.* **D2** 1285

18. Friedman J I 1991 *Rev. Mod. Phys.* **63** 615

19. Halzen F and Martin A D 1984 *Quarks and Leptons: An Introductory Course in Modern Particle Physics* (New York: John Wiley & Sons)

20. Peskin M E and Schroeder D V 2019 *An Introduction to Quantum Field Theory* (CRC Press)

21. Aitchison I J R and Hey A J G 2012 *Gauge Theories in Particle Physics: A Practical Introduction* (CRC Press)

22. Walecka J D 2004 *Theoretical Nuclear and Subnuclear Physics* (World Scientific)

23. Campbell J, Huston J, Krauss F 2018 *The Black Book of Quantum Chromodynamics: A Primer for the LHC Era* (Oxford University Press)

24. Gross D J and Wilczek F 1973 *Phys. Rev. Lett.* **30** 1343; Politzer H D 1973 *Phys. Rev. Lett.* **30** 1346

25. Landau L D and Lifshitz E M 1981 *Quantum Mechanics: Nonrelativistic Theory* (Butterworth-Heinemann)

26. Nielsen N K 1981 *Am. J. Phys.* **49** 1171

27. Bethke S 2000 *J. Phys.* **G26** R27

28. Hanauske M, et al. 2017 *J. Phys: Conference Series* **878** 012031

29. Steinheimer J, Schramm S and Stöcker H 2011 *Phys. Rev.* **C84** 045208

30. Wong C Y 1994 *Introduction to high energy heavy ion collisions* (Singapore: World Scientific)

31. Bertulani C A and Baur G 1988 *Phys. Rep.* **163** 299

32. Bertulani C A, Klein S, Nystrand J 2005 *Ann. Rev. Nuc. Part. Sci.* **55** 271

33. Adeluyi A, Bertulani C A 2011 *Phys. Rev.* **C84** 024916

34. Bertulani C A 2009 *Phys. Rev.* **C79** 047901

35. Bertulani C A, Baur G 1994 *Relativistic heavy ion physics without nuclear contact* (Physics Today, AIP)

Index

Printed in the United States
By Bookmasters